Stem Cell Genetics for Biomedical Research

Raul Delgado-Morales
Editor

Stem Cell Genetics for Biomedical Research

Past, Present, and Future

Springer

Editor
Raul Delgado-Morales
Duran i Reynals Hospital
Bellvitge Biomedical Research Institute (IDIBELL)
Barcelona, Spain

ISBN 978-3-030-08076-1 ISBN 978-3-319-90695-9 (eBook)
https://doi.org/10.1007/978-3-319-90695-9

Softcover re-print of the Hardcover 1st edition 2018

Printed on acid-free paper

This Springer imprint is published by the registered company Springer International Publishing AG part of Springer Nature.
The registered company address is: Gewerbestrasse 11, 6330 Cham, Switzerland

Contents

Part I Stem Cells for Brain Disorders

1 **Human-Induced Pluripotent Stem Cell-Derived Neurons to Model and Gain Insights into Alzheimer's Disease Pathogenesis.** 3
Jessica E. Young and Raul Delgado-Morales

2 **Modeling Schizophrenia with Human Stem Cells** 13
Juliana Minardi Nascimento, Veronica M. Saia-Cereda, Giuliana S. Zuccoli, Danielle Gouvêa-Junqueira, and Daniel Martins-de-Souza

3 **Rett Syndrome and Stem Cell Research** 27
Keita Tsujimura and Kinichi Nakashima

4 **Stem Cell Applications in Spinal Cord Injury: A Primer** 43
James Hong, Christopher E. Rodgers, and Michael G. Fehlings

5 **Cell-Based Therapy for Retinal Degenerative Disease** 73
Marco A. Zarbin

6 **Past, Present and Future of Cell-Based Therapy in Progressive Multiple Sclerosis** 87
Giulio Volpe, Joshua D. Bernstock, Luca Peruzzotti-Jametti, and Stefano Pluchino

Part II Stem Cells for Cardiovascular Diseases

7 **Cardiac Stem Cells: A Plethora of Potential Therapies for Myocardial Regeneration Within Reach** 135
Enrique Gallego-Colon and Eduardo Oliver

8 **Human-Induced Pluripotent Stem Cell-Derived Cardiomyocytes in the Evaluation of Cardiotoxic Potential of Drugs** 173
Yee Xiang Chang, and Christine L. Mummery

Part III Stem Cells for General Medicine

9 **Regenerative Medicine for Diabetes Treatment: New β-Cell Sources** 197
Rosa Gasa and Meritxell Rovira

10 **Dental Pulp Stem Cells Promote Wound Healing and Muscle Regeneration** 221
E. Martínez-Sarrà, S. Montori, C. Gil-Recio, R. Núñez-Toldrà, N. Carrio Bertran, A. Al Madhoun, M. Sampaolesi, and M. Atari

11 **From Bench to Bedside of Mesenchymal Stem Cells Use for Rheumatoid Arthritis Treatment** 241
R. A. Contreras, N. Luque, F. Djouad, Roberto Elizondo-Vega, and Patricia Luz-Crawford

12 **Stem Cells and Cancer** 271
Meredith Davis, Katja Gassner, Ruth Rodriguez-Barrueco, and David Llobet-Navas

13 **The Relevance of Induced Pluripotent Stem Cells for the Study of Physiological and Premature Aging** 311
Dido Carrero and Carlos López-Otín

Part IV Technical Challenges and Future

14 **Pluripotent Stem Cell Banks** 337
Begoña Aran, Ángel Raya, and Anna Veiga

15 **Engineering of Human-Induced Pluripotent Stem Cells for Precise Disease Modeling** 369
P. Lisowski

Index 413

Contributors

A. Al Madhoun Research Division, Dasman Diabetes Institute, Dasman, Kuwait

Begoña Aran Center of Regenerative Medicine in Barcelona (CMRB), Barcelona, Spain

National Stem Cell Bank-Barcelona Node, Biomolecular and Bioinformatics Resources Platform PRB2, ISCIII, CMRB, Barcelona, Spain

M. Atari Regenerative Medicine Research Institute, Universitat Internacional de Catalunya, Barcelona, Spain

Joshua D. Bernstock Department of Clinical Neurosciences and NIHR Biomedical Research Centre, University of Cambridge, Cambridge, UK

Stroke Branch, National Institute of Neurological Disorders and Stroke, National Institutes of Health (NINDS/NIH), Bethesda, MD, USA

N. Carrio Bertran Regenerative Medicine Research Institute, Universitat Internacional de Catalunya, Barcelona, Spain

Dido Carrero Departamento de Bioquímica y Biología Molecular, Facultad de Medicina, Instituto Universitario de Oncología (IUOPA), Universidad de Oviedo, Oviedo, Spain

Yee Xiang Chang University of Utrecht, Utrecht, The Netherlands

R. A. Contreras Laboratorio de Inmunología Celular y Molecular, Centro de Investigaci6n Biomédica, Facultad de Medicina, Universidad de los Andes, Santiago, Chile

Inserm, U 1183, Montpellier, France

Meredith Davis Institute of Genetic Medicine, Newcastle University, Newcastle upon Tyne, UK

Raul Delgado-Morales Cancer Epigenetics and Biology Program (PEBC), Bellvitge Biomedical Research Institute (IDIBELL), L'Hospitalet de Llobregat, Barcelona, Catalonia, Spain

Department of Psychiatry & Neuropsychology, School for Mental Health and Neuroscience (MHeNs), Maastricht University, Maastricht, The Netherlands

F. Djouad Inserm, U 1183, Montpellier, France

Université de Montpellier, Montpellier, France

Roberto Elizondo-Vega Laboratorio de Inmunología Celular y Molecular, Centro de Investigación Biomédica, Facultad de Medicina, Universidad de los Andes, Santiago, Chile

Michael G. Fehlings University of Toronto, Toronto, ON, Canada

University Health Network, Toronto, ON, Canada

Enrique Gallego-Colon Department of Biochemistry, Medical University of Silesia, School of Medicine in Katowice, Katowice, Poland

Rosa Gasa Centro de Investigación Biomédica en Red de Diabetes y Enfermedades Metabólicas Asociadas (CIBERDEM), Barcelona, Spain

Diabetes and Obesity Research Laboratory, August Pi i Sunyer Biomedical Research Institute (IDIBAPS), Barcelona, Spain

Katja Gassner Institute of Genetic Medicine, Newcastle University, Newcastle upon Tyne, UK

Bellvitge Biomedical Research Institute (IDIBELL), Gran via de l'Hospitalet, L'Hospitalet de Llobregat, Spain

C. Gil-Recio Regenerative Medicine Research Institute, Universitat Internacional de Catalunya, Barcelona, Spain

Danielle Gouvêa-Junqueira Lab of Neuroproteomics, Department of Biochemistry and Tissue Biology, Institute of Biology, University of Campinas (UNICAMP), Campinas, Brazil

James Hong University of Toronto & Krembil Research Institute, Toronto, ON, Canada

P. Lisowski Department of Molecular Biology, Institute of Genetics and Animal Breeding, Polish Academy of Sciences, Jastrzebiec, Magdalenka, Poland

Department of Medical Genetics, Centre for Preclinical Research and Technology (CePT), Warsaw Medical University, Warsaw, Poland

Mitochondria and Cell Fate Reprogramming Group, Department of Proteomics and Molecular Mechanisms of Neurodegenerative Diseases, Max Delbrück Center for Molecular Medicine (MDC) in the Helmholtz Association, Berlin, Germany

David Llobet-Navas Institute of Genetic Medicine, Newcastle University, Newcastle upon Tyne, UK

Bellvitge Biomedical Research Institute (IDIBELL), Gran via de l'Hospitalet, L'Hospitalet de Llobregat, Spain

Carlos López-Otín Departamento de Bioquímica y Biología Molecular, Facultad de Medicina, Instituto Universitario de Oncología (IUOPA), Universidad de Oviedo, Oviedo, Spain

Centro de Investigación en Red de Cáncer, Madrid, Spain

N. Luque Laboratorio de Inmunología Celular y Molecular, Centro de Investigación Biomédica, Facultad de Medicina, Universidad de los Andes, Santiago, Chile

Patricia Luz-Crawford Laboratorio de Inmunología Celular y Molecular, Centro de Investigación Biomédica, Facultad de Medicina, Universidad de los Andes, Santiago, Chile

E. Martínez-Sarrà Regenerative Medicine Research Institute, Universitat Internacional de Catalunya, Barcelona, Spain

Translational Cardiomyology Laboratory, Stem Cell Biology and Embryology Unit, Department of Development and Regeneration, Leuven, Belgium

Daniel Martins-de-Souza Instituto Nacional de Biomarcadores em Neuropsiquiatria (INBION) Conselho Nacional de Desenvolvimento Cientifico e Tecnologico, São Paulo, Brazil

UNICAMP's Neurobiology Center, Campinas, Brazil

Lab of Neuroproteomics, Department of Biochemistry and Tissue Biology, Institute of Biology, University of Campinas (UNICAMP), Campinas, Brazil

S. Montori Regenerative Medicine Research Institute, Universitat Internacional de Catalunya, Barcelona, Spain

Christine L. Mummery Department of Anatomy and Embryology, Leiden University Medical Centre, Leiden, The Netherlands

Kinichi Nakashima Department of Stem Cell Biology and Medicine, Graduate School of Medical Sciences, Kyushu University, Higashi-ku, Fukuoka, Fukuoka, Japan

Juliana Minardi Nascimento Lab of Neuroproteomics, Department of Biochemistry and Tissue Biology, Institute of Biology, University of Campinas (UNICAMP), Campinas, Brazil

D'Or Institute for Research and Education (IDOR), Rio de Janeiro, Brazil

R. Núñez-Toldrà Regenerative Medicine Research Institute, Universitat Internacional de Catalunya, Barcelona, Spain

Eduardo Oliver Centro Nacional de Investigaciones Cardiovasculares Carlos III (CNIC), Madrid, Spain

Luca Peruzzotti-Jametti Department of Clinical Neurosciences and NIHR Biomedical Research Centre, University of Cambridge, Cambridge, UK

Stefano Pluchino Department of Clinical Neurosciences and NIHR Biomedical Research Centre, University of Cambridge, Cambridge, UK

Ángel Raya Institució Catalana de Recerca I Estudis Avançats (ICREA), Barcelona, Spain

Center of Regenerative Medicine in Barcelona (CMRB), Barcelona, Spain

Center for Networked Biomedical Research on Bioengineering, Biomaterials and Nanomedicine (CIBER-BBN), Madrid, Spain

Christopher E. Rodgers University of Toronto & Krembil Research Institute, Toronto, ON, Canada

Ruth Rodriguez-Barrueco Institute of Genetic Medicine, Newcastle University, Newcastle upon Tyne, UK

Bellvitge Biomedical Research Institute (IDIBELL), Gran via de l'Hospitalet, L'Hospitalet de Llobregat, Spain

Universitat de Barcelona, Facultat de Medicina i Ciències de la Salut. Departament d'Anatomia i Terapèutica Experimental. Bellvitge Biomedical Research Institute (IDIBELL), Gran via de l'Hospitalet, L'Hospitalet de Llobregat, Spain

Meritxell Rovira Center of Regenerative Medicine in Barcelona (CMRB), Hospital Duran i Reynals, Barcelona, Spain

Veronica M. Saia-Cereda Lab of Neuroproteomics, Department of Biochemistry and Tissue Biology, Institute of Biology, University of Campinas (UNICAMP), Campinas, Brazil

M. Sampaolesi Translational Cardiomyology Laboratory, Stem Cell Biology and Embryology Unit, Department of Development and Regeneration, Leuven, Belgium

Human Anatomy Unit, Department of Public Health, Experimental and Forensic Medicine, University of Pavia, Pavia, Italy

Keita Tsujimura Department of Psychiatry, Nagoya University Graduate School of Medicine, Showa, Nagoya, Aichi, Japan

Institute for Advanced Research, Nagoya University, Chikusa, Nagoya, Aichi, Japan

Department of Stem Cell Biology and Medicine, Graduate School of Medical Sciences, Kyushu University, Higashi-ku, Fukuoka, Fukuoka, Japan

Anna Veiga Center of Regenerative Medicine in Barcelona (CMRB), Barcelona, Spain

National Stem Cell Bank-Barcelona Node, Biomolecular and Bioinformatics Resources Platform PRB2, ISCIII, CMRB, Barcelona, Spain

Center for Networked Biomedical Research on Bioengineering, Biomaterials and Nanomedicine (CIBER-BBN), Madrid, Spain

Giulio Volpe Department of Clinical Neurosciences and NIHR Biomedical Research Centre, University of Cambridge, Cambridge, UK

Jessica E. Young Department of Pathology, University of Washington, Seattle, WA, USA

Institute for Stem Cell and Regenerative Medicine, University of Washington, Seattle, WA, USA

Marco A. Zarbin Institute of Ophthalmology and Visual Science, Rutgers-New Jersey Medical School, Rutgers University, Newark, NJ, USA

Giuliana S. Zuccoli Lab of Neuroproteomics, Department of Biochemistry and Tissue Biology, Institute of Biology, University of Campinas (UNICAMP), Campinas, Brazil

Part I
Stem Cells for Brain Disorders

Chapter 1
Human-Induced Pluripotent Stem Cell-Derived Neurons to Model and Gain Insights into Alzheimer's Disease Pathogenesis

Jessica E. Young and Raul Delgado-Morales

Abstract Alzheimer's disease (AD) is the most common adult neurodegenerative disorder with the projected number of patients increasing to over 100 million in the next decades. Currently, there is no treatment that stops the progression of AD. Like other complex disorders, modeling AD in the laboratory is challenging due to inaccessibility of relevant living tissue, unknown interactions between genetics and environment, and species-specific differences between animal models and human subjects. Human induced pluripotent stem cell (hiPSC) technology has revolutionized the field of complex disease modeling. With this system, unique patient genetic backgrounds are captured in the dish and the cells can be directly differentiated to any cell type of an organism, including central nervous system cells affected in AD. This methodology allows for the design of genetic, molecular and biochemical experiments to decipher the complexity of AD using human neurons, or other CNS cell types, in the laboratory. Here we review current models of AD using hiPSCs, discuss advantages and disadvantages to the system, and propose future directions for this technology.

Keywords Alzheimer's disease · Human Induced Pluripotent Stem Cells · Neurons · Disease Modeling · Gene-editing · Therapeutic screening

---

J. E. Young (✉)
Department of Pathology, University of Washington, Seattle, WA, USA

Institute for Stem Cell and Regenerative Medicine, University of Washington, Seattle, WA, USA
e-mail: jeyoung@uw.edu

R. Delgado-Morales
Cancer Epigenetics and Biology Program (PEBC), Bellvitge Biomedical Research Institute (IDIBELL), L'Hospitalet de Llobregat, Barcelona, Catalonia, Spain

Department of Psychiatry & Neuropsychology, School for Mental Health and Neuroscience (MHeNs), Maastricht University, Maastricht, The Netherlands

R. Delgado-Morales (ed.), *Stem Cell Genetics for Biomedical Research*,
https://doi.org/10.1007/978-3-319-90695-9_1

1.1 Introduction

Alzheimer's disease (AD), the worldwide leading cause of dementia, is an age-related neurodegenerative disorder currently affecting approximately 46 million people worldwide. The global estimate cost of dementia is US$818 billion for 2016 [1], without taking into account inestimable social and health costs to relatives and other caregivers. Drug development for AD has proven to be difficult despite huge investment and research. Currently, only two types of drugs have been approved for AD treatment (acetylcholinesterase inhibitors and modulators of N-methyl-D-aspartate (NMDA) receptor), but none are extremely effective. Due to the lack of preventative therapeutics, the estimated number of patients by 2050 is more than 130 million [1].

Although clinical manifestations typically occur in patients over 65 years of age, AD develops over decades and it is believed that the first pathogenic processes start several decades before typical symptoms manifest [2, 3]. At the molecular level, AD is characterized by the appearance of two neuropathological hallmarks: the senile plaques comprised of insoluble aggregates of amyloid-beta (Aβ) and neurofibrillary tangles comprised of abnormally hyperphosphorylated Tau [4, 5]. Although the scientific community has amassed enormous knowledge of the molecular alterations and temporal progression of the disease, the driving events originating the disorder are still unknown. Genetic and nongenetic factors contribute to the etiopathology of AD. To date, almost 300 different mutations have been associated with familial AD, mainly appearing in APP, PSEN1, or PSEN2 genes [6]. However, approximately only 5% of all AD cases are attributed to deterministic genetic factors, pointing out the necessity of deciphering the other contributive factors of late-onset AD (LOAD) and sporadic AD (SAD). These factors include genetic variants, epigenetic changes, and environmental stimuli that lead to the development of the disorder.

The success rate of developing a disease-modifying therapeutic for AD has been poor and is impacted by an incomplete understanding of AD in a preclinical stage, thereby resulting in administration of treatment too late in the disease progression. Existing models should take this into account, and new models need to be developed to better understand the early stages of disease pathogenesis. However, to date, AD studies have mainly relied on the use of transgenic mice models, the use of nonneural human cell cultures, and human postmortem tissue analyses [7, 8]. Although the significant impact of transgenic mice models on progress in understanding various aspects of the disorder is undeniable, they only reproduce specific AD hallmarks and do not reflect clinical phenotypes completely [9]. Moreover, the use of nonneural cell lines omits unique neural features and therefore may fail to capture essential biological processes. It is these issues that make AD, and other age-related neurodegenerative disorders, challenging to study in the laboratory. The brain is a relatively inaccessible organ making sampling living tissue while a patient is alive challenging. Even when samples are available, the use of postmortem tissue does not allow differentiation between molecular hallmarks that are involved in the causes or consequences of the disease. For these reasons, there has been an ongoing demand for innovative and predictive model systems that closely resemble unique human neural features and which allow to study cause-effect relationships in a controlled setting.

The advent 10 years ago of generating induced pluripotent stem cells from mouse and human fibroblasts [10, 11] resulted in great excitement in the neurodegenerative research community. These stem cells can be directed to differentiate into all cellular lineages of an adult and can, in the case of complex disorders, capture complex human genetic background in a cell line, providing an infinitely renewable source of material for mechanistic in vitro experiments. This is particularly important for AD research because although basic mechanisms of neurodegeneration in AD, primarily those related to amyloid-beta generation, metabolism, and clearance, have been elucidated in other laboratory models, the cellular mechanisms leading to AD pathogenesis in late-onset sporadic AD are largely not well understood.

In this section, we will review the published models of AD using hiPSCs and discuss the advantages and disadvantages of these experiments. This topic has also been recently reviewed [12].

1.2 Familial AD Models

1.2.1 Patient Cell Lines

The first published report of hiPSCs from AD patient cells was in 2011 when Yagi et al. reprogrammed fibroblasts from patients harboring the presenilin 1 A246E mutation and the presenilin 2 N141I mutation, also known as the Volga German mutation [13]. Presenilins 1 and 2 comprise the γ-secretase complex which enzymatically cleaves the β-C-terminal fragment of APP to generate varying lengths of the Aβ peptide [14]. Yagi et al. reported increased Aβ 42 and decreased Aβ 40 from neurons derived from the presenilin 2 (PS2) mutant cell lines increasing the Aβ 42:40 ratio, similar to what has been observed in patient tissue. Pharmacological agents, which inhibit γ-secretase (Compound E) or selectively lower Aβ 42 (Compound W), had significant effects in these cells. The other main neuropathological hallmark of AD, phosphorylated tau protein, was examined, but no significant differences between the control cell lines and the FAD cell lines were determined, suggesting that perhaps a longer culture time was needed to observe significant tauopathy from hiPSC-derived neurons.

Israel et al. (2012) reported hiPSC lines from patients with a duplication of the APP gene, which leads to early-onset AD [15], as well as from two patients with SAD. Interestingly, increased phosphorylated tau protein was detected from the two APPdp patients and from one of the two SAD patients. Treatment with a β-secretase inhibitor, but not a γ-secretase inhibitor, reduced pTau in these lines, suggesting that the βCTF fragment of APP may be driving tau phosphorylation to a greater extent than Aβ peptides itself. Intriguingly, the authors also observed enlarged Rab5-positive early endosomes in neurons from the APP duplication and one SAD patient cell line, corroborating earlier reports of enlarged endosomes in SAD and FAD patient tissue [15–18].

In 2013, Kondo et al. reported iPSCs derived from APPV717lL and APPE693d mutation patients as well as two sporadic patients [19]. After generating cortical neurons from hiPSCs, they were the first group to report the detection of intracellular Aβ oligomers from one APPE693d mutant and one sporadic patient. Oligomeric Aβ is thought to be the most toxic species of the peptide [20], and Kondo et al. report increased neural cell stress responses, including increased amounts of the ER stress proteins BiP and caspase-4. Gene expression analysis uncovered increased expression of oxidative stress pathway genes, and treatment with docosahexaenoic acid (DHA), an ER stress inhibitor, alleviated these phenotypes. Interestingly, DHA was tested and was unsuccessful in AD clinical trials, but this work suggests that it may be beneficial in a subset of patients, arguing for a personalized or precision medicine approach to AD treatment. This type of approach can be validated using individual patient cells derived from hiPSCs.

Muratore and colleagues derived hiPSC lines from patients harboring the London mutation in APP, APPV717I [21]. Notably, this study documented an increase in total and phosphorylated levels of Tau, a phenotype reversed by treatment with Aβ-specific antibodies. In addition to observation of tauopathy in these patient-derived cell lines, this group also demonstrated altered APP cleavage, with increased Aβ 38 and Aβ 42 and a higher Aβ 42:40 ratio in the FAD cells, suggesting that altered APP cleavage may also contribute to phospho-tau phenotypes. More recently, additional work by this group demonstrates that hiPSCs from the APPV7171I patients can be patterned to different neuronal subtypes [22]. Intriguingly, they found that neurons directed toward rostral vs. caudal fates had differing responses to Aβ peptide species. In particular, they demonstrate that caudally directed neurons with APPV717I mutations have a reduced Aβ 40:42 ratio and no elevation of total and phospho-tau as compared to rostral neurons. Taken together, these elegant studies demonstrate the ability to begin to dissect the mechanisms of why certain cell types and brain regions are more susceptible to AD pathology than others.

Finally, hiPSC technology allows for identification of novel gene expression patterns. Sproul et al. generated hiPSC-derived neural progenitors from PS1 patient fibroblasts and characterized global gene expression [23]. They found several hundred differentially expressed genes and further narrowed the analysis to a handful of genes that appeared to be strongly differentially regulated in PS1 mutant cells. Several of these genes also showed significant differences in late-onset brains demonstrating that using iPSC-derived progenitors may be a valid method for identifying novel molecular targets in AD that are common between early- and late-onset forms of the disorder.

1.2.2 Gene Editing

One of the first descriptions of gene editing in hiPSCs was in 2013 where Woodruff et al. introduced the PS1 ΔE9 mutation into WT hiPSCs using TALEN-mediated genome engineering [24]. This study resulted in an allelic series of PS2 ΔE9 genotypes so that the mutant allele could be directly compared to null alleles in the same genetic background. In hiPSC-derived purified neurons harboring heterozygous null mutations and homozygous and heterozygous ΔE9 mutations, the authors observed impaired γ-secretase activity, leading to increased Aβ 42:40 ratio in ΔE9 mutant cell lines, but no impact on γ-secretase-independent functions of PS1, such as nicastrin maturation. Additionally, haploinsufficiency of PS1 had no effect on the Aβ 42:40 ratio in these neurons. This data demonstrated that PS1 loss of function and the ΔE9 mutation are not equivalent in terms of Aβ generation. Conversely, nicastrin maturation, a γ-secretase-independent function of PS1, was only impaired in the PS1 haploinsufficient cells. Together, this work suggests that PS1 mutations do not act as a simple loss of PS1 functions, an idea that has been controversial in AD biology [25].

The relative ease of gene editing with the CRISPR/Cas9 system is a powerful method to confirm the effects of dominant AD mutations and represents a novel avenue to specifically target disease-causing genetic variants. Recently, Ortiz-Virumbrales et al. went back to the PS2 N141I mutation to more specifically look at its effects in basal forebrain cholinergic neurons (BFCNs), one of the first vulnerable cell types in AD associated with loss of short-term memory [26]. In addition to corroborating the previous work by Yagi et al. and reporting an increase in the Aβ 42:40 ratio, the authors generated electrophysiologically active BFCNs and observed fewer maximum spikes in response to depolarization from the PS2 N141I mutant lines. They then generated isogenic lines in which the PS2 N141I mutation was corrected by CRISPR/Cas9 gene editing and demonstrated that the correction of the mutation ameliorated the observed functional deficit and restored spike activity to wild-type levels. As phenotypic variation is an issue in human cell lines from diverse genetic backgrounds, experiments such as this represent a powerful way to pinpoint an observed cellular phenotype to a specific mutation in an isogenic genetic background.

1.2.3 3D Models

One drawback to two-dimensional cultures for AD modeling is the lack of the neuropathology in terms of aggregated protein species (i.e., plaques and tangles) in two-dimensional cell culture. A plausible hypothesis is that a more three-dimensional structure is needed to observe this pathology and further mature the neural cells. In 2013, Choi et al. developed a culture paradigm in which ReN cells were transfected with APP and PS1 overexpression constructs and embedded in layers of Matrigel, resulting in three-dimensional cultures that could be sectioned and analyzed [27]. These studies demonstrated elevated aggregations of extracellular Aβ deposits and phospho-tau accumulation in processes reminiscent of dystrophic neurites. Interestingly, both of these phenotypes were attenuated by treatment with β- or γ-secretase inhibitors.

In 2016, Raja et al. demonstrated similar results in terms of the presence of aggregated proteins in self-organizing 3D neural organoids derived from patient hiPSCs [28]. Importantly, these models were derived from patient cell lines and thus expressed FAD mutations at endogenous levels, without the need for overexpression. In addition to extracellular aggregates of Aβ and increased aggregated, hyperphosphorylated Tau protein, they also report the presence of enlarged endosomes, a phenotype that has been reported in postmortem tissue [18, 29]. Taken together, this suggests that this system can recapitulate many aspects of AD neuropathology in vitro.

1.2.4 SAD Models

Due to complex human genetics and variable contributions from epigenetic and environmental factors, modeling late-onset sporadic AD (SAD) is understandably more variable than purely genetic forms of the disorder. In their 2012 work, Israel et al. examined hiPSC-derived neurons from two sporadic patients as well as from familial APP duplication carriers [15]. Interestingly, they found that cellular phenotypes from one of the two sporadic patients were consistently similar to what they observed from the APP duplication cells while the other patient line phenotypically segregated with the control cell lines. This data, although with a small number of genomes, implicated that diversity in patient genetic background may strongly impact cellular readouts. Kondo et al. (2013) also included some SAD patients along with FAD patients in their study and found that, again, one of two SAD patient cell lines exhibited phenotypes similar to FAD [19]. In this case, they report one SAD patient cell line accumulated intracellular Aβ oligomers, suggesting that a genetic factor may lead to an intracellular environment conducive to the formation of the oligomeric Aβ.

The apolipoprotein E4 allele (ApoE4) is the most associated risk factor for AD in the absence of APP or presenilin mutations. There are three genotypes of ApoE—E2, E3, and E4—and the E4 variant increases AD risk 3- to greater than 12-fold in heterozygous and homozygous carriers, respectively [30]. In 2014, Duan et al. generated neurons BFCNs from AD patients with ApoE3/E4 genotypes. They documented altered Aβ 42:40 ratio and increased neuronal cell death in response to excitotoxic stimuli in the E4 carriers [31].

Despite the well-documented association with ApoE, the genetics and heritability of SAD is complex, multiple genome-wide association studies (GWAS) have consistently identified around 20 loci associated with AD risk. Although the effect size of these risk variants in a population is low, it is unknown what the effect is in a specific individual, with a variable burden of risk alleles, or even in a specific cell type, such as purified neurons, astrocyte, or microglial cells. In 2015, Young et al. attempted to address this question by analyzing a cohort of patients genotyped for risk variants in the *SORL1* gene [32]. *SORL1* encodes the protein SORLA, which is involved in endocytic trafficking, a pathway established to be dysregulated in AD [33, 34]. In this study, the authors found that patient cell lines harboring two copies of an AD-associated risk haplotype did not induce *SORL1* expression above baseline level when treated with the neurotrophic factor BDNF. Furthermore, while BDNF treatment lowered secreted Aβ levels from purified neurons with protective variants, it did not have this protective effect in cells with two copies of the risk haplotype. Interestingly, this phenotype correlated only with variant genotype, not disease state, showing what might be expected of a genetic factor that contributes to disease risk but is not deterministic factor for AD.

In 2015, another report of SAD patient hiPSCs was published by Hossini et al. While this study was limited by a small number of cell lines and differentiations, transcriptome analysis suggested upregulation of AD-related pathways in the AD-iPSC-derived neurons [35]. Also, by comparing neuronal cells derived from AD-iPSCs to H9 hES-derived neurons, the authors constructed a network to model the association of AD-related proteins in hiPSC-derived neurons.

While this early work points to hiPSC-derived neurons as useful tools to interrogate complex human genetic backgrounds and individual transcriptomic profiles, it is clear that studies which will begin to elucidate the molecular basis of late-onset SAD will require large numbers of cell lines from a diverse population of patients. Also, as an AD diagnosis is only truly confirmed by neuropathological analysis upon autopsy, generation of cell lines from large patient cohorts with a bank of clinical and neuropathological data will be necessary to determine the accuracy of in vitro models.

1.2.5 hiPSC-Derived Cells for Therapeutic Screening

One of the most powerful aspects of hiPSC technology is the ability to generate an infinite number of cells that harbor an individual patient genetic background, differentiated into a disease-appropriate cell type. In terms of developing screening paradigms for new or existing therapeutics, this aspect of hiPSC disease modeling has the potential to be incredibly impactful for translatable, personalized medicine. Multiple studies, summarized above, have determined that key, detectable, and modifiable AD cellular phenotypes are generated from hiPSC-derived neurons. Most notably these include the detection of extracellular and intracellular Aβ and various phosphorylated forms of tau protein. As hiPSC-derived neurons can be differentiated in large numbers and plated in high-throughput screening (HTS) formats, this provides a biologically relevant platform to test novel small molecules or repurpose drugs from existing FDA-approved libraries. Recently, a study by Brownjohn et al. demonstrated the feasibility of this approach by using cortical neurons derived from trisomy 21 models (which have an extra copy of APP, thereby producing increased Aβ peptides) to identify small molecule modifiers of Aβ production in human neurons [36]. After a screen of a chemical library, they identified the avermectins, a class of macrocyclic lactones that decreases the production of longer Aβ fragments and increases shorter forms of the peptide. This effect is a phenocopy of γ-secretase-modulating drugs, which have also been proposed to be beneficial for AD [37]. An added advantage of avermectins, however, is that they exert no adverse effects on the γ-secretase enzymatic complex, which has many other substrates important for cellular health.

Additional work is moving quickly to develop phenotypic screens in a three-dimensional format, and recent work by Medda et al. describes the detection of tau aggregates from hiPSC-derived cortical neurons in a three-dimensional format in a 384-well plate [38]. Further work shows that mature neurons can be produced from hiPSCs in a large-scale format in three dimensions and that these cells can survive for months in culture, further increasing their maturity [39]. Taken together these studies provide a strong groundwork for developing screening systems that have the ability to assay multiple aspects of AD pathogenesis, such as endosome abnormalities, mitochondrial dysfunction, and synaptic aberrations, in addition to the defined neuropathological hallmarks of Aβ and phospho-tau.

1.3 Future Directions

The ability to generate AD-specific neural cell types has provided, and will continue to provide, unique insights into AD pathogenesis. As with any model system, there are drawbacks, one being the relative immaturity of the hiPSC-derived neurons. However, advances in culture technologies, especially 3D modeling, and in other methods to derive human neurons, such as transdifferentiation [40–42], can overcome some of these limitations. Now that living human neurons and other critical CNS cell types, including astrocytes, oligodendrocytes, and microglia [43–45], can be generated in reproducible and sufficient quantities, it will be important to integrate the data generated from human cells in vitro with the wealth of data collected in elegant animal models and from meticulous longitudinal and postmortem studies to gain new insights into the molecular pathogenesis of AD, in particular in ways that can productively develop effective therapeutics.

References

1. Wimo A et al (2017) The worldwide costs of dementia 2015 and comparisons with 2010. Alzheimers Dement J Alzheimers Assoc 13(1):1–7
2. Jack CR Jr et al (2010) Brain beta-amyloid measures and magnetic resonance imaging atrophy both predict time-to-progression from mild cognitive impairment to Alzheimer's disease. Brain 133(11):3336–3348
3. Reiman EM et al (2012) Brain imaging and fluid biomarker analysis in young adults at genetic risk for autosomal dominant Alzheimer's disease in the presenilin 1 E280A kindred: a case-control study. Lancet Neurol 11(12):1048–1056
4. Mandelkow EM, Mandelkow E (1998) Tau in Alzheimer's disease. Trends Cell Biol 8(11):425–427
5. Selkoe DJ (1991) The molecular pathology of Alzheimer's disease. Neuron 6(4):487–498
6. Cacace R, Sleegers K, Van Broeckhoven C (2016) Molecular genetics of early-onset Alzheimer's disease revisited. Alzheimers Dement J Alzheimers Assoc 12(6):733–748
7. Sullivan SE, Young-Pearse TL (2017) Induced pluripotent stem cells as a discovery tool for Alzheimers disease. Brain Res 1656:98–106

8. Wojda U, Kuznicki J (2013) Alzheimer's disease modeling: ups, downs, and perspectives for human induced pluripotent stem cells. J Alzheimers Dis 34(3):563–588
9. Franco R, Cedazo-Minguez A (2014) Successful therapies for Alzheimer's disease: why so many in animal models and none in humans? Front Pharmacol 5:146
10. Takahashi K et al (2007) Induction of pluripotent stem cells from adult human fibroblasts by defined factors. Cell 131(5):861–872
11. Takahashi K, Yamanaka S (2006) Induction of pluripotent stem cells from mouse embryonic and adult fibroblast cultures by defined factors. Cell 126(4):663–676
12. Mungenast AE, Siegert S, Tsai LH (2016) Modeling Alzheimer's disease with human induced pluripotent stem (iPS) cells. Mol Cell Neurosci 73:13–31
13. Yagi T et al (2011) Modeling familial Alzheimer's disease with induced pluripotent stem cells. Hum Mol Genet 20(23):4530–4539
14. De Strooper B, Iwatsubo T, Wolfe MS (2012) Presenilins and gamma-secretase: structure, function, and role in Alzheimer disease. Cold Spring Harb Perspect Med 2(1):a006304
15. Israel MA et al (2012) Probing sporadic and familial Alzheimer's disease using induced pluripotent stem cells. Nature 482(7384):216–220
16. Cataldo AM et al (2008) Down syndrome fibroblast model of Alzheimer-related endosome pathology: accelerated endocytosis promotes late endocytic defects. Am J Pathol 173(2):370–384
17. Cataldo AM et al (2004) Presenilin mutations in familial Alzheimer disease and transgenic mouse models accelerate neuronal lysosomal pathology. J Neuropathol Exp Neurol 63(8):821–830
18. Cataldo AM et al (2000) Endocytic pathway abnormalities precede amyloid beta deposition in sporadic Alzheimer's disease and down syndrome: differential effects of APOE genotype and presenilin mutations. Am J Pathol 157(1):277–286
19. Kondo T et al (2013) Modeling Alzheimer's disease with iPSCs reveals stress phenotypes associated with intracellular Abeta and differential drug responsiveness. Cell Stem Cell 12(4):487–496
20. Shankar GM et al (2008) Amyloid-beta protein dimers isolated directly from Alzheimer's brains impair synaptic plasticity and memory. Nat Med 14(8):837–842
21. Muratore CR et al (2014) The familial Alzheimer's disease APPV717I mutation alters APP processing and Tau expression in iPSC-derived neurons. Hum Mol Genet 23(13):3523–3536
22. Muratore CR et al (2017) Cell-type dependent Alzheimer's disease phenotypes: probing the biology of selective neuronal vulnerability. Stem Cell Reports 9(6):1868–1884
23. Sproul AA et al (2014) Characterization and molecular profiling of PSEN1 familial Alzheimer's disease iPSC-derived neural progenitors. PLoS One 9(1):e84547
24. Woodruff G et al (2013) The presenilin-1 DeltaE9 mutation results in reduced gamma-secretase activity, but not total loss of PS1 function, in isogenic human stem cells. Cell Reports 5(4):974–985
25. Anonymous (2007) Alzheimer research forum discussion: gain or loss of function – time to shake up assumptions on gamma-secretase in Alzheimer disease? J Alzheimers Dis 11(3):399–417
26. Ortiz-Virumbrales M et al (2017) CRISPR/Cas9-correctable mutation-related molecular and physiological phenotypes in iPSC-derived Alzheimer's PSEN2 (N141I) neurons. Acta Neuropathol Commun 5(1):77
27. Choi SH et al (2014) A three-dimensional human neural cell culture model of Alzheimer's disease. Nature 515(7526):274–278
28. Raja WK et al (2016) Self-organizing 3D human neural tissue derived from induced pluripotent stem cells recapitulate Alzheimer's disease phenotypes. PLoS One 11(9):e0161969
29. Cataldo AM, Hamilton DJ, Barnett JL, Paskevich PA, Nixon RA (1996) Abnormalities of the endosomal-lysosomal system in Alzheimer's disease: relationship to disease pathogenesis. Adv Exp Med Biol 389:271–280
30. Avramopoulos D (2009) Genetics of Alzheimer's disease: recent advances. Genome Med 1(3):34

31. Duan L et al (2014) Stem cell derived basal forebrain cholinergic neurons from Alzheimer's disease patients are more susceptible to cell death. Mol Neurodegener 9:3
32. Young JE et al (2015) Elucidating molecular phenotypes caused by the SORL1 Alzheimer's disease genetic risk factor using human induced pluripotent stem cells. Cell Stem Cell 16(4):373–385
33. Small SA, Simoes-Spassov S, Mayeux R, Petsko GA (2017) Endosomal traffic jams represent a pathogenic hub and therapeutic target in Alzheimer's disease. Trends Neurosci 40(10):592–602
34. Willnow TE, Andersen OM (2013) Sorting receptor SORLA – a trafficking path to avoid Alzheimer disease. J Cell Sci 126(Pt 13):2751–2760
35. Hossini AM et al (2015) Induced pluripotent stem cell-derived neuronal cells from a sporadic Alzheimer's disease donor as a model for investigating AD-associated gene regulatory networks. BMC Genomics 16:84
36. Brownjohn PW et al (2017) Phenotypic screening identifies modulators of amyloid precursor protein processing in human stem cell models of Alzheimer's disease. Stem Cell Reports 8(4):870–882
37. Liu Q et al (2014) Effect of potent gamma-secretase modulator in human neurons derived from multiple presenilin 1-induced pluripotent stem cell mutant carriers. JAMA Neurol 71(12):1481–1489
38. Medda X et al (2016) Development of a scalable, high-throughput-compatible assay to detect tau aggregates using iPSC-derived cortical neurons maintained in a three-dimensional culture format. J Biomol Screen 21(8):804–815
39. Rigamonti A et al (2016) Large-scale production of mature neurons from human pluripotent stem cells in a three-dimensional suspension culture system. Stem Cell Reports 6(6):993–1008
40. Huh CJ et al (2016) Maintenance of age in human neurons generated by microRNA-based neuronal conversion of fibroblasts. Elife 5:pii: e18648
41. Mertens J et al (2015) Directly reprogrammed human neurons retain aging-associated transcriptomic signatures and reveal age-related nucleocytoplasmic defects. Cell Stem Cell 17(6):705–718
42. Vierbuchen T et al (2010) Direct conversion of fibroblasts to functional neurons by defined factors. Nature 463(7284):1035–1041
43. Tcw J et al (2017) An efficient platform for astrocyte differentiation from human induced pluripotent stem cells. Stem Cell Reports 9(2):600–614
44. Ehrlich M et al (2017) Rapid and efficient generation of oligodendrocytes from human induced pluripotent stem cells using transcription factors. Proc Natl Acad Sci U S A 114(11):E2243–E2252
45. Abud EM et al (2017) iPSC-derived human microglia-like cells to study neurological diseases. Neuron 94(2):278–293. e279

Chapter 2
Modeling Schizophrenia with Human Stem Cells

Juliana Minardi Nascimento, Veronica M. Saia-Cereda, Giuliana S. Zuccoli, Danielle Gouvêa-Junqueira, and Daniel Martins-de-Souza

Abstract The complexity of psychiatric disorders is a challenge still to overcome, and schizophrenia has been the most prevalent yet little understood. Several studies have used knowledge from postmortem brain tissue and other models, to address difficult questions regarding diagnostic and treatment. An improvement in the translational capacity of molecular profiling studies of psychiatric disorders was achieved with the development of human-induced pluripotent stem cells (iPSCs), through provision of human neuronal-like tissue. The finding that iPSCs can recapitulate the phenotype of the donor also affords the possibility of using this approach to study both the disease and control states in a given medical area.

Keywords Brain · iPSC · Neural cell · Neurodevelopment · Psychiatric disorders

---

J. Minardi Nascimento (✉)
Lab of Neuroproteomics, Department of Biochemistry and Tissue Biology, Institute of Biology, University of Campinas (UNICAMP), Campinas, Brazil

D'Or Institute for Research and Education (IDOR), Rio de Janeiro, Brazil
e-mail: juminardi@gmail.com

V. M. Saia-Cereda · G. S. Zuccoli · D. Gouvêa-Junqueira
Lab of Neuroproteomics, Department of Biochemistry and Tissue Biology, Institute of Biology, University of Campinas (UNICAMP), Campinas, Brazil

D. Martins-de-Souza
Instituto Nacional de Biomarcadores em Neuropsiquiatria (INBION) Conselho Nacional de Desenvolvimento Cientifico e Tecnologico, São Paulo, Brazil

UNICAMP's Neurobiology Center, Campinas, Brazil

Lab of Neuroproteomics, Department of Biochemistry and Tissue Biology, Institute of Biology, University of Campinas (UNICAMP), Campinas, Brazil

R. Delgado-Morales (ed.), *Stem Cell Genetics for Biomedical Research*,
https://doi.org/10.1007/978-3-319-90695-9_2

2.1 Introduction

Neuropsychiatric disorders, such as schizophrenia (SCZ), bipolar disorder (BPD), major depressive disorder (MDD), and autism spectrum disorders (ASD), are devastating and complex disorders of the brain, with a strong personal, social, and economical burden in modern society. Those disorders are responsible for a lifelong disability and decrease in life expectancy [68]. Schizophrenia is one the most prevalent among them and shows a variety of symptoms, which overlap to other neuropsychiatric diseases, such as MDD and manic traits, affecting precise diagnosis and the design of treatment strategies [24]. In general, the initial symptoms are present at the beginning of adulthood; however, molecular roots of the disease can be linked to the neurodevelopment [52]. As a poorly understood disorder regarding its molecular mechanisms, schizophrenia research claims for novel discoveries, and characterization of new individual factors, contributing to onset or progression, is necessary [32].

Such complexity is only partially explained by genetics and its polygenic etiology [49, 59] and reflects the intricate relationship with environmental conditions, from prenatal neurodevelopment to the onset of the disease during early adult life. The complex cross talk of gene and protein expression in schizophrenia leads to abnormal processes during neurodevelopment [20, 52]. The multifactorial aspects of the disorder have impaired the understanding of the basis of the disease in cellular and molecular levels. Nevertheless, several genetic variants have been related to increased risk, yet not sufficient as a cause of the diseases, keeping the study of complex psychiatric disorders challenging [11–13].

On this matter, technological development of next-generation sequencing allowed deeper sample coverage and higher throughput in neuropsychiatric studies [12, 21, 54]. Genetic variations are being addressed by international collaborations, such as the Psychiatric Genomics Consortium (PGC), to provide a comprehensive number of subjects and samples analyzed. Several independent SCZ-associated loci (>100) and structural variation of genomic copy number variations (CNVs) were found, which contribute to disease susceptibility [11–14]. Those strongly support the polygenic contribution of small allelic effects on schizophrenia [6, 50].

An often-appointed limitation of genetic approaches in psychiatric disorders is the extrapolation to function, where transcription (of coding and noncoding RNAs) and protein expression are not taken into account. Proteins most of all still undergo several modifications from translation to posttranslation. Recognizing disturbances within the proteome of neuropsychiatric models of study enables a more complex cellular and molecular analysis of a human brain disorder in vitro [76].

Advances from the last decade directed research of postmortem brains and animal models to a broader spectrum of tools, from brain imaging to cell-based studies, including induced pluripotent stem cells (iPSCs). This approach consists of generation of pluripotent cells by genetically reprogramming somatic cells to produce the multiple neural cell types as observed within the nervous tissue. Nowadays, iPSCs are reprogrammed from several somatic cell types (skin, blood, urinary tract epithelium) and enable the study of patient-specific and disease-related specific cells

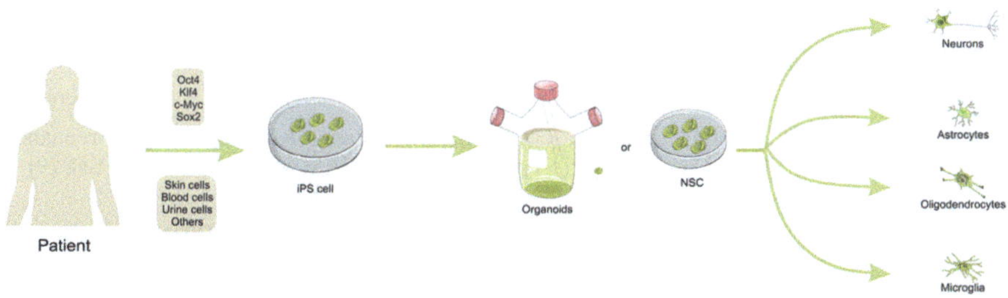

Fig. 2.1 Strategy for modeling the pathophysiology of human diseases using specific patient human iPS cells

in vitro (Fig. 2.1). Theoretically human-induced pluripotent stem cells (iPSCs) can differentiate into any cell type within the human organism, thus supplying a vast amount of cells of any fate, including brain cells. As an exciting tool supporting biomedical development, human iPSCs provide the basis to study the relationship between genotype and phenotype, key to understand some mechanisms underlying schizophrenia.

During the past decades, there have been several advances in understanding the molecular basis of neural development and disturbance consequences during disease. Furthermore, human iPSCs overcome the unpractical and poor accessibility of human brain cell types [47] and are providing the possibility to challenge and question neural cell fates, according to the disease in question; allowing detection of molecular signatures of the disease in question in living cells, prior or during onset; and providing a platform for identifying new molecular targets or novel biomarkers of the disease.

Therefore, as more models of study are available, the closer we are to untangle difficulties and challenges of schizophrenia research, such as genetic backgrounds and environmental cues involved. Taken into account the ever-growing interest in using stem cells to model human complex diseases, here we give an overview (Table 2.1) and discuss recent advances in the study of schizophrenia.

2.2 From Postmortem Brain Findings to Translational Possibilities

Several hypotheses for the pathophysiology of schizophrenia have emerged from studies of patients' postmortem brains. Despite the utmost significance of those studies, there are still ongoing efforts to concatenate and thus better understand the pathophysiology of the disease as a whole. Among those is the quest to understand the cell types involved in those deregulations observed. Previous studies, focusing on the role of neurons as the protagonist on the disease, were precedent, and most recent years showed several discussions about the role of glia cells in schizophrenia [4].

Table 2.1 Summary of studies using iPS cells for studying schizophrenia

Genetic information	iPSC cases	Neural differentiation	Findings	Reference
Sporadic	4 SCZ, 6 CTR	NPCs and neurons	Changes in Wnt and cAMP signaling pathways and NCAM expression in SCZ neurons; with decreased neuronal connectivity and neurite number. Loxapine rescued neuronal connectivity in patients	Brennand et al. [9]
DISC1 microdeletion	2 SCZ, 1 CTR	–	First iPSCs derived from adult patients with SCZ, carrying DISC1	Chiang et al. [10]
22q11.2 microdeletion and sporadic	3 SCZ, 2 CTR	Glutamatergic neurons	Delayed Oct4 and Nanog downregulation during neural differentiation in SCZ patients with 22q11.2 deletion	Pedrosa et al. [47]
Sporadic	1 SCZ, 1 CTR	NPC	NPCs from SCZ patients present higher ROS levels (which is reverted by valproic acid) and higher extramitochondrial oxygen consumption	Paulsen et al. [45]
Sporadic clozapine-treated	3 SCZ, 2 CTR	Neurons	Reduced efficiency in differentiation to dopaminergic neurons compared to glutamatergic neurons in SCZ	Robicsek et al. [51]
Sporadic	3 SCZ, 3 CTR	Neurons	SCZ-derived neurons have changes in secreted catecholamines and increased percentage of TH-positive neurons	Hook et al. [28]
DISC1 4 bp deletion	1 SCZ, 1 MDD, 3 CTR	Forebrain hNPCs and neurons	DISC1-mutated has impairment in synaptic processes, such as vesicle release	Wen et al. [66]
15q11.2 or DISC1 microdeletion	5 SCZ, 3 CTR	NPC and rosettes	Impairment in adherens junction and apical polarity on 15q11.2 microdeletion	Yoon et al. [73]
Sporadic	4 SCZ, 6 CTR	Forebrain NPCs	Aberrant migration; increase in oxidative stress; reduced neuronal connectivity, neurite outgrowths, and PSD95 levels	Brennand et al. [8]
CNTNAP2 deletions	1 SCZ, 7 CTR	Forebrain NPC, neurons and OPCs	Exon- and allele-specific CNTNAP2 expression in NPCs, neurons, and OPCs correlates with aberrant migration in NPCs	[35]
Sporadic	4 SCZ, 6 CTR	NPC	WNT signaling differentially expressed, showed increased canonical WNT signaling	Topol et al. [62]
22q11.2 microdeletion	6 SCZ, 6 CTR	Neurons	Neurons with 22q11.2 del recapitulate the miRNA expression patterns expected of 22q11.2 haploinsufficiency. Expression of several miRNAs that are similar to those found in clinical samples	Zhao et al. [75]

(continued)

Table 2.1 (continued)

Genetic information	iPSC cases	Neural differentiation	Findings	Reference
CNTNAP2 deletions	1 SCZ, 3 CTR	NPC, astrocyte, neurons	Intragenic CNTNAP2 deletions affect genes involved in neuronal development and neuronal activity	Flaherty et al. [19]
22q11.2 microdeletion	8 SCZ, 7 CTR	Neurons	Correlates the 22q11.2 deletion with impaired neuronal development, pointing possible pathways as drug and treatment targets	Lin et al. [36]
Sporadic miR-9	14 SCZ, 16 CTR	NPCs	Changes in the expression of miR-9, and its indirect targets, are related to risk of developing schizophrenia	Topol et al. [61]
DISC1 4 bp deletion	1 SCZ, 2 CTR	Forebrain organoids	Delayed cell-cycle progression during mitosis of radial glia in organoids of individual with DISC1 mutation (with a defect in its interaction with Nde1/Ndel1)	Ye et al. [71]

Although postmortem brain samples are able to provide information on different neural cell types and pathophysiology of the disease, those samples are influenced by a number of factors, such as the difference in age, lifestyle, cause of death, and especially the action of different types of medication that a patient receives throughout their life. Besides, samples are collected from chronically ill patients, which will reflect the final phases of the disorder. Consequently, those are confounding factors preventing deeper conclusions on mechanisms and deregulations of genes, proteins, or metabolites of those patients, broadening the puzzle regarding disease-related phenotypes.

In order to address these immediate concerns, several research groups have started to use iPSC models on issues related to CNS diseases, and most have a central focus on the role of neurons. First reports studied iPSC-derived neural progenitors from patients with disrupted in schizophrenia 1 (DISC1) mutation [10]. Some others were using more complex and sporadic genetic cases, which were then taken to neuronal fate [9, 45–47]. And as presented in Table 2.1, several others followed, with focus on NPCs and neurons from patients with DISC1 mutation. Unregulated or altered DISC1 has been implicated in several neurodevelopmental processes, and recent advances have focused on functional interaction between DISC1 and Ndel1/Nde1, including proliferation and disruption in neurogenesis by delaying cell-cycle progression of iPSC-derived cells [71]. In addition, synaptic maturation and formation of synaptic processes, such as vesicle release, axon and dendrite outgrowth, and neuronal migration and cell-to-cell adhesion, were observed [66].

Further studies, using more patient-derived iPSCs, have exhibited abnormal gene and protein expression as compared to controls, in addition to impaired migration of progenitors [7, 61, 62], deficits associated with adherens junctions and polarity [73], and reduced connectivity and low synaptic activity of SCZ-derived neurons [9, 74]. Taken together, they have observed similar phenotypes, such as altered protein levels of synaptic proteins, altered oxygen metabolism, and increase in oxidative stress, which were comparable to postmortem brain findings [42, 53, 67].

As previously indicated, schizophrenia is a multifactorial disease; nonetheless, regarding the genetic aspect several groups have focused on small studies of rare mutations associated with schizophrenia. The 22q11.2 microdeletions, considered one of the potential high risks for developing schizophrenia and other schizoaffective disorder or ASD, affect mostly genes expressed in the nervous system [36, 47, 75], and the individuals can show cognitive and behavioral symptoms associated with the development of schizophrenia [30]. Other observed phenotypes have been associated with delayed neuronal development, including those related to microRNA processing genes, thus impairing miRNA synthesis and, consequently, translation [75].

Alterations in miRNA have been correlated to increased risk to develop schizophrenia including miR-137 and miR-9 [15, 50, 61]. miR-137 is found to be highly expressed in several brain regions, such as the hippocampus, thalamus, and different regions of the cortex [72]. Associated with processes like proliferation and differentiation in neural stem cells, its target genes are related to pathways on axon guidance and long-term potentiation, affecting information process and transmission [56, 69, 72]. While manipulation of miR-9 impacts neural migration, changes of indirect targets are related to risk of developing schizophrenia [61]. Those miRNAs can be found in exosomes; nanovesicles have an essential role in cell signaling and communication between neurons and glial cells [3, 22, 23].

In addition to a complex genetic background, most studies have generated mixed neural populations, and thus, finding a precise functional gene/protein which is disease-associated is as ever challenging. However, several protocols are being developed taking into account cell diversity.

2.3 A Sea of Neural Diversity: Looking Beyond Neurons

The possibility to test different hypothesis in neuropsychiatric research, which were generated from postmortem brain studies, became possible with the introduction of stem cell biology. Both embryonic and induced pluripotent stem cells have shown possibilities to generate functional cells. Currently, challenges faced are to develop differentiation strategies that yield disease-relevant cell types, from different subtypes of neurons to glial cells.

Last decade has also witnessed an increase in evidences of glia being implicated in neuropsychiatric disorders [4, 53]. Since a seminal study 16 years ago [64], numerous studies point to the role of oligodendrocytes in the pathophysiology of schizophrenia, since molecules characteristic of this cellular type, such as MPB and MOG, have been often found deregulated in patients [25, 39, 43]. The main role of these molecules is in the formation of the myelin sheath and, within the scope of schizophrenia, the possible deregulation in the production of myelin in axons of neurons. Still, the models of analysis of this type of in vitro function are still precarious. Thus, the rise in cellular models, such as iPSC differentiation into oligodendrocytes is required, has open possibilities to understand the mechanisms of interaction and the possibility to study drugs that can act in the reversion of schizophrenia phenotype.

Mature oligodendrocytes have not yet been differentiated from iPSCs of patients with schizophrenia, let alone cocultivation of these cells with neurons, in order to attest to deregulation in the formation of the myelin sheath. However, recent studies have started to look toward oligodendrocyte progenitors (OPCs) derived from iPSCs [34] in schizophrenia patients, to understand specific function of proteins in different cell types. Nonetheless, differentiation of mature oligodendrocytes from iPSC was shown to be possible; and myelination deficiency was reversed by the injection of these cells in both mutant shiverer mice, which has a low neonatal myelination of brains [65], as well as in young brains of rats treated with drugs that decreases the number of OPCs [48]. Thus, it is only a matter of time to observe more neuropsychiatric disorders from the point of view of in vitro myelin formation and to better understand how impairment in myelination may contribute to reduce white matter integrity, their function, and outcomes in schizophrenia, mostly related to cognitive impairment and brain aging.

Astrocytes, on the other hand, are the most numerous cells present in the brain and have a great morphological and spatial variety. The past decade has seen an increased interest in the biology of those cells, beyond the scope of giving support to neurons. In addition to supplying neurons and oligodendrocytes with substrate for energy metabolism, astrocytes have been known to participate in electrolyte homeostasis; production and control of neuromodulators, ions, and neurotransmitters; modulation of immune response; synapsis; migration; and development and maintenance of the myelin sheath (reviewed in [31]). GFAP protein, a known marker of the cytoskeleton of astroglia, has been often found in postmortem tissue studies of various brain regions of patients with schizophrenia [42].

Since 2013, there are methods to differentiate astrocytes from iPSCs, but protocol was long and expensive, and more alternatives were explored [55]. Nowadays, there are numerous studies and different methods to differentiate pluripotent stem cells to astrocytes, both in 2D and 3D culture, as well as in co-culture with neurons [17, 44]. iPSC-derived astrocytes are both gene functionally and phenotypically comparable to human primary astrocyte cells, with also a strong correlation of transcriptome and proteome analyses of astrocytes [17, 60]. In addition, mature astrocytes from iPSCs were also derived from human cortical spheroids cultured for up to 590 days [57]. Those astrocytes were derived in several periods along the 3D differentiation, showing the course of maturation and differentiation those cells would take in vitro. In summary, several groups were able to differentiate functional astrocytes, which were derived following the pattern of development, and behaved similarly to primary human astrocytes in culture. Yet, understanding better how neuropsychiatric disorders behave during the course of differentiation and which steps are modulated by the disease are still only beginning to be answered.

To date, no study has been reported analyzing iPSC-derived astrocytes from patients with schizophrenia. However, when it comes to neurodevelopment, one of the major focuses of astrocyte deregulation in schizophrenia is related to DISC1 protein. Mice studies have shown DISC1 affects deregulation of functions as dendritic support and synapse maturation [1, 70], functions known to be altered in patients with schizophrenia. In addition, there are several studies relating new pharmacological treatments with action on astrocytes, such as D-serine and glycine [37], which mainly reverse the negative symptoms of schizophrenia (reviewed in [33]). And here, iPSC-derived astrocytes from DISC1-mutated cells might be key, not only to attest the reversal of deregulations but also to better understand mechanisms of action of these compounds.

The complexity of schizophrenia has stimulated several theories on disease pathogenesis, one of them associates the immune system influence. And some light was shed to the influence of microglia on schizophrenia brain. Compared to other types of glial cells, microglia has fewer studies when it comes to iPSCs, which are still under development. However, recently some studies started to show and characterize those cell types [2, 18, 41]. These cells play an important role in brain immunity and also actively participate in neurodevelopmental functions such as maturation of neurons and astrocytes, proliferation of astrocytes, and differentiation and maintenance of oligodendrocytes [40], and microglia-conditioned medium can stimulate proliferation of neural progenitor cells [5]. Regarding schizophrenia, there are still no studies using iPSC-derived microglia from patients, and the neurodevelopment-microglia-schizophrenia relationship is still to be elucidated.

While many advances have been obtained with methods of preparation and analysis of glial cells from iPSCs, as elucidated above, these still have the potential to be more explored in schizophrenia research, particularly with regard to alternative answers related to neurodevelopment and cell-specific analysis.

2.4 Genetic Editing on Pluripotent Stem Cells

Current developments in human pluripotent stem cell technology and techniques for genetic editing have provided new possibilities for the investigation of human biology in health and disease as well as the development of therapeutic applications. Different studies have accomplished the successful editing of disease-linked genes in animal models as well as patient-derived iPSCs [16].

Clustered Regularly Interspaced Short Palindromic Repeats (CRISPR), together with CRISPR-associated proteins (Cas), are a part of the adaptive immune system found in bacteria and archaea responsible for the promotion of immune response and elimination of invading genetic material [29]. This molecular machinery has recently been adapted for use in higher organisms, which revolutionized the field of gene editing and expression. The technique enables an efficient deletion and/or substitution of whole genes as well as their silencing or activation in almost any type of organism and cell, opening a wide range of possibilities to study [63]. One of the most important advantages of the CRISPR-Cas9 system is that it does not rely on protein engineering, differently from zinc-finger nucleases (ZFNs) and transcription activator-like effector nucleases (TALENs) [63]. Recently, this approach has been used to repress or activate gene expression and to gain-of-function or loss-of-function in several organisms and cell types, with high fidelity [38].

This ability to specifically control disease-related genes can be even more powerful when combined to iPSC-derived cells, mostly on learning the contribution to key risk genes related to a disease. In molecular psychiatry, that combination might be a real promise for joining the potential to study the neurodevelopmental component and to profile and test the genetic risk involved. Regarding schizophrenia, many gene loci have been implicated as contributors to the development of the condition. Therefore, employing TALENs to reverse a frameshift mutation on DISC1 in iPSC, a study generated different types of isogenic iPSC of the SCZ-associated characteristic. Results have shown synaptic vesicle formation and release deficit in forebrain neurons of those patients containing DISC1 mutation, but also on DISC1-mutated isogenic cells [66]. In addition, using CRISPR/Cas9 and TALENs to disrupt the DISC1 gene in iPSCs derived from control subjects, others have shown main alteration in baseline Wnt signaling and an increased DISC1 nonsense-mediated decay of splice transcripts, altering neural cell fate via an established patterning pathway but not neuronal differentiation itself [58].

More recent studies aimed to use CRISPR/Cas9 to edit open chromatin peaks in patient-derived iPSC line carrying alleles of SCZ risk variants, since those regions of open chromatin are usually enriched for disease-associated variants. This genetic editing at the miR-137 locus promoted alterations in several neurodevelopmental features of iPSC neurons in parallel with lower expression of miR-137, such as reduced neuronal cell proliferation, higher dendrite arborization, and increased neurite outgrowth [26]. Others showed CRISPR-mediated gene-activation/repression platform on human iPSC-derived NPCs, neurons, and astrocytes, manipulating endogenous expression levels of candidate neuropsychiatric risk genes recently pointed [12, 13], such as channel tetramerization domain containing 13 (KCTD13), thousand and one amino acid protein kinase 2 (TOAK2), neurexin 1 (NRXN1), synaptosome-associated protein 91 (SNAP91) and chloride voltage-gated channel (CLCN3) [27]. As recent and fast-growing field, CRISPR tools associated to iPSC are only beginning to address molecular psychiatry demands, and several groups are prone to investigate those concerns, in addition to developing new and better methods to study brain disorders.

2.5 Final Remarks

A considerable number of molecular studies on psychiatric disorders conducted in human samples have been performed on postmortem tissues. Moreover, intrinsic limitations of animal models to psychiatric studies claim for new strategies to study biological mechanisms associated to these disorders. The introduction of iPSC as a new model approach in recent years has increased the possibilities to inquire about mechanisms involved in cellular connection in the human brain, since they can differentiate into disease-affected cell types and recapitulate the early steps of neuronal development. The recent year has seen an increase in studies using iPSC from schizophrenia patients, from sporadic manifestations and specific familial gene mutations.

So far, this model has shown promising results, and the more it is known, the better is the outcome to understand this complex disorder. In addition, proteins previously identified as potential biomarkers can be verified and possibly used to rapidly increase drug development. So required when it comes to understand the mechanisms of the disease, like schizophrenia, which has a very low number of good responders to currently available drugs, and thus helping with diagnosis and treatment.

Acknowledgments JMN, VMSC, GSZ, and DMS are supported by the São Paulo Research Foundation (FAPESP) grants 14/21035-0, 16/07332-7, 16/04912-2, 13/08711-3, and 14/10068-4.

References

1. Abazyan S, Yang EJ, Abazyan B et al (2014) Mutant disrupted-in-schizophrenia 1 in astrocytes: focus on glutamate metabolism. J Neurosci Res 92:1659–1668. https://doi.org/10.1042/BST20130220
2. Abud EM, Ramirez RN, Martinez ES et al (2017) iPSC-derived human microglia-like cells to study neurological diseases. Neuron 94:278–293.e9. https://doi.org/10.1016/j.neuron.2017.03.042
3. Banigan MG, Kao PF, Kozubek JA et al (2013) Differential expression of exosomal microRNAs in prefrontal cortices of schizophrenia and bipolar disorder patients. PLoS One 8:e48814. EP –. https://doi.org/10.1371/journal.pone.0048814
4. Bernstein H-G, Steiner J, Guest PC et al (2015) Glial cells as key players in schizophrenia pathology: recent insights and concepts of therapy. Schizophr Res 161:4–18. https://doi.org/10.1016/j.schres.2014.03.035
5. Beumer W, Gibney SM, Drexhage RC et al (2012) The immune theory of psychiatric diseases: a key role for activated microglia and circulating monocytes. J Leukoc Biol 92:959–975. https://doi.org/10.1189/jlb.0212100
6. Bigdeli TB, Ripke S, Bacanu S-A et al (2015) Genome-wide association study reveals greater polygenic loading for schizophrenia in cases with a family history of illness. Am J Med Genet 171:276–289. https://doi.org/10.1016/j.ajhg.2010.11.011
7. Brennand K, Savas JN, Kim Y et al (2014) Phenotypic differences in hiPSC NPCs derived from patients with schizophrenia. Mol Psychiatry 20:361–368. https://doi.org/10.1038/mp.2014.22
8. Brennand KJ, Marchetto MC, Benvenisty N et al (2015) Creating patient-specific neural cells for the in vitro study of brain disorders. Stem Cell Reports 5:933–945. https://doi.org/10.1016/j.stemcr.2015.10.011
9. Brennand KJ, Simone A, Jou J et al (2011) Modelling schizophrenia using human induced pluripotent stem cells. Nature 473:221–225. https://doi.org/10.1038/nature09915
10. Chiang C-H, Su Y, Wen Z et al (2011) Integration-free induced pluripotent stem cells derived from schizophrenia patients with a DISC1 mutation. Mol Psychiatry 16:358–360. https://doi.org/10.1126/science.1172482
11. Committee TPGCS (2008) A framework for interpreting genome-wide association studies of psychiatric disorders. Mol Psychiatry 14:10–17. https://doi.org/10.1038/mp.2008.126

12. Consortium CASWGOTPG (2016) Contribution of copy number variants to schizophrenia from a genome-wide study of 41,321 subjects. Nat Genet 49:27–35. https://doi.org/10.1038/ng.3725
13. Consortium SWGOTPG (2014) Biological insights from 108 schizophrenia-associated genetic loci. Nature 511:421–427. https://doi.org/10.1038/nature13595
14. Consortium TSPG-WASG (2011) Genome-wide association study identifies five new schizophrenia loci. Nat Genet 43:969–976. https://doi.org/10.1038/ng.940
15. Consortium TSPG-WASG, Ripke S, Sanders AR et al (2011) Genome-wide association study identifies five new schizophrenia loci. Nat Genet 43:969. https://doi.org/10.1038/ng.940
16. Deleidi M, Yu C (2016) Genome editing in pluripotent stem cells: research and therapeutic applications. Biochem Biophys Res Commun 473:665–674. https://doi.org/10.1016/j.bbrc.2016.02.113
17. Dezonne RS, Sartore RC, Nascimento JM et al (2017) Derivation of functional human astrocytes from cerebral organoids. Sci Rep:1–14. https://doi.org/10.1038/srep45091
18. Douvaras P, Sun B, Wang M et al (2017) Directed differentiation of human pluripotent stem cells to microglia. Stem Cell Reports 8:1516–1524. https://doi.org/10.1016/j.stemcr.2017.04.023
19. Flaherty E, Deranieh RM, Artimovich E et al (2017) Patient-derived hiPSC neurons with heterozygous CNTNAP2 deletions display altered neuronal gene expression and network activity. npj Schizophrenia 3:35. https://doi.org/10.1038/s41537-017-0033-5
20. Freedman R (2003) Schizophrenia. N Engl J Med 349:1738–1749. https://doi.org/10.1056/NEJMra035458
21. Fromer M, Pocklington AJ, Kavanagh DH et al (2014) De novo mutations in schizophrenia implicate synaptic networks. Nature 506:179–184. https://doi.org/10.1038/nature12929
22. Frühbeis C, Fröhlich D, Krämer-Albers E-M (2012) Emerging roles of exosomes in neuron–glia communication. Front Physiol. https://doi.org/10.3389/fphys.2012.00119
23. Frühbeis C, Fröhlich D, Kuo WP et al (2013) Neurotransmitter-triggered transfer of exosomes mediates oligodendrocyte–neuron communication. PLoS Biol 11:e1001604. https://doi.org/10.1371/journal.pbio.1001604
24. Gonzalez-Pinto A, Gutierrez M, Mosquera F et al (1998) First episode in bipolar disorder: misdiagnosis and psychotic symptoms. J Affect Disord 50:41–44. https://doi.org/10.1016/S0165-0327(98)00032-9
25. Hakak Y, Walker JR, Li C et al (2001) Genome-wide expression analysis reveals dysregulation of myelination-related genes in chronic schizophrenia. Proc Natl Acad Sci U S A 98:4746. https://doi.org/10.1073/pnas.081071198
26. Hauberg ME, Fullard J, Giambartolomei C et al (2017) Cell-type specific open chromatin profiling in human postmortem brain infers functional roles for non-coding schizophrenia LOCI. Eur Neuropsychopharmacol 27:S428–S429. https://doi.org/10.1016/j.euroneuro.2016.09.483
27. Ho S-M, Hartley BJ, Flaherty E et al (2017) Evaluating synthetic activation and repression of neuropsychiatric-related genes in hiPSC-derived NPCs, neurons, and astrocytes. Stem Cell Reports. https://doi.org/10.1016/j.stemcr.2017.06.012
28. Hook V, Brennand KJ, Kim Y et al (2014) Human iPSC neurons display activity-dependent neurotransmitter secretion: aberrant catecholamine levels in schizophrenia neurons. Stem Cell Reports 3:531–538. https://doi.org/10.1016/j.stemcr.2014.08.001
29. Jinek M, Chylinski K, Fonfara I et al (2012) A programmable dual-RNA–guided DNA endonuclease in adaptive bacterial immunity. Science 337:816–821. https://doi.org/10.1126/science.1225829
30. Karayiorgou M, Simon TJ, Gogos JA (2010) 22q11.2 microdeletions: linking DNA structural variation to brain dysfunction and schizophrenia. Nat Rev Neurosci 11:402–416. https://doi.org/10.1038/nrn2841
31. Kimelberg HK (2010) Functions of mature mammalian astrocytes: a current view. Neuroscientist 16:79–106. https://doi.org/10.1177/1073858409342593
32. Kohane IS, Masys DR, Altman RB (2006) The Incidentalome: a threat to genomic medicine. JAMA 296:212–215. https://doi.org/10.1001/jama.296.2.212

33. Koyama Y (2015) Functional alterations of astrocytes in mental disorders: pharmacological significance as a drug target. Front Cell Neurosci. https://doi.org/10.3389/fncel.2015.00261
34. Lee IS, Carvalho CMB, Douvaras P et al (2015) Characterization of molecular and cellular phenotypes associated with a heterozygous CNTNAP2 deletion using patient-derived hiPSC neural cells. npj Schizophrenia 1:171. https://doi.org/10.1371/journal.pone.0044017
35. Lee IS, Carvalho CMB, Douvaras P et al (2015) Characterization of molecular and cellular phenotypes associated with a heterozygous CNTNAP2 deletion using patient-derived hiPSC neural cells. npj Schizophrenia 1:15019. https://doi.org/10.1038/npjschz.2015.19
36. Lin M, Pedrosa E, Hrabovsky A et al (2016) Integrative transcriptome network analysis of iPSC-derived neurons from schizophrenia and schizoaffective disorder patients with 22q11.2 deletion. BMC Syst Biol 10:105. https://doi.org/10.1186/s12918-016-0366-0
37. Ma TM, Abazyan S, Abazyan B et al (2013) Pathogenic disruption of DISC1-serine racemase binding elicits schizophrenia-like behavior via D-serine depletion. Mol Psychiatry 18:557–567. https://doi.org/10.1007/s00213-003-1582-z
38. Mandegar MA, Huebsch N, Frolov EB et al (2016) CRISPR interference efficiently induces specific and reversible gene silencing in human iPSCs. Stem Cell 18:541–553. https://doi.org/10.1016/j.stem.2016.01.022
39. Martins-de-Souza D, Maccarrone G, Wobrock T et al (2010) Proteome analysis of the thalamus and cerebrospinal fluid reveals glycolysis dysfunction and potential biomarkers candidates for schizophrenia. J Psychiatr Res 44:1176–1189. https://doi.org/10.1016/j.jpsychires.2010.04.014
40. Muffat J, Li Y, Omer A et al (2017) A possible role of microglia in Zika virus infection of the fetal human brain. bioRxiv. https://doi.org/10.1101/142497
41. Muffat J, Li Y, Yuan B et al (2016) Efficient derivation of microglia-like cells from human pluripotent stem cells. Nat Med 22:1358–1367. https://doi.org/10.1016/j.cell.2011.06.019
42. Nascimento JM, Martins-de-Souza D (2015) The proteome of schizophrenia. npj Schizophrenia 1:14003. EP –. https://doi.org/10.1038/npjschz.2014.3
43. Novikova SI, He F, Cutrufello NJ, Lidow MS (2006) Identification of protein biomarkers for schizophrenia and bipolar disorder in the postmortem prefrontal cortex using SELDI-TOF-MS ProteinChip profiling combined with MALDI-TOF-PSD-MS analysis. Neurobiol Dis 23:61–76. https://doi.org/10.1016/j.nbd.2006.02.002
44. Paşca AM, Sloan SA, Clarke LE et al (2015) Functional cortical neurons and astrocytes from human pluripotent stem cells in 3D culture. Nat Methods 12:671–678. https://doi.org/10.1038/nmeth.3415
45. Paulsen BDS, Maciel R d M, Galina A et al (2012) Altered oxygen metabolism associated to neurogenesis of induced pluripotent stem cells derived from a schizophrenic patient. Cell Transplant 21:1547–1559. https://doi.org/10.3727/096368911X600957
46. Paulsen BS, Souza CS, Chicaybam L et al (2011) Agathisflavone enhances retinoic acid-induced neurogenesis and its receptors α and β in pluripotent stem cells. Stem Cells Dev 20:1711–1721. https://doi.org/10.1089/scd.2010.0446
47. Pedrosa E, Sandler V, Shah A et al (2011) Development of patient-specific neurons in schizophrenia using induced pluripotent stem cells. J Neurogenet 25:88–103. https://doi.org/10.3109/01677063.2011.597908
48. Piao J, Major T, Auyeung G et al (2015) Human embryonic stem cell-derived oligodendrocyte progenitors remyelinate the brain and rescue behavioral deficits following radiation. Stem Cell 16:198–210. https://doi.org/10.1016/j.stem.2015.01.004
49. Purcell SM, Moran JL, Fromer M et al (2014) A polygenic burden of rare disruptive mutations in schizophrenia. Nature 506:185–190. https://doi.org/10.1038/nature12975
50. Ripke S, O'Dushlaine C, Chambert K et al (2013) Genome-wide association analysis identifies 13 new risk loci for schizophrenia. Nat Genet 45:1150–1159. https://doi.org/10.1038/ng.2742
51. Robicsek O, Karry R, Petit I et al (2013) Abnormal neuronal differentiation and mitochondrial dysfunction in hair follicle-derived induced pluripotent stem cells of schizophrenia patients. Mol Psychiatry 18:1067–1076. https://doi.org/10.1038/mp.2013.67

52. Schmidt MJ, Mirnics K (2015) Neurodevelopment, GABA system dysfunction, and schizophrenia. Neuropsychopharmacology 40:190–206. https://doi.org/10.1038/npp.2014.95
53. Schmitt A, Martins-de-Souza D, Akbarian S et al (2016) Consensus paper of the WFSBP Task Force on Biological Markers: criteria for biomarkers and endophenotypes of schizophrenia, part III: molecular mechanisms. World J Biol Psychiatry:1–27. https://doi.org/10.1080/15622975.2016.1224929
54. Schreiber M, Dorschner M, Tsuang D (2013) Next-generation sequencing in schizophrenia and other neuropsychiatric disorders. Am J Med Genet B Neuropsychiatr Genet 162B:671–678. https://doi.org/10.1002/ajmg.b.32156
55. Shaltouki A, Peng J, Liu Q et al (2013) Efficient generation of astrocytes from human pluripotent stem cells in defined conditions. Stem Cells 31:941–952. https://doi.org/10.1016/j.cell.2008.10.029
56. Silber J, Lim DA, Petritsch C et al (2008) miR-124 and miR-137 inhibit proliferation of glioblastoma multiforme cells and induce differentiation of brain tumor stem cells. BMC Med 6:1. 6:14. https://doi.org/10.1186/1741-7015-6-14
57. Sloan SA, Darmanis S, Huber N et al (2017) Human astrocyte maturation captured in 3D cerebral cortical spheroids derived from pluripotent stem cells. Neuron 95:779–790.e6. https://doi.org/10.1016/j.neuron.2017.07.035
58. Srikanth P, Han K, Callahan DG et al (2015) Genomic DISC1 disruption in hiPSCs alters Wnt signaling and neural cell fate. Cell Rep 12:1414–1429. https://doi.org/10.1016/j.celrep.2015.07.061
59. Sullivan PF, Daly MJ, O'Donovan M (2012) Genetic architectures of psychiatric disorders: the emerging picture and its implications. Nat Rev Genet 13:537–551. https://doi.org/10.1038/nrg3240
60. Tcw J, Wang M, Pimenova AA et al (2017) An efficient platform for astrocyte differentiation from human induced pluripotent stem cells. Stem Cell Reports 9:600–614. https://doi.org/10.1016/j.stemcr.2017.06.018
61. Topol A, Zhu S, Hartley BJ et al (2016) Dysregulation of miRNA-9 in a subset of schizophrenia patient-derived neural progenitor cells. Cell Rep 15:1024–1036. https://doi.org/10.1016/j.celrep.2016.03.090
62. Topol A, Zhu S, Tran N et al (2015) Correspondence. Biol Psychiatry:1–6. https://doi.org/10.1016/j.biopsych.2014.12.028
63. Torres-Ruiz R, Rodriguez-Perales S (2017) CRISPR-Cas9 technology: applications and human disease modelling. Brief Funct Genomics 16:4–12. https://doi.org/10.1093/bfgp/elw025
64. Uranova N, Orlovskaya D, Vikhreva O et al (2001) Electron microscopy of oligodendroglia in severe mental illness. Brain Res Bull 55:597–610. https://doi.org/10.1016/S0361-9230(01)00528-7
65. Wang S, Bates J, Li X et al (2013) Human iPSC-derived oligodendrocyte progenitor cells can myelinate and rescue a mouse model of congenital hypomyelination. Cell Stem Cell 12:252–264. https://doi.org/10.1016/j.stem.2012.12.002
66. Wen Z, Nguyen HN, Guo Z et al (2014) Synaptic dysregulation in a human iPS cell model of mental disorders. Nature. epub ahead of print. https://doi.org/10.1038/nature13716:414, https://doi.org/10.1038/nature13716
67. Wong AHC, Van Tol HHM (2003) Schizophrenia: from phenomenology to neurobiology. Neurosci Biobehav Rev 27:269–306. https://doi.org/10.1016/S0149-7634(03)00035-6
68. World Health Organization (2008) The global burden of disease. World Health Organization, Geneva
69. Wright C, Turner JA, Calhoun VD, Perrone-Bizzozero N (2013) Potential impact of miR-137 and its targets in schizophrenia. Front Genet 4:58. https://doi.org/10.3389/fgene.2013.00058
70. Xia M, Zhu S, Shevelkin A et al (2016) DISC1, astrocytes and neuronal maturation: a possible mechanistic link with implications for mental disorders. J Neurochem 138:518–524. https://doi.org/10.1111/jnc.13663

71. Ye F, Kang E, Yu C et al (2017) DISC1 regulates neurogenesis via modulating kinetochore attachment of Ndel1/Nde1 during mitosis. Neuron 96:1041–1054.e5. https://doi.org/10.1016/j.neuron.2017.10.010
72. Yin J, Lin J, Luo X et al (2014) miR-137: a new player in schizophrenia. Int J Mol Sci 15:3262–3271. https://doi.org/10.1016/j.biopsych.2013.06.016
73. Yoon K-J, Nguyen HN, Ursini G et al (2014) Modeling a genetic risk for schizophrenia in iPSCs and mice reveals neural stem cell deficits associated with Adherens junctions and polarity. Stem Cell 15:79–91. https://doi.org/10.1016/j.stem.2014.05.003
74. Yu DX, Di Giorgio FP, Yao J et al (2014) Modeling hippocampal neurogenesis using human pluripotent stem cells. Stem Cell Reports 2:295–310. https://doi.org/10.1016/j.stemcr.2014.01.009
75. Zhao D, Lin M, Chen J et al (2015) MicroRNA profiling of neurons generated using induced pluripotent stem cells derived from patients with schizophrenia and schizoaffective disorder, and 22q11.2 del. PLoS One 10:e0132387. EP –. https://doi.org/10.1371/journal.pone.0132387
76. Zuccoli GS, Martins-de-Souza D, Guest PC et al (2017) Combining patient-reprogrammed neural cells and proteomics as a model to study psychiatric disorders. In: Guest PC (ed) Proteomic methods in neuropsychiatric research. Springer International Publishing, Cham, pp 279–287

Chapter 3
Rett Syndrome and Stem Cell Research

Keita Tsujimura and Kinichi Nakashima

Abstract Rett syndrome (RTT) is a devastating neurodevelopmental disorder resulting from mutations in the X-linked gene encoding methyl-CpG-binding protein 2 (MECP2). *MECP2* mutations are also associated with other neurodevelopmental diseases, including autism and schizophrenia. Therefore, elucidating the mechanism of RTT can contribute to understanding the pathogenesis of a wide range of neurodevelopmental diseases. Despite its importance, however, the RTT pathogenesis remains unclear, and effective therapeutic treatment has not been developed. Offering an opportunity to move toward this goal, however, is the recent advance in the stem cell research field of the development of induced pluripotent stem cell (iPSC) technology. This technology can yield important insights into disease pathophysiology and has the potential to provide disease models for screening new drugs. Here, we discuss applications of recent stem cell technology to the field of research on RTT and describe the stem cell biology of RTT pathogenesis.

Keywords MeCP2 · Rett syndrome · Neurodevelopmental disorders · Induced pluripotent stem cells (iPSCs) · Neural stem cells (NSCs)

K. Tsujimura (✉)
Department of Psychiatry, Nagoya University Graduate School of Medicine, Showa, Nagoya, Aichi 466-8550, Japan

Institute for Advanced Research, Nagoya University, Chikusa, Nagoya, Aichi 464-8601, Japan

Department of Stem Cell Biology and Medicine, Graduate School of Medical Sciences, Kyushu University, Higashi-ku, Fukuoka, Fukuoka 812-8582, Japan
e-mail: tsujimura@med.nagoya-u.ac.jp

K. Nakashima (✉)
Department of Stem Cell Biology and Medicine, Graduate School of Medical Sciences, Kyushu University, Higashi-ku, Fukuoka, Fukuoka 812-8582, Japan
e-mail: kin1@scb.med.kyushu-u.ac.jp

R. Delgado-Morales (ed.), *Stem Cell Genetics for Biomedical Research*,
https://doi.org/10.1007/978-3-319-90695-9_3

3.1 Introduction

Stem cells are specialized cells that have the unique dual capability of self-renewal and differentiation into a variety of lineage-specific cell types. Because of these potentials, stem cells are of great interest in cell therapy research. Moreover, research on stem cells not only helps to elucidate the development and function of organs but also offers a way of treating diseases. However, human stem cell research also raises ethical issues and limitation of research source materials.

A major technological breakthrough in stem cell research was achieved in 2006 when Yamanaka's group reported that cells with a gene expression profile and developmental potential similar to embryonic stem cells (ESCs) can be produced from mouse somatic cells through the ectopic expression of four transcription factors [74]. They named these cells induced pluripotent stem cells (iPSCs). Similarly to ESCs, iPSCs can be expanded indefinitely and are capable of differentiating into all three germ layers. In the following year, two independent groups, including Yamanaka's group, reported generation of iPSCs from human somatic cells [73, 82].

This human iPSC technology has led to a new era in research and therapy. Advantages of this technology include the human origin of the iPSCs, easy accessibility, expandability, potential to produce almost any cell type(s) desired, avoidance of ethical concerns associated with human ESCs, and the potential to develop personalized medicine using patient-specific iPSCs. This technology has tremendous potential for cell-based drug discoveries, cell therapy, and disease modeling (Fig. 3.1).

Studying diseases in the nervous system presents certain unique challenges due to the complexity of the systems involved and the relative lack of accessibility of the human brain. iPSC technology has allowed scientists to establish iPSC lines from patients and direct their differentiation to various types of nervous system cells, such as neurons and glial cells. Thus, iPSC technology provides promising tools for the study of disease mechanisms and development of new drugs for neurodevelopmental diseases, including RTT.

3.2 Rett Syndrome

Rett syndrome (RTT, OMIM 312750) is a severe progressive neurodevelopmental disorder that affects approximately 1 in 10,000 females. Andreas Rett first described this syndrome in 1966 [65]. RTT is one of the most common causes of mental retardation in females [9]. Affected individuals appear to develop normally for the first 6–18 months but then regress, with the onset of various neurological symptoms including impaired motor function, mental retardation, seizures, autistic features, and stereotypical behaviors [6]. RTT is predominantly caused by de novo mutations in the X-linked gene encoding methyl-CpG-binding protein 2 (MECP2) [1].

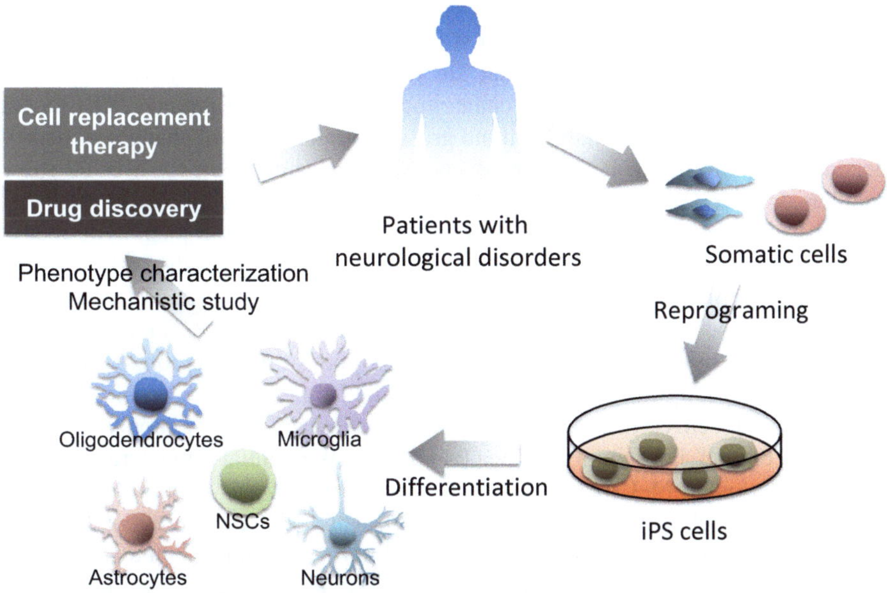

Fig. 3.1 Application of iPSCs for neurodevelopmental disorders. The use of human-induced pluripotent stem cells (iPSCs) typically involves the following steps. iPSCs are derived from individual patients. The iPSCs are then differentiated into specific cell types, such as neural cells, and the resultant cells are studied for phenotype characterization and identification of new pathological mechanisms, providing opportunities for cell replacement therapy and drug discovery

Because of random X chromosome inactivation, most affected individuals are females who are heterozygous for the *MECP2* mutation and thus mosaic in MeCP2 deficiency. Since males have only one X chromosome, male individuals with an *MECP2* mutation are more severely affected than females and generally die earlier [9]. RTT is not considered to be a neurodegenerative disease. The brain of RTT patients exhibits no obvious atrophy or degeneration, and there is no abnormal neuronal migration [34, 64]. RTT patients show decreases in overall brain volume which are linked with abnormally small, densely packed neurons. Reduced dendritic complexity and synaptic density are also observed in RTT brains [4].

3.3 RTT Mouse Model

Since the discovery that RTT is caused by loss-of-function mutations in *MECP2*, two groups have successfully generated MeCP2-null mouse models [11, 27]. MeCP2-null mice (both male hemizygotes and female homozygotes) are viable and undergo a period of normal development, followed by severe progressive neurological dysfunction, including uncoordinated motor behavior, reduced activity, and breathing irregularities, leading to death of hemizygous males at 8–10 weeks of age.

Female heterozygous mice are, in contrast, viable and fertile and seem normal even in early adulthood. However, at a later stage (at about 6 months of age), they begin to show neurological symptoms that are reminiscent of those in RTT patients. Consistent with the findings of neuroanatomical studies in RTT patients, RTT model mice also exhibit microcephaly, with no gross neuropathological changes or neurodegeneration. Similar to RTT neurons, mouse neurons deficient for MeCP2 show smaller somas and reduced dendritic complexity [11, 23, 40]. Importantly, mice that have the null mutation only in the CNS were found to show a phenotype that is indistinguishable from that of mice with the mutation throughout the body [11, 27], which strongly indicated that MeCP2 dysfunction in the CNS is sufficient to cause the disease. In addition, various conditional MeCP2-knockout (KO) mouse lines in which the gene has been deleted in specific cell types or brain region(s), or at specific developmental stage(s), have been reported. [10, 11, 24, 25, 83]. These studies revealed that localized loss of MeCP2 causes a subset of the commonly observed symptoms of RTT and support the idea that MeCP2 is essential for general neuronal health. Deletion of MeCP2 from glia such as astrocytes and microglia results in relatively mild behavioral phenotypes [19, 46], while MeCP2 expression in astrocytes partially ameliorates the symptoms of abnormal breathing, motor activities, and anxiety levels of mice lacking MeCP2 in all other cell types [46]. In contrast, expression of MeCP2 only in postmitotic neurons prevents reduction in brain weight and activity and shorter lifespan [26, 48]. Taken together, this body of evidence suggests that dysfunction of MeCP2 in neurons is the dominant driver of the pathogenesis of RTT.

3.4 MECP2 Gene, Protein Functions, and Mutations

3.4.1 MECP2 Gene

The MECP2 gene consists of four exons that encode two different isoforms (MeCP2_e1 and MECP2_e2) due to alternate splicing of exon 2 (Fig. 3.2). These MeCP2 splice variants differ only at their extreme N-terminal region. The more abundant MeCP2_e1 isoform (encoded by *MECP2α* or *MECP2B*) includes 21 amino acids encoded by exon 1 and lacks 9 amino acids encoded by exon 2, while the MeCP2_e2 isoform (encoded by *MECP2β* or *MECP2A*) is expressed from exon 2 [20, 42, 55]. Although MeCP2_e1 is the predominant isoform in the brain, the two isoforms are thought to be largely functionally equivalent [38]. Moreover, *MECP2* contains a long highly conserved 3′-untranslated region (3′UTR), and the two isoforms are differentially polyadenylated, producing four different 3’UTR lengths [70]. These different *MECP2* transcripts might be subject to tissue- and cell type-specific and developmental stage-specific regulation. MeCP2 protein is especially abundant in the brain, primarily in postmitotic neurons [69, 71].

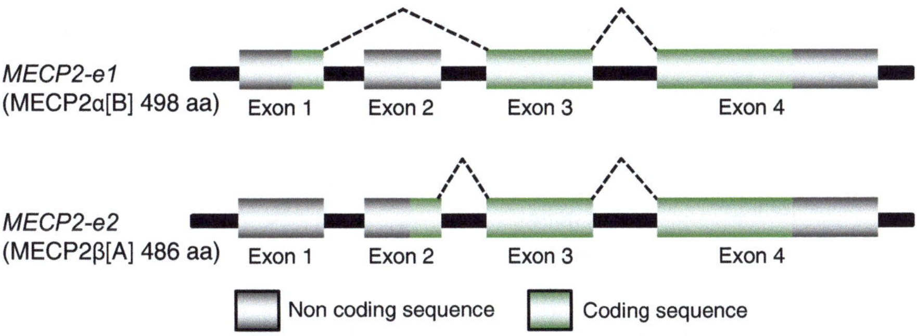

Fig. 3.2 Gene structure of *MECP2*. The *MECP2-e1* isoform is expressed from an ATG in exon1, while the *MECP2-e2* isoform is expressed from an ATG in exon 2. Alternative splicing of the *MECP2-e1* transcript excludes exon 2

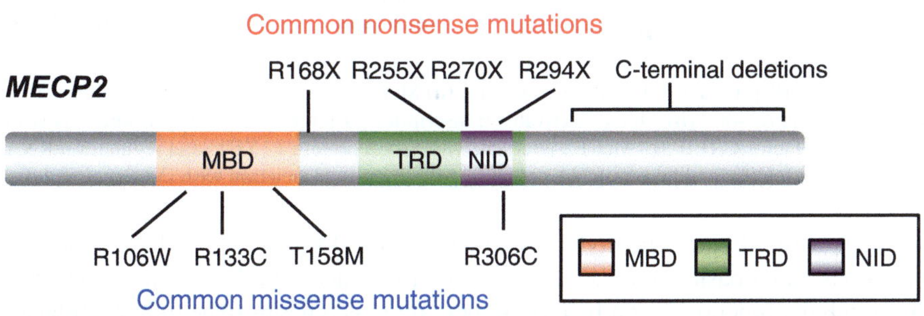

Fig. 3.3 MeCP2 protein domains and the locations of RTT-causing mutations. The MeCP2 protein contains distinct functional domains that are pertinent to RTT pathology: the methyl-CpG-binding domain (MBD) (yellow), the transcriptional repression domain (TRD) (green), and the NCoR/SMRT interaction domain (NID) (purple). The locations of common RTT-causing point mutations are indicated

3.5 MECP2 Protein Functions

MeCP2 is a member of the methyl-CpG-binding protein family [30] and comprises three domains: the methyl-CpG-binding domain (MBD), the transcriptional repression domain (TRD), and the C-terminal domain (Fig. 3.3). MeCP2 was originally identified as a methylated-DNA-binding transcriptional repressor [37, 57]. MeCP2 induces repression of gene expression by recruiting a corepressor complex that can include histone deacetylases (HDACs), Sin3A, NCoR, and SMRT [28, 37, 49]. When MeCP2 binds to methylated DNA of target genes via its MBD, its TRD recruits corepressors. A recent study suggested that the MBD of MeCP2 also

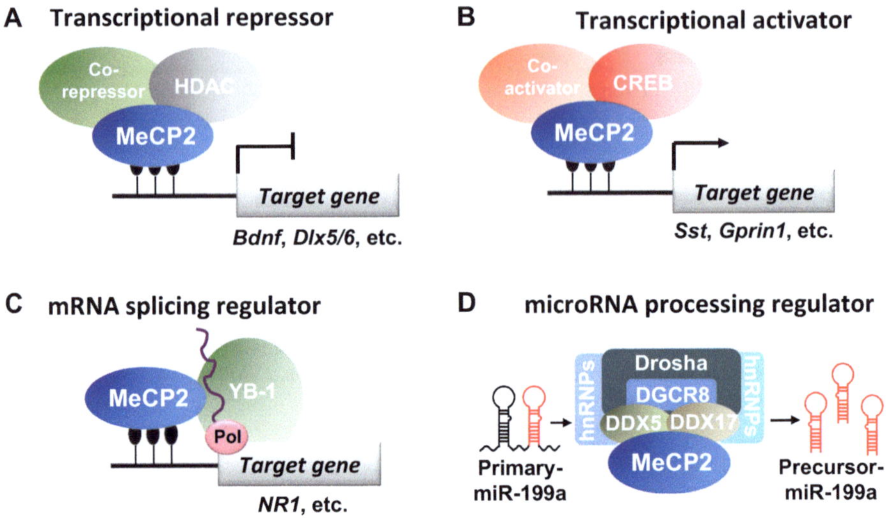

Fig. 3.4 MeCP2 protein exerts various functions. (**a**) MeCP2 acts as a transcriptional repressor by recruiting corepressors. (**b**) MeCP2 activates transcription of target genes when it interacts with transcription factor CREB. (**c**) MeCP2 interacts with Y-box protein 1 (YB1) and regulates mRNA splicing. (**d**) MeCP2 facilitates miRNA processing through its association with Drosha complex

interacts with 5-hydroxymethylcytosine-containing DNA [53]. In addition to MBD, AT-hooks and a basic amino acid cluster are associated with DNA binding [5, 29]. Although the functional significance of these regions remains to be fully elucidated, all of them may play a role in the regulation of chromatin states. Interestingly, a recent study demonstrated that truncated MeCP2 that contains only the MBD and NCoR/SMRT interaction domain (NID) can rescue RTT phenotypes of MeCP2-deficient mice [75].

Several other functions of MeCP2 have also been reported. Based on functional studies, MeCP2 has been proposed to act as an activator of transcription [8], a regulator of mRNA splicing [52, 81] or miRNA processing [13, 78], and a facilitator of chromosome looping or compaction [7] (Fig. 3.4). This evidence taken altogether suggests that MeCP2 might be an important multifunctional hub for many pathways that support brain function.

3.6 *MECP2* Mutations

Mutations in *MECP2* are found in more than 95% of sporadic cases of classic RTT and result in loss of function due to truncated, unstable, or abnormally folded proteins. A variety of mutation types, such as missense, nonsense, and frameshift mutations, as well as deletions encompassing whole exons, have been reported [16, 61, 63]. Eight common missense and nonsense mutations arise at CpG hotspots in

MECP2 [76, 80] and account for approximately 60–70% of RTT cases [58]. In addition to these hot-spot mutations, other common mutations, including small insertions or deletions in the C-terminal domain and large DNA deletions that lack the majority of the coding sequencing, of *MECP2* have been described [58].

3.7 Stem Cell Biology of Rett Syndrome

A number of studies using postmortem brains of RTT patients and RTT model mice, as described above, have revealed various neurological phenotypes. However, it remains unknown how developmental defects occur in the RTT brain. The mammalian CNS comprises three major neural cell types: neurons, astrocytes, and oligodendrocytes. These neural cell types are produced from common multipotent neural stem cells (NSCs) [56, 72]. Several lines of evidence suggest that defects in fate determination of NSCs are one of the common causes of neurodevelopmental disorders, including RTT [22]. Although MeCP2 has been considered to be involved in neuronal maturation rather than neural fate decision [40], several studies showed that MeCP2 regulates expression of astroglial genes [17, 41, 68]. Astrocytes derived from MeCP2-null mouse embryonic stem cells showed higher expression levels of astrocyte-specific genes such as *glial fibrillary acidic protein* (*Gfap*) and *S100β* than those derived from wild-type mouse embryonic stem cells [59, 60]. Other evidence showed that expression of MeCP2 in neural stem cells not only suppresses astrocytic differentiation but also promotes neuronal differentiation in mouse [77]. Moreover, a major truncated form of RTT-associated MeCP2 (R168X) fails to induce neuronal differentiation and instead abnormally enhances a high degree of astrocytic differentiation [77]. A recent study also showed that cell cycle-linked phosphorylation at serine 421 on MeCP2 regulates the balance between proliferation and differentiation through the Notch signaling pathway in adult NSCs [44]. Thus, abnormally enhanced astrocytic differentiation might occur in RTT brains, and it is possible that aberrant fate determination of neural stem cells contributes to the pathogenesis of RTT (Fig. 3.5). Regarding an explanation of the enhanced astrocytic differentiation in MeCP2-deficient NSCs, astrocytic differentiation-inducing signaling, such as signaling by the interleukin-6 family and bone morphogenetic protein (BMP) family of cytokines [56], might be enhanced due to MeCP2 dysfunction in NSCs. Further studies will be required to achieve a better understanding of the detailed mechanism that produces this phenotype.

3.8 iPSC Technology in RTT Research

Stem cell research has the potential to significantly contribute to the development of disease-modifying therapy for neurodevelopmental disorders, and considerable advances in this field have been made in creating patient-derived neural cells from

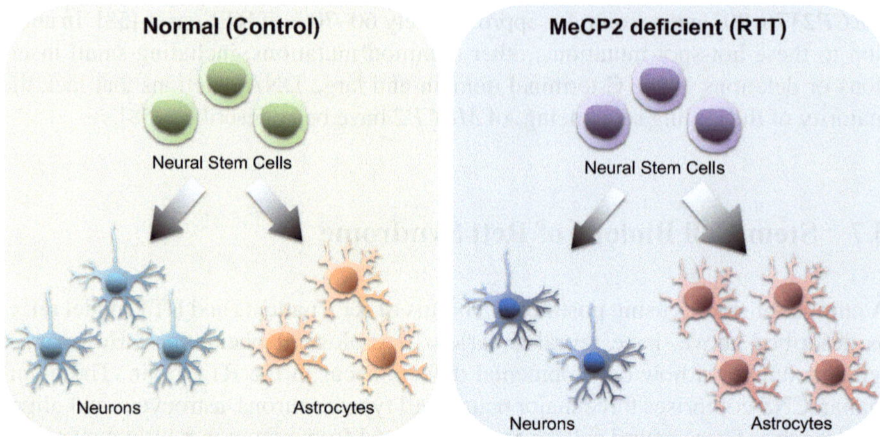

Fig. 3.5 Abnormal fate determination of MeCP2-deficient neural stem cells (NSCs). Astrocytic differentiation is increased and neuronal differentiation is decreased in MeCP2-deficient NSCs. Tsujimura et al. demonstsrated that MeCP2 inhibits astrocytic differentiation and promotes neuronal differentiation of NSCs

somatic cells. The development of new cell models of neurodevelopmental diseases is a particularly promising area of stem cell research. Cellular models of RTT generated from stem cells could help researchers to screen drugs more efficiently than currently available animal models and to study underlying biological mechanisms of RTT in cells isolated from the patients.

Ellis' group was the first to generate iPSCs from patients with RTT (RTT-iPSC) [33]. Since that report, several laboratories have established RTT-iPSCs and studied phenotypes of RTT-iPSC-derived neural cells (Table 3.1). So far, studies using RTT-iPSCs have revealed that neurons derived from RTT-iPSCs show reduced soma/nuclear size, fewer synapses, decreased dendritic spine density, altered calcium signaling, and lower expression levels of neuronal markers compared with non-affected controls [2, 14, 39, 50]. An electrophysiological study also demonstrated that RTT neurons show a decrease in excitatory synaptic transmission [50] (Fig. 3.6). Regarding these phenotypes of MECP2-deficient neurons, the likely involvement of the mechanistic target of rapamycin (mTOR), which functions as a hub for various signaling pathways, has been suggested [45, 66]: abnormal attenuation of mTOR signaling may contribute to these phenotypes, including decreased excitatory synaptic transmission. In support of this, we showed that activation of mTOR signaling in cultured neurons from MeCP2-deficient mice ameliorates abnormal phenotypes such as smaller soma size, decreased excitatory synaptic density, and impaired excitatory synaptic transmission [78]. In addition to abnormal neuronal phenotypes, abnormal fate determination of NSCs is also observed in RTT-iPSC-derived NSCs. A study of an isogenic RTT-iPSC model demonstrated that MeCP2 deficiency induces accelerated astrocyte differentiation from RTT-iPSC-derived NSCs [3]. A recent study also revealed that RTT-iPSC-derived NSCs show decreased neurogenesis [54]. These results are consistent with the studies using RTT murine models.

Table 3.1 Reports on derivation and characterization of iPSCs derived from RTT patients

Mutations	Cell types	Phenotypes	Therapeutics	Reference
R306C in MECP2	N/A	N/A	N/A	[33]
T158 M, R306C, Q244X, or 1155del32 in MECP2	Neurons	Smaller soma size Fewer synapses Reduced spine density Altered calcium transient Electrophysiological deficits	IGF1 Gentamicin	[50]
Deletion in exon 3 and 4 in MECP2	Neurons	Reduction in soma size	N/A	[14]
T158 M, E235fs, Q244X, R306C, or X487W in MECP2	Neurons	Reduction in neuronal marker positive cells	N/A	[39]
T158, V247X, R306C, or R294X in MECP2	Neurons	Smaller soma size	N/A	[2]
G269AfsX288 in MECP2	Neural stem cells	Accelerated astrocyte differentiation	N/A	[3]
R106W, or V247X in MECP2	Neural stem cells	Decreased neurogenesis	N/A	[54]

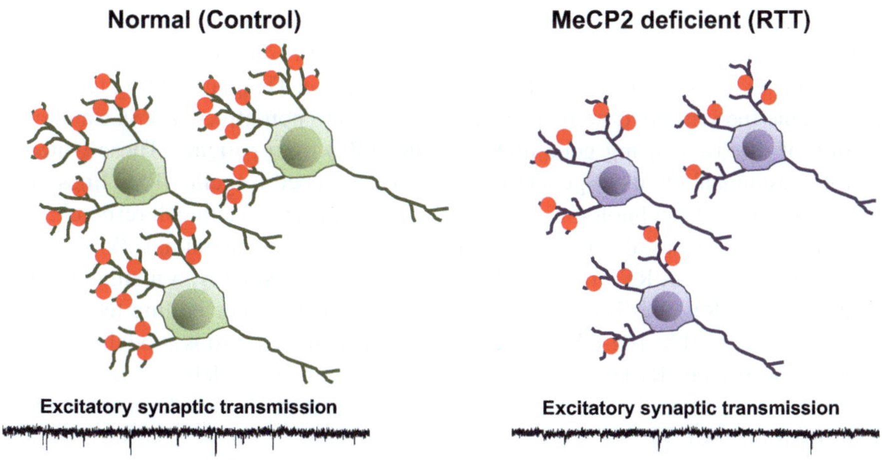

Fig. 3.6 Neuronal phenotypes found in RTT patients' iPSC-derived neurons. RTT-iPSC-derived neurons show smaller soma/nuclear size, fewer synapses, and decreased excitatory synaptic transmission when compared with neurons from non-affected individuals. The red puncta on the neuritis represent excitatory synapses

These in vitro recapitulations of RTT phenotypes provide proof of principle that RTT patient-derived iPSCs can be used to model RTT pathophysiology and serve as potential drug-screening platforms.

3.9 Conclusions and Future Prospects

Since the establishment of iPS technology a decade ago, great progress has been made in stem cell biology. Thus far, patient-derived iPSCs have been widely produced and used for drug discovery, cell therapy development, and disease modeling. Recent studies have further made iPSCs technology an even more powerful platform by combining it with other new technologies such as 3D organoids and gene editing technologies. The organoid is a self-organizing 3D structure grown from stem cells which recapitulates the in vivo architecture and multi-lineage differentiation of the original tissues in mammals [21]. Brain organoids have been generated from hiPSCs and appear to recapitulate the brain's 3D cytoarchitectural arrangement [36, 43]. Such studies have successfully recapitulated pathological phenotypes using patients' iPSC-derived brain organoids [43, 51]. Genome editing technologies, including zinc-finger nucleases (ZFN) [31, 84], transcription activator-like effector nucleases (TALENs) [15, 32, 67], and the CRISPR-Cas9 system [18, 35, 62], are also rapidly being developed and applied. Among them, in particular, the CRISPR-Cas9 system is currently most widely used to edit the genome of human iPSCs owing to its simplicity in design and ease of use. The gene editing technologies enable the introduction of specific mutations into non-affected iPSCs and correction of disease-causing gene mutations in iPSCs from patients. These genome editing systems allow us to generate genetically matched isogenic iPSC lines, and the emerging iPSC technologies will accelerate advances in stem cell research.

Rapid progress has also been made in RTT modeling using animals. In addition to mouse models, an RTT rat model was reported [79]. Notably, a nonhuman primate model of RTT has been generated by using TALEN gene editing technology [12, 47], and the TALEN-edited MeCP2 mutant monkeys in this model recapitulated phenotypes of RTT patients [12]. These new animal models will be valuable for further unraveling the disease pathophysiology and for the development of therapeutic strategies for RTT.

In this chapter we have described current advances in the RTT field, focusing on stem cell research. Many powerful tools to elucidate disease pathophysiology have emerged in recent years. Future studies exploiting new technologies will further reveal connections between the phenotypes of MeCP2-deficient stem cells and RTT disease mechanisms.

Acknowledgments This study was in part supported by JSPS KAKENHI Grant Number 16 K18391 to K.T. and MEXT KAKENHI Grant Number 17H01390 to K.N., Foundation of Synapse and Neurocircuit Pathology, and Intramural Research Grant 27-7 for Neurological and

Psychiatric Disorders of the National Center of Neurology and Psychiatry. We thank Elizabeth Nakajima for critical reading of the manuscript.

References

1. Amir RE, Van den Veyver IB, Wan M, Tran CQ, Francke U, Zoghbi HY (1999) Rett syndrome is caused by mutations in X-linked MECP2, encoding methyl-CpG-binding protein 2. Nat Genet 23:185–188
2. Ananiev G, Williams EC, Li H, Chang Q (2011) Isogenic pairs of wild type and mutant induced pluripotent stem cell (iPSC) lines from Rett syndrome patients as in vitro disease model. PLoS One 6:e25255
3. Andoh-Noda T, Akamatsu W, Miyake K, Matsumoto T, Yamaguchi R, Sanosaka T et al (2015) Differentiation of multipotent neural stem cells derived from Rett syndrome patients is biased toward the astrocytic lineage. Mol Brain 8:31
4. Armstrong D, Dunn JK, Antalffy B, Trivedi R (1995) Selective dendritic alterations in the cortex of Rett syndrome. J Neuropathol Exp Neurol 54:195–201
5. Baker SA, Chen L, Wilkins AD, Yu P, Lichtarge O, Zoghbi HY (2013) An AT-hook domain in MeCP2 determines the clinical course of Rett syndrome and related disorders. Cell 152:984–996
6. Bienvenu T, Chelly J (2006) Molecular genetics of Rett syndrome: when DNA methylation goes unrecognized. Nat Rev Genet 7:415–426
7. Brero A, Easwaran HP, Nowak D, Grunewald I, Cremer T, Leonhardt H et al (2005) Methyl CpG-binding proteins induce large-scale chromatin reorganization during terminal differentiation. J Cell Biol 169:733–743
8. Chahrour M, Jung SY, Shaw C, Zhou X, Wong ST, Qin J et al (2008) MeCP2, a key contributor to neurological disease activates and represses transcription. Science 320:1224–1229
9. Chahrour M, Zoghbi HY (2007) The story of Rett syndrome: from clinic to neurobiology. Neuron 56:422–437
10. Chao HT, Chen H, Samaco RC, Xue M, Chahrour M, Yoo J et al (2010) Dysfunction in GABA signalling mediates autism-like stereotypies and Rett syndrome phenotypes. Nature 468:263–269
11. Chen RZ, Akbarian S, Tudor M, Jaenisch R (2001) Deficiency of methyl-CpG binding protein-2 in CNS neurons results in a Rett-like phenotype in mice. Nat Genet 27:327–331
12. Chen Y, Yu J, Niu Y, Qin D, Liu H, Li G et al (2017) Modeling Rett syndrome using TALEN-edited *MECP2* mutant cynomolgus monkeys. Cell 169:945–955
13. Cheng TL, Wang Z, Liao Q, Zhu Y, Zhou WH, Xu W et al (2014) MeCP2 suppresses nuclear microRNA processing and dendritic growth by regulating the DGCR8/Drosha complex. Dev Cell 28:547–560
14. Cheung AY, Horvath LM, Grafodatskaya D, Pasceri P, Weksberg R, Hotta A et al (2011) Isolation of MECP2-null Rett syndrome patient hiPS cells and isogenic controls through X-chromosome inactivation. Hum Mol Genet 20:2103–2115
15. Christian M, Cermak T, Doyle EL, Schmidt C, Zhang F, Hummel A et al (2010) Targeting DNA double-strand breaks with TAL effector nucleases. Genetics 186:757–761
16. Christodoulou J, Grimm A, Maher T, Bennetts B (2003) RettBASE: the IRSA MECP2 variation database-a new mutation database in evolution. Hum Mutat 21:466–472
17. Colantuoni C, Jeon OH, Hyder K, Chenchik A, Khimani AH, Narayanan V et al (2001) Gene expression profiling in postmortem Rett syndrome brain: differential gene expression and patient classification. Neurobiol Dis 8:847–865
18. Cong L, Ran FA, Cox D, Lin S, Barretto R, Habib N et al (2013) Multiplex genome engineering using CRISPR-Cas systems. Science 339:819–823

19. Cronk JC, Derecki NC, Ji E, Xu Y, Lampano AE, Smirnov I et al (2015) Methyl-CpG binding protein 2 regulates microglia and macrophage gene expression in response to inflammatory stimuli. Immunity 42:679–691
20. Dragich JM, Kim YH, Arnold AP, Schanen NC (2007) Differential distribution of the MeCP2 splice variants in the postnatal mouse brain. J Comp Neurol 501:526–542
21. Dutta D, Heo I, Clevers H (2017) Disease modeling in stem cell-derived 3D organoid systems. Trends Mol Med 23:393–410
22. Ernst C (2016) Proliferation and differentiation deficits are a major convergence point for neurodevelopmental disorders. Trends Neurosci 39:290–299
23. Fukuda T, Itoh M, Ichikawa T, Washiyama K, Goto Y (2005) Delayed maturation of neuronal architecture and synaptogenesis in cerebral cortex of Mecp2-deficient mice. J Neuropathol Exp Neurol 64:537–544
24. Fyffe SL, Neul JL, Samaco RC, Chao HT, Ben-Shachar S, Moretti P et al (2008) Deletion of Mecp2 in Sim1-expressing neurons reveals a critical role for MeCP2 in feeding behavior, aggression, and the response to stress. Neuron 59:947–958
25. Gemelli T, Berton O, Nelson ED, Perrotti LI, Jaenisch R, Monteggia LM (2006) Postnatal loss of methyl-CpG binding protein 2 in the forebrain is sufficient to mediate behavioral aspects of Rett syndrome in mice. Biol Psychiatry 59:468–476
26. Giacometti E, Luikenhuis S, Beard C, Jaenisch R (2007) Partial rescue of MeCP2 deficiency by postnatal activation of MeCP2. Proc Natl Acad Sci U S A 104:1931–1936
27. Guy J, Hendrich B, Holmes M, Martin JE, Bird A (2001) A mouse Mecp2-null mutation causes neurological symptoms that mimic Rett syndrome. Nat Genet 27:322–326
28. Harikrishnan KN, Chow MZ, Baker EK, Pal S, Bassal S, Brasacchio D et al (2005) Brahma links the SWI/SNF chromatin-remodeling complex with MeCP2-dependent transcriptional silencing. Nat Genet 37:254–264
29. Heckman LD, Chahrour MH, Zoghbi HY (2014) Rett-causing mutations reveal two domains critical for MeCP2 function and for toxicity in MECP2 duplication syndrome mice. Elife 3:e02676
30. Hendrich B, Bird A (1998) Identification and characterization of a family of mammalian methyl-CpG binding proteins. Mol Cell Biol 18:6538–6547
31. Hockemeyer D, Soldner F, Beard C, Gao Q, Mitalipova M, DeKelver RC et al (2009) Efficient targeting of expressed and silent genes in human ESCs and iPSCs using zinc-finger nucleases. Nat Biotechnol 27:851–857
32. Hockemeyer D, Wang H, Kiani S, Lai CS, Gao Q, Cassady JP et al (2011) Genetic engineering of human pluripotent cells using TALE nucleases. Nat Biotechnol 29:731–734
33. Hotta A, Cheung AY, Farra N, Vijayaragavan K, Seguin CA, Draper JS et al (2009) Isolation of human iPS cells using EOS lentiviral vectors to select for pluripotency. Nat Methods 6:370–376
34. Jellinger K, Armstrong D, Zoghbi HY, Percy AK (1988) Neuropathology of Rett syndrome. Acta Neuropathol 76:142–158
35. Jinek M, Chylinski K, Fonfara I, Hauer M, Doudna JA, Charpentier E (2012) A programmable dual-RNA-guided DNA endonuclease in adaptive bacterial immunity. Science 337:816–821
36. Jo J, Xiao Y, Sun AX, Cukuroglu E, Tran HD, Goke J et al (2016) Midbrain-like organoids from human pluripotent stem cells contain functional dopaminergic and neuromelanin-producing neurons. Cell Stem Cell 19:248–257
37. Jones PL, Veenstra GJ, Wade PA, Vermaak D, Kass SU, Landsberger N et al (1998) Methylated DNA and MeCP2 recruit histone deacetylase to repress transcription. Nat Genet 19:187–191
38. Kerr B, Soto CJ, Saez M, Abrams A, Walz K, Young JI (2012) Transgenic complementation of MeCP2 deficiency: phenotypic rescue of Mecp2-null mice by isoform-specific transgenes. Eur J Hum Genet 20:69–76
39. Kim KY, Hysolli E, Park IH (2011) Neuronal maturation defect in induced pluripotent stem cells from patients with Rett syndrome. Proc Natl Acad Sci U S A 108:14169–14174

40. Kishi N, Macklis JD (2004) MECP2 is progressively expressed in post-migratory neurons and is involved in neuronal maturation rather than cell fate decisions. Mol Cell Neurosci 27:306–321
41. Kohyama J, Kojima T, Takatsuka E, Yamashita T, Namiki J, Hsieh J et al (2008) Epigenetic regulation of neural cell differentiation plasticity in the adult mammalian brain. Proc Natl Acad Sci U S A 105:18012–18017
42. Kriaucionis S, Bird A (2004) The major form of MeCP2 has a novel N-terminus generated by alternative splicing. Nucleic Acids Res 32:1818–1823
43. Lancaster MA, Renner M, Martin CA, Wenzel D, Bicknell LS, Hurles ME et al (2013) Cerebral organoids model human brain development and microcephaly. Nature 501:373–379
44. Li H, Zhong X, Chau KF, Santistevan NJ, Guo W, Kong G et al (2014) Cell cycle-linked MeCP2 phosphorylation modulates adult neurogenesis involving the Notch signalling pathway. Nat Commun 5:5601
45. Li Y, Wang H, Muffat J, Cheng AW, Orlando DA, Loven J et al (2013) Global transcriptional and translational repression in human-embryonic-stem-cell-derived Rett syndrome neurons. Cell Stem Cell 13:446–458
46. Lioy DT, Garg SK, Monaghan CE, Raber J, Foust KD, Kaspar BK et al (2011) A role for glia in the progression of Rett's syndrome. Nature 475:497–500
47. Liu H, Chen Y, Niu Y, Zhang K, Kang Y, Ge W et al (2014) TALEN-mediated gene mutagenesis in rhesus and cynomolgus monkeys. Cell Stem Cell 14:323–328
48. Luikenhuis S, Giacometti E, Beard CF, Jaenisch R (2004) Expression of MeCP2 in postmitotic neurons rescues Rett syndrome in mice. Proc Natl Acad Sci U S A 101:6033–6038
49. Lyst MJ, Ekiert R, Ebert DH, Merusi C, Nowak J, Selfridge J et al (2013) Rett syndrome mutations abolish the interaction of MeCP2 with the NCoR/SMRT co-repressor. Nat Neurosci 16:898–902
50. Marchetto MC, Carromeu C, Acab A, Yu D, Yeo GW, Mu Y et al (2010) A model for neural development and treatment of Rett syndrome using human induced pluripotent stem cells. Cell 143:527–539
51. Mariani J, Coppola G, Zhang P, Abyzov A, Provini L, Tomasini L et al (2015) FOXG1-dependent dysregulation of GABA/glutamate neuron differentiation in autism spectrum disorders. Cell 162:375–390
52. Maunakea AK, Chepelev I, Cui K, Zhao K (2013) Intragenic DNA methylation modulates alternative splicing by recruiting MeCP2 to promote exon recognition. Cell Res 23:1256–1269
53. Mellen M, Ayata P, Dewell S, Kriaucionis S, Heintz N (2012) MeCP2 binds to 5hmC enriched in the nervous system. Cell 151:1417–1430
54. Mellios N, Feldman DA, Sheridan SD, Ip JPK, Kwok S, Amoah SK et al (2018) MeCP2-regulated miRNAs control early human neurogenesis through differential effects on ERK and AKT signaling. Mol Psychiatry 23:1051-1065
55. Mnatzakanian GN, Lohi H, Munteanu I, Alfred SE, Yamada T, MacLeod PJ et al (2004) A previously unidentified MECP2 open reading frame defines a new protein isoform relevant to Rett syndrome. Nat Genet 36:339–341
56. Namihira M, Nakashima K (2013) Mechanisms of astrocytogenesis in the mammalian brain. Curr Opin Neurobiol 23:921–927
57. Nan X, Ng HH, Johnson CA, Laherty CD, Turner BM, Eisenman RN (1998) Transcriptional repression by the methyl-CpG-binding protein MeCP2 involves a histone deacetylase complex. Nature 393:386–389
58. Neul JL, Fang P, Barrish J, Lane J, Caeg EB, Smith EO et al (2008) Specific mutations in methyl-CpG-binding protein 2 confer different severity in Rett syndrome. Neurology 70:1313–1321
59. Okabe Y, Kusaga A, Takahashi T, Mitsumasu C, Murai Y, Tanaka E et al (2010) Neural development of methyl-CpG-binding protein 2 null embryonic stem cells: a system for studying Rett syndrome. Brain Res 1360:17–27

60. Okabe Y, Takahashi T, Mitsumasu C, Kosai K, Tanaka E, Matsuishi T (2012) Alterations of gene expression and glutamate clearance in astrocytes derived from an MeCP2-null mouse model of Rett syndrome. PLoS One 7:e35354
61. Pan H, Li MR, Nelson P, Bao XH, Wu XR, Yu S (2006) Large deletions of the MECP2 gene in Chinese patients with classical Rett syndrome. Clin Genet 70:418–419
62. Perez-Pinera P, Kocak DD, Vockley CM, Adler AF, Kabadi AM, Polstein LR et al (2013) RNA-guided gene activation by CRISPR-Cas9-based transcription factors. Nat Methods 10:973–976
63. Ravn K, Nielsen JB, Skjeldal OH, Kerr A, Hulten M, Schwartz M (2005) Large genomic rearrangements in MECP2. Hum Mutat 25:324
64. Reiss AL, Faruque F, Naidu S, Abrams M, Beaty T, Bryan RN et al (1993) Neuroanatomy of Rett syndrome: a volumetric imaging study. Ann Neurol 34:227–234
65. Rett A (1966) On a unusual brain atrophy syndrome in hyperammonemia in childhood. Wien Med Wochenschr 116:723–726
66. Ricciardi S, Boggio EM, Grosso S, Lonetti G, Forlani G, Stefanelli G et al (2011) Reduced AKT/mTOR signaling and protein synthesis dysregulation in a Rett syndrome animal model. Hum Mol Genet 20:1182–1196
67. Sanjana NE, Cong L, Zhou Y, Cunniff MM, Feng G, Zhang F (2012) A transcription activator-like effector toolbox for genome engineering. Nat Protoc 7:171–192
68. Setoguchi H, Namihira M, Kohyama J, Asano H, Sanosaka T, Nakashima K (2006) Methyl-CpG binding proteins are involved in restricting differentiation plasticity in neurons. J Neurosci Res 84:969–979
69. Shahbazian MD, Antalffy B, Armstrong DL, Zoghbi HY (2002) Insight into Rett syndrome: MeCP2 levels display tissue- and cell-specific differences and correlate with neuronal maturation. Hum Mol Genet 11:115–124
70. Singh J, Saxena A, Christodoulou J, Ravine D (2008) MECP2 genomic structure and function: insight from ENCODE. Nucleic Acids Res 36:6035–6047
71. Skene PJ, Illingworth RS, Webb S, Kerr AR, James KD, Turner DJ et al (2010) Neuronal MeCP2 is expressed at near histone-octamer levels and globally alters the chromatin state. Mol Cell 37:457–468
72. Svendsen CN, ter Borg MG, Armstrong RJ, Rosser AE, Chandran S, Ostenfeld T et al (1998) A new method for the rapid and long term growth of human neural precursor cells. J Neurosci Methods 85:141–152
73. Takahashi K, Tanabe K, Ohnuki M, Narita M, Ichisaka T, Tomoda K et al (2007) Induction of pluripotent stem cells from adult human fibroblasts by defined factors. Cell 131:861–872
74. Takahashi K, Yamanaka S (2006) Induction of pluripotent stem cells from mouse embryonic and adult fibroblast cultures by defined factors. Cell 126:663–676
75. Tillotson R, Selfridge J, Koerner MV, Gadalla KKE, Guy J, De Sousa D et al (2017) Radically truncated MeCP2 rescues Rett syndrome-like neurological defects. Nature 550:398–401
76. Trappe R, Laccone F, Cobilanschi J, Meins M, Huppke P, Hanefeld F et al (2001) MECP2 mutations in sporadic cases of Rett syndrome are almost exclusively of paternal origin. Am J Hum Genet 68:1093–1101
77. Tsujimura K, Abematsu M, Kohyama J, Namihira M, Nakashima K (2009) Neuronal differentiation of neural precursor cells is promoted by the methyl-CpG-binding protein MeCP2. Exp Neurol 219:104–111
78. Tsujimura K, Irie K, Nakashima H, Egashira Y, Fukao Y, Fujiwara M et al (2015) miR-199a links MeCP2 with mTOR signaling and its dysregulation leads to Rett syndrome phenotypes. Cell Rep 12:1887–1901
79. Veeraragavan S, Wan YW, Connolly DR, Hamilton SM, Ward CS, Soriano S et al (2016) Loss of MeCP2 in the rat models regression, impaired sociability and transcriptional deficits of Rett syndrome. Hum Mol Genet 25:3284–3302

80. Wan M, Lee SS, Zhang X, Houwink-Manville I, Song HR, Amir RE et al (1999) Rett syndrome and beyond: recurrent spontaneous and familial MECP2 mutations at CpG hotspots. Am J Hum Genet 65:1520–1529
81. Young JI, Hong EP, Castle JC, Crespo-Barreto J, Bowman AB, Rose MF et al (2005) Regulation of RNA splicing by the methylation-dependent transcriptional repressor methyl-CpG binding protein 2. Proc Natl Acad Sci U S A 102:17551–17558
82. Yu J, Vodyanik MA, Smuga-Otto K, Antosiewicz-Bourget J, Frane JL, Tian S et al (2007) Induced pluripotent stem cell lines derived from human somatic cells. Science 318:1917–1920
83. Zhang W, Peterson M, Beyer B, Frankel WN, Zhang ZW (2014) Loss of MeCP2 from forebrain excitatory neurons leads to cortical hyperexcitation and seizures. J Neurosci 34:2754–2763
84. Zou J, Maeder ML, Mali P, Pruett-Miller SM, Thibodeau-Beganny S, Chou BK et al (2009) Gene targeting of a disease-related gene in human induced pluripotent stem and embryonic stem cells. Cell Stem Cell 5:97–110

Chapter 4
Stem Cell Applications in Spinal Cord Injury: A Primer

James Hong, Christopher E. Rodgers, and Michael G. Fehlings

Abstract Despite advances in perioperative care, spinal cord injury (SCI) continues to be a devastating neurological condition. As SCI pathophysiology is multifaceted and time-dependent, the adaptive nature of cell therapy has been considered a promising approach for addressing these issues. In this chapter, we review the primary therapeutic targets of stem cell therapy including neuroprotection and neuroregeneration. Further, we present an in-depth review of the primary stem cell candidates (neural stem/progenitor and mesenchymal stem cells), a brief discussion of other cell types, and the status of combinatorial therapies for SCI. In closing, we discuss the translational challenges at the frontier and the progress made at the trial stage.

Keywords Spinal Cord Injury · Stem cells · Transplantation

4.1 Introduction

4.1.1 Spinal Cord Injury

Traumatic spinal cord injury (SCI) is a debilitating condition that severely restricts the mobility, freedom, and livelihoods of approximately 40 out of every million people in North America (Fig. 4.1, [6]). The temporal phases of SCI are classified as acute (<48 h), subacute (48 h–14 days), intermediate (14 days–6 months), and chronic (>6 months) (Fig. 4.2, [6]).

Initially, primary traumatic injury to the spinal cord damages and destroys neurons and glia at the injury site. Subsequently, a complex secondary injury cascade

J. Hong · C. E. Rodgers
University of Toronto & Krembil Research Institute, Toronto, ON, Canada
e-mail: jms.hong@mail.utoronto.ca; chris.rodgers@mail.utoronto.ca

M. G. Fehlings (✉)
University of Toronto, Toronto, ON, Canada

University Health Network, Toronto, ON, Canada

R. Delgado-Morales (ed.), *Stem Cell Genetics for Biomedical Research*,
https://doi.org/10.1007/978-3-319-90695-9_4

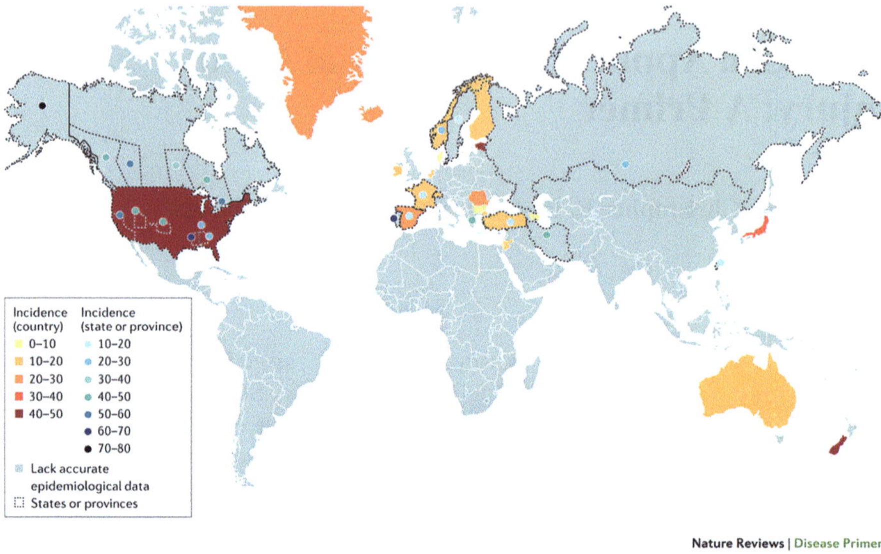

Fig. 4.1 Annual incidence of spinal cord injury across reported countries, states or provinces, and regions [6]

characterized by blood-spinal cord barrier (BSCB) disruption, ischemia, inflammation, fibrosis, and gliosis results in further tissue loss [6]. The formation of the glial scar involves the secretion of chondroitin sulfate proteoglycans (CSPGs) by astrocytes and surrounding cells and forms a barrier around the lesion core. Although the glial scar prevents the spread of the hemorrhage and inflammation, it also blocks perilesional axonal regeneration, thus limiting endogenous repair and recovery [6].

With the lifetime treatment cost of SCI being $1.1–4.6 million per patient, research into therapeutic treatments is of vital importance. Stem cells, with their potential to regenerate and/or protect central nervous system (CNS) tissue, are currently an active therapeutic candidate for traumatic SCI. While it has yet to show definitive success in clinical trials, it has presented significant promise in basic science research with animal models [6].

4.1.2 *Stem Cells*

Stem cells (Fig. 4.3) refer broadly to cell types possessing the fundamental properties of both (1) unlimited self-renewal and proliferation through mitotic divisions and (2) multipotency, the ability to have multiple downstream differentiation fates. The two principal stem cells studied for SCI therapy are (1) neural stem and progenitor cells (or the heterogeneous mixture known as neural precursor cells (NPCs)) and (2) mesenchymal stem cells, also known as mesenchymal stromal cells [90, 91, 95, 100].

Fig. 4.2 Pathophysiology of traumatic spinal cord injury [6]

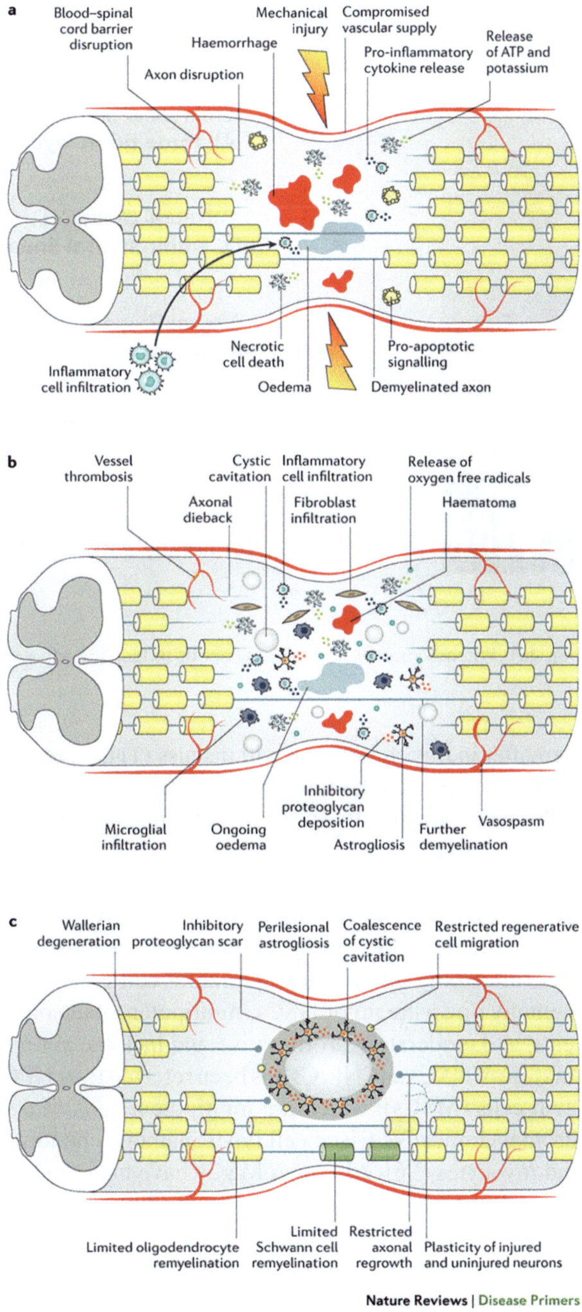

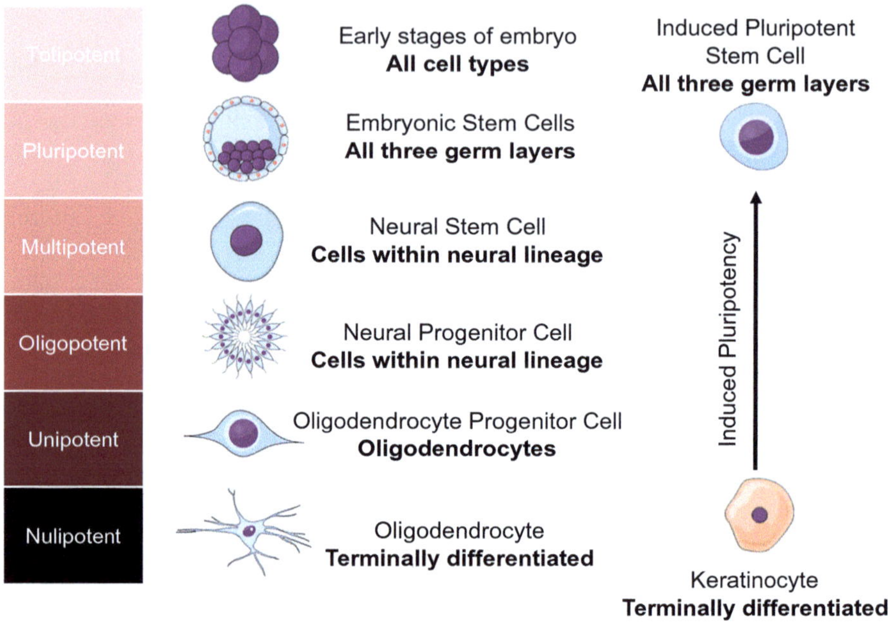

Fig. 4.3 Potency of cells and the concept of induced pluripotency

Hereafter in this chapter, we will discuss (1) the preclinical applications of stem cells in SCI therapy, (2) the translational challenges and the generation of stem cells for SCI, and (3) the results and status of clinical trials involving stem cell treatment for traumatic SCI.

Although the terms neural stem cells (NSCs) and neural progenitor cells (NPrCs) have occasionally been used interchangeably, there are differences between these stem cell types. NSCs have unlimited self-renewal, proliferation, and multipotency toward forming all downstream neural cell types. Conversely, NPrCs are developmentally downstream of NSCs, multipotent (but sometimes only bipotent or unipotent), and moderately proliferative and lack the ability for self-renewal. Finally, the mixture of NSCs and NPrCs has been referred to as NPCs [22, 23, 116]. Mesenchymal stem cells (MSCs), which are multipotent cells of a stromal origin, can only form stromal/connective tissue cell types, namely, chondrocytes, osteoblasts, adipocytes, and myocytes [96]. Although MSCs have been traditionally isolated from the bone marrow of adult mammals, they have additionally been discovered to reside in the stromal tissue of various organs [96].

In order to fully comprehend and appreciate the importance and function of stem cell therapies, a brief historical background is necessary. Over 120 years before the first isolation of adult NSCs/NPCs, Swiss neurologist Wilhelm His Sr. demonstrated that the embryonic human brain is derived from cells of an initially multipotent origin. During development, these *germinal cells* divide in the ventricles prior to acquiring neuronal fates; after which, mitosis ceases and migration to the cortex or other terminal destinations ensues [23, 44–46]. A second principal observation by His and supported by the Swedish anatomist Gustaf Retzius lays in the idea that neurons and glial cells are derived from different branches of progenitor cells. We now understand these to be neuronal precursors and oligodendrocyte progenitor cells (OPCs), which develop downstream of NPCs [23, 103, 104].

Moving forward to the mid-twentieth century, James Till and Ernest McCulloch discovered the existence of adult multipotent stem cells, understood now to have been both mesenchymal and hematopoietic stem cells (HSCs) [19, 83, 120, 132]. With regard to adult NSCs/NPCs, they were first postulated to exist and contribute to neurogenesis in the adult CNS in 1965, a change from the previously believed notion that this occurred only in neurodevelopment. It would not be until over two decades later that such self-renewing multipotent adult neural cells would be discovered in the subventricular zone by Sally Temple. Three years afterward, NSCs/NPCs would be isolated from the striatum of adult mice by B.A. Reynolds and S. Weiss [9, 77, 105, 130].

4.2 Stem Cells: Potential Mechanisms of Action in Spinal Cord Injury

The two principal mechanisms of action for experimental stem cell therapy treatments for traumatic SCI are (1) neuroprotection, referring to direct or indirect increases in the sparing and survival of existing neurons and glia, and (2) neuroregeneration/neurorepair, referring to the regrowth of endogenous neurons and oligodendrocytes or the integration of grafted cells. This regrowth and integration leads to the regeneration of axons and reestablishment of synaptic connectivity of damaged neural tracts and/or the remyelination of surviving, demyelinated axons [4, 6, 55]. During the acute phase of traumatic SCI, neuroprotective strategies such as surgical decompression, blood pressure modulation, and neuroprotective drug treatments are moderately effective in reducing cell and tissue death and can ultimately mitigate motor deficit [4, 6]. However, in the chronic phase characterized by a fully developed glial scar, neuroregeneration is understood to be the only possible avenue for accomplishing any significant functional recovery [4, 6].

Neuroprotection and neuroregeneration provided by stem cell therapies post-SCI are understood to be aided and accelerated via the processes of (1) trophic support, such as the secretion of ciliary neurotrophic factor (CNTF), epidermal growth factor (EGF), basic fibroblast growth factor (BFGF), and platelet-derived growth factor subunit A (PDGF-A); (2) immunomodulation, such as the upregulation of interleukin 6 (IL-6) to promote neurogenesis; and (3) scaffolding, referring to the ability of transplanted NPCs (or downstream NPC-derived cells)/MSCs to provide physical structural support for remyelination and axonal regeneration to occur [4, 28, 51, 52, 70, 145].

4.2.1 Neuroprotection

After traumatic SCI, neuroprotection can mitigate cell and tissue damage, thereby reducing functional motor and sensory loss [5]. With regard to stem cell-facilitated neuroprotection, MSCs can modify the negative and inhibitory microenvironment of the SCI lesion area (and of other injury microenvironments) by secreting immune and growth factors [15, 40, 50, 59, 85, 90, 96, 100, 101, 119]. Using a rat model of SCI and quantitative real-time polymerase chain reaction, MSCs have been shown to induce a neuroprotective effect by upregulating interleukin 10 (IL-10), glial-derived neurotrophic factor (GDNF), and vascular endothelial growth factor (VEGF) at 24 h, 48 h, 8 days, and 21 days posttreatment [110]. Simultaneously, MSCs have been shown to reduce acute expression of TNF-α and transforming growth factor-beta (TGF-β) [110].

Additionally, studies have demonstrated that transplanted MSCs can reduce recruitment and infiltration of reactive astrocytes and macrophages, subsequently ameliorating the inflammatory microenvironment and reducing the formation of the glial scar [81, 95]. A possible mechanism of MSC-facilitated protection is the switch from inflammatory macrophages (M1) to repair-promoting macrophages (M2) infiltrating the SCI. This reconciles the aforementioned IL-10 upregulation due to the ability of M2 macrophages to secrete it [95]. Apoptosis of neurons and glia in the SCI lesion is known to be reduced by transplanted MSCs via the upregulation of the stromal cell-derived factor-1 (SDF-1)/CXC chemokine receptor 4 (CXCR4) axis in response to inflammatory cues such as TNF-α and interleukin 1-β [100, 144]. MSCs can also promote neuronal/glial sparing and improve the microenvironment through secretion of such factors as hepatocyte growth factor (HGF) and SDF-1α/CXCR4 axis.

Macroscopically, the neuroprotective effects of MSCs have promoted the functional recovery of both bladder and hind limb motor functions in rat SCI models [81, 93, 94, 100]. This coincides with physical regenerative improvements as corticospinal tract sprouting, remyelination of surviving demyelinated axons, microvasculature repair, and BSCB integrity improvements [86]. As an example of the neuroprotective potency of MSC grafts, an intravenous infusion of MSCs significantly reduced hemorrhaging and lesion volume, increased IL-10 systemically in the spleen, and improved BSCB stability after traumatic SCI in the rat [15]. Additionally, significant neuronal and glial tissue sparing was found up to 10 weeks post-injury, which coincided with improved hind limb motor function [15]. Various efforts have been undertaken to further enhance the neuroprotective effects offered by MSCs. This includes genetic modifications to secrete additional neural growth/sparing-promoting factors such as brain-derived neurotrophic factor (BDNF). In chronic canine SCI models, BDNF secretion was shown to enhance functional recovery after 6–8 weeks *by* upregulating expression of neurogenesis markers (β3-tubulin, neurofilament medium, and nestin) while downregulating inflammatory/gliosis markers (TNF-α, IL-6, COX-2, and GFAP). Additionally, Wnt3a, which promoted neurogenic markers and axonal regeneration, led to functional motor score improvements 7 weeks posttreatment in a rat SCI model study [68, 117]. Although neuroprotection offered by strategies such as MSC transplantation has proven beneficial for providing both physical repair and functional recovery following traumatic SCI, future therapeutic treatments may require more complex combinatorial approaches to truly yield complete therapeutic restoration [5].

4.2.1.1 Remyelination

A major hallmark of SCI is the loss of myelination and death of oligodendrocytes [14]. This leads to an interruption of action potential progression and neuronal degeneration. As 10% of axons are required to regain substantial function post-SCI [37], a promising strategy is to use stem cell therapy to remyelinate the damaged axons. This strategy achieves two therapeutic targets: (1) neuroprotection, by preventing further damage to spared axons, and (2) cell replacement, by replenishing dead oligodendrocytes. Several preclinical studies have demonstrated that the transplantation of OPCs and NPCs is effective for axon remyelination and improved functional recovery. These cells generated de novo myelin sheets and oligodendrocytes, which resulted in neurobehavioral recovery [14, 54, 97, 98].

If left untreated, demyelination will result in elevated intracellular levels of sodium and increased ATP expenditure [97]. This leads to reversal of $Na+/Ca2+$ exchanger and calcium regulation in neurons, interfering with neurotransmitter release [89, 97]. Demyelination will also cause mitochondrial dysfunction, free radical formation, and increased protease activity, ultimately culminating in axonal dysfunction [97]. In parallel, ROS, glutamate, and ATP affect the surrounding oligodendrocytes, which lead to cell death [97]. Remyelination induced by stem cell therapy is able to stop this pathogenesis and to provide beneficial effects after SCI [14, 54, 98].

4.2.2 Neuroregeneration and Combinatorial Approaches

Stem cells can only provide a certain level of repair and restoration for CNS injury [6]. To aid in the regenerative capacity of stem cell transplants, the use of biomaterials, such as agarose hydrogels and various forms of organic polymers, has shown remarkable success in preclinical studies [3]. Biomaterials can provide scaffolding support and enable the delivery of antibodies, hormones, regulatory factors, and other supportive capacities alongside stem cell strategies [3, 118, 151]. In this section, we will briefly introduce some of the intriguing possibilities offered by the combinatorial approaches of biomaterials and stem cells for CNS damage and SCI.

One biomaterial strategy that has shown potential is the use of an agarose hydrogel to deliver thyroid hormone and to induce neuroregeneration in a rat model of SCI [118]. In the study by Shultz et al., a screening test was performed to determine the most effective substance for promoting oligodendro gliogenesis and functional remyelination. The substances screened included a few different microinjected transcription factors and hormones: Sonic Hedgehog, BDNF, neurotrophin 3, and thyroid hormone 3 (T3) [118]. From the screen, T3 was selected as the best option, demonstrating 4-week post-microinjection to have promoted significant increases in MBP expression and axon ensheathment [118]. The agarose hydrogel biomaterial component played a vital role in the study of Shultz et al., as it allowed specific and direct administration of T3 to the SCI lesion, lessening the dangers of creating complications and side effects from systemic T3 treatment [118].

Another area of neuroregeneration in which biomaterials have shown potential is in the supportive role of promoting endogenous neuroregeneration via inhibiting cytoskeletal actin and actomyosin modifying GTPases of the Ras homolog gene family (Rho). In particular, RhoA inhibits growth cone formation by acting on lamellipodia and filopodia and to activate collaspin-1, preventing neurite outgrowth/axonal regrowth [41, 69, 84, 129]. A number of studies have used RhoA inhibitor C3 transferase to prevent ADP-ribosylation, inducing neurite outgrowth and axonal regeneration, both in vitro and in vivo. Biomaterials have acted as integral components of such research by supporting efficient delivery of C3 [41, 84, 129].

One such study using biomaterials to promote RhoA inhibition utilized PLGA microspheres with a virally delivered C3 transferase gene [129]. The PLGA microsphere-coated C3 transferase displayed rapid and effective penetration of neighboring cells within 6 h and remarkably remained active for up to 3 weeks [129]. During this time, significant increases in growth cone formation, neurite outgrowth, and axonal regeneration were shown for neuroblastoma cells in vitro [129]. Another study that recently displayed regenerative success used a cationic, amphiphilic copolymer, termed PgP, as a delivery vehicle for an siRNA targeting RhoA mRNA. After only 4 weeks, PgP + siRNA-RhoA injected into the lesion site of SCI rats was capable of inhibiting RhoA, thus attenuating neuronal apoptosis and cavity volume and ultimately promoting axonal regrowth [41].

Another avenue of the biomaterials-stem cell application in neuroregeneration studies has been the promotion of neurogenesis and differentiation of NPCs through adhesive surfaces [35]. Through the use of bifunctionalized hydrogels comprising of polylysine and a peptide containing a laminin motif, Farrukh et al. demonstrated rapid and increased differentiation of embryonic NPCs toward cortical neurons with β1-integrin-containing focal adhesions and significantly increased proliferation and expansion of adult NPCs via the promotion of neurogenesis [35].

Polyethylene glycol (PEG) hydrogels and chitosan spheroids present promising options for combinatorial biomaterials-stem cell approaches for neuroregeneration [63]. PEG hydrogels have been demonstrated to promote migration, proliferation, and differentiation of NPC transplants in models of contusive SCI in rodents [63]. Furthermore, evidence suggests that PEG hydrogels mitigate glial scar formation, promote axonal regeneration and synaptic reestablishment, reduce neuroinflammation, and provide neuroprotection by reducing axonal/neuronal degeneration from spreading necrosis/apoptosis [63]. Chitosan, in the form of spheroids as an encapsulation and delivery vehicle for a combined treatment of NPCs and MSCs, is capable of increasing the survival of transplanted NPCs and MSCs [42]. Additionally, chitosan spheroids co-encapsulate NPCs and MSCs and support positive interactions between the two cell types [42]. A proof of concept study that injected adult and embryonic zebra fish injured brains with chitosan spheroid displayed increased survival for chitosan spheroid-encased NPCMSC combinations and neurological recovery [42].

A critical obstacle of SCI repair is the inhibitory microenvironment, which contains inflammatory factors and myelin debris, which inhibit neuronal differentiation of transplanted NPCs [5, 6, 151]. One downstream target is epidermal growth factor receptor (EGFR). A number of studies have attempted to target EGFR using EGFR antibodies and inhibitors delivered using biomaterial scaffolds [34, 43, 72]. In the aforementioned studies, transplanted and endogenous NPC exhibited enhanced differentiation, neurite outgrowth, axonal regeneration, and functional synapse formation. A recent study used the EGFR antibody cetuximab implanted via a collagen scaffold in a dog model of traumatic SCI [73]. The antibody significantly increased endogenous neurogenesis/neuronal differentiation, functional reestablishment of severed synaptic connections, and remyelination [73]. Further studies are necessary to assess the regenerative and neuronal differentiation-promoting capacity of biomaterial scaffold-delivered EGFR inhibitors in contusive models of SCI, where the more inflammatory microenvironment may play an additional inhibitory role on preventing neurogenesis and axonal regeneration [6].

All in all, biomaterials offer a unique supportive role in combination with NPC and MSC transplantation or endogenous activation for post-SCI neuroregeneration, and their use may prove to be vital toward providing options for future translational repair strategies.

4.3 Clinical Translation and Trials

4.3.1 Species-Specific Differences

To better understand and promote the use of NPCs for SCI, it is necessary to recognize the genetic differences between human and animal stem cells [5, 6, 137]. Although neural development generally occurs along the same path among mammals, there are nevertheless differences at the tissue/organ and the genetic and epigenetic regulatory levels [137]. One such species-specific difference exists in the development of the neocortex of rodents. In rodents, it is smooth (lissencephalic) and simplistic, and in humans, it is rough and elaborate (gyrencephalic) owing to its importance in higher-level thought processing [88, 137].

Differences at the cellular level underpin and explain the neocortex macroscopic contrasts between rodents and humans. Firstly, the outer subventricular zone which contains NPcCs and specific progenitors is more complex in humans. These cells are essential in establishing the neocortex [137]. The visual cortex is yet another CNS region in which differences exist between humans and rodents. Of the 1000 genes that regulate visual cortex development in humans and mice, 21% differ in spatiotemporal expression patterns [137]. At the genetic regulatory/epigenetic level, there are species-specific differences in differential binding of the transcription factor repressor element 1 silencing transcription factor (REST) to ESCs which affect neural lineage fate differentiations [107, 137]. Additionally, human ESC genomes contain 8000 binding sites for REST, whereas those of mice possess half as many sites [107, 137]. Despite such differences, REST binding was found to be even more highly conserved between humans and mice than any other traditional ESC markers (e.g., Sox2, OCT4, NANOG). Many of the non-conserved/human-specific REST binding sites were additionally shown to possess repressor element 1 (RE1) motifs, with many being primate-exclusive. Thus, they have been postulated to have more recent evolutionary developments, perhaps initiated by the integration of such transposable elements as long interspersed nuclear elements or short interspersed nuclear elements [2, 108, 137].

When examining other vertebrates, zebra fish represent an additional common model for neurodevelopment and neuroregeneration studies. Therefore, it is also important to recognize the specific difference in neurobiology and neurodevelopment between humans and zebra fish [17]. In mammals, relatively few regions of NPC genesis and proliferation exist, including the subventricular zone (SVZ), the dentate gyrus of the hippocampus, and quite possibly the central canal of the spinal cord. These are found deep within the surface of the brain and spinal cord and are usually inaccessible for live in vivo imaging. Conversely, in zebra fish there are several surface brain regions of neurogenesis, which prominently includes the surface gray and white matter (termed pallium) of the dorsal telencephalon [17].

NPCs of the zebra fish pallium and the mouse hippocampal dentate gyrus are understood to undergo only 1–2 rounds of proliferation prior to becoming terminally specified as neurons. Conversely, NPCs of the mouse ventral telencephalon experience significant proliferation prior to neuronal differentiation [17]. The rate of neurogenesis is also understood to be considerably higher in zebra fish than rodents or other mammals [10, 114]. Furthermore, a novel phenomenon of direct NPC-neuron conversion without initial cell division was recently detected in zebra fish [17].

One shared feature of neurogenesis between zebra fish, mammals, and other vertebrates is the gradual depletion of the adult NPC pool with increased organism age, accompanied by a shift toward increased gliogenic differentiation fates later in life [17, 21]. With regard to gene expression, signaling from Notch1 is understood to be pivotal for maintaining the NPC pools of zebra fish pallium and mouse SVZ. However, Id1 and Fezf2 play opposite roles between these two species, promoting quiescence of NPCs in zebra fish while suppressing it in mice [10, 20]. Perhaps the most critical among the mammal-teleost differences in neurogenesis (specifically between mice and zebra fish), and most important for neuroregeneration studies, is understanding that neuroinflammation plays a solely positive role after injury for the zebra fish CNS. Specifically, neuroinflammation stimulates neurogenesis from specific zebra fish neural progenitors, known as radial-glial cells, which immediately promote, repair, and regenerate [10, 60–62, 66]. Additionally, unlike mammals, there is no formation of an inhibitory glial scar after CNS injury in zebra fish [10, 62, 80]. As such, zebra fish studies have indicated significant endogenous regeneration following SCI by activating and recruiting latent progenitors, known as ependymal glia, which reside in the central canal [10, 102]. Similar evidence from rodent SCI has indicated activation of central canal ependymal NPCs with tripotency post-injury. However, perhaps due to the restricted proliferation and self-renewal capacity in comparison with zebra fish, endogenous repair is greatly limited and insufficient for full recovery in rodents [10, 18]. Acquiring a stronger understanding of the cellular and molecular pathways involved in the endogenous neuroregeneration of lower vertebrates is important for developing techniques to address the current challenges of translational neural repair in humans. This knowledge may allow us to move closer toward manipulating similar mechanisms and pathways that can lead to future therapeutics for SCI, along with other CNS injuries and neurodegenerative conditions.

4.3.2 Overcoming Immune Rejection

As described previously, using autologous iPSCs is a promising strategy for stem cell therapy as it minimizes the host's immune rejection [74] and avoids ethical concerns associated with embryonic stem cell use. However, recent studies have provided evidence of immunogenicity upon administration of iPSC-derived cells [150]. Although this is an active area of research, and multiple contradictory results have been demonstrated, it is believed that differential expression of certain proteins interferes with the immune system's recognition [74, 122]. The major evidence for immunogenicity of iPSCs is based on the transplantation of autologous iPSCs in C57BL/6 mice [150]. By performing a teratoma formation assay, Zhao et al. examined the host organism immune reaction to the teratomas formed by iPSCs introduced via subcutaneous injection [150]. They observed that embryonic stem cells were capable of forming teratomas, whereas iPSCs were lacking that capability, suggesting that iPSCs are targeted by the immune system. This was due to the changes in the expression of immunogenic proteins culminating in termination of peripheral tolerance [150].

There has been major concern regarding the efficacy of teratoma formation assay as a test for immunogenicity. Most notably, this method of examining immunogenicity has been challenged by Araki et al. due to the indication of tumor-related genes as a source of immunogenicity [13]. When iPSCs were differentiated in vivo into the skin and bone marrow, and transplanted into recipient mice, no immune reaction was observed. This was also followed by no observable changes in the expression of immunogenic genes [13]. Other studies have echoed these observations by demonstrating that the immune reaction elicited by undifferentiated iPSCs is different from differentiated iPSCs [8, 13]. In particular, the immune reaction to undifferentiated iPSC involves infiltration of lymphocytes and increases in interferon-gamma, granzyme B, and perforin intragrafts, whereas immune reaction to differentiated iPSCs resembles self-tolerance [8]. However, it is important to note that in vitro differentiation of iPSCs has been shown to result in immune reaction and T-cell infiltration [13]. For instance, when iPSCs were transformed into cardiomyocytes in vitro, significant immune reaction was observed after transplantation [13].

Three main factors have been investigated for their potential impact on the iPSC immunogenicity: (1) reprogramming technique, (2) epigenetic and genetic instability, and (3) transplantation site [122]. Various methods are available to reprogram somatic cells to iPSCs and change the immune reaction [150]. Furthermore, reprogramming can cause genetic and epigenetic changes in the cell, subsequently changing the immune signature of the cells. The site at which the iPSCs are transplanted also affects the body's immune reaction. For instance, Todorova et al. demonstrated that iPSC-derived hepatocytes are immune-tolerated when transplanted into the kidney capsule but are rejected in subcutaneous injection. These considerations indicate that further experimentation is required to better understand the immunological response to iPSCs and their derivatives before the clinical implications of stem cell therapy can be evaluated.

Interestingly, efforts are under way to apply immunogenicity as a control mechanism in order to prevent tumor formation in stem cell therapy [53]. Undifferentiated cells after transplantation lead to the formation of teratomas, a significant concern for stem cell therapy. Kawamura et al. illustrated that allogenic iPSCs combined with immunosuppressants are able to form teratoma in the host, but upon cessation of immunosuppressants, the teratoma disappears [53]. Alternatively, Kruse et al. have used IL-2 to activate autologous natural killer (NK) cells to kill human iPSCs [64].

4.3.3 Generation of iPSC-Derived Cells for SCI

Stem cell therapies in SCI generally consist of three main steps: (1) cells are obtained for reprogramming, (2) induced pluripotent stem cells that are capable of producing the desired downstream cells are generated, and (3) cells are differentiated into the desired state. These three steps have been investigated in order to optimize the therapeutic efficacy of stem cell therapy [56].

4.3.4 Source Cells

Various types of adult cells have been utilized for generating iPSCs. These cell types possess different characteristics in their capacity to transform into iPSCs and the properties that are expressed once pluripotency is achieved. In general, there is a trade-off between the availability of source cells and their reprogramming capacity [56]. The further differentiated cells are more abundant and easily available but are more difficult to be turned into iPSCs. In contrast, less differentiated cells, such as multipotent stem cells, are scarcer but require less effort to transform into iPSCs [56].

Multiple factors should be considered when determining source cells for iPSC generation, including reprogramming efficiency, differentiation potential, tumorigenicity, and phenotypic status [56]. Reprogramming efficiency quantifies the difficulty of transforming source cells into pluripotent cells. This is influenced by endogenous levels of pluripotent proteins, such as klf4 and c-Myc, which vary among different cell types [56]. Once pluripotency is achieved, different sources vary in their tendency to differentiate into a downstream state. In some cases, cells retain certain epigenetic modifications during reprogramming (known as epigenetic memory), resulting in a preferred path of differentiation [99]. For instance, iPSCs derived from keratinocytes are more likely to differentiate into NPCs than CD34+, possibly because of a common ectodermal germ layer origin. This phenomenon is referred to as source memory, and it limits our capability to produce the desired cells. Another important factor that varies among cell sources is the tendency to form a tumor after transplantation. These factors limit the domain used for extracting cell sources.

Variation among cell sources arises due to epigenetics, single nucleotide polymorphism (SNP), copy number variations (CNVs), and aneuploidy [67, 125]. Epigenetics include nucleotide and histone modifications, such as methylation and acetylation. These modifications result in the silencing or activation of certain regions of chromatin. One important component of the epigenome that varies among source cells is random X chromosome inactivation [24]. In addition to epigenetic modifications, genetic mutations in somatic tissue might also influence the properties of source cells. For instance, CNVs have been observed in pluripotent genes, such as NANOG [67]. Aneuploidy – abnormal chromosome number – is estimated to be present in 13% of iPSCs [125].

Investigations are underway to select the most suitable source cell for stem cell therapy in order to optimize this treatment approach for SCI. Currently, numerous source cells have been proposed for SCI treatment, with some source candidates for the efficient and safe production of iPSCs including skin fibroblasts, keratinocytes, umbilical cord cells, and adipose stem cells.

4.3.4.1 Skin Fibroblast

Skin fibroblasts are a promising source cell and commonly used cell type for reprogramming. Due to their accessibility, these cells are ideal candidates for stem cell therapy. The downside of these cells is that they have low efficiency for reprogramming and require an extensive amount of time [146]. Previous studies have demonstrated that the amplification process required for skin fibroblasts is 3–4 weeks in duration. Afterward, an additional 3–4 weeks are required for the colonies to appear [92]. Using Yamanaka's four factors, only about 0.01% of skin fibroblasts achieve pluripotency [47]. It is important to note that the time requirement and low yield of skin fibroblast significantly increase the cost and resources required for transition of these cells. Also, in the case of extraction of these cells from the patient, the patient can miss the time period in which stem cells are most effective.

4.3.4.2 Keratinocytes

Similar to skin fibroblasts, keratinocytes are widely available and can be extracted from a patient's skin. However, compared to the skin fibroblasts, keratinocytes achieve a higher efficiency [1]. This could be due to the elevated endogenous levels of klf4 and c-Myc in keratinocytes. It has been demonstrated that keratinocyte cells can achieve a 2.8% yield for transitioning into pluripotency [56].

4.3.4.3 Umbilical Cord Blood Cells

Different cell types from the umbilical cord can be used to generate iPSCs. These cells are easy to obtain using a minimally invasive procedure and possess epigenetic characteristics that are similar to pluripotent cells. However, they are restricted to autologous transfer only if the patient's umbilical cord was preserved at the time of birth. The most efficacious type of cell is the CD133+, which has a success rate of 0.45% using only SOX2 and OCT4. The reprogramming efficiency of CD34+ cells isolated from peripheral blood is 0.01–0.02% with Yamanaka's factors [56].

4.3.4.4 Adipose Stem Cells

Adipose stem cells are a promising choice for stem cell therapy. These cells are classified as multipotent and have endogenously elevated levels of klf4. Using liposuction, adipose stem cells can be easily extracted from the patient's body. These cells can achieve pluripotency in 10–15 days with 0.2% yield [56].

4.3.5 Inducing Pluripotency

Reprogramming cells to a pluripotent state is a crucial step that requires a safe and clinically applicable method (Fig. 4.4). Introduction of Takahashi and Yamanaka's factors – Oct4, Sox2, Klf2, and c-Myc – has opened a new frontier in stem cell therapy, as it allows for the efficient production of iPSCs without sacrificing a human embryo [127, 128]. However, this application of cell reprogramming is limited in clinical settings due to safety concerns. Particularly, the use of proto-oncogenes, such as c-Myc, can lead to increased risk of tumor formation [56, 87]. Therefore, derivations to the four aforementioned factors have been proposed. Two of these alternatives include either the use of Nanog and Lin28 as a replacement for Klf4 and

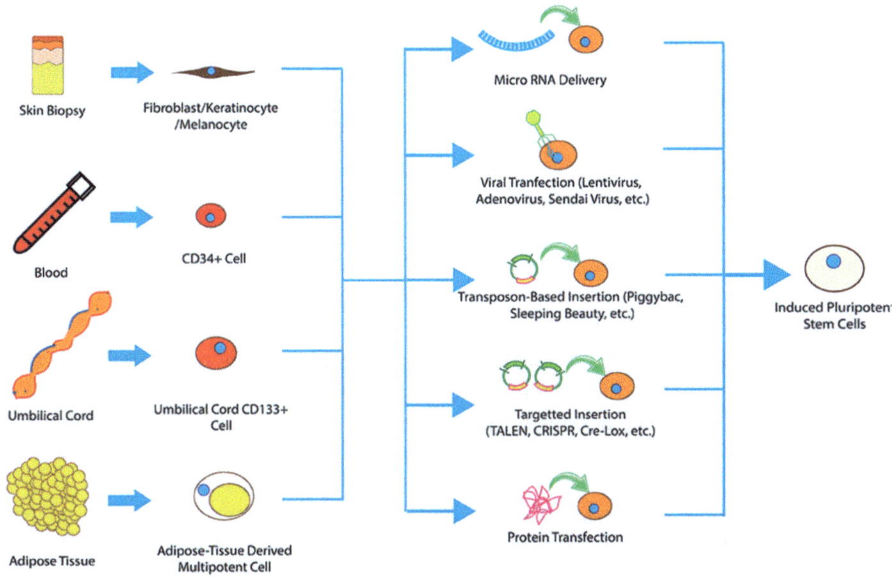

Fig. 4.4 Different types of cells have been used to produce iPSCs, including fibroblasts, keratinocytes, melanocytes, CD34+ cells, cord blood cells, and adipose stem cells. (Khazaei et al. [56] (open access))

c-Myc or Glis1 as a replacement for c-Myc [79, 148]. Although safety is improved, these alternatives could potentially comprise reprogramming efficiency [56, 87]. In addition, viral delivery of these factors to the cell affects its clinical practicality. Lentivirus, adenovirus, and Sendai virus can be used; however, adenovirus and Sendai viruses are safer than lentiviruses, as they do not cause mutagenesis by integrating into the genome [56]. Other approaches include removable transposons [143], microRNAs [123], mRNAs [138], episomal vectors [149], and recombinant proteins [57]. In the future, technologies such as blind CRISPR-Cas9 may be used for transcriptional control by bringing fused transcriptional regulators to the site of interest in the genome [32]. Furthermore, investigations are underway examining the use of small molecules that resemble the effects of transcriptional factors. For instance, valproic acid and histone deacetylase inhibitors are capable of producing the effects of c-Myc during reprogramming [56].

4.3.6 Differentiated Cells

Following the induction of pluripotency, the next step involves differentiating iPSCs into the appropriate cell type in order to exert their therapeutic effect. Various differentiation paths have been investigated for treating primary and secondary SCI [16]. These pathways produce NPcCs, mesenchymal stromal cells (MSC), astrocytes, and oligodendrocyte progenitor cells, among others [16].

4.3.6.1 Neural Progenitor Cells (NPrCs)

NPrCs are one the most promising cells for treatment of SCI, as they are capable of differentiating into both neurons and glia. In vivo studies demonstrated that these cells can regenerate and remyelinate axons and improve the survivability of neurons. Several protocols are available for differentiating iPSCs into NPCs, such as dual SMAD inhibition. Alternatively, NPCs can be developed by NOTCH pathway induction using nonviral transposons. Upon transplantation, these cells are able to integrate into the host tissue. An important observation for NPCs is their ability to migrate large distances from the source of injury and act as interneurons to reconstruct local neural circuits. In addition, due to their neural commitment, NPCs lack tumorigenicity and are safe candidates for SCI treatment. Interestingly, NPC can also be directly generated from the source cells without transforming it into a pluripotent state. This can be achieved by using Sox2, FoxG1, and Brn2 [76, 140].

4.3.6.2 Mesenchymal Stromal Cells (MSCs)

Intravenous injection of MSCs has been shown to mitigate inflammation and decrease the level of IL10 – a potent anti-inflammatory cytokine. MSCs also reduce BSCB permeability, hemorrhage, and acute lesion volume. One potential downside of mesenchymal stem cells is that they are multipotent and can differentiate into unwanted cells [16].

4.3.6.3 Astrocytes

Astrocytes play a crucial role in providing nutritional and trophic support and controlling synapse formation. SCI results in the loss of astrocytes in both white and gray matter. Although reactive astrocytes play a role in the pathobiology of glial scarring and chronic secondary SCI, transplanting astrocytes immediately post-injury is postulated to have beneficial effects. Several studies have examined the efficacy of transplanting iPSC-derived astrocytes in SCI. Astrocytes have shown to promote axonal regeneration, tissue reorganization, and functional recovery [56].

4.3.6.4 Oligodendrocyte Progenitor Cells (OPCs)

The use of OPCs to treat SCI has gained popularity due to the involvement of these cells in myelinating damaged axons [71]. Following SCI, endogenous OPCs will proliferate and increase spontaneous remyelination of axons [139]. However, due to the presence of inhibitory molecules and glial scaring, its frequency is limited and rare. By administrating exogenous OPCs, myelination of damaged axons is enhanced, leading to improved functional recovery after SCI [16]. In addition, OPCs are involved in the production of neurotrophic factors that promote axonal regeneration [16].

4.3.7 Direct Reprogramming/Transdifferentiation

Direct reprogramming has emerged recently as a result of the ethical and rejection-related issues associated with embryonic stem cells and the general inefficiency and slow pace of induced pluripotent reprogramming. Direct reprogramming between adult cell types across different developmental tissue lineages is a strategy for sourcing NPCs, downstream neural cell types, and their derivatives [11, 29, 33, 49, 91, 113, 124, 135, 136, 147]. The first successful cross-lineage reprogramming or transdifferentiation into a neural cell type was achieved through the use of lentiviral-mediated transgenic expression of the three neuronal fate-specifying transcription factors termed BAM factors: *Ascl1*, *Brn2*, and *Mytl1* [136]. This work demonstrated the conversion of mouse embryonic fibroblasts into functional, Tuj1-expressing adult neurons (termed induced neurons, iNs), in only 12 days [136]. These iNs were shown to also express the general neuronal markers (MAP2, NeuN, and synapsin), generate functional action potentials (APs), form synapses with each other, and is a heterogeneous mix of GABAnergic and glutamatergic neurons [136].

Building on the initial success of converting mouse fibroblasts into iNs, the same research group reproduced their results using human fibroblasts of both embryonic and postnatal origin [91]. One notable difference required to generate iNs from human fibroblasts is the requirement of the transcription factor, *NeuroD1*, and the aforementioned BAM factors – the new combination is termed the BAMN factors [91]. The study discovered that while antiviral-transfected expression of the BAM factors alone was sufficient to produce Tuj1-expressing cells from human fibroblasts within 7–10 days, these cells were found to be immature neurons/neuronal progenitors and were unable to generate APs or form synapses [91]. Additionally, the combination of *Brn2 and NeuroD1* was found to be capable of directly reprogramming human fibroblasts into a Tuj1-expressing state, although similarly to the BAM factors, they could not produce mature functional iNs [91].

Expression of the BAMN factors in human fibroblasts was subsequently shown to induce expression of the pan-neuronal markers Tuj1, NeuN, PSA-NCAM, and MAP2 after only 2 weeks [91]. After 4–5 weeks of culture, the iNs expressed the synaptic vesicle proteins, synapsin and synaptotagmin, in addition to other synaptic markers including the microtubule proteins – β-III-tubulin, doublecortin (DCX), and (MAP2) – as well as neural cell adhesion molecule (NCAM) [91]. The iNs produced from human fibroblasts using the BAMN factors were found to be predominantly glutamatergic, expressing vGLUT1 and/or vGLUT2 [91]. Further, these iNs were a mixture of forebrain (expressing T-box brain 1, Tbr1) and peripheral neurons (expressing Peripherin) [91]. In order to further refine the iN production procedure, later experiments by a different team demonstrated that the transcription factors *Brn2* and *Mytl1*, in tandem with microRNA 124 (miR-124), were capable of producing functional adult neurons from adult human fibroblasts [11].

Generating NPCs from fibroblasts through direct reprogramming was discovered to be possible through several methods [58, 76, 82]. Fibroblast to NPC direct reprogramming methods have included (1) the expression of the iPS reprogramming factors, Oct4, Sox2, Klf4, and c-Myc, for 12 days using the doxycycline (dox)-inducible secondary mouse embryonic fibroblast (MEF) system or (2) the expression of the three transcription factors – *Brn2, Sox2*, and *FoxG1* – to produce tripotent NPCs capable of forming functional and mature neurons, oligodendrocytes, and astrocytes [58, 76].

The fibroblast-NPC reprogramming method using iPS factors was shown to involve the cells transitioning through a temporary "plastic" developmental intermediate fate (separate from a total pluripotent iPS fate), before becoming specified as NPCs through induction by the RepM neural media containing FGF2, EGF, and FGF4 [58]. Additionally, iPS factor directly reprogramed NPCs were found to be heterogeneous in nature, additionally containing colony-forming cells that expressed Prominin-1, PSA-NCAM, and A2B5, and separate Tuj1+, DCX+, and tyrosine-hydroxylase neurons [58]. Upon differentiation in N2 media for 2 weeks, Kim et al.'s NPCs were able to produce GABAergic, NeuN, and Map2 neurons, as well as astrocytes expressing GFAP. However, in contrast to the fibroblast-NPC direct reprogramming method using *Brn2*, *Sox2*, and *FoxG1*, the aforementioned study of Kim et al. did not detect mature, MBP-expressing oligodendrocytes [58, 76]. Lujan et al. were similarly unable to produce mature oligodendrocytes without the addition of *Brn2* to *Box2* and *Foxg1*, although the latter factors alone were demonstrated to be capable of producing Tau-EGFP+, Tuj1+, and MAP2+ neurons and GFAP+ astrocytes [76]. In contrast, the omission of Sox2 restricted the reprogrammed NPCs from differentiating into mature neurons [76].

The critical, notable differences between the lineage-specific, three-factor fibroblast NPC reprogramming method used by *Lujan* et al. and the iPS factor reprogramming strategy employed by *Kim* et al. were that (1) *Lujan* et al.'s NPCs were shown to be tripotent (while *Kim* et al.'s NPCs being unable to differentiate into oligodendrocytes), (2) *Lujan* et al.'s NPCs were able to proliferate beyond 3–5 passages, and (3) *Lujan* et al.'s NPCs did not undergo a transition through a semi-pluripotent/plastic intermediate state before establishing an NPC fate [58, 76]. A recent genetic study into the brain tumor, glioblastoma multiforme (GBM), elucidated the role in which two of the three factors (*Sox2 and FoxG1*), used by *Lujan* et al. to drive fibroblast-NPC reprogramming, may accomplish their task. *Sox2* and *FoxG1* were found to affect the regulation of the downstream cell cycle and epigenetic effectors including *Plk1*, *Mycn*, *Dnmt1*, *Dnmt3b*, *Tet3*, and particularly *Foxo3* [26, 76].

Reprogrammed NPCs from adult mouse and human fibroblasts have been successfully generated using Sox2 alone [106]. Sox2-only reprogrammed NPCs are tripotent and resemble the morphology and gene expression profile of endogenous NPCs (*Sox2*, *Nestin*, *Sox1*, *Zbtb17 gene).* They are also capable of self-renewal and neurosphere formation and can integrate safely into mouse brains without tumor formation [106]. Despite the promise of single-factor fibroblast-NPC reprogramming, more recent work suggests that Sox2-alone NPC reprogramming may also yield only partially reprogrammed and senescent NPCs [141]. This is demonstrated by the study's NPCs being incapable of both astrocytic differentiation and full self-renewal, in addition to reduced neuronal differentiation [141].

4.3.8 Non-stem Autologous Cell Transplantation for SCI

4.3.8.1 Schwann Cells

Under normal conditions, Schwann cells are responsible for the production of myelin sheaths in the peripheral nervous system. Transplantation of these cells after SCI has demonstrated promising results [16]. Potential mechanisms through which Schwann cells improve SCI recovery include repairing and remyelinating the damaged axons and secreting neurotrophic factors in order to enhance the survivability of damaged neurons [97, 131]. During the first clinical trial, transplantation of Schwann cells did not show any adverse effects, and improved motor and sensory skills were observed in one out of four patients [112]. Further clinical studies are currently underway to better understand and assess the efficacy and mechanism of Schwann cell transplantation [27].

4.3.8.2 Olfactory Ensheathing Cells (OECs)

OECs are considered to be a suitable choice for SCI treatment because of their ability to induce axon regeneration and integrate into the spinal cord of the host organism [31]. Notably, these cells are capable of remyelinating and scaffolding to repair the damaged axons [56]. In addition, OECs are involved in the secretion of neurotrophic factors and other pro-regenerative factors, which assist recovery post-SCI [109]. Due to easy accessibility of these cells from olfactory mucosa, they are well suited for autologous transplantation [30]. Accordingly, in preclinical rat models, OECs have demonstrated promising results in both axonal repair and structural and functional recovery. However, in clinical trials, OECs have failed to demonstrate significant improvements in functional recovery, measured using the ASIA (American Spinal Injury Association) scale [39, 78]. The lack of ASIA score improvement could be linked to the cohort used, which was an entirely thoracic paraplegic population. More recent clinical studies have demonstrated functional improvement after OEC transplantation [126].

4.3.9 Impact of Injury *Level* Milieu *on Stem Cell Fate*

In addition to the other barriers of translating cell therapy to SCI, there is the potential barrier of injury level differences in the injury niche. While over 50% of SCIs occur in the cervical (neck) region [121], the majority of preclinical cell therapy evidence is in thoracic SCI models. The preference for using thoracic models is due to reduced cardiovascular and respiratory complications as compared to cervical models. Thus, a critical knowledge gap exists regarding the efficacy and timing of cell therapy in cervical SCI [65]. Overall, the cervical region is anatomically different compared to the thoracic region, in that it has a greater central and peripheral blood supply, gray-to-white matter ratio, cord diameter, and smaller vertebra. The cervical cord is more susceptible to BSCB disruption and immune cell infiltration post-SCI due to greater vascular supply and reduced pericyte coverage of its gray-matter capillaries [142]. Additionally, there are distinct level-specific secondary injury pathologies in the chronic injury phase of cervical injuries due to immunosuppression after the loss of supraspinal sympathetic input [25, 48, 75, 133]. Further, injury level stratification in clinical trials of neuroprotective therapies has also indicated level-dependent heterogeneity in terms of their therapeutic efficacy [38, 101]. Taken together, it is difficult to ignore the potential level-dependent impact of the injury niche on stem cell fate as NPCs, MSCs, and OPCs are all known to respond to injury and inflammation differently [7, 111, 134]. As such, it is of paramount importance that we seek to understand the level- and time-dependent changes to the injury niche and to tailor cell therapy accordingly.

4.3.10 Clinical Trials

In order to realize the therapeutic potential of stem cell treatments and translate them for use in SCI patients, clinical trials are necessary to evaluate the safety and efficacy of these treatments [5, 6, 12, 36]. A number of SCI clinical trials involving a variety of stem cell transplantation treatments have already been conducted, and some are currently underway (Table 4.1) [5, 6, 12, 36].

A clinical trial by StemCells, Inc. began in March 2011 and involved the use of human central nervous system stem cells (HuCNS-SCs, [12]). The trial began with one center in Switzerland and soon added two sites in Canada [12]. Twelve thoracic injury (T2–T11) patients took part in the trial, receiving one dose of intramedullary delivered HuCNS-SCs followed by subsequent immunosuppression for 9 months [12]. The trial evaluated safety and efficacy and presented findings at the fourth Joint International Spinal Cord Society and American Spinal Injury Association meeting in Montreal in 2015. However, as of October 2017, the final evaluations for this trial have yet to appear in an official scientific peer-reviewed publication [12].

A similar phase II trial, which set out to evaluate intramedullary delivered HuCNS-SCs for safety and efficacy in cervical (C5–C7) SCI patients, began in October 2014 [12]. Thirty-one cervical SCI patients with ASIA Impairment Scale levels of B–C were recruited across 13 sites in North America [12]. Despite showing promise after 6 months with 80% of patients reportedly showing motor improvements, trial NCT02163876 was ultimately terminated in May of 2016 due to financial concerns and failure to show significant functional motor improvements or a trend for improvement over time [12].

In addition to stem and progenitor cells, Schwann cells have also been evaluated as a possible therapeutic in SCI clinical trials [5]. A phase I trial, which was conducted on nine thoracic (T3–T11) injured with motor levels of A on the International Standards for Neurological Classification of Spinal Cord Injury (ISNCSCI), was conducted in November 2012 [5]. Patients received autologous sural nerve-derived

Table 4.1 List of completed and ongoing clinical trials involving stem cells for SCI

Trial name	Date started	ClinicalTrials.gov Identifier	Treatment evaluated	Phase	# of participants	Status	Date completed
Study of human central nervous system stem cells (HuCNS-SC) in patients with thoracic spinal cord injury	March 2011	NCT01321333	HuCNS-SC cells – T2–T11 injuries, ASIA A–C	I/II	12	Completed	April 2015
Study of human central nervous system (CNS) stem cell transplantation in cervical spinal cord injury	October 2014	NCT02163876	HuCNS-SC cells – C5–C7 injuries, ASIA B–C	II	31	Terminated (financial reasons)	May 2016
Safety of autologous human Schwann cells (ahSC) in subjects with subacute SCI	November 2012	NCT01739023	Autologous (sural nerve) human Schwann cells – T3–T11 injuries, ISNCSCI grade A	I	9	Completed	August 2016
Administration of expanded autologous adult bone marrow mesenchymal cells in established chronic spinal cord injuries	July 2015	NCT02570932	Bone marrow-derived MSCs – all injury levels, ASIA levels A–D	II	10	Ongoing (not recruiting)	February 2018
Umbilical cord mesenchymal stem cell transplantation to patients with spinal cord injury	January 2014	NCT02481440	Umbilical cord MSCs – all injury levels, ASIA levels A–D	I/II	44	Ongoing (actively recruiting)	December 2018
Dose escalation study of AST-OPC1 in spinal cord injury	March 2015	NCT02302157	AST-OPC1 (ESC-derived OPCs, Asterias Biotherapeutics, Inc.) Last fully preserved single neurological level (SNL) C4–C7 ASIA A (cohorts #1, 2, 3) ASIA B (cohorts #4, 5)	I/II	35	Ongoing (actively recruiting)	December 2018
Olfactory ensheathing cells in complete human spinal cord injury	November 2010	NCT01231893	Patient sampled autologous OECs Single injury C5–L5 ASIA A	I	10	Unknown	Unknown

Schwann cell grafts, with sural nerve harvests taking place 30 days post-injury and transplantations up to 72 days post-injury [5]. In August 2016, the trial concluded and successfully demonstrated safety and a lack of adverse effects after 1 year [5].

MSCs, as mentioned in previous sections in this chapter, also represent a promising stem cell treatment for SCI and have thus been the primary subject of several clinical trials [5]. Two recent MSC-SCI trials, NCT02481440 (phase I/II) and NCT02570932 (phase II), started in January 2014 and July 2015, respectively, to test the safety and efficacy of MSCs [5]. Both ongoing trials are being conducted in patients of all injury levels and overall ASIA impairment scale levels ranging from A to D [5]. Forty-four patients are currently enrolled in trial NCT02481440 at one site (Guangdong, China), which is evaluating intrathecally delivered human umbilical cord-derived MSCs for safety and efficacy over a 24-month period [5]. Trial NCT02570932 on the other hand (Madrid, Spain) is utilizing expanded autologous bone marrow-derived MSCs delivered intrathecally into each patient's subarachnoid space using lumbar punctures. The punctures are performed three times, with 3-month intervals between each treatment. Similarly, results will report on the safety and efficacy of MSCs over a 24-month period [5]. Both ongoing MSC trials are expected to be completed before the end of 2018 and will hopefully provide support for safe and effective MSC treatments and eventual phase III MSC trials for SCI.

In preclinical studies, OPCs have been demonstrated to be a potentially useful cell-based therapy for SCI and are currently the subject of a phase I/II trial, NCT02302157 [5]. OPC clinical trials originally began in 2010 with a phase I trial, NCT01217008, evaluating the safety of human ESC-derived OPCs for thoracic-injured (T3–T11) ASIA level A patients [115]. This trial was unfortunately canceled after 1 year, with only five patients receiving transplants due to funding concerns [115]. In 2014, Asterias Biotherapeutics, Inc. revived the clinical trial usage of OPCs when it began sponsoring the aforementioned NCT02302157, this time using their proprietary cell line termed "Asterias OPC1" (AST-OPC1). As of October 2017, the Asterias trial is currently ongoing, evaluating the dose escalation of OPC treatments (one injection of either 2 or 10 million cells or two injections of 10 million cells) among 5 cohorts of 35 patients, with ASIA levels of A or B [5, 6]. Results of the Asterias trial, specifically patient upper extremity motor scores and ISCCSCI motor levels evaluated at six time points (30, 60, 90, 180, 270, and 365 days), are expected to be released by December 2018.

Despite their significant therapeutic potential, iPS-derived NPCs (or other cells) have yet to be used in any clinical trial for SCI [36]. Currently the main roadblocks for translating iPS-derived cells have been safety, particularly the danger of tumorigenicity, and low efficiency of induced pluripotency [36]. Perhaps in the near future, as iPS protocols continue to improve, there may still be hope for iPS-based clinical trials, perhaps involving allogenic NPCs for the treatment of traumatic SCI [36]. Recent advancements, such as the optimization of transcription factor cocktails and continued study of nonviral reprogramming factor delivery and transformation methods, have increased the likelihood to clinical trials for iPS-based treatments in the future [36].

References

1. Aasen T, Raya A, Barrero MJ, Garreta E, Consiglio A, Gonzalez F, Vassena R, Bilić J, Pekarik V, Tiscornia G, Edel M, Boué S, Belmonte JCI (2008) Efficient and rapid generation of induced pluripotent stem cells from human keratinocytes. Nat Biotechnol 26(11):1276–1284
2. Abrajano JJ, Qureshi IA, Gokhan S, Molero AE, Zheng D, Bergman A, Mehler MF (2010) Corepressor for element-1-silencing transcription factor preferentially mediates gene networks underlying neural stem cell fate decisions. Proc Natl Acad Sci U S A 107(38):16685–16690
3. Agbay, Andrew; Edgar, John M; Robinson, Meghan; Styan, Tara; Wilson, Krista; Schroll, Julian; Ko, Junghyuk; Khadem Mohtaram, Nima, ; Jun, Martin Byung-Guk; Willerth, Stephanie M.(2016). Biomaterial strategies for delivering stem cells as a treatment for spinal cord injury. Cells Tissues Organs 202(1–2): 42–51
4. Ahuja CS, Fehlings M (2016) Concise review: bridging the gap: novel neuroregenerative and neuroprotective strategies in spinal cord injury. Stem Cells Transl Med 5(7):914–924
5. Ahuja CS, Nori S, Tetreault L, Wilson J, Kwon B, Harrop J, Choi D, Fehlings MG (2017) Traumatic spinal cord injury-repair and regeneration. Neurosurgery 80(3S):S9–S22
6. Ahuja CS, Wilson JR, Nori S, Kotter MRN, Druschel C, Curt A, Fehlings MG (2017) Traumatic spinal cord injury. Nat Rev Dis Primers 3:17018
7. Akkermann R, Beyer F, Küry P (2017) Heterogeneous populations of neural stem cells contribute to myelin repair. Neural Regen Res 12(4):509–517
8. de Almeida PE, Meyer EH, Kooreman NG, Diecke S, Dey D, Sanchez-Freire V, Hu S, Ebert A, Odegaard J, Mordwinkin NM, Brouwer TP, Lo D, Montoro DT, Longaker MT, Negrin RS, Wu JC (2014) Transplanted terminally differentiated induced pluripotent stem cells are accepted by immune mechanisms similar to self-tolerance. Nat Commun 5:3903
9. Altman J, Das GD (1965) Autoradiographic and histological evidence of postnatal hippocampal neurogenesis in rats. J Comp Neurol 124(3):319–335
10. Alunni A, Bally-Cuif L (2016) A comparative view of regenerative neurogenesis in vertebrates. Development 143(5):741–753
11. Ambasudhan R, Talantova M, Coleman R, Yuan X, Zhu S, Lipton SA, Ding S (2011) Direct reprogramming of adult human fibroblasts to functional neurons under defined conditions. Cell Stem Cell 9(2):113–118
12. Anderson AJ, Piltti KM, Hooshmand MJ, Nishi RA, Cummings BJ (2017) Preclinical efficacy failure of human CNS-derived stem cells for use in the pathway study of cervical spinal cord injury. Stem Cell Reports 8(2):249–263
13. Araki R, Uda M, Hoki Y, Sunayama M, Nakamura M, Ando S, Sugiura M, Ideno H, Shimada A, Nifuji A, Abe M (2013) Negligible immunogenicity of terminally differentiated cells derived from induced pluripotent or embryonic stem cells. Nature 494(7435):100–104
14. Assinck P, Duncan GJ, Plemel JR, Lee MJ, Stratton JA, Manesh SB, Liu J, Ramer LM, Kang SH, Bergles DE, Biernaskie J, Tetzlaff W (2017) Myelinogenic plasticity of oligodendrocyte precursor cells following spinal cord contusion injury. J Neurosci 37(36):8635–8654
15. Badner A, Vawda R, Laliberte A, Hong J, Mikhail M, Jose A, Dragas R, Fehlings M (2016) Early intravenous delivery of human brain stromal cells modulates systemic inflammation and leads to vasoprotection in traumatic spinal cord injury. Stem Cells Transl Med 5(8):991–1003
16. Badner A, Siddiqui AM, Fehlings MG (2017) Spinal cord injuries: how could cell therapy help? Expert Opin Biol Ther 17(5):529–541
17. Barbosa JS, Ninkovic J (2016) Adult neural stem cell behavior underlying constitutive and restorative neurogenesis in zebrafish. Neurogenesis (Austin, TX) 3(1):e1148101
18. Barnabé-Heider F, Göritz C, Sabelström H, Takebayashi H, Pfrieger FW, Meletis K, Frisén J (2010) Origin of new glial cells in intact and injured adult spinal cord. Cell Stem Cell 7(4):470–482
19. Becker AJ, McCulloch EA, Till JE (1963) Cytological demonstration of the clonal nature of spleen colonies derived from transplanted mouse marrow cells. Nature 197:452–454

20. Berberoglu MA, Dong Z, Li G, Zheng J, Trejo Martinez L d CG, Peng J, Wagle M, Reichholf B, Petritsch C, Li H, Pleasure SJ, Guo S (2014) Heterogeneously expressed fezf2 patterns gradient Notch activity in balancing the quiescence, proliferation, and differentiation of adult neural stem cells. J Neurosci Off J Soc Neurosci 34(42):13911–13923
21. Bonaguidi MA, Wheeler MA, Shapiro JS, Stadel RP, Sun GJ, Ming G-l, Song H (2011) In vivo clonal analysis reveals self-renewing and multipotent adult neural stem cell characteristics. Cell 145(7):1142–1155
22. Bond AM, Ming G-L, Song H (2015) Adult mammalian neural stem cells and neurogenesis: five decades later. Cell Stem Cell 17(4):385–395
23. Breunig JJ, Haydar TF, Rakic P (2011) Neural stem cells: historical perspective and future prospects. Neuron 70(4):614–625
24. Briggs SF, Reijo Pera RA (2014) X chromosome inactivation: recent advances and a look forward. Curr Opin Genet Dev 28:78–82
25. Brommer B, Engel O, Kopp MA, Watzlawick R, Müller S, Prüss H, Chen Y, DeVivo MJ, Finkenstaedt FW, Dirnagl U, Liebscher T, Meisel A, Schwab JM (2016) Spinal cord injury-induced immune deficiency syndrome enhances infection susceptibility dependent on lesion level. Brain J Neurol 139(Pt 3):692–707
26. Bulstrode H, Johnstone E, Marques-Torrejon MA, Ferguson KM, Bressan RB, Blin C, Grant V, Gogolok S, Gangoso E, Gagrica S, Ender C, Fotaki V, Sproul D, Bertone P, Pollard SM (2017) Elevated FOXG1 and SOX2 in glioblastoma enforces neural stem cell identity through transcriptional control of cell cycle and epigenetic regulators. Genes Dev 31(8):757–773
27. Bunge MB (2016) Efficacy of Schwann cell transplantation for spinal cord repair is improved with combinatorial strategies: improving Schwann cell transplantation. J Physiol 594(13):3533–3538
28. Cao Q, He Q, Wang Y, Cheng X, Howard RM, Zhang Y, DeVries WH, Shields CB, Magnuson DSK, Xu X-M, Kim DH, Whittemore SR (2010) Transplantation of ciliary neurotrophic factor-expressing adult oligodendrocyte precursor cells promotes remyelination and functional recovery after spinal cord injury. J Neurosci Off J Soc Neurosci 30(8):2989–3001
29. Chambers SM, Studer L (2011) Cell fate plug and play: direct reprogramming and induced pluripotency. Cell 145(6):827–830
30. Chhabra HS, Lima C, Sachdeva S, Mittal A, Nigam V, Chaturvedi D, Arora M, Aggarwal A, Kapur R, Khan TAH (2009) Autologous mucosal transplant in chronic spinal cord injury: an Indian Pilot Study. Spinal Cord 47:887–895
31. Dlouhy BJ, Awe O, Rao RC, Kirby PA, Hitchon PW (2014) Autograft-derived spinal cord mass following olfactory mucosal cell transplantation in a spinal cord injury patient. J Neurosurg Spine 21:618–622
32. Dominguez AA, Lim WA, Qi LS (2015) Beyond editing: repurposing CRISPR–Cas9 for precision genome regulation and interrogation. Nat Rev Mol Cell Biol 17(1):5–15
33. Efe JA, Hilcove S, Kim J, Zhou H, Ouyang K, Wang G, Chen J, Ding S (2011) Conversion of mouse fibroblasts into cardiomyocytes using a direct reprogramming strategy. Nat Cell Biol 13(3):215–222
34. Fan C, Li X, Xiao Z, Zhao Y, Liang H, Wang B, Han S, Li X, Xu B, Wang N, Liu S, Xue W, Dai J (2017) A modified collagen scaffold facilitates endogenous neurogenesis for acute spinal cord injury repair. Acta Biomater 51:304–316
35. Farrukh A, Ortega F, Fan W, Marichal N, Paez JI, Berninger B, Campo AD, Salierno MJ (2017) Bifunctional hydrogels containing the laminin motif IKVAV promote neurogenesis. Stem Cell Reports 9(5):1432–1440
36. Faulkner SD, Vawda R, Fehlings MG (2014) Adult-derived pluripotent stem cells. World Neurosurg 82(3–4):500–508
37. Fehlings MG, Tator CH (1995) The relationships among the severity of spinal cord injury, residual neurological function, axon counts, and counts of retrogradely labeled neurons after experimental spinal cord injury. Exp Neurol 132(2):220–228
38. Fehlings MG, Theodore N, Harrop J, Maurais G, Kuntz C, Shaffrey CI, Kwon BK, Chapman J, Yee A, Tighe A, McKerracher L (2011) A phase I/IIa clinical trial of a recombinant Rho protein antagonist in acute spinal cord injury. J Neurotrauma 28(5):787–796

39. Féron F, Perry C, Cochrane J et al (2005) Autologous olfactory ensheathing cell transplantation in human spinal cord injury. Brain 128(Pt 12):2951–2960
40. Forostyak S, Jendelova P, Sykova E (2013) The role of mesenchymal stromal cells in spinal cord injury, regenerative medicine and possible clinical applications. Biochimie 95(12):2257–2270
41. Gwak S-J, Macks C, Jeong DU, Kindy M, Lynn M, Webb K, Lee JS (2017) RhoA knockdown by cationic amphiphilic copolymer/siRhoA polyplexes enhances axonal regeneration in rat spinal cord injury model. Biomaterials 121:155–166
42. Han H-W, Hsu S-H (2017) Chitosan derived co-spheroids of neural stem cells and mesenchymal stem cells for neural regeneration. Colloids Surf B Biointerfaces 158:527–538
43. Han Q, Jin W, Xiao Z, Ni H, Wang J, Kong J, Wu J, Liang W, Chen L, Zhao Y, Chen B, Dai J (2010) The promotion of neural regeneration in an extreme rat spinal cord injury model using a collagen scaffold containing a collagen binding neuroprotective protein and an EGFR neutralizing antibody. Biomaterials 31(35):9212–9220
44. His W Sr (1874) Unsere Körperform und das physiologische Problem ihrer Entstehung : Briefe an einen befreundeten Naturforscher. F.C.W. Vogel, Leipzig
45. His W Sr (1886) Zur Geschichte der menschlichen Rückenmarkes und der Nervenwurzeln. Bei S. Hirzel, Leipzig, p 13
46. His W Sr (1904) Die Entwicklung des menschlichen Gehirns während der ersten Monate : Untersuchungsergebnisse. Leipzig, Hirzel
47. Huangfu D, Osafune K, Maehr R et al (2008) Induction of pluripotent stem cells from primary human fibroblasts with only Oct4 and Sox2. Nat Biotechnol 26(11):1269–1275
48. Ibarra A, Jiménez A, Cortes C, Correa D (2007) Influence of the intensity, level and phase of spinal cord injury on the proliferation of T cells and T-cell-dependent antibody reactions in rats. Spinal Cord 45(5):380–386
49. Ieda M, Fu J-D, Delgado-Olguin P, Vedantham V, Hayashi Y, Bruneau BG, Srivastava D (2010) Direct reprogramming of fibroblasts into functional cardiomyocytes by defined factors. Cell 142(3):375–386
50. Kamada T, Koda M, Dezawa M, Anahara R, Toyama Y, Yoshinaga K, Hashimoto M, Koshizuka S, Nishio Y, Mannoji C, Okawa A, Yamazaki M (2011) Transplantation of human bone marrow stromal cell-derived Schwann cells reduces cystic cavity and promotes functional recovery after contusion injury of adult rat spinal cord. Neuropathology Off J Japanese Soc Neuropathol 31(1):48–58
51. Karimi-Abdolrezaee S, Eftekharpour E, Wang J, Morshead CM, Fehlings MG (2006) Delayed transplantation of adult neural precursor cells promotes remyelination and functional neurological recovery after spinal cord injury. J Neurosci Off J Soc Neurosci 26(13):3377–3389
52. Karimi-Abdolrezaee S, Eftekharpour E, Wang J, Schut D, Fehlings MG (2010) Synergistic effects of transplanted adult neural stem/progenitor cells, chondroitinase, and growth factors promote functional repair and plasticity of the chronically injured spinal cord. J Neurosci Off J Soc Neurosci 30(5):1657–1676
53. Kawamura A, Miyagawa S, Fukushima S, Kawamura T, Kashiyama N, Ito E, Watabe T, Masuda S, Toda K, Hatazawa J, Morii E, Sawa Y (2016) Teratocarcinomas arising from allogeneic induced pluripotent stem cell-derived cardiac tissue constructs provoked host immune rejection in mice. Sci Rep 6:19464
54. Keirstead HS (2005) Human embryonic stem cell-derived oligodendrocyte progenitor cell transplants remyelinate and restore locomotion after spinal cord injury. J Neurosci 25(19):4694–4705
55. Khazaei M, Siddiqui AM, Fehlings MG (2014) The potential for iPS-derived stem cells as a therapeutic strategy for spinal cord injury: opportunities and challenges. J Clin Med 4(1):37–65
56. Khazaei M, Ahuja CS, Fehlings MG (2017) Induced pluripotent stem cells for traumatic spinal cord injury. Front Cell Dev Biol 4:152
57. Kim D, Kim C-H, Moon J-I, Chung Y-G, Chang M-Y, Han B-S, Ko S, Yang E, Cha KY, Lanza R, Kim K-S (2009) Generation of human induced pluripotent stem cells by direct delivery of reprogramming proteins. Cell Stem Cell 4(6):472–476

58. Kim J, Efe JA, Zhu S, Talantova M, Yuan X, Wang S, Lipton SA, Zhang K, Ding S (2011) Direct reprogramming of mouse fibroblasts to neural progenitors. Proc Natl Acad Sci U S A 108(19):7838–7843
59. Kim YC, Kim YH, Kim JW, Ha KY (2016) Transplantation of mesenchymal stem cells for acute spinal cord injury in rats: comparative study between intralesional injection and scaffold based transplantation. J Korean Med Sci 31(9):1373–1382
60. Kizil C, Dudczig S, Kyritsis N, Machate A, Blaesche J, Kroehne V, Brand M (2012) The chemokine receptor cxcr5 regulates the regenerative neurogenesis response in the adult zebrafish brain. Neural Dev 7:27
61. Kizil C, Kyritsis N, Dudczig S, Kroehne V, Freudenreich D, Kaslin J, Brand M (2012) Regenerative neurogenesis from neural progenitor cells requires injury-induced expression of Gata3. Dev Cell 23(6):1230–1237
62. Kizil C, Kyritsis N, Brand M (2015) Effects of inflammation on stem cells: together they strive? EMBO Rep 16(4):416–426
63. Kong X-B, Tang Q-Y, Chen X-Y, Tu Y, Sun S-Z, Sun Z-L (2017) Polyethylene glycol as a promising synthetic material for repair of spinal cord injury. Neural Regen Res 12(6):1003–1008
64. Kruse V, Hamann C, Monecke S, Cyganek L, Elsner L, Hübscher D, Walter L, Streckfuss-Bömeke K, Guan K, Dressel R, Björkström NK (2015) Human induced pluripotent stem cells are targets for allogeneic and autologous natural killer (NK) cells and killing is partly mediated by the activating NK receptor DNAM-1. PLoS One 10(5):e0125544
65. Kwon BK, Hillyer J, Tetzlaff W (2010) Translational research in spinal cord injury: a survey of opinion from the SCI community. J Neurotrauma 27(1):21–33
66. Kyritsis N, Kizil C, Brand M (2014) Neuroinflammation and central nervous system regeneration in vertebrates. Trends Cell Biol 24(2):128–135
67. Laurent LC, Ulitsky I, Slavin I, Tran H, Schork A, Morey R, Lynch C, Harness JV, Lee S, Barrero MJ, Ku S, Martynova M, Semechkin R, Galat V, Gottesfeld J, Belmonte JCI, Murry C, Keirstead HS, Park H-S, Schmidt U, Laslett AL, Muller F-J, Nievergelt CM, Shamir R, Loring JF (2011) Dynamic changes in the copy number of pluripotency and cell proliferation genes in human ESCs and iPSCs during reprogramming and time in culture. Cell Stem Cell 8(1):106–118
68. Lee SH, Kim Y, Rhew D, Kim A, Jo KR, Yoon Y, Choi KU, Jung T, Kim WH, Kweon O-K (2016) Impact of local injection of brain-derived neurotrophic factor-expressing mesenchymal stromal cells (MSCs) combined with intravenous MSC delivery in a canine model of chronic spinal cord injury. Cytotherapy. https://doi.org/10.1016/j.jcyt.2016.09.014
69. Lehmann M, Fournier A, Selles-Navarro I, Dergham P, Sebok A, Leclerc N, Tigyi G, McKerracher L (1999) Inactivation of Rho signaling pathway promotes CNS axon regeneration. J Neurosci 19(17):7537–7547
70. Leibinger M, Müller A, Gobrecht P, Diekmann H, Andreadaki A, Fischer D (2013) Interleukin-6 contributes to CNS axon regeneration upon inflammatory stimulation. Cell Death Dis 4:e609
71. Levine JM, Reynolds R, Fawcett JW (2001) The oligodendrocyte precursor cell in health and disease. Trends Neurosci 24:39–47
72. Li X, Han J, Zhao Y, Ding W, Wei J, Han S, Shang X, Wang B, Chen B, Xiao Z, Dai J (2015) Functionalized collagen scaffold neutralizing the myelin-inhibitory molecules promoted neurites outgrowth in vitro and facilitated spinal cord regeneration in vivo. ACS Appl Mater Interfaces 7(25):13960–13971
73. Li X, Zhao Y, Cheng S, Han S, Shu M, Chen B, Chen X, Tang F, Wang N, Tu Y, Wang B, Xiao Z, Zhang S, Dai J (2017) Cetuximab modified collagen scaffold directs neurogenesis of injury-activated endogenous neural stem cells for acute spinal cord injury repair. Biomaterials 137:73–86
74. Liu X, Li W, Fu X, Xu Y (2017) The immunogenicity and immune tolerance of pluripotent stem cell derivatives. Front Immunol 8:645

75. Lucin KM, Sanders VM, Jones TB, Malarkey WB, Popovich PG (2007) Impaired antibody synthesis after spinal cord injury is level dependent and is due to sympathetic nervous system dysregulation. Exp Neurol 207(1):75–84
76. Lujan E, Chanda S, Ahlenius H, Südhof TC, Wernig M (2012) Direct conversion of mouse fibroblasts to self-renewing, tripotent neural precursor cells. Proc Natl Acad Sci U S A 109(7):2527–2532
77. Mackay-Sim A, Kittel P (1991) Cell dynamics in the adult mouse olfactory epithelium: a quantitative autoradiographic study. J Neurosci Off J Soc Neurosci 11(4):979–984
78. Mackay-Sim A, Féron F, Cochrane J et al (2008) Autologous olfactory ensheathing cell transplantation in human paraplegia: a 3-year clinical trial. Brain J Neurol 131:2376–2386
79. Maekawa M, Yamaguchi K, Nakamura T, Shibukawa R, Kodanaka I, Ichisaka T, Kawamura Y, Mochizuki H, Goshima N, Yamanaka S (2011) Direct reprogramming of somatic cells is promoted by maternal transcription factor Glis1. Nature 474(7350):225–229
80. März M, Schmidt R, Rastegar S, Strähle U (2011) Regenerative response following stab injury in the adult zebrafish telencephalon. Dev Dyn Off Publ Am Assoc Anatomists 240(9):2221–2231
81. Matyas JJ, Stewart AN, Goldsmith A, Nan Z, Skeel RL, Rossignol J, Dunbar GL (2017) Effects of bone-marrow-derived MSC transplantation on functional recovery in a rat model of spinal cord injury: comparisons of transplant locations and cell concentrations. Cell Transplant 26(8):1472–1482
82. Maucksch C, Jones KS, Connor B (2013) Concise review: the involvement of SOX2 in direct reprogramming of induced neural stem/precursor cells. Stem Cells Transl Med 2(8):579–583
83. McCulloch EA, Till JE (1960) The radiation sensitivity of normal mouse bone marrow cells, determined by quantitative marrow transplantation into irradiated mice. Radiat Res 13:115–125
84. McKerracher L, Higuchi H (2006) Targeting Rho to stimulate repair after spinal cord injury. J Neurotrauma 23(3–4):309–317
85. Melo FR, Bressan RB, Forner S, Martini AC, Rode M, Delben PB, Rae GA, Figueiredo CP, Trentin AG (2017) Transplantation of human skin-derived mesenchymal stromal cells improves locomotor recovery after spinal cord injury in rats. Cell Mol Neurobiol 37(5):941–947
86. Morita T, Sasaki M, Kataoka-Sasaki Y, Nakazaki M, Nagahama H, Oka S, Oshigiri T, Takebayashi T, Yamashita T, Kocsis JD, Honmou O (2016) Intravenous infusion of mesenchymal stem cells promotes functional recovery in a model of chronic spinal cord injury. Neuroscience 335:221–231
87. Nakagawa M, Koyanagi M, Tanabe K, Takahashi K, Ichisaka T, Aoi T, Okita K, Mochiduki Y, Takizawa N, Yamanaka S (2007) Generation of induced pluripotent stem cells without Myc from mouse and human fibroblasts. Nat Biotechnol 26(1):101–106
88. Namba T, Huttner WB (2017) Neural progenitor cells and their role in the development and evolutionary expansion of the neocortex. Wiley Interdiscip Rev Dev Biol 6(1):1–16
89. Nashmi R, Fehlings MG (2001) Changes in axonal physiology and morphology after chronic compressive injury of the rat thoracic spinal cord. Neuroscience 104(1):235–251
90. Oliveri RS, Bello S, Biering-Sørensen F (2014) Mesenchymal stem cells improve locomotor recovery in traumatic spinal cord injury: systematic review with meta-analyses of rat models. Neurobiol Dis 62:338–353
91. Pang ZP, Yang N, Vierbuchen T, Ostermeier A, Fuentes DR, Yang TQ, Citri A, Sebastiano V, Marro S, Südhof TC, Wernig M (2011) Induction of human neuronal cells by defined transcription factors. Nature 476(7359):220–223
92. Park I-H, Lerou PH, Zhao R, Huo H, Daley GQ (2008) Generation of human-induced pluripotent stem cells. Nat Protoc 3(7):1180–1186
93. Park H-W, Lim M-J, Jung H, Lee S-P, Paik K-S, Chang M-S (2010) Human mesenchymal stem cell-derived Schwann cell-like cells exhibit neurotrophic effects, via distinct growth factor production, in a model of spinal cord injury. Glia 58(9):1118–1132

94. Park WB, Kim SY, Lee SH, Kim H-W, Park J-S, Hyun JK (2010) The effect of mesenchymal stem cell transplantation on the recovery of bladder and hindlimb function after spinal cord contusion in rats. BMC Neurosci 11:119
95. Peng Z, Gao W, Yue B, Jiang J, Gu Y, Dai J, Chen L, Shi Q (2016) Promotion of neurological recovery in rat spinal cord injury by mesenchymal stem cells loaded on nerve-guided collagen scaffold through increasing alternatively activated macrophage polarization. J Tissue Eng Regen Med 12(3):e1725–e1736
96. Phinney DG, Prockop DJ (2007) Concise review: mesenchymal stem/multipotent stromal cells: the state of transdifferentiation and modes of tissue repair–current views. Stem Cells (Dayton, Ohio) 25(11):2896–2902
97. Plemel JR, Keough MB, Duncan GJ, Sparling JS, Yong VW, Stys PK, Tetzlaff W (2014) Remyelination after spinal cord injury: is it a target for repair? Prog Neurobiol 117:54–74
98. Priest CA, Manley NC, Denham J, Wirth ED, Lebkowski JS (2015) Preclinical safety of human embryonic stem cell-derived oligodendrocyte progenitors supporting clinical trials in spinal cord injury. Regen Med 10(8):939–958
99. Puri MC, Nagy A (2012) Concise review: embryonic stem cells versus induced pluripotent stem cells: the game is on. Stem Cells 30(1):10–14
100. Qu J, Zhang H (2017) Roles of mesenchymal stem cells in spinal cord injury. Stem Cells Int 2017:5251313
101. Rahimi-Movaghar V (2009) Clinical trials for the treatment of spinal cord injury: cervical and lumbar enlargements versus thoracic area. Brain J Neurol 132(Pt 7):e115
102. Reimer MM, Sörensen I, Kuscha V, Frank RE, Liu C, Becker CG, Becker T (2008) Motor neuron regeneration in adult zebrafish. J Neurosci Off J Soc Neurosci 28(34):8510–8516
103. Retzius G (1893) Die Cajal'schen Zellen der Grosshirnrinde beim Menschen und bei Säugetieren. Biologische Untersuchungen 5:1–9
104. Retzius G (1893) Studien uber Ependym and Neuroglia. Biologische Untersuchungen. 5:2–26
105. Reynolds BA, Weiss S (1992) Generation of neurons and astrocytes from isolated cells of the adult mammalian central nervous system. Science (New York, NY) 255(5052):1707–1710
106. Ring KL, Tong LM, Balestra ME, Javier R, Andrews-Zwilling Y, Li G, Walker D, Zhang WR, Kreitzer AC, Huang Y (2012) Direct reprogramming of mouse and human fibroblasts into multipotent neural stem cells with a single factor. Cell Stem Cell 11(1):100–109
107. Rockowitz S, Zheng D (2015) Significant expansion of the REST/NRSF cistrome in human versus mouse embryonic stem cells: potential implications for neural development. Nucleic Acids Res 43(12):5730–5743
108. Rockowitz S, Lien W-H, Pedrosa E, Wei G, Lin M, Zhao K, Lachman HM, Fuchs E, Zheng D (2014) Comparison of REST cistromes across human cell types reveals common and context-specific functions. PLoS Comput Biol 10(6):e1003671
109. Roet KCD, Verhaagen J (2014) Understanding the neural repair-promoting properties of olfactory ensheathing cells. Exp Neurol 261:594–609
110. Rosado IR, Carvalho PH, Alves EGL, Tagushi TM, Carvalho JL, Silva JF, Lavor MSL, Oliveira KM, Serakides R, Goes AM, Melo EG (2017) Immunomodulatory and neuroprotective effect of cryopreserved allogeneic mesenchymal stem cells on spinal cord injury in rats. Genet Mol Res 16(1):1–24
111. Ruddy RM, Morshead CM (2017) Home sweet home: the neural stem cell niche throughout development and after injury. Cell Tissue Res 371(1):125–141
112. Saberi H, Moshayedi P, Aghayan H-R, Arjmand B, Hosseini S-K, Emami-Razavi S-H, Rahimi-Movaghar V, Raza M, Firouzi M (2008) Treatment of chronic thoracic spinal cord injury patients with autologous Schwann cell transplantation: an interim report on safety considerations and possible outcomes. Neurosci Lett 443(1):46–50
113. Sandler VM, Lailler N, Bouhassira EE (2011) Reprogramming of embryonic human fibroblasts into fetal hematopoietic progenitors by fusion with human fetal liver CD34+ cells. PLoS One 6(4):e18265

114. Schmidt R, Strähle U, Scholpp S (2013) Neurogenesis in zebrafish – from embryo to adult. Neural Dev 8:3
115. Scott CT, Magnus D (2014) Wrongful termination: lessons from the Geron clinical trial. Stem Cells Transl Med 3(12):1398–1401
116. Seaberg RM, van der Kooy D (2003) Stem and progenitor cells: the premature desertion of rigorous definitions. Trends Neurosci 26(3):125–131
117. Seo DK, Kim JH, Min J, Yoon HH, Shin E-S, Kim SW, Jeon SR (2017) Enhanced axonal regeneration by transplanted Wnt3a-secreting human mesenchymal stem cells in a rat model of spinal cord injury. Acta Neurochir 159(5):947–957
118. Shultz RB, Zhong Y (2017) Minocycline targets multiple secondary injury mechanisms in traumatic spinal cord injury. Neural Regen Res 12(5):702–713
119. Siddiqui AM, Khazaei M, Fehlings MG (2015) Translating mechanisms of neuroprotection, regeneration, and repair to treatment of spinal cord injury. Prog Brain Res 218:15–54
120. Siminovitch L, Mcculloch EA, Till JE (1963) The distribution of colony-forming cells among spleen colonies. J Cell Comp Physiol 62:327–336
121. Singh A, Tetreault L, Kalsi-Ryan S, Nouri A, Fehlings MG (2014) Global prevalence and incidence of traumatic spinal cord injury. Clin Epidemiol 6:309–331
122. Sordi V, Pellegrini S, Piemonti L (2017) Immunological issues after stem cell-based β cell replacement. Curr Diab Rep 9:17
123. Subramanyam D, Lamouille S, Judson RL, Liu JY, Bucay N, Derynck R, Blelloch R (2011) Multiple targets of miR-302 and miR-372 promote reprogramming of human fibroblasts to induced pluripotent stem cells. Nat Biotechnol 29(5):443–448
124. Szabo E, Rampalli S, Risueño RM, Schnerch A, Mitchell R, Fiebig-Comyn A, Levadoux-Martin M, Bhatia M (2010) Direct conversion of human fibroblasts to multilineage blood progenitors. Nature 468(7323):521–526
125. Taapken SM, Nisler BS, Newton MA, Sampsell-Barron TL, Leonhard KA, McIntire EM, Montgomery KD (2011) Karyotypic abnormalities in human induced pluripotent stem cells and embryonic stem cells. Nat Biotechnol 29(4):313–314
126. Tabakow P, Jarmundowicz W, Czapiga B et al (2013) Transplantation of autologous olfactory ensheathing cells in complete human spinal cord injury. Cell Transplant 22:1591–1612
127. Takahashi K, Yamanaka S (2006) Induction of pluripotent stem cells from mouse embryonic and adult fibroblast cultures by defined factors. Cell 126(4):663–676
128. Takahashi K, Tanabe K, Ohnuki M, Narita M, Ichisaka T, Tomoda K, Yamanaka S (2007) Induction of pluripotent stem cells from adult human fibroblasts by defined factors. Cell 131(5):861–872
129. Tan EYM, Law JWS, Wang C-H, Lee AYW (2007) Development of a cell transducible RhoA inhibitor TAT-C3 transferase and its encapsulation in biocompatible microspheres to promote survival and enhance regeneration of severed neurons. Pharm Res 24(12):2297–2308
130. Temple S (1989) Division and differentiation of isolated CNS blast cells in microculture. Nature 340(6233):471–473
131. Tetzlaff W, Okon EB, Karimi-Abdolrezaee S, Hill CE, Sparling JS, Plemel JR, Plunet WT, Tsai EC, Baptiste D, Smithson LJ (2011) A systematic review of cellular transplantation therapies for spinal cord injury. J Neurotrauma 28(8):1611–1682
132. Till JE, McCulloch EA (1961) A direct measurement of the radiation sensitivity of normal mouse bone marrow cells. Radiat Res 14:213–222
133. Ulndreaj A, Tzekou A, Mothe AJ, Siddiqui AM, Dragas R, Tator CH, Torlakovic EE, Fehlings MG (2017) Characterization of the antibody response after cervical spinal cord injury. J Neurotrauma 34(6):1209–1226
134. Vidane AS, Zomer HD, Oliveira BMM, Guimarães CF, Fernandes CB, Perecin F, Silva LA, Miglino MA, Meirelles FV, Ambrósio CE (2013) Reproductive stem cell differentiation: extracellular matrix, tissue microenvironment, and growth factors direct the mesenchymal stem cell lineage commitment. Reprod Sci (Thousand Oaks, CA) 20(10):1137–1143
135. Vierbuchen T, Wernig M (2011) Direct lineage conversions: unnatural but useful? Nat Biotechnol 29(10):892–907

136. Vierbuchen T, Ostermeier A, Pang ZP, Kokubu Y, Südhof TC, Wernig M (2010) Direct conversion of fibroblasts to functional neurons by defined factors. Nature 463(7284):1035–1041
137. Wang P, Zhao D, Rockowitz S, Zheng D (2016) Divergence and rewiring of regulatory networks for neural development between human and other species. Neurogenesis (Austin, TX) 3(1):e1231495
138. Warren L, Manos PD, Ahfeldt T, Loh Y-H, Li H, Lau F, Ebina W, Mandal PK, Smith ZD, Meissner A, Daley GQ, Brack AS, Collins JJ, Cowan C, Schlaeger TM, Rossi DJ (2010) Highly efficient reprogramming to pluripotency and directed differentiation of human cells with synthetic modified mRNA. Cell Stem Cell 7(5):618–630
139. Watanabe M, Toyama Y, Nishiyama A (2002) Differentiation of proliferated NG2-positive glial progenitor cells in a remyelinating lesion. J Neurosci Res 69:826–836
140. Wilcox JT, Satkunendrarajah K, Zuccato JA, Nassiri F, Fehlings MG (2014) Neural precursor cell transplantation enhances functional recovery and reduces astrogliosis in bilateral compressive/contusive cervical spinal cord injury. Stem Cells Transl Med 3(10):1148–1159
141. Winiecka-Klimek M, Smolarz M, Walczak MP, Zieba J, Hulas-Bigoszewska K, Kmieciak B, Piaskowski S, Rieske P, Grzela DP, Stoczynska-Fidelus E (2015) SOX2 and SOX2-MYC reprogramming process of fibroblasts to the neural stem cells compromised by senescence. PLoS One 10(11):e0141688
142. Winkler EA, Sengillo JD, Bell RD, Wang J, Zlokovic BV (2012) Blood-spinal cord barrier pericyte reductions contribute to increased capillary permeability. J Cereb Blood Flow Metab 32(10):1841–1852
143. Woltjen K, Michael IP, Mohseni P, Desai R, Mileikovsky M, Hämäläinen R, Cowling R, Wang W, Liu P, Gertsenstein M, Kaji K, Sung H-K, Nagy A (2009) piggyBac transposition reprograms fibroblasts to induced pluripotent stem cells. Nature 458(7239):766–770
144. Xu Q, Wang J'A, He J, Zhou M, Adi J, Webster KA, Yu H (2011) Impaired CXCR4 expression and cell engraftment of bone marrow-derived cells from aged atherogenic mice. Atherosclerosis 219(1):92–99
145. Xu B, Zhao Y, Xiao Z, Wang B, Liang H, Li X, Fang Y, Han S, Li X, Fan C, Dai J (2017) A dual functional scaffold tethered with EGFR antibody promotes neural stem cell retention and neuronal differentiation for spinal cord injury repair. Adv Healthc Mater 6(9):1–12
146. Yamanaka S (2009) Elite and stochastic models for induced pluripotent stem cell generation. Nature 460(7251):49–52
147. Yang L (2011) From fibroblast cells to cardiomyocytes: direct lineage reprogramming. Stem Cell Res Ther 2(1):1
148. Yu J, Vodyanik MA, Smuga-Otto K, Antosiewicz-Bourget J, Frane JL, Tian S, Nie J, Jonsdottir GA, Ruotti V, Stewart R (2007) Induced pluripotent stem cell lines derived from human somatic cells. Science 318(5858):1917–1920
149. Yu J, Hu K, Smuga-otto K et al (2009) Human induced pluripotent stem cells free of vector and transgene sequences. Science 324(5928):797–801
150. Zhao T, Zhang Z-N, Rong Z, Xu Y (2011) Immunogenicity of induced pluripotent stem cells. Nature 474(7350):212–215
151. Zhao Y, Xiao Z, Chen B, Dai J (2017) The neuronal differentiation microenvironment is essential for spinal cord injury repair. Organogenesis 13(3):63–70

Chapter 5
Cell-Based Therapy for Retinal Degenerative Disease

Marco A. Zarbin

Abstract The availability of non-invasive high-resolution imaging technology, the immune suppressive nature of the subretinal space, and the existence of surgical techniques that permit transplantation surgery to be a safe procedure all render the eye an ideal organ in which to begin cell-based therapy in the central nervous system. A number of early stage clinical trials are underway to assess the safety and feasibility of cell-based therapy for retinal blindness. Cell-based therapy using embryonic stem cell derived differentiated cells (e.g., retinal pigment epithelium (RPE)), neural progenitor cells, photoreceptor precursors, and bone marrow-derived hematopoietic stem/progenitor cells all have demonstrated successful rescue and/or replacement in preclinical models of human retinal degenerative disease. Additional research is needed to identify the mechanisms that control synapse formation/disjunction (to improve photoreceptor transplant efficacy), to identify factors that limit RPE survival in areas of geographic atrophy (to improve RPE transplant efficacy in eyes with age-related macular degeneration), and to identify factors that regulate immune surveillance of the subretinal space (to improve long-term photoreceptor and RPE transplant survival).

Keywords Stem Cells · Retina · Transplantation · Retinitis Pigmentosa · Macular Degeneration · Blindness

Supported in part by the Joseph J. & Marguerite DiSepio Retina Research Fund, the New Jersey Lions Eye Research Foundation, the Eng Family Foundation, and the National Eye Institute (EY021542)

M. A. Zarbin (✉)
Institute of Ophthalmology and Visual Science, Rutgers-New Jersey Medical School, Rutgers University, Newark, NJ, USA

R. Delgado-Morales (ed.), *Stem Cell Genetics for Biomedical Research*,
https://doi.org/10.1007/978-3-319-90695-9_5

5.1 Introduction

A number of treatments for blindness arising from retinal dysfunction are under development including gene therapy, photochemical therapy, drug therapy, and cell-based therapy. Each modality has strengths and limitations that may result in different approaches being selected for different causes of retina-based blindness. In the case of gene therapy, if the therapeutic gene is intended to provide functional protein that addresses a specific protein deficiency, then the target patient population is restricted to those patients who have inherited the specific genetic abnormality [1–3]. Alternatively, gene therapy can provide a generic therapeutic benefit if the goal is to inhibit or stimulate a particular biochemical pathway [4]. Photochemical therapy provides a generic approach to visual restoration in which cells that are not photosensitive normally (e.g., retinal ganglion cells) are rendered photosensitive either via genetic modification (an approach termed optogenetic therapy) or through the use of photosensitive molecules that alter the membrane potential of these cells in a light-sensitive manner [5, 6]. Cell transplants have the potential to replace cells that are dying or dead and restore lost vision (an approach termed *replacement*) even in end-stage disease. Cell transplants also may provide needed trophic factors that prevent or delay cell death, thus promoting sight preservation (an approach termed *rescue*). *Replacement* and *rescue* are not mutually exclusive goals, as some cell transplants have the capacity to replace lost cells as well as provide trophic factors (e.g., RPE transplants [7]). Blindness due to photoreceptor (PR) death, such as occurs with retinitis pigmentosa (RP) or retinal detachment, could be treated with replacement of lost PRs. Blindness due to retinal ganglion cell death, such as occurs with glaucoma or traumatic optic neuropathy, could be treated with replacement of retinal ganglion cells. Cell-based therapy may provide a generic approach to treating retinal degenerative disease, i.e., therapeutic benefit independent of the specific mutation(s) or environmental insult(s) that caused blindness.

5.2 Disease Targets

As noted elsewhere, a number of early-stage clinical trials of cell-based therapy for degenerative retinal disease are in progress [8]. The main diseases under study are geographic atrophy (GA) associated with age-related macular degeneration (AMD), Stargardt macular dystrophy (SD), myopic macular degeneration, and retinitis pigmentosa (RP). Age-related macular degeneration is the leading cause of blindness in the industrialized world, affecting approximately 1.5% of persons over age 40 years and 15% of women over age 80 [9]. Patients lose vision through the growth of abnormal blood vessels under the fovea, which is the central part of the retina supporting high acuity vision, or due to atrophy of the foveal PRs. The atrophic manifestation of the disease, termed geographic atrophy (GA), involves apoptotic degeneration of the PRs and subjacent support tissue, i.e., the retinal pigment epithelium (RPE) and choriocapillaris. Retinitis pigmentosa afflicts one in 2500–7000 persons [10]. Regardless of the causative mutation, the first clinical manifestation of RP is nyctalopia associated with primary loss of rod PRs. As the disease progresses, visual field loss progresses in a centripetal manner, with central visual loss, associated with secondary cone PR death, occurring in the later stages of the disease. Stargardt macular dystrophy (SMD) is the most common retinal dystrophy in children with a prevalence of 1:10,000 births [11]. This disease typically arises from a mutation in PRs that results in lipofuscin accumulation in RPE cells, with premature RPE death, secondary loss of foveal PRs, and loss of central vision. Myopic macular degeneration is a major cause of visual disability in Asia [12]. Genetic and environmental factors contribute to the development of pathologic myopia, which is associated with abnormal elongation of the globe, thinning of the posterior choroid and retina, and development of choriocapillaris-RPE-PR atrophy as well as choroidal neovascularization.

The conditions mentioned above have different causes and different demographics, but at their end-stage, they are associated with central visual loss, including damage to foveal PRs. While there is effective therapy for the abnormal choroidal vessel growth (termed choroidal neovascularization) in AMD [13–15], there is no effective treatment for GA [16]. At present there is no proved sight-preserving or sight-restoring treatment for the vast majority of patients with RP although, as noted above, gene therapy may be effective for some patients [17, 18], and small molecule therapy has shown promise in early phase studies [19]. Cell-based therapy could be an important therapeutic option for patients with RP and GA who cannot be treated with gene or drug therapy [20, 21]. Gene therapy and pharmacological therapy (https://www.businesswire.com/news/home/20170105006500/en/Acucela-Receives-Orphan-Drug-Designation-FDA-Treatment) for SMD are being explored as is cell-based therapy [20, 22]. Currently, there is no effective treatment for the blinding atrophic complications of myopic macular degeneration.

5.3 Treatment Approaches

Advantages of the Eye as a Target Organ for Cell-Based Therapy As pointed out elsewhere, the eye has unique advantages as a target for cell-based therapy in the central nervous system [8]. Retinal anatomy and physiology are well described [23, 24]. Surgical techniques to gain access to the vitreous cavity and subretinal space are established and have an excellent safety record (Fig. 5.1) [25]. The subretinal space is an immune privileged site although it is not entirely sequestered from immune surveillance [26]. Remarkably, the number of PRs required to restore high acuity vision may be relatively few, perhaps in the range of 3000–250,000, depending on the number of cone vs. rod PRs that integrate with the host retina [27]. Finally, the human retina can be imaged in vivo with high resolution (e.g., 3 μm) using noninvasive technologies [28–31], which enable one to monitor a transplant's anatomy closely after surgery. Electrodiagnostic tests,

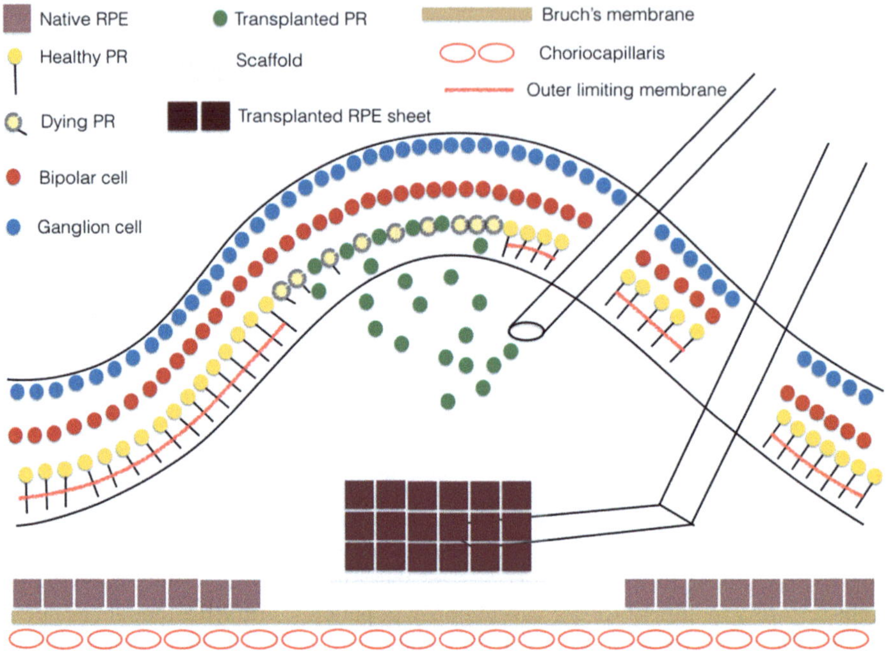

Fig. 5.1 Surgical approach to subretinal cell transplantation. Schematic drawing illustrating subretinal injection of a suspension of rod photoreceptor precursor cells as might be done for a patient with photoreceptor degeneration due to a retinal dystrophy. The cells integrate into the retina preferentially in areas of external limiting membrane breakdown. Also shown is subretinal delivery of an RPE sheet on a scaffold to replace a localized RPE defect on Bruch's membrane as could occur in patients with geographic atrophy. (Reproduced with permission from Zarbin [8])

such as the electroretinogram (ERG), and psychophysical tests, such as microperimetry, enable assessment of functional recovery in ways that are more incremental than the gross psychophysical measurement of visual acuity. Preclinical experiments in animal models of human RP indicate that approximately 150,000 functioning rods are required to generate a reproducible dark-adapted (scotopic) ERG signal [32].

Sources of Cells The costs associated with bringing a therapeutic cell line through the regulatory process are very high [33]. Currently, embryonic stem cells (ESCs) and induced pluripotent stem cells (iPSCs) as well as adult stem cells are preferred as sources of cells for ophthalmic cell-based therapeutics because of their unlimited proliferative potential. Once a cell line has received regulatory approval, it can provide an unlimited source of cells for any number of patients requiring the cell-based therapy derived from the cell line's progeny. In principle, pluripotent cells can be differentiated into any needed cell type, and RPE, PRs, an optic cup, and neurons have been derived from pluripotent stem cells [34–41].

Multipotent retinal stem cells may have a role in cell replacement therapy. Mammalian Müller glial cells may be able to differentiate into PRs [42–45]. A subpopulation of adult human RPE cells can be activated in vitro to become multipotent stem cells that produce RPE cells [46, 47]. If the signaling pathways that regulate differentiation of these cells into desired phenotypes (e.g., PRs, RPE) could be identified, then it might not be necessary to undertake transplantation surgery. Cell replacement could be initiated endogenously via gene reprogramming and differentiation of adult stem cells in situ.

Sanges et al. [48] used cell fusion to induce PR replacement in the rd10 mouse model, which has a mutation in the enzyme phosphodiesterase-6-beta that is present in some human patients with RP [48, 49]. These investigators exploited the fact that transplanted bone marrow-derived stem cells can fuse with somatic cells and generate hybrids, which in turn can differentiate into neurons and other cell types in vivo. Cell fusion, which is essential for fertilization, viral entry, muscle differentiation, and placental development [50], enables Müller glia to reenter the cell cycle, proliferate, become reprogrammed to a pluripotent/neural progenitor-like state, and differentiate into PRs. Sanges et al. [48] stimulated Wnt signaling in bone marrow-derived hematopoietic stem/progenitor cells and transplanted these cells into the subretinal space of rd10 mice to regenerate lost PRs. Wnt ligand is a secreted glycoprotein that binds to Frizzled receptors, leading to the formation of a larger cell surface complex with LRP5/6 [51]. The Wnt/β-catenin pathway regulates stem cell pluripotency and cell fate decisions during development [51]. Wnt/β-catenin signaling strongly enhances efficiency of reprogramming of somatic cells upon cell fusion with pluripotent stem cells, and the Wnt signaling pathway can be activated by the GSK-3 inhibitor, 6 bromoindirubin-3′-oxime (BIO) [48]. Transplantation of BIO-treated wild-type bone marrow-derived hematopoietic stem/progenitor cells into the subretinal space of rd10 mice at postnatal day 18 resulted in a 300% increase in the thickness of the outer nuclear layer (which contains PR cell bodies) compared to sham-treated controls in 2-month-old recipients [48]. The transplants showed marked recovery of the ERG, including both the A- and B-waves [48]. B-wave recovery indicates that the newly formed PRs made functional synaptic connections with the host second-order neurons in the visual pathway, i.e., bipolar cells.

Methods of Cell Delivery Cells can be delivered as a dispersed preparation or as anatomically organized sheets. Each approach has advantages and disadvantages. Delivery of dispersed cell preparations generally involves a relatively simple surgical procedure. The simplest example involves intravitreal injection of cells for the purpose of trophic factor delivery and rescue therapy [21, 52]. Subretinal injection of dispersed cells can be done using a small (e.g., 33-gauge), self-sealing retinotomy in conjunction with pars plana vitrectomy and is associated with minimal surgical risk [20].

As noted elsewhere [8], delivery of cells as an anatomically organized preparation has several potential advantages over dispersed cell preparations. In the case of RPE transplantation, the transplanted cells can be differentiated and organized anatomically as a monolayer on a scaffold, which resembles the in situ configuration. Thus, hexagonal morphology, apical-basal polarization, intact cell-cell junctions (e.g., zonula adherens), and normal transepithelial electrical resistance (i.e., trans-RPE permeability) exist at the time of RPE-scaffold transplantation. Transplanting differentiated RPE may prove advantageous because once foveal PRs are detached, they begin to degenerate. This process is arrested by re-apposition of the retina against functional RPE cells. When delivered as a suspension in organ culture, approximately 1–2 weeks is required for the RPE to attach to Bruch's membrane and to reacquire features of differentiation [53]. This time interval can be shortened or eliminated by the use of a scaffold delivering differentiated RPE cells. In addition, scaffold delivery may be associated with a lower antigen load if fewer cells are delivered with scaffolds than with cell suspensions. Although the subretinal space is an immune-suppressive environment [26], the immune privilege can be abrogated by disruption of the RPE monolayer [54], and reducing the antigen load might provide less stimulus to immune surveillance of the transplanted cells [55]. Furthermore, it may be possible to integrate growth factors, immunomodulatory molecules, or other useful moieties into a scaffold, thus prolonging RPE graft survival as well as PR survival. Lu and coworkers [56] nanoengineered parylene C scaffolds for RPE transplantation. The scaffolds are 6 μm thick to provide mechanical support and have 0.3 μm thick diffusion zones to facilitate cell survival. These diffusion zones occupy 58% of the surface area of the scaffold and are 40 μm in diameter. One can also transplant sheets of PRs and RPE cells without an underlying scaffold [57–59].

Sheets of pure PRs can be manufactured [60], and sheet PR transplants have been reported in preclinical [61–64] and human studies [65]. Experiments with documented vision improvement have used full-thickness retina, which leads to interposition of the transplanted inner retina (e.g., ganglion cells) between the transplanted PRs and their synaptic target, the host bipolar cells. Transplantation of glial tissue, an integral component of full-thickness grafts, may lead to an immune response to the transplant [66, 67].

Scaffold-based approaches generally will require a larger retinotomy for cell delivery to the subretinal space compared with dispersed cell delivery. Thus, the use of scaffolds may be associated with increased risk of cell egress into the vitreous cavity where the cells might undergo transdifferentiation or uncontrolled proliferation, scar tissue formation (depending on the cell type), and development of epiretinal scar tissue and/or postoperative retinal detachment. Cell sheets can migrate postoperatively to locations distant from the fovea, and such migration has been documented clinically [58]. If postoperative migration occurs, then the trophic effect of the transplant may be diminished, and the replacement effect may be lost entirely. Intraoperative use of devices such as heavier-than-water liquids may fix the scaffold/sheet in the proper location at the time of retinal reattachment (following cell delivery to the subretinal space), which may mitigate the scaffold migration risk.

5.4 Potential Obstacles to Success

Immune Response The combination of the relative immune privilege of the subretinal space (or, to a lesser degree, the vitreous cavity) and the immune privileged nature of the transplanted tissue may result in the induction of ignorance, which is the failure of the immune system to recognize transplanted tissue [68]. Transplantation of PRs to the subretinal space might serve as an example, as PRs have low major histocompatibility (MHC) class I expression and express MHC class 1b antigens (HLA-G and HLA-E) that bind CD94-NKG2 and block NK cell-mediated lysis [68]. Preliminary clinical studies in patients with RP found no signs of immune rejection of allogeneic PR sheet transplants [57]. Nonetheless, preclinical data indicate that long-term survival of transplanted PRs requires immune suppression therapy [67]. (Cellular contaminants may have played a role in stimulating immune surveillance of transplanted PRs in these experiments.)

Although RPE cells can express MCH class II antigens [69], they also express Fas ligand [70] and may be able to modulate the host immune response. Evidence from human studies, however, indicates that subretinal RPE transplants can be rejected [58], and initial experience with elderly patients receiving RPE transplants indicates that many cannot tolerate sustained immune suppression [58, 71].

Differentiated progeny of ESCs expresses major histocompatibility (MHC) class I antigens [72, 73]. Stem cells generated by somatic cell nuclear transfer are syngeneic to the nuclear donor, but the mitochondrial genes, which are of oocyte origin, are a source of minor histocompatibility antigens [74, 75]. Disparities at minor histocompatibility loci can provoke rejection of ESC-derived tissue [76]. If transplanted into the donor, cells derived from iPSCs should be devoid of alloreactivity, but if the iPSC harbors a genetic abnormality and if this abnormality is corrected before transplantation into the iPSC donor, then an immune response may occur [77]. Donor iPSC banks are being created to match donors and recipients who are homozygous at some of the MHC loci [78–82]. Because disparities at minor histocompatibility loci can provoke immune rejection, it is not clear that this approach will be useful for many patients. Although MHC matching could be supplemented with immune-suppressive therapy, this approach might be accompanied by an increased risk of ESC-derived tumor formation.

Induction of tolerance (i.e., absence of a destructive immune response to transplanted tissue without immune suppression) might prevent immune surveillance of the transplant [68]. Tolerance can be achieved via mixed chimerism or by inducing anergy through blockade of co-stimulatory signals that activate T cells (e.g., with belatacept). Minimizing surgical trauma may be another important consideration in minimizing immune surveillance of the transplant because surgical trauma can induce activation of the innate immune system, which in turn can activate the adaptive immune system [66]. For this reason, one should minimize the use of materials (e.g., scaffold components) that activate the innate immune response and stimulate inflammation.

Retinal Trauma Delivery of transplanted cells to the subretinal space requires creation of a localized retinal detachment. Wang et al. [83] demonstrated that the first synapse of the visual pathway can be disrupted quickly after localized retinal detachment. Specifically, rod axon terminals withdraw from their bipolar postsynaptic partners toward their cell somata in the outer nuclear layer within 2 h after a surgically induced retinal detachment in pigs. Of perhaps even greater significance is that synaptic disjunction occurs not only in the detached retina but also in regions many millimeters away [83]. Thus, detachment affects large areas of the retina outside the area of iatrogenic detachment even if the PRs are otherwise healthy. These findings may mean that even extrafoveal delivery of cells can compromise foveal function. Fontainhas and Townes-Anderson [84] demonstrated that activation of the RhoA pathway is involved in PR synaptic retraction. Townes-Anderson et al. [83, 85] found that using a Rho kinase (ROCK) inhibitor at the time of detachment can reduce the amount of axon retraction of rod photoreceptors by 40–50%, which suggests a potential strategy to avoid this complication in patients receiving subretinal cell transplants [86].

Cancer Risk Transplantation of pluripotent stem cells or cells derived from pluripotent stem cells might be associated with an increased risk of cancer [87–89]. Protocols that do not require c-Myc, Sox-2, and/or Klf4 [90–95] and generation of iPSCs without viral vectors [96–102] probably reduce the cancer risk. As part of clinical grade manufacturing, a number of steps are taken to monitor for malignant potential after cell manufacture including G-band karyotyping to detect chromosomal abnormalities; genomic DNA polymerase chain reaction (PCR) to confirm the absence of episomal plasmid genes; assessment of pluripotency by immunostaining with surface and nuclear pluripotency markers for subsequent flow cytometry quantification, by quantitative reverse PCR of endogenous pluripotency genes, and by gene chip and bioinformatics PluriTest assays; and in vivo transplant experiments of the differentiated therapeutic cells in immune compromised animals (e.g., athymic mice) to detect cancer (e.g., teratoma) formation.

5.5 Conclusions

A number of different approaches to treating blindness secondary to retinal degenerative disease and trauma are under study including gene therapy, photochemical therapy, drug therapy, and cell-based therapy. One advantage of cell-based therapy is that it can provide a generic approach to sight preservation and restoration and so may be directed to a broader patient population compared to therapies that are more disease-specific (e.g., gene therapy). Current disease targets for cell-based therapy include conditions associated with PR death and/or RPE death (e.g., GA, RP, SMD, and myopic macular degeneration). The eye is well suited for cell-based therapy due to well-established, safe surgical delivery procedures, the relative immune-suppressive environment of the intraocular space, the ability to image transplanted cells in vivo noninvasively at high resolution, and the ability to measure incremental improvement in retinal function following surgery. A variety of cell sources are being explored in human and preclinical studies including pluripotent stem cells, multipotent retinal stem cells, and bone marrow-derived hematopoietic stem cells that can fuse with Müller glial cells in the retina. The latter approach, which stimulates Müller cells to differentiate into PRs, seems to hold great promise for generating large numbers of PRs that integrate with the host retina. Cells have been delivered in preclinical and clinical studies as dispersed preparations and as sheets of organized cells with or without an underlying scaffold. More research is needed to identify the best method of cell delivery for retinal therapy. Potential obstacles to successful cell-based therapy include immune surveillance of the transplanted tissue, trauma-induced reaction of the host retina to subretinal cell delivery, and the possibility of tumor formation (in the case of stem cell-derived products). A number of early-stage clinical trials of cell-based therapy are in progress. It seems likely the retinal cell-based therapy will be among the first function-restoring cell-based treatments in the central nervous system.

References

1. Maguire AM, Simonelli F, Pierce EA et al (2008) Safety and efficacy of gene transfer for Leber's congenital amaurosis. N Engl J Med 358(21):2240–2248
2. Cideciyan AV, Hauswirth WW, Aleman TS et al (2009) Human RPE65 gene therapy for Leber congenital amaurosis: persistence of early visual improvements and safety at 1 year. Hum Gene Ther 20(9):999–1004
3. Bainbridge JW, Smith AJ, Barker SS et al (2008) Effect of gene therapy on visual function in Leber's congenital amaurosis. N Engl J Med 358(21):2231–2239
4. Rakoczy EP, Lai CM, Magno AL et al (2015) Gene therapy with recombinant adeno-associated vectors for neovascular age-related macular degeneration: 1 year follow-up of a phase 1 randomised clinical trial. Lancet 386(10011):2395–2403
5. Van Gelder RN (2015) Photochemical approaches to vision restoration. Vis Res 111(Pt B):134–141
6. Francis PJ, Mansfield B, Rose S (2013) Proceedings of the first international optogenetic therapies for vision symposium. Transl Vis Sci Technol. 2(7):4
7. Lund RD, Adamson P, Sauve Y et al (2001) Subretinal transplantation of genetically modified human cell lines attenuates loss of visual function in dystrophic rats. Proc Natl Acad Sci U S A 98(17):9942–9947
8. Zarbin M (2016) Cell-based therapy for degenerative retinal disease. Trends Mol Med 22(2):115–134
9. Friedman DS, O'Colmain BJ, Munoz B et al (2004) Prevalence of age-related macular degeneration in the United States. Arch Ophthalmol 122(4):564–572
10. Parmeggiani F (2011) Clinics, epidemiology and genetics of retinitis pigmentosa. Curr Genomics 12(4):236–237
11. Genead MA, Fishman GA, Stone EM, Allikmets R (2009) The natural history of stargardt disease with specific sequence mutation in the ABCA4 gene. Invest Ophthalmol Vis Sci 50(12):5867–5871
12. Wong TY, Loon SC, Saw SM (2006) The epidemiology of age related eye diseases in Asia. Br J Ophthalmol 90(4):506–511

13. Rosenfeld PJ, Brown DM, Heier JS et al (2006) Ranibizumab for neovascular age-related macular degeneration. N Engl J Med 355(14):1419–1431
14. Brown DM, Kaiser PK, Michels M et al (2006) Ranibizumab versus verteporfin for neovascular age-related macular degeneration. N Engl J Med 355(14):1432–1444
15. Heier JS, Brown DM, Chong V et al (2012) Intravitreal aflibercept (VEGF trap-eye) in wet age-related macular degeneration. Ophthalmology 119(12):2537–2548
16. Zarbin MA, Rosenfeld PJ (2010) Pathway-based therapies for age-related macular degeneration: an integrated survey of emerging treatment alternatives. Retina 30(9):1350–1367
17. Ashtari M, Zhang H, Cook PA et al (2015) Plasticity of the human visual system after retinal gene therapy in patients with Leber's congenital amaurosis. Sci Transl Med 7(296):296ra110
18. Jacobson SG, Cideciyan AV, Roman AJ et al (2015) Improvement and decline in vision with gene therapy in childhood blindness. N Engl J Med 372(20):1920–1926
19. Koenekoop RK, Sui R, Sallum J et al (2014) Oral 9-cis retinoid for childhood blindness due to Leber congenital amaurosis caused by RPE65 or LRAT mutations: an open-label phase 1b trial. Lancet 384(9953):1513–1520
20. Schwartz SD, Tan G, Hosseini H, Nagiel A (2016) Subretinal transplantation of embryonic stem cell-derived retinal pigment epithelium for the treatment of macular degeneration: an assessment at 4 years. Invest Ophthalmol Vis Sci 57(5):ORSFc1–ORSFc9
21. Klassen H (2016) Stem cells in clinical trials for treatment of retinal degeneration. Expert Opin Biol Ther 16(1):7–14
22. Binley K, Widdowson P, Loader J et al (2013) Transduction of photoreceptors with equine infectious anemia virus lentiviral vectors: safety and biodistribution of StarGen for Stargardt disease. Invest Ophthalmol Vis Sci 54(6):4061–4071
23. Sung CH, Chuang JZ (2010) The cell biology of vision. J Cell Biol 190(6):953–963
24. Ramsden CM, Powner MB, Carr AJ, Smart MJ, da Cruz L, Coffey PJ (2014) Neural retinal regeneration with pluripotent stem cells. Dev Ophthalmol 53:97–110
25. Binder S, Stanzel BV, Krebs I, Glittenberg C (2007) Transplantation of the RPE in AMD. Prog Retin Eye Res 26(5):516–554
26. Streilein JW (2003) Ocular immune privilege: the eye takes a dim but practical view of immunity and inflammation. J Leukoc Biol 74(2):179–185
27. Geller AM, Sieving PA (1993) How many cones are required to "see?": lessons from Stargardt's macular dystrophy and from modeling with degenerate photoreceptor arrays. In: Hollyfield JG, Anderson RE, La Vail MM (eds) Retinal degeneration. Plenum Press, New York, pp 25–34
28. Zarbin MA, Casaroli-Marano RP, Rosenfeld PJ (2014) Age-related macular degeneration: clinical findings, histopathology, imaging techniques. In: Casaroli-Marano RP, Zarbin MA (eds) Cell-based therapy for retinal degenerative disease. Karger Medical and Scientific Publishers, Basel, pp 1–32
29. Menghini M, Duncan JL (2014) Diagnosis and complementary examinations. Dev Ophthalmol 53:53–69
30. Scoles D, Flatter JA, Cooper RF et al (2016) Assessing photoreceptor structure associated with ellipsoid zone disruptions visualized with optical coherence tomography. Retina 36(1): 91–103
31. Zayit-Soudry S, Duncan JL, Syed R, Menghini M, Roorda AJ (2013) Cone structure imaged with adaptive optics scanning laser ophthalmoscopy in eyes with nonneovascular age-related macular degeneration. Invest Ophthalmol Vis Sci 54(12):7498–7509
32. Pearson RA, Barber AC, Rizzi M et al (2012) Restoration of vision after transplantation of photoreceptors. Nature 485(7396):99–103
33. Sheu J, Klassen H, Bauer G (2014) Cellular manufacturing for clinical applications. Dev Ophthalmol 53:178–188
34. Zhong X, Gutierrez C, Xue T et al (2014) Generation of three-dimensional retinal tissue with functional photoreceptors from human iPSCs. Nat Commun 5:4047
35. Phillips MJ, Wallace KA, Dickerson SJ et al (2012) Blood-derived human iPS cells generate optic vesicle-like structures with the capacity to form retinal laminae and develop synapses. Invest Ophthalmol Vis Sci 53(4):2007–2019

36. Klimanskaya I, Hipp J, Rezai KA, West M, Atala A, Lanza R (2004) Derivation and comparative assessment of retinal pigment epithelium from human embryonic stem cells using transcriptomics. Cloning Stem Cells 6(3):217–245
37. Osakada F, Ikeda H, Sasai Y, Takahashi M (2009) Stepwise differentiation of pluripotent stem cells into retinal cells. Nat Protoc 4(6):811–824
38. Reichman S, Terray A, Slembrouck A et al (2014) From confluent human iPS cells to self-forming neural retina and retinal pigmented epithelium. Proc Natl Acad Sci U S A 111(23):8518–8523
39. Lamba DA, McUsic A, Hirata RK, Wang PR, Russell D, Reh TA (2010) Generation, purification and transplantation of photoreceptors derived from human induced pluripotent stem cells. PLoS One 5(1):e8763
40. Ohlemacher SK, Iglesias CL, Sridhar A, Gamm DM, Meyer JS (2015) Generation of highly enriched populations of optic vesicle-like retinal cells from human pluripotent stem cells. Curr Protoc Stem Cell Biol 32:1H 8 1–1H 8 20
41. Tucker BA, Anfinson KR, Mullins RF, Stone EM, Young MJ (2013) Use of a synthetic xeno-free culture substrate for induced pluripotent stem cell induction and retinal differentiation. Stem Cells Transl Med 2(1):16–24
42. Karl MO, Reh TA (2010) Regenerative medicine for retinal diseases: activating endogenous repair mechanisms. Trends Mol Med 16(4):193–202
43. Pollak J, Wilken MS, Ueki Y et al (2013) ASCL1 reprograms mouse Muller glia into neurogenic retinal progenitors. Development (Cambridge, England) 140(12):2619–2631
44. Del Debbio CB, Balasubramanian S, Parameswaran S, Chaudhuri A, Qiu F, Ahmad I (2010) Notch and Wnt signaling mediated rod photoreceptor regeneration by Muller cells in adult mammalian retina. PLoS One 5(8):e12425
45. Jayaram H, Jones MF, Eastlake K et al (2014) Transplantation of photoreceptors derived from human Muller glia restore rod function in the P23H rat. Stem Cells Transl Med 3(3):323–333
46. Salero E, Blenkinsop TA, Corneo B et al (2012) Adult human RPE can be activated into a multipotent stem cell that produces mesenchymal derivatives. Cell Stem Cell 10(1):88–95
47. Blenkinsop TA, Salero E, Stern JH, Temple S (2013) The culture and maintenance of functional retinal pigment epithelial monolayers from adult human eye. Methods Mol Biol 945:45–65
48. Sanges D, Simonte G, Di Vicino U et al (2016) Reprogramming Muller glia via in vivo cell fusion regenerates murine photoreceptors. J Clin Invest 126(8):3104–3116
49. Chang B, Hawes NL, Pardue MT et al (2007) Two mouse retinal degenerations caused by missense mutations in the beta-subunit of rod cGMP phosphodiesterase gene. Vis Res 47(5):624–633
50. Ogle BM, Cascalho M, Platt JL (2005) Biological implications of cell fusion. Nat Rev Mol Cell Biol 6(7):567–575
51. Clevers H, Nusse R (2012) Wnt/beta-catenin signaling and disease. Cell 149(6):1192–1205
52. Mizumoto H, Mizumoto K, Whiteley SJ, Shatos M, Klassen H, Young MJ (2001) Transplantation of human neural progenitor cells to the vitreous cavity of the Royal College of Surgeons rat. Cell Transplant 10(2):223–233
53. Gullapalli VK, Sugino IK, Van Patten Y, Shah S, Zarbin MA (2005) Impaired RPE survival on aged submacular human Bruch's membrane. Exp Eye Res 80(2):235–248
54. Wenkel H, Streilein JW (1998) Analysis of immune deviation elicited by antigens injected into the subretinal space. Invest Ophthalmol Vis Sci 39(10):1823–1834
55. Zhang X, Bok D (1998) Transplantation of retinal pigment epithelial cells and immune response in the subretinal space. Invest Ophthalmol Vis Sci 39(6):1021–1027
56. Lu B, Tai YC, Humayun MS (2014) Microdevice-based cell therapy for age-related macular degeneration. Dev Ophthalmol 53:155–166
57. Kaplan HJ, Tezel TH, Berger AS, Wolf ML, Del Priore LV (1997) Human photoreceptor transplantation in retinitis pigmentosa. A safety study. Arch Ophthalmol 115(9):1168–1172
58. Tezel TH, Del Priore LV, Berger AS, Kaplan HJ (2007) Adult retinal pigment epithelial transplantation in exudative age-related macular degeneration. Am J Ophthalmol 143(4):584–595

59. Mandai M, Watanabe A, Kurimoto Y et al (2017) Autologous induced stem-cell-derived retinal cells for macular degeneration. N Engl J Med 376(11):1038–1046
60. Huang JC, Ishida M, Hersh P, Sugino IK, Zarbin MA (1998) Preparation and transplantation of photoreceptor sheets. Curr Eye Res 17(6):573–585
61. Seiler MJ, Aramant RB, Thomas BB, Peng Q, Sadda SR, Keirstead HS (2010) Visual restoration and transplant connectivity in degenerate rats implanted with retinal progenitor sheets. Eur J Neurosci 31(3):508–520
62. Assawachananont J, Mandai M, Okamoto S et al (2014) Transplantation of embryonic and induced pluripotent stem cell-derived 3D retinal sheets into retinal degenerative mice. Stem Cell Reports 2(5):662–674
63. Seiler MJ, Lin RE, McLelland BT et al (2017) Vision recovery and connectivity by fetal retinal sheet transplantation in an immunodeficient retinal degenerate rat model. Invest Ophthalmol Vis Sci 58(1):614–630
64. Mandai M, Fujii M, Hashiguchi T et al (2017) iPSC-derived retina transplants improve vision in rd1 end-stage retinal-degeneration mice. Stem Cell Reports 8(1):69–83
65. Radtke ND, Aramant RB, Petry HM, Green PT, Pidwell DJ, Seiler MJ (2008) Vision improvement in retinal degeneration patients by implantation of retina together with retinal pigment epithelium. Am J Ophthalmol 146(2):172–182
66. Xian B, Huang B (2015) The immune response of stem cells in subretinal transplantation. Stem Cell Res Ther 6:161
67. West EL, Pearson RA, Barker SE et al (2010) Long-term survival of photoreceptors transplanted into the adult murine neural retina requires immune modulation. Stem Cells 28(11):1997–2007
68. Tena A, Sachs DH (2014) Stem cells: immunology and immunomodulation. Dev Ophthalmol 53:122–132
69. Kohen L, Enzmann V, Faude F, Wiedemann P (1997) Mechanisms of graft rejection in the transplantation of retinal pigment epithelial cells. Ophthalmic Res 29:298–304
70. Kaplan HJ, Leibole MA, Tezel T, Ferguson TA (1999) Fas ligand (CD95 ligand) controls angiogenesis beneath the retina. Nat Med 5(3):292–297
71. Schwartz SD, Regillo CD, Lam BL et al (2015) Human embryonic stem cell-derived retinal pigment epithelium in patients with age-related macular degeneration and Stargardt's macular dystrophy: follow-up of two open-label phase 1/2 studies. Lancet 385(9967):509–516
72. Boyd AS, Wood KJ (2009) Variation in MHC expression between undifferentiated mouse ES cells and ES cell-derived insulin-producing cell clusters. Transplantation 87(9):1300–1304
73. Tian L, Catt JW, O'Neill C, King NJ (1997) Expression of immunoglobulin superfamily cell adhesion molecules on murine embryonic stem cells. Biol Reprod 57(3):561–568
74. Wakayama T, Tabar V, Rodriguez I, Perry AC, Studer L, Mombaerts P (2001) Differentiation of embryonic stem cell lines generated from adult somatic cells by nuclear transfer. Science 292(5517):740–743
75. Fairchild PJ, Nolan KF, Cartland S, Waldmann H (2005) Embryonic stem cells: a novel source of dendritic cells for clinical applications. Int Immunopharmacol 5(1):13–21
76. Robertson NJ, Brook FA, Gardner RL, Cobbold SP, Waldmann H, Fairchild PJ (2007) Embryonic stem cell-derived tissues are immunogenic but their inherent immune privilege promotes the induction of tolerance. Proc Natl Acad Sci U S A 104(52):20920–20925
77. Fairchild PJ (2010) The challenge of immunogenicity in the quest for induced pluripotency. Nat Rev Immunol 10(12):868–875
78. Boyd AS, Fairchild PJ (2010) Approaches for immunological tolerance induction to stem cell-derived cell replacement therapies. Expert Rev Clin Immunol 6(3):435–448
79. Nakatsuji N, Nakajima F, Tokunaga K (2008) HLA-haplotype banking and iPS cells. Nat Biotechnol 26(7):739–740
80. Zimmermann A, Preynat-Seauve O, Tiercy JM, Krause KH, Villard J (2012) Haplotype-based banking of human pluripotent stem cells for transplantation: potential and limitations. Stem Cells Dev 21(13):2364–2373

81. Taylor CJ, Peacock S, Chaudhry AN, Bradley JA, Bolton EM (2012) Generating an iPSC bank for HLA-matched tissue transplantation based on known donor and recipient HLA types. Cell Stem Cell 11(2):147–152
82. Turner M, Leslie S, Martin NG et al (2013) Toward the development of a global induced pluripotent stem cell library. Cell Stem Cell 13(4):382–384
83. Wang J, Zarbin M, Sugino I, Whitehead I, Townes-Anderson E (2016) RhoA signaling and synaptic damage occur within hours in a live pig model of CNS injury, retinal detachment. Invest Ophthalmol Vis Sci 57(8):3892–3906
84. Fontainhas AM, Townes-Anderson E (2011) RhoA inactivation prevents photoreceptor axon retraction in an in vitro model of acute retinal detachment. Invest Ophthalmol Vis Sci 52(1):579–587
85. Townes-Anderson E, Wang J, Halasz E et al (2017) Fasudil, a clinically used ROCK inhibitor, stabilizes rod photoreceptor synapses after retinal detachment. Transl Vis Sci Technol 6(3):22
86. Townes-Anderson E, Sugino I, Zarbin M (2017) Using Rho kinase inhibitors for retinal detachment. JAMA Ophthalmol 135(8):895
87. Arnhold S, Klein H, Semkova I, Addicks K, Schraermeyer U (2004) Neurally selected embryonic stem cells induce tumor formation after long-term survival following engraftment into the subretinal space. Invest Ophthalmol Vis Sci 45(12):4251–4255
88. Okita K, Ichisaka T, Yamanaka S (2007) Generation of germline-competent induced pluripotent stem cells. Nature 448(7151):313–317
89. Yamanaka S (2007) Strategies and new developments in the generation of patient-specific pluripotent stem cells. Cell Stem Cell 1(1):39–49
90. Nakagawa M, Koyanagi M, Tanabe K et al (2008) Generation of induced pluripotent stem cells without Myc from mouse and human fibroblasts. Nat Biotechnol 26(1):101–106
91. Li W, Zhou H, Abujarour R et al (2009) Generation of human-induced pluripotent stem cells in the absence of exogenous Sox2. Stem Cells 27(12):2992–3000
92. Zhu S, Li W, Zhou H et al (2010) Reprogramming of human primary somatic cells by OCT4 and chemical compounds. Cell Stem Cell 7(6):651–655
93. Kolomeyer AM, Zarbin MA (2014) Trophic factors in the pathogenesis and therapy for retinal degenerative diseases. Surv Ophthalmol 59(2):134–165
94. Huangfu D, Osafune K, Maehr R et al (2008) Induction of pluripotent stem cells from primary human fibroblasts with only Oct4 and Sox2. Nat Biotechnol 26(11):1269–1275
95. Shi Y, Desponts C, Do JT, Hahm HS, Scholer HR, Ding S (2008) Induction of pluripotent stem cells from mouse embryonic fibroblasts by Oct4 and Klf4 with small-molecule compounds. Cell Stem Cell 3(5):568–574
96. Okita K, Nakagawa M, Hyenjong H, Ichisaka T, Yamanaka S (2008) Generation of mouse induced pluripotent stem cells without viral vectors. Science 322(5903):949–953
97. Yu J, Hu K, Smuga-Otto K et al (2009) Human induced pluripotent stem cells free of vector and transgene sequences. Science 324(5928):797–801
98. Kaji K, Norrby K, Paca A, Mileikovsky M, Mohseni P, Woltjen K (2009) Virus-free induction of pluripotency and subsequent excision of reprogramming factors. Nature 458(7239):771–775
99. Warren L, Manos PD, Ahfeldt T et al (2010) Highly efficient reprogramming to pluripotency and directed differentiation of human cells with synthetic modified mRNA. Cell Stem Cell 7(5):618–630
100. Zhou H, Wu S, Joo JY et al (2009) Generation of induced pluripotent stem cells using recombinant proteins. Cell Stem Cell 4(5):381–384
101. Kim D, Kim CH, Moon JI et al (2009) Generation of human induced pluripotent stem cells by direct delivery of reprogramming proteins. Cell Stem Cell 4(6):472–476
102. Balasubramanian S, Babai N, Chaudhuri A et al (2009) Non cell-autonomous reprogramming of adult ocular progenitors: generation of pluripotent stem cells without exogenous transcription factors. Stem Cells 27(12):3053–3062

Chapter 6
Past, Present and Future of Cell-Based Therapy in Progressive Multiple Sclerosis

Giulio Volpe, Joshua D. Bernstock, Luca Peruzzotti-Jametti, and Stefano Pluchino

Abstract Although substantial progress has been made in the development of effective drugs for relapsing-remitting multiple sclerosis, no convincingly treatment exists to slow and/or reverse progressive forms of the disease. Advances in the field of cell-based therapies have opened to novel possible application to treat inflammatory and degenerative disease of the central nervous system including progressive MS. Major expectation arises from the use of (i) oligodendrocyte progenitor cells for directly replacing the damaged myelin and (ii) non-haematopoietic stem cells for the potential of influencing host immune response and endogenous mechanisms of repair. Each approach has potential advantages but also safety concerns and unresolved questions that are to be addressed in order to witness further advance in clinics.

In this chapter, we summarize the preclinical research leading to the current status of cell-based therapies to treat progressive form of MS and discuss the current obstacle and main objective for prospective clinical application.

Keywords Multiple sclerosis · Cell-based therapy · Oligodendrocyte progenitor cells · Mesenchymal stem cells · Neural stem cells · Non-haematopoietic stem cells · Regenerative medicine · Stem cells

---

G. Volpe · L. Peruzzotti-Jametti · S. Pluchino (✉)
Department of Clinical Neurosciences and NIHR Biomedical Research Centre, University of Cambridge, Cambridge, UK
e-mail: spp24@cam.ac.uk

J. D. Bernstock
Department of Clinical Neurosciences and NIHR Biomedical Research Centre, University of Cambridge, Cambridge, UK

Stroke Branch, National Institute of Neurological Disorders and Stroke, National Institutes of Health (NINDS/NIH), Bethesda, MD, USA

R. Delgado-Morales (ed.), *Stem Cell Genetics for Biomedical Research*,
https://doi.org/10.1007/978-3-319-90695-9_6

6.1 Introduction

The last 20 years have seen major progress in the treatment of multiple sclerosis (MS) introducing drugs with differing mechanisms of action that target the dysfunctional immunological response leading to inflammation in the central nervous system (CNS). Disease-modifying treatments (DMTs) have revolutionized the treatment of relapsing-remitting multiple sclerosis (RRMS) patients by reducing the frequency of relapses and accumulation of relapse-related disability [245]. Nonetheless, it has soon become evident that none of the currently available DMTs are capable of slowing disease progression or directly promoting repair of preexisting CNS damage [47]. In other words, all clinical trials of such agents in progressive form of MS (primary or secondary progressive) have failed or demonstrated limited efficacy [206]. Yet, the observation of a differing response to the same treatment in the different phases of MS has provided initial evidence that pathophysiological mechanisms change along the course of the disease and new therapeutic strategies are required for progressive forms. Following advances in histopathology and magnetic resonance imaging (MRI), neglected aspects of MS pathophysiology have been identified allowing to discriminate the pathological events that characterize disease activity during initial acute inflammatory relapses from those that occur during chronic progression. Despite the exact mechanism(s) of such evolution remains unknown, the prevailing hypothesis involves a systemic adaptive inflammatory autoimmune process in the early stages that eventually results in the chronic activation of the CNS intrinsic innate immune system that ultimately leads to neurodegeneration [164]. Thus, disease progression may derive from degenerative process driven by widespread activation of microglial cells and infiltrating blood-derived macrophages (MPs) that maintain a persistent background of mild-to-moderate inflammation compartmentalized within the CNS [152, 174]. As a matter of fact, any agent wishing to have an impact on the progressive phase of MS should (i) possess therapeutic proprieties capable of dealing with MS pathophysiology in later stages, (ii) be effective in slowing or preventing progression, (iii) have the potential to setback such evolution and hence reverse patient's disability and (iv) be used safely with fewer delivery and adherence concerns given the chronic course of the disease [51].

Advances in cell-based therapy research have generated substantial evidence as potential tools to address these gaps by working through various mechanisms, which include tissue regeneration, immunomodulation and neurotrophic support [266]. Exogenous stem cell therapy in MS has been pursued experimentally for at least four decades and has witnessed continuous development of therapeutic strategies in parallel to the ongoing understanding of stem/progenitor cell biology. Nowadays, the most investigated applications of cell-based therapy in MS include the use of (i) autologous haematopoietic stem cell (HSC) to reconstitute the immune system following therapeutic immune ablation [193], (ii) oligodendrocyte progenitor cells (OPCs) to replace damaged or lost myelin-making oligodendrocytes and (iii) non-haematopoietic stem cells (nHSCs) (i.e. mesenchymal stem cells [MSCs] and neural stem/progenitor cells [NSPCs]) to modulate both immune reactions and endogenous mechanisms of repair.

Immuno-ablation followed by autologous HSC transplantation has shown to stabilize patient outcomes in those suffering from severe forms of RRMS, which have ongoing disease activity (clinical relapses or MRI lesion activity), despite DMTs [240]. However similar to DMTs, limited efficacy has been demonstrated in patients with progressive course where inflammation becomes intrinsically sustained within the CNS [29].

In considering progressive MS pathology instead, transplantation of myelin-forming cells has long emerged as a promising therapeutic tool to repair damaged CNS and revert disability. Being the oldest approach involving application of cell-based therapy in MS, substantial preclinical findings have been collected over the past decades [24]. Nonetheless, serious practical issues have concurred to delay clinical experimentation with the first trial testing feasibility and safety of OPCs transplantation in progressive MS patients announced only recently [103]. In striking contrast, a large number of small clinical trials have already tested the transplantation of different sources of nHSCs (essentially MSCs) in MS patients on the basis of an extensive literature describing the complex crosstalk between nHSCs and the affected microenvironment present in the inflamed CNS [138, 264]. Preclinical studies have recognized in nHSCs the ability to influence both the activity of the host immune system and efficacy of endogenous regenerative responses in a way that could be helpful in patients with progressive forms of MS where both aspects are involved in regulating disease progression.

In this chapter, we will describe the preclinical research leading to the current status of cell-based therapeutic strategies for progressive MS. In particular, we will discuss the remyelinating potential of stem/progenitor cell transplantation in preclinical model of demyelinating/dysmyelinating disease, before focusing on the immunomodulatory and neuroprotective potential of MSCs and NSCs as major representative of nHSC-based therapy. In describing each therapeutic approach, we will review the potential for clinical translation summarizing the practical, scientific, clinical, regulatory and ethical concerns influencing the advance of such therapies from the bench to the bedside.

6.2 Cell-Based Remyelinating Strategies in MS

Widespread demyelinated areas represent one of the most prominent pathological findings observed in post-mortem evaluation of MS patients [161]. The loss of myelin has serious functional consequences in MS and underlies many of its clinical and laboratory features by disrupting nerve conduction and placing demyelinated axons at risk of subsequent degeneration [198]. Even though there is no direct evidence of pathogenic mechanisms triggering demyelination in MS, the similarities observed among experimental models [142] and the histological examination of biopsies and autopsies derived from acute/active MS lesions have led to the

Table 6.1 Core pathological features of progressive MS

Progressive MS pathology: slowly expanding/inactivated CNS lesions		
Inflammation	Demyelination	Neurodegeneration
CNS trapped with: Rare infiltrating T or B lymphocytes Microglial density within the lesion generally lower than in surrounding normal-appearing tissue Lesion edge including a small rim of activated microglia and blood-borne derived MPs Meningeal B-cell aggregates	Different patterns with: Lesion centre generally devoid of myelin Lack of signs of remyelination despite presence of endogenous OPCs Subpial cortical demyelination Remyelinated shadow plaques present in a subset of aged (>50 years) patients with long-lasting disease	Brain atrophy with: Profound axonal loss embedded in astrocytic scar tissue Diffuse white matter injury

MPs mononuclear phagocytes, *OPCs* oligodendrocyte progenitor cells

assumption that in MS demyelination occurs as result to a primary damage of the myelin sheaths and/or to the death of oligodendrocytes themselves [161]. Hence, remyelination has two major consequences: it may restore and speed impulse conduction [253] and may protect axons against degeneration, thus halting disease progression [121, 142] (Table 6.1).

In experimental models of demyelination, the adult CNS has shown to possess remarkable ability to be remyelinated by an endogenous response and to be restored in function [69, 121]. In human MS, however, such endogenous mechanism of repair appears to be not effective and exhaust despite signs of remyelination can be observed in patient with longer disease course [86]. The recognition of remyelination as spontaneous response to demyelination, which might potentially bring to a complete tissue healing and functional recover [69, 88], has given rise to novel expectation for the identification of the underpinned causes leading to the human inefficient response to the demyelinating insult [67]. In MS the failure of remyelination has been partially related to the inability of OPCs' local population to differentiate in mature myelinating oligodendrocytes [89, 145], the sole source of myelin in the adult CNS. Therefore, therapeutic strategies have aimed at either identifying molecular targets to foster the endogenous process of remyelination [119] or at replacing damaged or lost myelin-making oligodendrocytes by transplantation of exogenous glial cells. While the field of study looking at the pharmacological manipulation of the endogenous process of myelin repair currently holds major interest [89], and in some manner also employs glial progenitor's culture to perform high-throughput drugs screening, in considering cell-based remyelinating therapy in MS, the replacement of damaged or lost myelin-making oligodendrocytes by transplantation of exogenous glial cells occupies, beyond any doubt a substantial part of the past and present of stem/progenitor cell research, having brought many insights both in progenitor glial cell biology and pathology.

The initial idea of providing an external source of myelin-forming cells for treating demyelinating/dysmyelinating disorders derives from the early evidence which showed that transplanted glial cells (i.e. mixed glia or Schwann cells) were capable of ensheathing naked axons into the CNS of rodents with myelin mutations [148] or with toxin-induced focal demyelination [25, 68]. Following these studies, over the last 30 years, different types of cells with the potential of committing towards oligodendrocytic fate have demonstrated or confirmed to possess similar proprieties in experimental models of focal demyelination. In this regard cell types which have attracted more attention include embryonic stem cell (ESC) [31], neural stem/precursor cells (NSPCs) [4, 55], primary OPCs [271] and Schwann cells (SCs) [14, 115]. Interestingly, limited remyelinating potential has been recognized in postmitotic oligodendrocytes [53], thus underlining the superiority of mitotic glial precursors/progenitors in achieving remyelination. Despite many of these cell types have demonstrated the potential of differentiating towards an oligodendroglial lineage in vivo, several obstacles have limited their clinical translation with OPCs remaining the foremost representative cell type of current cell-based approach for replacing damaged myelin in MS. For instance, clinical translation of ESCs has been rightfully halted by the fear of potential tumorigenesis [113]. Indeed, human embryonic stem cell-based therapies might give rise to teratomas growing from persistent undifferentiated ES cells in the graft [23]. While NSPCs will be discussed later in this chapter, on account of the recognition of the prevalence of therapeutic potentials unrelated to their commitment for oligodendroglial differentiation, it is interesting to note that graft environment has shown to significantly influence NSPCs' ability to differentiate in oligodendroglial lineage in vivo with considerably better results in model where inflammation is not consistent such as in model of toxin-induced demyelination [33, 132, 200], rather than in model where inflammation becomes predominant like in the experimental autoimmune/allergic encephalomyelitis (EAE) [19, 215]. A possible explanation for this difference may be found in the abundance of bone morphogenetic protein (BMP)-4 and the BMP-4 antagonist noggin in the inflammatory environment [217]. With regard to Schwann cell transplantation instead, it is noteworthy to mention that a clinical trial was performed into single demyelinating lesions in three MS patients in 2001 at Yale University [127]. Though the surgical procedure proved to be safe, the study was discontinued in early 2003, after follow-up brain biopsies could not demonstrate either Schwann cell survival or new myelin formation. It is not known whether this failure resulted from poor cell preparation and autoimmune attack or if it was due to a lack of integration of peripheral myelinating cells in the adult human brain environment; nonetheless this study has dampened further clinical experimentation with Schwann cell transplants.

6.2.1 Oligodendrocyte Progenitor Cells (OPCs)

The first characterization of population of oligodendrocyte precursors was conducted in vitro by culturing developing neuronal tissue [223]. A better insight in their physiology was then achieved following isolation from rat optic nerve during the late 1980s [221]. In culture, oligodendrocyte precursors (OPCs) have a characteristic bipolar morphology, are highly migratory and express a number of specific cell surface characteristics which can indicate stages of development. For instance, immature oligodendrocyte precursors express a range of gangliosides that bind the monoclonal antibody A2B5 [222] and express the proteoglycan NG2 epitope [201] and platelet-derived growth factor receptors (PDGF alpha receptors) [235], whereas once differentiated into more mature phenotypes, they express galactocerebroside [223], adenomatous polyposis coli (CC1) [149], myelin basic protein (MBP), proteolipid protein (PLP) and myelin oligodendrocyte glycoprotein (MOG) on their surface. It is important to note that there is extensive overlap in the expression of these markers and they roughly mark stages of early and late progenitor development [32].

In mammals glial progenitors develop during embryogenesis and early postnatal life from restricted periventricular germinal regions and disperse widely throughout the CNS, pervading both grey and white matter [262]. In adult human brain, glial progenitors comprise roughly 3% of all cells in the adult forebrain and are the major mitotic neural phenotype [238]. Similar to rodents, it is supposed that during adult life, glial progenitors continue to arise from neural stem cells of the subventricular zone (SVZ) and generate both major macroglial phenotypes, astrocytes and oligodendrocyte precursors [177]. While one of the first characteristics of OPCs observed in vitro was that in response to specific environmental cues, they have the potential to generate a subpopulation of astrocytes [222]; more recent evidence has suggested that OPCs may hold further neuroectodermal competences when removed from autocrine and paracrine influences [17]. These findings have challenged the previous concept of OPCs as cell lineage purely committed towards the production of oligodendrocytes, thereby raising some doubts on the appropriation of the term oligodendrocyte precursor cells to define this cell line [87].

6.2.2 Evidence from Disease/Pathway Models

The isolation of OPCs from foetal, neonatal and adult rat brain has allowed the study of their remyelinating potential in experimental models of focal demyelination [124] and in genetically hypomyelinated animals as model of dysmyelinating diseases [13, 85, 280]. While early experiments were conducted by transplanting mixed glial cell populations, more recent advances in culturing and sorting methodology have allowed to isolate nearly pure population by means of fluorescence-activated cell sorting (FACS), immunopanning or magnetic bead sorting with antibodies recognizing OPCs or immature oligodendrocytes [71, 250, 251, 273]. In discussing the major findings related to these studies, it is important to clarify that none of these experimental settings accurately reproduce the environment exhibited in chronic/progressive MS lesions. Nonetheless, the collected data have contributed to evaluate many aspects that constitute the theoretical and practical frame of modern cell-based remyelinating strategies in MS including its putative advantages and drawbacks.

The adoption of genetically hypomyelinated animals presenting impaired primary development of myelin (i.e. myelin-deficient [md] rat and C3HeB/FeJ-shiverer [shi] mouse) has demonstrated the possibility of achieving repair across the brain and spinal cord [140]. In general transplantation of OPCs in rodents, CNS has been conducted locally showing successful myelination at the site of engraftment, thus suggesting that different strategic sites both in the brain and spinal cord could be remyelinated by more than a single transplant. In myelin mutant animals, transplant-derived myelin is recognized by "white" areas in a myelin-deficient background and confirmed by the presence of the missing myelin proteins and by electron microscopy (EM) analysis [70]. Myelin sheaths made by transplanted cells are frequently thin (as also occurring in endogenous remyelination); however this may relate to the length of time post grafting given that normal thickness myelin sheaths in myelin mutant pup can be observed when transplanted as neonates and studied after longer periods of engraftment [11]. In terms of function, myelin made by the grafted cells is associated with the formation of normal nodes of Ranvier that is critical for restoration of impulse conduction [236]. Though transplantation in neonatal animals has provided important insight in the potential of adopting cell-based therapy to myelinate developmental CNS, the pathobiology observed in human MS imposes to differently weigh such findings. Indeed, genetically hypomyelinated animals are models which mostly resemble children's dysmyelinating diseases like leukodystrophies in which "primary" myelination is impaired and transplanted cells have not competed with endogenous myelinating cells to sheathe naked axons. Moreover, during neonatal development, the CNS presents stimuli related to the undergoing process of primary myelination occurring in early age; on the contrary in MS, the disease is driven by an autoimmune response affecting an already mature CNS which may lack such developmental clues. To test myelin-making oligodendrocytes' graft in an environment closer to human MS, another myelin mutant model has been adopted in research involving rodents which undergo progressive and chronic loss of the myelin sheath while ageing (i.e. homozygous taiep rats) [85]. By 1 year of age, in taiep rats, the CNS tissue has many features of chronic areas of demyelination possibly resembling MS: chronically demyelinated axons present in an astrocytic environment in the absence of acute/active inflammation. Using the taiep rat and a combination of X-irradiation, in order to suppress the endogenous response for myelin repair, it has been possible to address a number of questions concerning remyelination failure in chronic MS lesions, such as whether chronically demyelinated axons have undergone changes that render them refractory to remyelination and why remyelination is absent when oligodendrocyte progenitor cells (OPCs) are present [275]. In this study, Foote and Blakemore have disclosed the possibility to remyelinate axons that have been depleted of their myelin sheath for a considerable time and have provided important information regarding the influence of the environment in which the cells are transplanted [85]. It appears that adult CNS may influence the myelinogenic potential of OPCs, for the presence of endogenous OPCs could reduce the graft's ability to remyelinate axons. Such aspect has to be taken in consideration on account of the evidence showing a large population of endogenous OPCs in chronic demyelinated lesion of MS patients [275].

A more comprehensive understanding regarding the transplantation of OPCs in the adult demyelinated CNS has been achieved through the study of models of focal demyelination induced by the injection of myelin-toxic chemicals into the brain or spinal cord. Focal demyelination can be induced through diet (i.e. with the copper chelator cuprizone) or injection of toxins (i.e. ethidium bromide, lysophosphatidylcholine [lysolecithin]) with the potential of isolating demyelination and remyelination as discrete events with spatiotemporal predictability. In this experimental setting, endogenous OPC involvement can be either suppressed by x-ray irradiation or kept, mimicking in such aspect MS but differently from the human disease; chemical-induced demyelination does not associate with an undergoing adaptive immune inflammation [225]. The adoption of these models has proven that OPCs can remyelinate axons also in the adult CNS and that progenitors are a better source of remyelination than adult oligodendrocytes [53]. In addition, toxin-induced model of demyelination has confirmed the importance of the environment in influencing graft's beneficial effects. When OPCs are transplanted in demyelinated lesions, the presence of activated astrocytes appears to favour successful remyelination [27], yet gliosis can inhibit the migration of exogenous cells. Such aspects should be accurately evaluated when considering timing and route of cell injection given that acute and chronic lesion of MS patients shows different status of astrocytes activation [196]. In addition, while comparing the exogenous and endogenous response to repair, it was noted that transplanted glial cells migrated over greater distances and remyelinated axons faster than endogenous cells [26]; however, when OPCs were transplanted in normal myelinated tissue, they did not seem to migrate or survive well [203]. These observations may suggest that inhibitory signals for endogenous OPCs in MS plaques may not interfere with transplanted exogenous cells but the presence of normal glia may obstacle cell migration.

In an attempt to better translate such results in human diseases, protocols to isolate glial progenitor cells expressing A2B5 from human foetal sources [251] and adult forebrain [271] have been developed. When tested in C3HeB/FeJ-shiverer (shi) mice, both human foetal- and adult-derived OPCs are capable to extensively remyelinate unsheathed axon [271, 273]. However, in comparing foetal- and adult-derived cells, the former becomes substantially more engrafted in the recipient brains than the adult counterpart [271, 273]. Further studies have then demonstrated that foetal-derived cells are significantly more proliferative and migratory when transplanted in C3HeB/FeJ-shiverer (shi) mice than mouse OPCs. Moreover, experiment adopting both C3HeB/FeJ-shiverer (shi) and wild-type mice has confirmed that a competent host myelinogenic compartment in wild-type mice can considerably reduce the migration of exogenous cells of both mouse and human origin [274]. In considering adult CNS, when human OPCs were transplanted directly into lysolecithin-demyelinated lesions in rat brain, the cells demonstrated to efficiently mature as oligodendrocytes and myelinate residual host axons [272]. However human OPCs revealed lower efficiency in remyelinate adult CNS than congenitally hypomyelinated brain. Thus similar to the rodent counterpart, human

OPC grafts appear to be consistently influenced by the presence of endogenous myelinogenic cells.

These results have clarified the need to improve our understanding of the molecular environment of demyelinating foci and the transcriptional response of OPCs to that environment. It is reasonable to speculate that only through such understanding it would be possible to maximize the efficiency of the graft and identify the best timing for OPC transplantation [188].

6.2.3 OPCs: Clinical Translation to Date

The first planned phase I trial in patients with secondary progressive multiple sclerosis will involve foetal OPCs purified by selecting for expression CD140a and platelet-derived growth factor alpha receptor. CD 140a + cell pool comprises a well-defined population of cells that uniformly retain oligodendrocytic lineage competence (comprises the entire pool of cells capable of generating myelinogenic oligodendrocytes) [251]. The trial planned by the New York State Consortium for Cell Therapy will aim to evaluate the feasibility and safety of intracerebral injections of escalating doses of OPCs. Intracerebral injection is assumed necessary because OPCs are not expected to be capable of trafficking from blood or CSF into the CNS parenchyma, yet this introduces an additional level of technical and safety concerns. To prevent rejection of allogeneic transplant, all patients will be treated with immunosuppressive therapy protocol (a combination of tacrolimus and mycophenolate mofetil for 6 months, followed by monotherapy with mofetil until the end of the study). Based on previous observation, immunosuppression should not affect OPC proliferation, survival or migratory [110].

Studies of the disease-specific tissue environment in genetically hypomyelinated newborn rodents and adult demyelinating disorders have suggested that the application of cell-based remyelinating therapy may be effective in treating dysmyelinating diseases and unique, limited area of demyelination within adult CNS. In considering OPC application to treat demyelination in progressive MS, exogenous graft should be able to remyelinate axons in chronic, non-repairing MS plaques in which endogenous, nondifferentiating OPC persists, perhaps associated with low-grade inflammation [42]. In such complex environment, the same factors negatively influencing endogenous OPC differentiation may somehow concur to limit the remyelinating potential of the graft such as the lack of factor(s) necessary to sustain remyelination and/or the presence of inhibitory clues in the lesion environment. Analogously, it is uncertain how already degenerated axons may support remyelination. Without accurately considering the timing of graft delivery, the administration of exogenous cells may not address the need. Even in the case that this trial may demonstrate safety, there will be many unanswered questions to address including the following: (i) What information on efficacy can be obtained from a phase 1 study focusing on feasibility and safety? (ii) How to monitor fate of injected cells? (iii) What is the most appropriate population within progressive patients to benefit from this approach? (iv) What is the best timing for its administration as so to stratify to those patients with a tissue environment permissive for donor cell integration and sufficient axonal preservation [51]? In addition, many of the obstacles that have long delayed the clinical investigation of OPC transplantation have yet to be overwhelmed.

While the development of protocols to isolate OPCs form foetal or adult human brain [271] has provided a potential source of cells for regenerative medicine, deriving glial progenitor from human foetuses 21- to 23-week-old or from patients undergoing temporal lobe resection for medication-refractory epilepsy clearly holds many limitations in terms of a putative clinical translation. Apart from the clear ethical controversy of obtaining cells from human foetuses, deriving glial tissue from surgical procedures involving the CNS neither can be considered an abundant source of cells for transplantation, given the limed possibility for in vivo expansion. Indeed, any practical source for cell-based therapy in myelin disorders should provide large numbers of progenitors biased to oligodendrocyte differentiation and myelinogenesis in order to cover the necessity of the clinical procedures. It is reasonable to assume that similar limitation involves OPCs derived from human embryonic stem cells (hESCs) [281] plus the additional concern related to the possibility of incidentally transplanted hESC remnants retaining the potential for undesired expansion after implantation [237]. In addition, both foetal/adult brain and hESC-derived OPCs are allotransplants, thereby requiring the need for immunosuppression in graft recipients. With this regard, the expansion of the field of cell reprogramming has provided novel promising alternatives through the adoption of patient autologous cells, thus obviating the need for immunosuppression and overcoming the ethics of embryonic/foetal sources. In 2006, the group of Yamanaka showed that mouse embryonic cells and adult fibroblasts may acquire properties similar to those of embryonic stem (ES) cells if transduced with four transcription factors (namely, Oct3/4, Sox2, Klf4 and c-Myc) [258]. This first generation of induced pluripotent stem (iPS) cells was similar to ES cells in morphology, proliferation and teratomas formation. However, iPS cells had a different global gene expression pattern from ES cells, and they failed to produce adult chimeric mice. In 2007, germline transmission was achieved with mouse iPS cells [205], and soon afterwards iPS cells could be efficiently generated from human fibroblasts [257]. Nowadays, iPS cell technology holds a tremendous promise for regenerative medicine, and research groups have efficiently developed human OPCs by differentiating somatic cell-derived iPS cells with the overarching aim to treat MS [261]. Nonetheless, the field of cellular reprogramming is one that is in its relative infancy presenting numerous limitations. These include major difficulties in obtaining differentiation into specific lineages and the required extensive in vitro passaging [84]. Further one cannot exclude the possibility of remaining neoplastic pluripotent cells within the final medical product [158]. Last but not least, there are no yet available clinical protocols [105, 120, 160]. Despite some preclinical studies have already shown that iPS-derived glial cells can efficiently myelinate unsheathed axons and enhance survival of congenitally hypomyelinated mouse [269] and remyelinate focal area of toxin-induced demyelination in the adult mouse spinal cord [189], further research will be required in particular concerning the mechanisms leading to genetic alterations in human iPSCs [97]. Of note, within the field of cell reprogramming, novel techniques continue to emerge tor the generation of autologous OPCs. Functional induced oligodendrocyte precursor cells (iOPCs) that exhibit morphological and molecular features consistent with OPCs have also been directly derived by delivering Sox10, Olig2 and Nkx6.2 in somatic fibroblasts [195]. Such avoidance of pluripotential intermediates in the generation of glial progenitors may mitigate the risk of tumorigenesis and circumvent many of the issues associated with the current remyelinating strategies involving cell therapy.

6.3 Modulation of Host Immune Responses and Promotion of Endogenous Repair Following Non-haematopoietic Stem Cell (nHSC) Transplantation

While the essential role of an inflammatory response for healing throughout the body has been known for a long time, it was until recently that inflammation in the CNS has been purely considered a synonym for a detrimental immune response that should be mitigated. This negative view evolved owing to the privileged immunological features of the CNS, protected behind the blood-brain barrier (BBB), and lack of awareness of the complexity of immune response during both acute and chronic inflammatory conditions. In the wake of the more recent understanding of CNS inflammation, however, such view has consistently expanded, in particular regarding the involvement of immune cells in CNS maintenance and repair [169, 183, 249, 284]. We now know that the entire process of recovery from injury involves a network encompassing both innate (microglia, monocytes, monocyte-derived macrophages) and adaptive immunity (T and B cells) at different stages and locations throughout the repair process. For instance, macrophage paracrine signal and removal of myelin debris allow migration and maturation of OPCs following demyelinating damage [85, 181], and analogously effector helper T cells and regulatory T cells contribute to CNS repair through the halting of harmful inflammatory response and promotion of tissue renewal [183, 283, 284]. With this regard, it has become clear that the innate and adaptive immune arms have variable influence on neurogenesis [76] and gliogenesis [34, 252]. This interplay with endogenous neural stem/progenitor cells (NSPCs) can be either detrimental or beneficial based on the context in which immune activation occurs. Thus, pro-inflammatory cytokines (e.g. IL-1β, TNF-α, IL-6) seem to impair endogenous NSPC proliferation [35, 141, 184], whereas anti-inflammatory cytokines (IL-4, IL-15) [34, 102, 107] and trophic factors (IGF-1, BDNF) [3, 118] appear to be pro-neuro and pro-gliogenic. The amount and the timing of these factors' release influence the net effect that the innate or adaptive immune cells, or both, exert on endogenous stem/progenitor cells; thereby the same cytokine or growth factor can be both pro- and anti-neurogenic depending on the microenvironment in which it operates [138]. Despite the precise cellular and molecular mechanisms regulating such relationship are still elusive, it can be considered a complex phenomenon in which the subtype of immune cells, the nature of the pathology and whether the inflammatory condition is acute or chronic will determine the overall outcome.

The possibility to modulate immune response in the CNS, in order to halt damage perpetuation and somehow reset immune cell action towards their essential role for healing, holds a great potential for treating chronic immune-mediated disease such as MS. Preclinical studies involving the experimental autoimmune/allergic encephalomyelitis [EAE], as immune-mediated model of MS, have provided overwhelming proofs for the capacity of different types of nHSCs to positively modulate immune response and improve animal phenotype. The understanding of these "non-canonical" properties has been at the centre of substantial stem cell research in MS aiming to develop cell-based therapy approaches to promote recovery by preventing tissue damage and/or sustaining the reestablishment of appropriately functioning neuronal circuits [170]. In this regard, MSCs and NSPCs have attracted the most attention collecting many lines of evidence for their preclinical efficacy when transplanted either directly in the CNS or systemically through a peripheral vein in EAE animals as immune-mediated model of MS.

6.3.1 MSCs (Including Multipotent Mesenchymal Stromal Cells)

The isolation of postnatal pluripotent cells within the connective tissue of the bone marrow was first described by Friedenstein et al. in the late 1960s (mid-1970s) [92]. During these pioneering studies, they established a colony of adherent fibroblast-like cells from single-cell suspensions of the bone marrow, which proliferated and survived several passages in vitro.

Defined as colony-forming unit fibroblasts (CFU-Fs), adherent cells were heterogeneous in appearance and give rise to stromal cells capable to serve as feeder layers for the culture of haematopoietic stem cells (HSCs) and to differentiate into adipocytes, chondrocytes and osteocytes both in vitro and in vivo [92, 93]. Bone marrow-derived stromal cells (BMSCs) were originally identified as specific progenitor for skeletal tissue; nonetheless, in the early 1990s, Caplan introduced the terminology "mesenchymal stem cells" (MSCs) extending their progenitor's role beyond the range of skeletal tissues [40] owing to the nomenclature "mesenchymal" referring to the histological term describing the transient embryonic connective tissue arising primarily from the mesoderm, a source of bone during embryological development but also of other tissues like muscle, tendons and ligaments.

In line with this assumption during the past decades, progenitor cells presenting phenotypical features analogous to the ones observed in the BMSCs, and accordingly termed MSCs, have been identified within the connective tissues of many adult and foetal organs (i.e. the foetal liver [37], the aorta-gonad-mesonephros

[176], the placenta [43], the umbilical cord [78], dental pulp [139] and adipose tissue [285]). As a consequence of the broad adoption of the term mesenchymal stem cells, the International Society for Cellular Therapy (ISCT), in an attempt to better clarify the nomenclature for MSCs, has suggested that the fibroblast-like plastic-adherent cells, regardless of the tissue from which they are isolated, should be termed "multipotent mesenchymal stromal cells", while the term mesenchymal stem cells has to be used only for cells that meet specified stem cell criteria [116]. In addition, ISCT has set minimal criteria for the identification of human MSCs as follows: (i) MSC must be plastic-adherent when maintained in standard culture conditions; (ii) MSC must express CD105, CD73 and CD90 and may lack expression of CD45, CD34, CD14 or CD11b, CD79a or CD19, and HLA-DR surface molecules; and (iii) MSC must differentiate to osteoblasts, adipocytes, and chondroblasts in vitro [65]. Nonetheless, the abbreviation MSCs is used to refer to either multipotent or mesenchymal stem cells and often indicate both categories given that no one of these surface markers (or a combination of them) is really specific for mesenchymal stem cells. Indeed, ISCT criteria do not support the purification of a homogenous stem cell population but rather outline heterogeneous population of cells that contain a mixture of progenitors at different stages of commitment to the mesodermal lineage and only a very small number of multipotential self-renewing stem cells.

While ISCT criteria have attempted to provide defined parameters to characterize MSCs in vitro purification, in vivo assay for MSCs identification has yet to be established. With this regard, Sacchetti et al. have indicated CD146 (also known as MCAM), as a marker for the identification and in vivo localization of bona fide stem cells within the subendothelial/perivascular zone of human bone marrow sinusoids [241]. Based on this evidence, recent studies have identified a ubiquitous population of highly myogenic and skeletogenic CD146+ cells, associated with microvessels of skeletal muscle and other tissues [54], thus supporting the ubiquity of MSCs within adult stromal tissue. However, while cells isolated from different tissues have demonstrated to grow under identical conditions and to express all the "MSC" markers, the analysis of the transcriptome has emphasized several important differences related to their tissue source. For instance, bone marrow-derived CD146+ cells expressed osteogenic transcription factors, whereas muscle-derived CD146+ cells expressed myogenic transcription factors [242]. The commitment of these cells to a specific lineage is confirmed by strict differentiation assays which showed bone marrow-derived CD146+ cells forming a bone/marrow organ upon in vivo transplantation but did not spontaneously form myotubes in vitro, and muscle-derived CD146+ cells did not form bone in vivo but did spontaneously form myotubes in vitro in the absence of exogenous myoblasts [242]. This data, while contrasting with the idea of a uniform class of "MSCs", have contributed to support the view that different mesoderm derivatives include distinct classes of tissue-specific committed progenitors that may have different developmental origins.

Despite the complexity involved in their biology, MSCs have received increasing interest following the recognition of biologic potentials which go behind their role in tissue renewal. Several lines of evidence suggest that MSCs possess additional proprieties which justify their use as a potential therapeutic tool for diseases in which tissue repair is needed and/or inflammation is extensive. By expressing a spectrum of signalling molecules, MSCs appear to exert modulatory effects on both adaptive and innate immunity and regulate tissue homeostasis responding to an extensive array of different environmental cues [264]. While these proprieties appear to be intrinsic in MSCs derived from varied stromal tissues, the putative translation for cell therapy has led to focus the attention on sources of cells which are easier to harvest and possess the potential for autologous application; these include MSCs derived from bone marrow (BM-MSCs), adipose tissues (AT-MSCs) and umbilical cord perivascular cells (UCPVCs). Nonetheless, the ideal source for cell-based therapy has yet to be determined [264].

6.3.2 *Evidence from Disease/Pathway Models*

6.3.2.1 Systemic Delivery of MSCs

Preclinical investigations of the impact of intravenous MSC (i.v.) delivery in acute [168] and chronic [147] EAE mice have mostly shown to ameliorate clinical manifestations, CNS inflammatory infiltration, demyelination and axonal damage. Interestingly, these studies have revealed that MSC efficiency was significantly influenced by the time of delivery [147], being the most effective when administered either before disease onset, at disease onset or at the peak of disease.

In line with the pathophysiology of EAE in which the immunization of animals with myelin-derived peptides (e.g. proteolipid protein [PLP], myelin oligodendrocyte glycoprotein [MOG] and myelin basic protein [MBP]) [50, 151] results in CNS infiltration of myelin-specific T cells yielding to the primary demyelination of axonal tracks, it has been suggested that the clinical and pathological benefit observed following systemic MSC transplantation may be a direct consequence of the inhibition of peripheral encephalitogenic T-cell proliferation and differentiation [279]. As a matter of fact, very few donor MSCs were detected in recipient CNS, and the majority of transplanted MSCs were found within lymph nodes and spleen [98, 279]. In the lymph nodes of BM-MSCs-treated mice, there is a significant decrease of T_H17 cell numbers and an increased percentage of $CD4^+CD25^+Foxp3^+$ Treg [163, 219], and when T cells from the lymph nodes of these mice are re-challenged with MOG peptide in vitro, they do not proliferate as expected [279]. Although a clear understanding of the mechanisms involved in MSC ability to inhibit T-cell proliferation has yet to be fully established, MSCs have proven to inhibit T-cell proliferation induced by multiple stimuli including mitogens [61, 62, 153], alloantigens [153, 263] and antibodies [263]. Several studies have suggested that both cell-to-cell contact mechanisms and secretion of paracrine factors may be involved. In a contact-dependent manner, MSCs appear to interact with T cells through adhesion molecules like ICAM-1 and VCAM-1 [144, 228]. Likewise, many MSC-derived products have been related to the inhibition of T-cell proliferation including truncated CCL-2 (tCCL-2) [220], hepatocyte growth factor (HGF) [61], indoleamine 2, 3-dioxygenase (IDO) [263], nitric oxide [244], prostaglandin E2 (PGE2) [1], galectin 1 [156] and semaphorin 3A [156]. Interestingly, inhibition of T-cell proliferation has been related to the arrest cell cycle in the G0/G1 phase rather than to the induction of apoptosis [21, 100]. The concept that MSCs ameliorate EAE through the induction of peripheral immune tolerance has been further nourished by the evidence that MSCs also appear to influence CD4 and CD8 T-cell differentiation and/or activity. It has emerged that MSCs induce an increase in CD4 T helper cell secretion of IL-4, whereas they decrease the production of interferon γ (IFN-γ), thus suggesting a shift from the prominence of pro-inflammatory T_H1/T_H17 cells towards an increase in anti-inflammatory T_H2 cells [1]. In considering CD8 cytotoxic T cells instead, MSCs suppress their cytotoxic potential, meanwhile not being targeted by such effect themselves [227]. Overall, these findings suggest that MSCs can modulate the intensity of an immune response by inhibiting antigen-specific T-cell proliferation and cytotoxicity while promoting their differentiation towards more tolerant phenotype for the sustainment of tissue regeneration.

According to the dependence of B-cell responses to T cell, systemic delivery of MSCs in mice with proteolipid protein (PLP)-induced EAE has proven to inhibit the in vivo production of pathogenic PLP-specific antibodies and to suppress the encephalitogenic potential of PLP-specific T cells following transfer in healthy recipient [98]. In considering the influence of MSCs on B-cell proliferation, contrasting views have emerged from in vitro coculture studies. While some have reported that MSCs inhibit the proliferation of B cells that are activated with anti-immunoglobulin antibodies, soluble CD40 ligand or cytokine [12, 100], others have reported that MSCs promoted proliferation and differentiation of transitional and naive B cells into immunoglobulin-secreting cells and strongly enhanced proliferation and differentiation of memory B-cell populations into plasma cells [226]. However, given the results observed in EAE mice, it is reasonable to emphasize that B-cell responses are mainly T-cell dependent and therefore the final outcome of the interaction between MSCs and B cells in vivo might be significantly influenced by the aforementioned MSC-mediated inhibition of T-cell functions.

In the wake of the increasing importance of innate immunity in many autoimmune diseases including MS, during the recent years, numerous studies have demonstrated the influence of MSCs on innate cells in particular like mononuclear macrophages (MPs), dendritic cells (DCs) and natural killer cells (NKs). Macrophages are an essential component in the orchestration and expression of innate immunity and adaptive immune responses. Depending on the microenvironment, macrophages can acquire distinct functional phenotypes playing a central role in both inflammation and tissue healing [167]. With this regard, MPs are typically divided into two phenotypes: classically activated macrophages (M1MPs) and alternatively activated MPs (M2MPs) [186]. M1MP polarization, induced by interferon-γ (IFN-γ) and lipopolysaccharide (LPS), is characterized by a high production of pro-inflammatory cytokines, nitric oxide and respiratory burst, whereas M2MP polarization, induced by T helper 2 (T_H2) cell cytokines (interleukin-4 [IL-4] and IL-13), presents low production of pro-inflammatory cytokines like IL-12 and high production of anti-inflammatory cytokines such as IL-10. In vitro studies have shown that MPs cocultured with MSCs consistently display an upregulation of M2 markers, thus resulting in high levels of IL-10 and low levels of IL-12 and TNF-α production [135]. MSC-mediated polarization towards M2 phenotypes depends on the secretion of soluble factors, including PGE2, TSG-6, IL-6, IDO and TGF-β1[175, 199, 282]. Recent findings have shown more complex mutual interaction between MSCs and MPs, in which MPs also appear to affect MSC immunomodulatory proprieties but in a contact-dependent manner and in correlation to MP phenotype. M1MPs interact with MSCs by functional synapse through CD54 [80], also known as intercellular adhesion molecule 1, a highly glycosylated immunoglobulin superfamily member that binds the leukocyte integrins "leukocyte function antigen 1" (LFA-1) or CD54 itself. The result of this interaction is that MSCs incubated with M1MPs inhibit T-cell proliferation is stronger manner than when incubated with M2MPs or MSCs alone [80]. While the in vivo relevance of these well-documented in vitro phenomena remains to be established, it is possible to speculate that such an effect may be involved in the higher efficacy observed in MSC i.v. injection at the peak of disease compared to later delivery, possibly suggesting a timing for optimizing the efficiency of MSCs in the treatment of inflammatory pathologies.

Interesting immunomodulatory proprieties have also been observed on DCs. BM-MSCs have shown to strongly inhibit DC maturation and to reduce the expression of surface stimulatory/co-stimulatory molecules once they become mature [125, 197]. Moreover, DC generated in coculture with MSCs favours the generation of regulatory T cells. The influence of MSCs on DC appears to be related to soluble factors such as IL-6 or TGF- α, and, accordingly, conditioned medium from MSC cultures shows similar immunomodulatory proprieties [157]. These studies suggest another putative mechanism with which MSCs can induce a peripheral immune tolerance following system delivery by direct DC maturation towards an anti-inflammatory or regulatory phenotype responsible for an attenuated T-cell response.

Analogously, BM-MSCs have shown to inhibit cytotoxic activity of resting NK cells by downregulating expression of NKp30 and natural killer group 2, member D (NKG2D), which are activating receptors that are involved in NK-cell activation and target cell killing [255]. Moreover, MSCs can inhibit the IL-2-induced proliferation of resting, inactivated NK cells, whereas they had only a partial inhibitory effect on proliferating NK cells [255]. Resting NK cells cocultured in the presence of MSCs have reduced proliferation rate and statistically significantly decreased (more than 80%) in IFN-γ levels following stimulation with IL-2 [1].

Despite little evidence exists demonstrating the presence of MSCs in the CNS following systemic delivery, the existence of beneficial effects independent from peripheral immunomodulation cannot be excluded. While in vitro studies have supported the idea that MSCs could transdifferentiate in cells of ectodermal lineage [276], the i.v. injection of MSCs in acute and chronic EAE mice has provided no evidence of transplanted MSCs expressing marker of neuroectodermal lineage within the CNS [98, 279]. However, the fact that i.v.-delivered MSCs can promote the regeneration of axons in the spinal cord of mice with myelin oligodendrocyte glycoprotein (MOG)-induced EAE [182] has contributed to support the view that

MSCs can exert antioxidant and neuroprotective activities [150] and/or promote endogenous reparative responses through bystander mechanisms autonomous from the induction of T-cell anergy [49].

Human MSCs have shown similar therapeutic effects in ameliorating clinical and pathological manifestation of acute and chronic EAE models [15, 104, 109]. Upon i.v. delivery, they influence host immune responses by reducing T-cell proliferation [66], IFN-γ production and T_H1-like and T_H17-like inflammatory cells (19191336 [16]) and upregulate CD1d^{high}CD5+ regulatory B cells [109]. In contrast to rodent MSCs, human MSCs apparently infiltrate the CNS in higher proportion, accumulating at sites of myelin damage and inflammation [104]. Interestingly, human MSCs are intrinsically capable of fusing with endogenous Purkinje cells in vitro. This event increases in frequency upon the presence of TNF-α and/or IFN-γ [133] and also occurs consistently in vivo in the EAE brain following the i.v. delivery of human MSCs, as suggested by the occurrence of heterokaryons, multinucleated cells containing genetic material from both the cellular graft and the host. Human MSCs also appear to support neurogenesis by secretion of neurotrophic factors. When injected into the dentate gyrus of healthy mice, they are able to promote proliferation and differentiation of neural stem cells [192]. This effect is attributed to the elevated secretion of NTFs such as nerve growth factor (NGF), VEGF, ciliary neurotrophic factor (CNTF) and FGF2 from transplanted cells. Analogously, MSCs injected into the SVZ and SGZ of healthy mice promote neurogenesis in the SVZ but not in SGZ associated with elevated BDNF secretion [128].

6.3.2.2 Local Delivery of MSCs

MSCs have shown preclinical efficacy in EAE models when delivered locally into the CNS. The intracerebroventricular (i.c.v.) administration has demonstrated to significantly reduce CNS inflammation in terms of the total number, size and cellularity of the lymphocytic infiltrates [130]. When compared to systemic i.v. injection, local transplant appears to provide higher neuroprotective effect in situ as demonstrated by a significant decrease in axonal loss, not observed following i.v. [130]. On the other hand, i.c.v. delivery has shown to be effective mainly in mild EAE phenotypes, displaying potentially adverse effects, including the formation of cellular masses, when transplanted in severe EAE mice [106]. These findings are of considerable relevance when one considers clinical trials in MS.

In order to assess the neuroprotective capacity of MSCs in a non-T-cell mediated setting, recent studies have employed the local delivery of MSCs in models of chemically induced demyelination, such as cuprizone and lysolecithin. In these models, demyelination is induced via direct toxicity to oligodendrocytes without involving a consistent immune activation; thus any observed beneficial effects may be reasonably due to the release of trophic factors by MSCs and/or to a bystander promotion of endogenous reparative responses. Accordingly, evidence have shown that grafting of MSCs led to an increase in OPC migration to the lesion, remyelination and axon conduction velocity [112, 123]. To support the hypothesis that MSCs may have a direct influence on OPCs and/or OLs without the interposition of the immune cells, similar beneficial effect is observed in NOD-SCID immunocompromised mice (which lack B and T cells) [99]. While the direct neurotrophic proprieties of MSCs have yet to be fully elucidated in vivo, it is possible to speculate that many of the beneficial effects observed on exogenous repair following MSC delivery may be a consequence of direct effect on endogenous cells as well as indirect consequence of their immunomodulatory proprieties.

Human MSCs have also proven to be efficacious in EAE, after i.c.v. transplantation. Human placental MSCs (PL-MSCs) transplanted i.c.v. in EAE mice decrease the disease severity and increase the animals' survival [83]. PL-MSCs express anti-inflammatory mediators in response to pro-inflammatory cytokines (e.g. TNF-α, IL1-β) released by activated microglia and astrocyte in the CNS of EAE mice. Further reports have shown that following exposure to conditioned media from LPS-activated astrocytes, PL-MSCs secrete TNF-α-stimulated gene/protein 6 (TSG-6). Such results again suggest that human MSCs may influence activated microglia, astrocytes and/or inflammatory cells in EAE models [83].

6.3.3 MSCs: Clinical Translation to Date

MSCs represent the most investigated cell-based approach to promote endogenous tissue repair. On the basis of their capacity to modulate immune system and promote tissue regeneration witnessed in preclinical studies, MSC transplantation has been tested in several neuro-inflammatory and degenerative diseases [194]. To date, over 700 clinical trials using human MSCs are listed in clinicaltrials.gov, out of which more than 40 have reached phase III. Nonetheless, in considering MS, clinical investigation has yet to reach such advanced stage.

Various groups have published small studies exploring feasibility and safety of MSC transplantation in MS [266] but those included very different cohorts of patients, cell products and routes of administration. In order to regulate this emerging field, in 2009 an international panel comprised of experts in the treatment of MS formed a consortium entitled the "International MSCT Study Group" with the aim to derive a consensus on the utilization of MSCs for the treatment of MS, along with protocols for the culture of the cells and the treatment of patients [91]. The consensus recognized that current evidence did support the i.v. administration of autologous MSCs as inhibitors of the autoimmune response in patients who continued to show inflammatory activity despite attempts to treat with DMTs [91]. Despite many completed trials have generally supported the feasibility and safety of MSC transplantation in MS patients, with some of them also reporting preliminary evidence of benefit on clinical, neurophysiological or imaging outcomes [46, 48, 232], the most recent consensus workshop held at the International Conference on Cell-Based Therapies for MS in November 2015 [246] has firmly agreed that further clinical trials are warranted. Indeed larger, controlled phase 2 studies of both unseparated, non-expanded bone marrow-derived cells [233, 234] and purified, culture-expanded

MSCs [MEsenchymal StEm Cells for Multiple Sclerosis (MESEMS)] [91] are currently underway.

Prudence on the results of future trials will be required anyway on the basis of the several unresolved issues involving both MSC-based therapy and MS pathology. For instance, since MSCs can be derived by various autologous sources, there may be potentially important differences between cells isolated from diverse sources (e.g. bone marrow vs adipose tissue vs umbilical cord) in terms of migratory abilities and/or impact on the clinical and pathological disease outcomes of the stem cell graft [209]. Moreover, variability is not limited to the tissue's source but also includes donor characteristics like ages and health [5, 256, 278]. With this regard, some studies have shown that BM- and AT-MSCs decrease in therapeutic potential with age, whereas neonatal MSCs display reduced signs of senescence following multiple passages in vitro [60]. In AT-MSCs, a certain range of heterogeneity has been also demonstrated with cells isolated from different regions of the body, and evidence that, e.g. subcutaneous-derived AT-MSCs grow slightly faster than omental-derived AT-MSCs provided [265].

Donor's characteristics also include whether cells are allogenic or autologous. In this context, the immunogenicity of MSCs has contributed to spread the idea that MSCs can be considered intrinsically immune privileged, thus overcoming the necessity for graft-host matching and considerably boosting the idea for allogenic MSC transplantation in patients. The substantial evidence of the immunomodulatory capacity of MSCs has driven the consideration that MSCs could induce anti-inflammatory responses sufficient to tame any pro-inflammatory T-cell responses that they might concurrently induce [239]. Thus, it has been suggested that MSCs may be considered immune evasive rather than immune privileged [10] and that such propriety is balanced between their expression of immunogenic and immunosuppressive factors, which is substantially affected by the nature of local inflammatory environment. In the case of autologous MSCs, it is unclear whether the underlying biology of MS might affect MSC function. Some studies have demonstrated similar growth in culture, differentiation potential, surface antigen expression and immunomodulatory properties of MSCs isolated from MS subjects versus non-MS controls [131, 165, 173, 207], whereas other studies reported notable functional differences [58], suggesting that autologous cell transplantation might not be appropriate.

The definition of optimal protocol for cell isolation is another aspect of major debate. While most of the previous studies used mixed mononuclear cell preparations [232] or purified culture-expanded MSCs [28, 46, 48, 129, 277], at present time it is not clear yet which cell among such population may be the most valuable therapeutically. On the other hand, we cannot exclude that the heterogeneity of MSCs itself may affect the biological properties of the total population, including its immunomodulatory effects. In this case, it would be difficult to produce heterogenic population of MSCs exhibiting similar therapeutic potential given that every collection of cells may contain a different percentage of stromal cells containing stem, progenitors and differentiated mesodermal cells [96]. Likewise, MSCs from different donors may vary in therapeutic potential. Such issues could obstacle any attempt to compare results obtained in different trials. With this regard, the International Society for Cellular Therapy (ISCT) has suggested that standardized immune assays should be performed on MSC samples that will be used in therapy to obtain reproducible and consistent data [143]. The ISCT has proposed the in vitro evaluation of IDO production following treatment of MSCs with interferon gamma (IFN-γ). This assay is of interest because the use of IFN-γ-induced IDO upregulation may be a surrogate marker of MSC suppressor function owing to the degree of the IDO response among individual donor correlates with MSC immunomodulatory activity [143].

In a chronic-relapsing disease like MS, timing and methodology of transplantation may consistently influence trial outcome. MSCs were reported safe for administration to MS patients through intravenous [46, 48, 232] and intrathecal transplantation [129, 277]. Yet, in the absence of a head-to-head comparison study, the optimal route of delivery remains uncertain; thus for the reduced complexity of peripheral cell delivery, intravenous administration is actually preferred [91]. In i.v. infusion, it is well recognized that MSCs are rapidly cleared by the lungs and rarely escape from the circulation [154]. With this regard, emerging evidence suggests potential immunomodulating effects may also result from the interaction of MSCs and immune cells in the lung [154, 204]. However, it still remains unclear as to whether the sustained vitality of the graft in vivo is indeed indispensable for the therapeutic potential of MSC transplantation. Though lack of evidence supporting sustained engraftment suggests that MSCs may mediate their function through a "hit-and-run" mechanism [267], therefore multiple infusions would be intuitively needed [234]. Yet, there is little or no evidence indicating how many cells might be optimal for any therapeutic use and current cell dose is entirely empirical. Common target is approximately 1–2 × 10^6 cells/kg body weight, an amount that makes using primary MSCs near impossible for clinical use, thereby requiring in vitro expansion. The propagation of primary cells in vitro is limited by senescence, which is triggered by DNA damage response pathways during growth in inappropriate growth conditions [96]. The presence of senescent cells in a batch of MSCs that will be delivered to patients may affect the therapeutic potential of the MSCs given that evidence exists showing the secretome profile of senescent MSC cultures differ from that of young, healthy cultures [248]. It should also be noted that even donor samples that will be used without ex vivo expansion might have different percentages of senescent cells due to interindividual variation. For such reasons new approaches to expansion continue to be explored [114] (Table 6.2).

6.3.4 Neural Stem/Precursor Cells (NSPCs)

In the adult CNS, proliferating neural cells represent a heterogeneous population of mitotically active, self-renewing and multipotent cells showing complex patterns of gene expression that vary in space and time [122]. The categorization of neural stem/precursor cells (NSPCs) is used as a generic term encompassing both stem and

Table 6.2 Comparison between major candidate strategies for cell-based therapy in progressive MS

	Inflammation	Demyelination	Neurodegeneration
Transplantation of OPCs for replacing damaged or lost myelin-making oligodendrocytes		Successful remyelination at the site of engraftment [271, 273] Grafted cells appear to remyelinate naked axon quicker than endogenous cells [26]	Restoration of myelin sheaths protects axons from degeneration [121, 142] Restore normal speed conduction [253] Possibility to remyelinate axons that have been depleted of myelin sheath for a considerable time [85]
Modulation of host immune response and enhancement of endogenous repair mechanisms using nHSCs (MSCs and NSPCs)	Inhibition of peripheral encephalitogenic T-cell proliferation and differentiation [216, 279] Promote shift from pro-inflammatory (T_H1/T_H17) to an anti-inflammatory (T_H2) environment [1, 217] Increase in CD25+ and CD25+/CD62L+ regulatory T cells [75, 217] Inhibition of pathogenic antibody production [98] Upregulate MPs M2 anti-inflammatory phenotypes, thus resulting in high levels of IL-10 and low levels of IL-12 and TNF-α production [135, 231] Inhibited DC maturation [125, 197]	Stimulate the proliferation and differentiation of endogenous OPCs, and upregulate remyelination through the secretion of platelet-derived growth factor (PDGF)-α and bFGF [73, 99, 215] Potential to directly differentiate in oligodendrocytic lineage following engraftment [19, 33, 132, 200, 215]	Promote the regeneration of axons [182] Promote proliferation and differentiation of endogenous neural stem cells [192] Increase in vivo bioavailability of major neurotrophins (e.g. nerve growth factor [NGF], brain-derived neurotrophic factor [BDNF], ciliary neurotrophic factor [CNTF], glial-derived neurotrophic factor [GDNF]) [41, 74, 215, 259]

progenitor cells and includes “bona fide” CNS stem cells, (intermediate stage) multipotent neural progenitors and lineage-oriented neural precursors. CNS stem cells display cardinal features such as (virtually) unlimited capacity for self-renewal and ability to proliferate in response to mitogens and multipotency for the different neuroectodermal lineages of the CNS (e.g. astrocytes, oligodendrocytes and neurons).

Multipotent progenitors are proliferative cells with only limited self-renewal that can differentiate into at least two different cell lineages [270], while lineage-oriented precursors are cells with restriction to one distinct neural lineage.

In the late 1960s, proliferating neural cells, possibly representing newly generated neurons, were identified in the adult rat brain [6, 7]. Since then, neurogenesis in the adult mammalian brain [8, 101, 202], including humans [79, 137, 211], has been shown to occur in specialized microenvironments, or niches, under a diverse set of physiologic/pathophysiologic conditions [59, 95, 162]. In the healthy adult CNS, the subventricular zone (SVZ) of the lateral ventricles and the subgranular zone (SGZ) of the hippocampal dentate gyrus (DG) have been the brain regions better characterized for continuously generating new neuronal cells [94]. Mammal neural stem cell niches possess a highly complex organization made up of a variety of cell subpopulations consisting in different stages of immature neural progeny accompanied by endothelial, astroglial and ependymal cells [126]. It is through the interaction of multiple specific signals that such microenvironment regulates NSPC quiescence, activation, self-renewal and ultimate survival.

Nowadays, the identification of NSPCs in vivo relies on the analysis of cell morphology, mitotic activity and gene and protein expression. Commonly used NSPC markers include nestin, glial fibrillary acidic protein (GFAP), Musashi 1/2 and the Shy-related high mobility group box transcription factor 2 (Sox2) [63, 155, 243]. Other various markers expressed at different stages of NSPC progression have been recognized and are currently used to indicate early or later phase of neural progenitor development like Prox1, NeuroD1, doublecortin (DCX) and polysialated form of neural cell adhesion molecule (PSA-NCAM) [180].

Soon after the in vivo identification of NSPCs within the CNS, different procedures were developed to allow us to safely expand and maintain these multipotent cells in chemically defined media for years. NSPCs obtained from either embryo or adult CNS can be cultured/expanded in vitro if supplemented with multipotent growth factors (i.e. EGF and FGF) as free-floating spherical aggregates (i.e. neurospheres) and/or as adherent monolayers (provided the provision of an artificial matrix). Upon growth factor withdrawal, they spontaneously differentiate into specific neuroectodermal lineages, ultimately becoming neurons, astrocytes and oligodendrocytes [229, 230].

The intrinsic ability of NSPCs to differentiate into cells of the oligodendroglial lineage cultivated hope for their adoption as clinically relevant therapies in a number of demyelinating/dysmyelinating disorders. Therefore, during the last decade, an extensive number of preclinical studies involving both rodents and non-human primates have been generated to investigate the therapeutic potential of exogenous delivery of NSPCs in different models of MS including toxin-induced demyelination [132, 187, 200], myelin mutant animals [111] and acute/chronic EAE [74, 75]. As a result, our comprehension regarding the predominant mechanisms involved into the beneficial effects observed upon transplantation has increased significantly. In 2003 Einstein et al. demonstrated that NSPCs were capable of migrating into inflamed white matter and differentiated into glial cells when intraventricularly transplanted into acute EAE rats [74]. The direct remyelinating potential of NSPCs has been then widely confirmed in mutant rodents and in model of focal demyelination in which grafted cells have shown potential to differentiate in both OPCs and SCs and remyelinate unsheathed axons [187]. These studies have demonstrated the ability of NSPCs to migrate and differentiate in oligodendrocyte lineage in both presence and absence of a host competent myelinating compartment [187].

Despite these findings have showed the possibility of NSPCs to differentiate and directly remyelinate axons in vivo, it should be noted that our current understanding of the prevailing mechanisms governing the beneficial effects displayed by transplanted NSPCs revolves primarily around a number of graft-to-host interactions as opposed to the direct replacement of damaged tissue [18, 52]; the aforementioned has come to be known as "therapeutic plasticity" [214].

6.3.5 *Evidence from Disease/Pathway Models*

6.3.5.1 Systemic Delivery of NSPCs

The observation that transplanted NSPCs were capable to attenuate the clinical course of acute [75], chronic [74, 215] and relapsing [217] EAE in rodents and primates [213] has significantly contributed to expand our comprehension regarding the therapeutical potential of nHSCs.

Following i.v. delivery in EAE rodents, NSPCs demonstrate distinct homing capacity by establishing close relationship with inflammatory areas within the CNS [170, 213, 216, 217]. This phenomenon depends on NSPCs' constitutive expression of an armamentarium of membrane receptors enabling them to follow gradients of chemoattractants, such as pro-inflammatory cytokines and chemokines and danger signals. In the injured CNS, the activation of microglia, astrocytes and/or endothelial cells [170, 191] produces chemokines such as stromal cell-derived factor (SDF)-1α/CXCL12, monocyte chemotactic protein (MCP)-1/CCL2 and vascular endothelial growth factor (VEGF) that collectively function as homing beacons, not only for inflammatory cells but also for transplanted NSPCs [64, 172, 179, 191, 212, 254]. NSPCs, similar to immune cells, express adhesion molecules (e.g. [CD44], integrins [α4 β1]) and chemokine receptors (e.g. CCR1, CCR2, CCR5, CXCR3, CXCR4) which allow them to extravasate into the CNS via tethering, rolling and adhering to endothelial cells followed by trans-endothelial migration [9, 57, 190, 215, 217]. Once in the CNS, NSPCs establish atypical perivascular niches whose architecture is reminiscent of their native prototypical germinal niches and microenvironment is supportive for undifferentiated stem survival [36, 75, 217]. NSPCs essentially maintain an undifferentiated phenotype in the inflamed CNS, accounting for very few percentages undergoing terminal differentiation (mostly in astrocytes or oligodendrocyte progenitor cells) [18, 19, 38, 56, 217]. The lack of functional integration can, at least in part, be explained by the fact that primary pro-inflammatory cytokines (e.g. TNF-α, IL-1β, IFN-γ) render transplanted NSPCs unable to fully differentiate by increasing the expression of inhibitors of the cell cycle [108]. It has been suggested that an environment rich in bone morphogenetic protein (BMP)-4 and the BMP-4 antagonist noggin could be considered the cause for the permanence in such state [217]. The presence of reactive glial cells also appears to influence transplanted NSPCs in multiple ways. Activated astrocytes support NSPC proliferation, survival and/or differentiation by producing factors like basic fibroblast growth factor (bFGF) and insulin-like growth factor (IGF)-1 [159, 191, 254], whereas hypertrophic glial fibrillary acidic protein (GFAP)-enriched astrocytes of the glial scar produce factors such as slit homologue (SLIT)-2, TNF-α and hyaluronan that repel NSPCs and limit the regenerative potential of their progeny (10742153) [20, 77, 191, 254].

Amelioration of EAE phenotype in murine and non-human primate correlates with pathological improvements in terms of the reduction of glial scar, area of demyelination, number of inflammatory infiltrates and evidence of axonal rescue [73, 215]. Systemic injection of NSPCs has shown to stimulate the proliferation and differentiation of endogenous OPCs and consequently upregulate remyelination through the secretion of platelet-derived growth factor (PDGF)-α and bFGF [73, 215]. Moreover, increased in vivo bioavailability of major neurotrophins (e.g. nerve growth factor [NGF], brain-derived neurotrophic factor [BDNF], ciliary neurotrophic factor [CNTF], glial-derived neurotrophic factor [GDNF]) is also involved into the trophic effects related to NSPC transplantation [41, 74, 215, 259]. This may be consequent to the culturing protocols which call for elevated amounts of epidermal growth factor (EGF) and fibroblast growth factor (FGF)-2, thus selecting growth factor-responsive cells that if transplanted into the inflamed CNS are more responsive to environmental signals and therefore release more neurotrophins and growth factors [170]. Thus, transplanted NSPCs may enhance the adult CNS capacity to repair itself by restoring the ability of endogenous progenitors and stem cells to both respond properly to disease state and replace damaged CNS cells and the ability of severed axons to regenerate. Yet, a current and comprehensive illustration of whether these and other (tissue)trophic properties of NSPCs are also relevant to EAE and MS is lacking.

The modulation of the host immune response is actually the most prioritized therapeutic propriety exerted by nHSC graft, and similar to other stem cell types, NSPCs have demonstrated to exert such effects in EAE as preclinical models of MS. Transplantation of NSPCs facilitates the switch to a more anti-inflammatory lesional environment [171, 212]. In vitro studies have shown that NPSCs:

1. Induce apoptosis of T_H1 and T_H17, but not T_H2 lymphocytes, through Fas ligand (FasL), TRAIL and Apo-3 ligand (APO3L) [217]
2. Reduce T-cell proliferation through nitric oxide and prostaglandin E (PGE)-2 [268]
3. Reduce T-cell receptor (TCR)-dependent activation [216]
4. Inhibit IL-2 (T cell) and IL-6 (B cell) signalling [81, 136]
5. Reduce local populations of monocytes and macrophages through cytotoxic TNF-α secretion [231]

In vitro coculture studies have confirmed that NSPCs can influence innate immunity. In particular, NSPCs exert an anti-inflammatory effect via inhibiting macrophage M1 activation and the secretion of inflammatory cytokines (i.e. TNF-α and IL-1β) [44]. This observation acquires relevance considering the important roles played by microglia/macrophages in progressive MS.

The ability of transplanted NSPCs to act as immunomodulatory cells is supported by evidence showing that NSPCs can exert such effects outside the CNS as well as after i.v. injection. In mouse models of EAE, i.v. NSPCs may inhibit the initiation and maintenance of the inflammatory events occurring in the secondary lymphoid organs (lymph nodes and spleen) [72, 208, 213]. Dendritic cell antigen presentation and antigen-specific T-cell proliferation [72] are impaired after i.v. injection of NSPCs into EAE mice. This has been attributed to the secretion by the NSPCs of LIF, which antagonizes the IL-6-mediated phosphorylation of signal transducer and activator of transcription 3 (STAT3), both required for encephalitogenic T_H17 cell differentiation [39]. Whether most of the immune regulatory effects of systemically injected NSPCs in EAE act directly in the CNS or in the periphery is still an unanswered question, and further studies are needed to establish the absolute relevance of these preclinical data in EAE, where peripheral lymphoid organs play an important role in the regulation of the immune responses to self-myelin antigens, and indeed the possibility that both sites of action may become interrelated and pathophysiologically relevant to the future applications of NSPCs in MS [171].

As soon as protocol to isolate human somatic and embryonic stem cell-derived NSPCs were developed [82, 281], similar immunomodulatory proprieties have been observed following the transplantation in non-human primates and mice with EAE, respectively [2, 213]. It is also worth noting that, along with immunomodulation, undifferentiated human NSPCs persisting in inflamed CNS areas secrete a plethora of trophic factors capable of protecting endogenous neural cells from programmed cell death, preventing glial scar formation, reestablishing neuron-glia functional interactions and enhancing endogenous remyelination.

6.3.5.2 Local Delivery of NSPCs

Although NSPCs have been successfully transplanted i.c.v. in both acute and chronic EAE, functional recovery obtained upon transplantation scarcely correlates with absolute numbers of transplant-derived newly generated terminally differentiated neuronal cells [56]. Following i.c. or i.t. transplantation of either neurospheres or single-cell-dissociated NSPCs, clinico-pathological amelioration of EAE in rodents has been achieved, with transplanted cells exhibiting targeted migration almost exclusively to the inflamed periventricular and perivascular white, but not grey, matter of the forebrain [75, 217]. Attenuation of brain inflammation occurs in correlation with a reduction of CD3+ T cells and an increase in CD25+ and CD25+/CD62L+ regulatory T cells along with a downregulation of inflammatory markers intercellular adhesion molecule 1 (ICAM-1) and lymphocyte function-associated antigen 1 (LFA-1) [75, 217].

Human NSPCs ameliorate EAE in non-human primates [2, 213]; however, in contrast to their mouse counterparts, they display a limited cytotoxicity towards T cells in vitro and a higher cytotoxic potential against MPs perhaps due to the elevated expression of TNF-α, TNF-like protein (TL) 1A and TRAIL [231]. Such effects on MPs may support the potential adoption of hNSCs as a treatment for progressive MS, which shows CNS inflammatory infiltrates principally characterized by macrophages and microglia [166].

6.3.6 NSPCs: Clinical Translation to Date

Based on the encouraging results collected preclinically during the last 10 years, several phase I and II trials have been conducted for the use of NSPCs in incurable diseases of both developing and mature CNS [266]. Nonetheless to date, no clinical trials have yet investigated NSPC application in MS [45].

Despite the strong preclinical evidence supporting the beneficial effects of NSPC transplantation, clinical translation for MS patients has remained indeed elusive. This "loss in translation" depends mostly on several preliminary and unresolved issues which have yet to be addressed that include (1) the ideal stem cell source for transplantation, (2) the most appropriate route of stem cell administration, (3) the best approach to achieve functional and long-lasting integration of transplanted stem cells into the host tissue and, finally, (4) the clear determination of both the short- and long-term beneficial and side effects of the transplantation procedure.

Unlike MSCs that can be isolated from various autologous tissue sources, somatic NSPCs are derived from either foetal [82] or embryonic sources [281]. While there are serious ethical issues associated with such methods, practical limitations also include the immunogenicity of the allogeneic grafts and the limited expandability/genotypic stability over extensive passaging in vitro [210, 224]. For the reasons outlined, generation of NSPCs from iPSCs generated from the recipient is an attractive alternative approach to circumvent histocompatibility barriers and the ethics associated with the source of the tissue [257]. However, difficulties in obtaining differentiation into specific lineages and the required extensive in vitro passaging are still limiting such development [84]. In addition, induced pluripotent stem (iPS) cell-derived NSPCs, similar to embryonic NSPCs, can be associated with a risk of tumor formation for the possibility of remaining neoplastic pluripotent cells within the final medical product [158]. Genetic instability is another relevant issue related to human iPSCs with some studies having detected frequent genetic modifications, including aberrant DNA methylation and mutations in genes implicated in cancer [105, 120, 160]. With this regard, the first trial employing an individualized iPSC-based therapy (i.e. in age-related macular degeneration [AMD]) was suspended for the identification of three single-nucleotide variations and three copy-number variants that were present in a second patient's iPSCs which were not detectable in the patient's original fibroblasts [97]. Novel techniques have emerged to guarantee the generation of autologous NSPCs with direct reprogramming. In 2011, it was shown that four Yamanaka reprogramming factors in combination with NSPC-permissive culture condition could be used to directly reprogram fibroblasts in NSPCs [134]. While these induced neural stem/precursor cells (iNSPCs) could initially be expanded and self-renewed for only 3–5 passages, later studies showed stably expandable iNSPCs (more than 50 passages) obtained by constitutively inducing Sox2, Klf4 and c-Myc while strictly limiting Oct4 activity [260]. These iNSPCs not only possess a similar genome-wide transcriptional profile compared to brain-derived NSPCs but can also differentiate into neurons, astrocytes and oligodendrocytes. Although human iNSPCs have been also generated [178], investigations are still ongoing to evaluate their overall potentials and risks.

For a multifocal disease like MS, the route of cell delivery becomes a primary consideration. Directly injecting cells into specific lesions would provide little benefit in the diffuse grey and white matter involvement that characterizes MS. In preclinical studies NSPCs have demonstrated beneficial effects in both systemic [170] and CNS deliveries [56]. In terms of clinical translation, i.v. delivery, similar to conventional drugs, may be safer. NSPCs have shown specific CNS homing potential, thus suggesting the possibility of successful peripheral application [64]. However, it has yet to be clarified whether NSPCs' main effects are exerted in the CNS or derived from interaction with immune cells in peripheral lymphoid organs [170]. On the other hand, i.c./i.t. delivery may target the disease process directly producing the most effective outcome. Indeed, recent studies indicate the transition of immune pathogenesis in MS from a peripheral initiated inflammatory injury during early phase of disease to a CNS compartmentalized immune and degenerative process during the chronic phase that is localized mainly in subarachnoid space/pial surface [117].

Last but not least, while in preclinical models, it is relatively easy to monitor graft integration and pathological outcomes, for clinical application there is still lack of valuable markers to track cells after transplant [218] and defined endpoints to evaluate and compare trial results [247]. The challenge to detect neuroprotection and repair in clinical trials for MS is substantial, and only comprehensive approaches involving clinical, laboratory and radiological findings are likely to achieve success.

6.4 Conclusion

More than one million of people are living with a progressive form of MS worldwide [30]. Identifying effective therapies for progressive MS represents a significant unmet need for disease treatment and a top priority for the global MS research community. While cell-based therapy for MS has been perused experimentally for at least four decades, it has been over the last years that major advances have occurred mostly based on the scientific progress which has expanded our view on immune responses in the CNS and their action in neural degeneration, plasticity and repair. The two cell-based approaches discussed in this chapter has entered (enhancement of endogenous repair through nHSCs – MSCs) or will enter (remyelinating strategies – OPCs) clinical trials soon. Enormous effort has been required to reach this stage, yet further work will be required to define both mechanisms causing and sustaining progressive MS pathology and to ultimately disclose the concrete potential of cell-based therapy. Without such studies, we will not be able to ascertain whether the clinical benefits of such therapies outweigh any potential harm. The economic burden of these trials cannot fall on single institution or agency. Recently two high-profile cellular therapy developers have fallen short in the highly challenging neural therapy space (i.e. StemCells Inc. and Geron) [90]. It is necessary that collaborative funding networks like the International Progressive Multiple Sclerosis Alliance or the New York State Consortium for Stem Cell Therapy for Progressive Multiple Sclerosis along with other government agencies share the efforts. Such networks can function beyond the organization of a single multicentre clinical trial and can establish protocols, undertake studies and importantly set up registries to record transplantation protocols and outcomes. This is fundamental to avoid uncontrolled use of cell-based therapy as unfortunately have occurred lately. In one recent case, a patient was treated with multiple injections of allogeneic stem cells from different sources (including mesenchymal, embryonic and foetal neural stem cells) that were intended to reduce neurologic deficits stemming from a middle cerebral artery stroke. The injections were associated with the development of a glioproliferative lesion, which led to paraplegia and ultimately required radiotherapy [22]. In another instance, autologous stem cells derived from adipose tissue and injected intravitreally into the eyes of people with macular degeneration were associated with worsening vision in three people, two of whom became legally blind [146]. The uncontrolled use of cell-based therapy represents possibly the most serious obstacle to the further development of these therapeutic approaches. It is reasonable to believe that in order to address the unique challenges of stem cell clinical research, ensuring the safety of patients must be the foundation of any stem cell experimentation. In all clinical cell therapy endeavours, there is need for rigorous and stringent quality and safety control in cell production with, in the case of artificially expanded cells, assessment of phenotype and karyotype, mutagenesis testing and microbiological analysis [65, 185]. This will provide an important pathway for ensuring that safe, effective stem cell therapies may ultimately become available to patients in need, but only if derived from solid and validated scientific evidence.

References

1. Aggarwal S, Pittenger MF (2005) Human mesenchymal stem cells modulate allogeneic immune cell responses. Blood 105(4):1815–1822. https://doi.org/10.1182/blood-2004-04-1559
2. Aharonowiz M, Einstein O, Fainstein N, Lassmann H, Reubinoff B, Ben-Hur T (2008) Neuroprotective effect of transplanted human embryonic stem cell-derived neural precursors

in an animal model of multiple sclerosis. PLoS One 3(9):e3145. https://doi.org/10.1371/journal.pone.0003145
3. Ahmed S, Reynolds BA, Weiss S (1995) BDNF enhances the differentiation but not the survival of CNS stem cell-derived neuronal precursors. J Neurosci Off J Soc Neurosci 15(8):5765–5778
4. Akiyama Y, Honmou O, Kato T, Uede T, Hashi K, Kocsis JD (2001) Transplantation of clonal neural precursor cells derived from adult human brain establishes functional peripheral myelin in the rat spinal cord. Exp Neurol 167(1):27–39. https://doi.org/10.1006/exnr.2000.7539
5. Altaner C, Altanerova V, Cihova M, Hunakova L, Kaiserova K, Klepanec A, Vulev I, Madaric J (2013) Characterization of mesenchymal stem cells of "no-options" patients with critical limb ischemia treated by autologous bone marrow mononuclear cells. PLoS One 8(9):e73722. https://doi.org/10.1371/journal.pone.0073722
6. Altman J, Das GD (1965) Autoradiographic and histological evidence of postnatal hippocampal neurogenesis in rats. J Comp Neurol 124(3):319–335
7. Altman J, Das GD (1966) Autoradiographic and histological studies of postnatal neurogenesis. I. A longitudinal investigation of the kinetics, migration and transformation of cells incorporating tritiated thymidine in neonate rats, with special reference to postnatal neurogenesis in some brain regions. J Comp Neurol 126(3):337–389. https://doi.org/10.1002/cne.901260302
8. Altman J, Das GD (1967) Postnatal neurogenesis in the guinea-pig. Nature 214(5093):1098–1101
9. Andres RH, Choi R, Pendharkar AV, Gaeta X, Wang N, Nathan JK, Chua JY, Lee SW, Palmer TD, Steinberg GK, Guzman R (2011) The CCR2/CCL2 interaction mediates the transendothelial recruitment of intravascularly delivered neural stem cells to the ischemic brain. Stroke 42(10):2923–2931. https://doi.org/10.1161/STROKEAHA.110.606368
10. Ankrum JA, Ong JF, Karp JM (2014) Mesenchymal stem cells: immune evasive, not immune privileged. Nat Biotechnol 32(3):252–260. https://doi.org/10.1038/nbt.2816
11. Archer DR, Cuddon PA, Lipsitz D, Duncan ID (1997) Myelination of the canine central nervous system by glial cell transplantation: a model for repair of human myelin disease. Nat Med 3(1):54–59
12. Augello A, Tasso R, Negrini SM, Amateis A, Indiveri F, Cancedda R, Pennesi G (2005) Bone marrow mesenchymal progenitor cells inhibit lymphocyte proliferation by activation of the programmed death 1 pathway. Eur J Immunol 35(5):1482–1490. https://doi.org/10.1002/eji.200425405
13. Avellana-Adalid V, Nait-Oumesmar B, Lachapelle F, Baron-Van Evercooren A (1996) Expansion of rat oligodendrocyte progenitors into proliferative "oligospheres" that retain differentiation potential. J Neurosci Res 45(5):558–570. https://doi.org/10.1002/(SICI)1097-4547(19960901)45:5<558::AID-JNR6>3.0.CO;2-B
14. Bachelin C, Lachapelle F, Girard C, Moissonnier P, Serguera-Lagache C, Mallet J, Fontaine D, Chojnowski A, Le Guern E, Nait-Oumesmar B, Baron-Van Evercooren A (2005) Efficient myelin repair in the macaque spinal cord by autologous grafts of Schwann cells. Brain 128(Pt 3):540–549. https://doi.org/10.1093/brain/awh406
15. Bai L, Lennon DP, Caplan AI, DeChant A, Hecker J, Kranso J, Zaremba A, Miller RH (2012) Hepatocyte growth factor mediates mesenchymal stem cell-induced recovery in multiple sclerosis models. Nat Neurosci 15(6):862–870. https://doi.org/10.1038/nn.3109
16. Bai L, Lennon DP, Eaton V, Maier K, Caplan AI, Miller SD, Miller RH (2009) Human bone marrow-derived mesenchymal stem cells induce Th2-polarized immune response and promote endogenous repair in animal models of multiple sclerosis. Glia 57(11):1192–1203. https://doi.org/10.1002/glia.20841
17. Belachew S, Chittajallu R, Aguirre AA, Yuan X, Kirby M, Anderson S, Gallo V (2003) Postnatal NG2 proteoglycan-expressing progenitor cells are intrinsically multipotent and generate functional neurons. J Cell Biol 161(1):169–186. https://doi.org/10.1083/jcb.200210110

18. Ben-Hur T (2008) Immunomodulation by neural stem cells. J Neurol Sci 265(1–2):102–104. https://doi.org/10.1016/j.jns.2007.05.007
19. Ben-Hur T, Einstein O, Mizrachi-Kol R, Ben-Menachem O, Reinhartz E, Karussis D, Abramsky O (2003) Transplanted multipotential neural precursor cells migrate into the inflamed white matter in response to experimental autoimmune encephalomyelitis. Glia 41(1):73–80. https://doi.org/10.1002/glia.10159
20. Benedetti S, Pirola B, Pollo B, Magrassi L, Bruzzone MG, Rigamonti D, Galli R, Selleri S, Di Meco F, De Fraja C, Vescovi A, Cattaneo E, Finocchiaro G (2000) Gene therapy of experimental brain tumors using neural progenitor cells. Nat Med 6(4):447–450. https://doi.org/10.1038/74710
21. Benvenuto F, Ferrari S, Gerdoni E, Gualandi F, Frassoni F, Pistoia V, Mancardi G, Uccelli A (2007) Human mesenchymal stem cells promote survival of T cells in a quiescent state. Stem Cells 25(7):1753–1760. https://doi.org/10.1634/stemcells.2007-0068
22. Berkowitz AL, Miller MB, Mir SA, Cagney D, Chavakula V, Guleria I, Aizer A, Ligon KL, Chi JH (2016) Glioproliferative lesion of the spinal cord as a complication of "stem-cell tourism". N Engl J Med 375(2):196–198. https://doi.org/10.1056/NEJMc1600188
23. Bjorklund LM, Sanchez-Pernaute R, Chung S, Andersson T, Chen IY, McNaught KS, Brownell AL, Jenkins BG, Wahlestedt C, Kim KS, Isacson O (2002) Embryonic stem cells develop into functional dopaminergic neurons after transplantation in a Parkinson rat model. Proc Natl Acad Sci U S A 99(4):2344–2349. https://doi.org/10.1073/pnas.022438099
24. Blakemore WF (1977) Remyelination of CNS axons by Schwann cells transplanted from the sciatic nerve. Nature 266(5597):68–69
25. Blakemore WF, Crang AJ (1985) The use of cultured autologous Schwann cells to remyelinate areas of persistent demyelination in the central nervous system. J Neurol Sci 70(2):207–223
26. Blakemore WF, Gilson JM, Crang AJ (2000) Transplanted glial cells migrate over a greater distance and remyelinate demyelinated lesions more rapidly than endogenous remyelinating cells. J Neurosci Res 61(3):288–294. https://doi.org/10.1002/1097-4547(20000801)61:3<288::AID-JNR6>3.0.CO;2-#
27. Blakemore WF, Gilson JM, Crang AJ (2003) The presence of astrocytes in areas of demyelination influences remyelination following transplantation of oligodendrocyte progenitors. Exp Neurol 184(2):955–963. https://doi.org/10.1016/S0014-4886(03)00347-9
28. Bonab MM, Sahraian MA, Aghsaie A, Karvigh SA, Hosseinian SM, Nikbin B, Lotfi J, Khorramnia S, Motamed MR, Togha M, Harirchian MH, Moghadam NB, Alikhani K, Yadegari S, Jafarian S, Gheini MR (2012) Autologous mesenchymal stem cell therapy in progressive multiple sclerosis: an open label study. Curr Stem Cell Res Ther 7(6):407–414
29. Bowen JD, Kraft GH, Wundes A, Guan Q, Maravilla KR, Gooley TA, McSweeney PA, Pavletic SZ, Openshaw H, Storb R, Wener M, McLaughlin BA, Henstorf GR, Nash RA (2012) Autologous hematopoietic cell transplantation following high-dose immunosuppressive therapy for advanced multiple sclerosis: long-term results. Bone Marrow Transplant 47(7):946–951. https://doi.org/10.1038/bmt.2011.208
30. Browne P, Chandraratna D, Angood C, Tremlett H, Baker C, Taylor BV, Thompson AJ (2014) Atlas of multiple sclerosis 2013: a growing global problem with widespread inequity. Neurology 83(11):1022–1024. https://doi.org/10.1212/WNL.0000000000000768
31. Brustle O, Jones KN, Learish RD, Karram K, Choudhary K, Wiestler OD, Duncan ID, McKay RD (1999) Embryonic stem cell-derived glial precursors: a source of myelinating transplants. Science 285(5428):754–756
32. Buchet D, Baron-Van Evercooren A (2009) In search of human oligodendroglia for myelin repair. Neurosci Lett 456(3):112–119. https://doi.org/10.1016/j.neulet.2008.09.086
33. Buchet D, Garcia C, Deboux C, Nait-Oumesmar B, Baron-Van Evercooren A (2011) Human neural progenitors from different foetal forebrain regions remyelinate the adult mouse spinal cord. Brain J Neurol 134(Pt 4):1168–1183. https://doi.org/10.1093/brain/awr030
34. Butovsky O, Ziv Y, Schwartz A, Landa G, Talpalar AE, Pluchino S, Martino G, Schwartz M (2006) Microglia activated by IL-4 or IFN-gamma differentially induce neurogenesis and

oligodendrogenesis from adult stem/progenitor cells. Mol Cell Neurosci 31(1):149–160. https://doi.org/10.1016/j.mcn.2005.10.006
35. Cacci E, Ajmone-Cat MA, Anelli T, Biagioni S, Minghetti L (2008) In vitro neuronal and glial differentiation from embryonic or adult neural precursor cells are differently affected by chronic or acute activation of microglia. Glia 56(4):412–425. https://doi.org/10.1002/glia.20616
36. Calabrese C, Poppleton H, Kocak M, Hogg TL, Fuller C, Hamner B, Oh EY, Gaber MW, Finklestein D, Allen M, Frank A, Bayazitov IT, Zakharenko SS, Gajjar A, Davidoff A, Gilbertson RJ (2007) A perivascular niche for brain tumor stem cells. Cancer Cell 11(1):69–82. https://doi.org/10.1016/j.ccr.2006.11.020
37. Campagnoli C, Roberts IA, Kumar S, Bennett PR, Bellantuono I, Fisk NM (2001) Identification of mesenchymal stem/progenitor cells in human first-trimester fetal blood, liver, and bone marrow. Blood 98(8):2396–2402
38. Cao QL, Howard RM, Dennison JB, Whittemore SR (2002) Differentiation of engrafted neuronal-restricted precursor cells is inhibited in the traumatically injured spinal cord. Exp Neurol 177(2):349–359
39. Cao W, Yang Y, Wang Z, Liu A, Fang L, Wu F, Hong J, Shi Y, Leung S, Dong C, Zhang JZ (2011) Leukemia inhibitory factor inhibits T helper 17 cell differentiation and confers treatment effects of neural progenitor cell therapy in autoimmune disease. Immunity 35(2):273–284. https://doi.org/10.1016/j.immuni.2011.06.011
40. Caplan AI (1991) Mesenchymal stem cells. J Orthopaed Res Off Publ Orthopaed Res Soc 9(5):641–650. https://doi.org/10.1002/jor.1100090504
41. Capone C, Frigerio S, Fumagalli S, Gelati M, Principato MC, Storini C, Montinaro M, Kraftsik R, De Curtis M, Parati E, De Simoni MG (2007) Neurosphere-derived cells exert a neuroprotective action by changing the ischemic microenvironment. PLoS One 2(4):e373. https://doi.org/10.1371/journal.pone.0000373
42. Chang A, Tourtellotte WW, Rudick R, Trapp BD (2002) Premyelinating oligodendrocytes in chronic lesions of multiple sclerosis. N Engl J Med 346(3):165–173. https://doi.org/10.1056/NEJMoa010994
43. Chang CJ, Yen ML, Chen YC, Chien CC, Huang HI, Bai CH, Yen BL (2006) Placenta-derived multipotent cells exhibit immunosuppressive properties that are enhanced in the presence of interferon-gamma. Stem Cells 24(11):2466–2477. https://doi.org/10.1634/stemcells.2006-0071
44. Cheng Z, Zhu W, Cao K, Wu F, Li J, Wang G, Li H, Lu M, Ren Y, He X (2016) Anti-inflammatory mechanism of neural stem cell transplantation in spinal cord injury. Int J Mol Sci 17(9). https://doi.org/10.3390/ijms17091380
45. Chiu AY, Rao MS (2011) Cell-based therapy for neural disorders – anticipating challenges. Neurotherapeutics J Am Soc Exp NeuroTherap 8(4):744–752. https://doi.org/10.1007/s13311-011-0066-9
46. Cohen JA, Imrey PB, Planchon SM, Bermel RA, Fisher E, Fox RJ, Bar-Or A, Sharp SL, Skaramagas TT, Jagodnik P, Karafa M, Morrison S, Reese Koc J, Gerson SL, Lazarus HM (2017) Pilot trial of intravenous autologous culture-expanded mesenchymal stem cell transplantation in multiple sclerosis. Mult Scler 24(4):501–511. https://doi.org/10.1177/1352458517703802
47. Comi G (2013) Disease-modifying treatments for progressive multiple sclerosis. Mult Scler 19(11):1428–1436. https://doi.org/10.1177/1352458513502572
48. Connick P, Kolappan M, Crawley C, Webber DJ, Patani R, Michell AW, Du MQ, Luan SL, Altmann DR, Thompson AJ, Compston A, Scott MA, Miller DH, Chandran S (2012) Autologous mesenchymal stem cells for the treatment of secondary progressive multiple sclerosis: an open-label phase 2a proof-of-concept study. Lancet Neurol 11(2):150–156. https://doi.org/10.1016/S1474-4422(11)70305-2
49. Constantin G, Marconi S, Rossi B, Angiari S, Calderan L, Anghileri E, Gini B, Bach SD, Martinello M, Bifari F, Galie M, Turano E, Budui S, Sbarbati A, Krampera M, Bonetti B

(2009) Adipose-derived mesenchymal stem cells ameliorate chronic experimental autoimmune encephalomyelitis. Stem Cells 27(10):2624–2635. https://doi.org/10.1002/stem.194
50. Constantinescu CS, Farooqi N, O'Brien K, Gran B (2011) Experimental autoimmune encephalomyelitis (EAE) as a model for multiple sclerosis (MS). Br J Pharmacol 164(4):1079–1106. https://doi.org/10.1111/j.1476-5381.2011.01302.x
51. Correale J, Gaitan MI, Ysrraelit MC, Fiol MP (2017) Progressive multiple sclerosis: from pathogenic mechanisms to treatment. Brain J Neurol 140(3):527–546. https://doi.org/10.1093/brain/aww258
52. Cossetti C, Alfaro-Cervello C, Donega M, Tyzack G, Pluchino S (2012) New perspectives of tissue remodelling with neural stem and progenitor cell-based therapies. Cell Tissue Res 349(1):321–329. https://doi.org/10.1007/s00441-012-1341-8
53. Crang AJ, Gilson J, Blakemore WF (1998) The demonstration by transplantation of the very restricted remyelinating potential of post-mitotic oligodendrocytes. J Neurocytol 27(7):541–553
54. Crisan M, Yap S, Casteilla L, Chen CW, Corselli M, Park TS, Andriolo G, Sun B, Zheng B, Zhang L, Norotte C, Teng PN, Traas J, Schugar R, Deasy BM, Badylak S, Buhring HJ, Giacobino JP, Lazzari L, Huard J, Peault B (2008) A perivascular origin for mesenchymal stem cells in multiple human organs. Cell Stem Cell 3(3):301–313. https://doi.org/10.1016/j.stem.2008.07.003
55. Cummings BJ, Uchida N, Tamaki SJ, Salazar DL, Hooshmand M, Summers R, Gage FH, Anderson AJ (2005) Human neural stem cells differentiate and promote locomotor recovery in spinal cord-injured mice. Proc Natl Acad Sci U S A 102(39):14069–14074. https://doi.org/10.1073/pnas.0507063102
56. Cusimano M, Biziato D, Brambilla E, Donega M, Alfaro-Cervello C, Snider S, Salani G, Pucci F, Comi G, Garcia-Verdugo JM, De Palma M, Martino G, Pluchino S (2012) Transplanted neural stem/precursor cells instruct phagocytes and reduce secondary tissue damage in the injured spinal cord. Brain 135(Pt 2):447–460. https://doi.org/10.1093/brain/awr339
57. Darsalia V, Kallur T, Kokaia Z (2007) Survival, migration and neuronal differentiation of human fetal striatal and cortical neural stem cells grafted in stroke-damaged rat striatum. Eur J Neurosci 26(3):605–614. https://doi.org/10.1111/j.1460-9568.2007.05702.x
58. de Oliveira GL, de Lima KW, Colombini AM, Pinheiro DG, Panepucci RA, Palma PV, Brum DG, Covas DT, Simoes BP, de Oliveira MC, Donadi EA, Malmegrim KC (2015) Bone marrow mesenchymal stromal cells isolated from multiple sclerosis patients have distinct gene expression profile and decreased suppressive function compared with healthy counterparts. Cell Transplant 24(2):151–165. https://doi.org/10.3727/096368913X675142
59. Decimo I, Bifari F, Krampera M, Fumagalli G (2012) Neural stem cell niches in health and diseases. Curr Pharm Des 18(13):1755–1783
60. Deuse T, Stubbendorff M, Tang-Quan K, Phillips N, Kay MA, Eiermann T, Phan TT, Volk HD, Reichenspurner H, Robbins RC, Schrepfer S (2011) Immunogenicity and immunomodulatory properties of umbilical cord lining mesenchymal stem cells. Cell Transplant 20(5):655–667. https://doi.org/10.3727/096368910X536473
61. Di Nicola M, Carlo-Stella C, Magni M, Milanesi M, Longoni PD, Matteucci P, Grisanti S, Gianni AM (2002) Human bone marrow stromal cells suppress T-lymphocyte proliferation induced by cellular or nonspecific mitogenic stimuli. Blood 99(10):3838–3843
62. Djouad F, Plence P, Bony C, Tropel P, Apparailly F, Sany J, Noel D, Jorgensen C (2003) Immunosuppressive effect of mesenchymal stem cells favors tumor growth in allogeneic animals. Blood 102(10):3837–3844. https://doi.org/10.1182/blood-2003-04-1193
63. Doetsch F (2003) The glial identity of neural stem cells. Nat Neurosci 6(11):1127–1134. https://doi.org/10.1038/nn1144
64. Doetsch F, Petreanu L, Caille I, Garcia-Verdugo JM, Alvarez-Buylla A (2002) EGF converts transit-amplifying neurogenic precursors in the adult brain into multipotent stem cells. Neuron 36(6):1021–1034

65. Dominici M, Le Blanc K, Mueller I, Slaper-Cortenbach I, Marini F, Krause D, Deans R, Keating A, Prockop D, Horwitz E (2006) Minimal criteria for defining multipotent mesenchymal stromal cells. The International Society for Cellular Therapy position statement. Cytotherapy 8(4):315–317. https://doi.org/10.1080/14653240600855905
66. Donders R, Vanheusden M, Bogie JF, Ravanidis S, Thewissen K, Stinissen P, Gyselaers W, Hendriks JJ, Hellings N (2015) Human Wharton's Jelly-derived stem cells display immunomodulatory properties and transiently improve rat experimental autoimmune encephalomyelitis. Cell Transplant 24(10):2077–2098. https://doi.org/10.3727/096368914X685104
67. Dubois-Dalcq M, Ffrench-Constant C, Franklin RJ (2005) Enhancing central nervous system remyelination in multiple sclerosis. Neuron 48(1):9–12. https://doi.org/10.1016/j.neuron.2005.09.004
68. Duncan ID, Aguayo AJ, Bunge RP, Wood PM (1981) Transplantation of rat Schwann cells grown in tissue culture into the mouse spinal cord. J Neurol Sci 49(2):241–252
69. Duncan ID, Brower A, Kondo Y, Curlee JF Jr, Schultz RD (2009) Extensive remyelination of the CNS leads to functional recovery. Proc Natl Acad Sci U S A 106(16):6832–6836. https://doi.org/10.1073/pnas.0812500106
70. Duncan ID, Hammang JP, Jackson KF, Wood PM, Bunge RP, Langford L (1988) Transplantation of oligodendrocytes and Schwann cells into the spinal cord of the myelin-deficient rat. J Neurocytol 17(3):351–360
71. Duncan ID, Paino C, Archer DR, Wood PM (1992) Functional capacities of transplanted cell-sorted adult oligodendrocytes. Dev Neurosci 14(2):114–122
72. Einstein O, Fainstein N, Vaknin I, Mizrachi-Kol R, Reihartz E, Grigoriadis N, Lavon I, Baniyash M, Lassmann H, Ben-Hur T (2007) Neural precursors attenuate autoimmune encephalomyelitis by peripheral immunosuppression. Ann Neurol 61(3):209–218. https://doi.org/10.1002/ana.21033
73. Einstein O, Friedman-Levi Y, Grigoriadis N, Ben-Hur T (2009) Transplanted neural precursors enhance host brain-derived myelin regeneration. J Neurosci 29(50):15694–15702. https://doi.org/10.1523/JNEUROSCI.3364-09.2009
74. Einstein O, Grigoriadis N, Mizrachi-Kol R, Reinhartz E, Polyzoidou E, Lavon I, Milonas I, Karussis D, Abramsky O, Ben-Hur T (2006) Transplanted neural precursor cells reduce brain inflammation to attenuate chronic experimental autoimmune encephalomyelitis. Exp Neurol 198(2):275–284. https://doi.org/10.1016/j.expneurol.2005.11.007
75. Einstein O, Karussis D, Grigoriadis N, Mizrachi-Kol R, Reinhartz E, Abramsky O, Ben-Hur T (2003) Intraventricular transplantation of neural precursor cell spheres attenuates acute experimental allergic encephalomyelitis. Mol Cell Neurosci 24(4):1074–1082
76. Ekdahl CT, Claasen JH, Bonde S, Kokaia Z, Lindvall O (2003) Inflammation is detrimental for neurogenesis in adult brain. Proc Natl Acad Sci U S A 100(23):13632–13637. https://doi.org/10.1073/pnas.2234031100
77. Emsley JG, Mitchell BD, Kempermann G, Macklis JD (2005) Adult neurogenesis and repair of the adult CNS with neural progenitors, precursors, and stem cells. Prog Neurobiol 75(5):321–341. https://doi.org/10.1016/j.pneurobio.2005.04.002
78. Erices A, Conget P, Minguell JJ (2000) Mesenchymal progenitor cells in human umbilical cord blood. Br J Haematol 109(1):235–242
79. Eriksson PS, Perfilieva E, Bjork-Eriksson T, Alborn AM, Nordborg C, Peterson DA, Gage FH (1998) Neurogenesis in the adult human hippocampus. Nat Med 4(11):1313–1317. https://doi.org/10.1038/3305
80. Espagnolle N, Balguerie A, Arnaud E, Sensebe L, Varin A (2017) CD54-mediated interaction with pro-inflammatory macrophages increases the immunosuppressive function of human mesenchymal stromal cells. Stem Cell Reports 8(4):961–976. https://doi.org/10.1016/j.stemcr.2017.02.008
81. Fainstein N, Vaknin I, Einstein O, Zisman P, Ben Sasson SZ, Baniyash M, Ben-Hur T (2008) Neural precursor cells inhibit multiple inflammatory signals. Mol Cell Neurosci 39(3):335–341. https://doi.org/10.1016/j.mcn.2008.07.007

82. Ferrari D, Binda E, De Filippis L, Vescovi AL (2010) Isolation of neural stem cells from neural tissues using the neurosphere technique. Curr Protoc Stem Cell Biol. Chapter 2:Unit2D 6. https://doi.org/10.1002/9780470151808.sc02d06s15
83. Fisher-Shoval Y, Barhum Y, Sadan O, Yust-Katz S, Ben-Zur T, Lev N, Benkler C, Hod M, Melamed E, Offen D (2012) Transplantation of placenta-derived mesenchymal stem cells in the EAE mouse model of MS. J Mol Neurosci 48(1):176–184. https://doi.org/10.1007/s12031-012-9805-6
84. Fong CY, Gauthaman K, Bongso A (2010) Teratomas from pluripotent stem cells: a clinical hurdle. J Cell Biochem 111(4):769–781. https://doi.org/10.1002/jcb.22775
85. Foote AK, Blakemore WF (2005) Inflammation stimulates remyelination in areas of chronic demyelination. Brain 128(Pt 3):528–539. https://doi.org/10.1093/brain/awh417
86. Franklin RJ (2002) Why does remyelination fail in multiple sclerosis? Nat Rev Neurosci 3(9):705–714. https://doi.org/10.1038/nrn917
87. Franklin RJ, Ffrench-Constant C (2008) Remyelination in the CNS: from biology to therapy. Nat Rev Neurosci 9(11):839–855. https://doi.org/10.1038/nrn2480
88. Franklin RJ, ffrench-Constant C, Edgar JM, Smith KJ (2012) Neuroprotection and repair in multiple sclerosis. Nat Rev Neurol 8(11):624–634. https://doi.org/10.1038/nrneurol.2012.200
89. Franklin RJ, Gallo V (2014) The translational biology of remyelination: past, present, and future. Glia 62(11):1905–1915. https://doi.org/10.1002/glia.22622
90. Frantz S (2012) Embryonic stem cell pioneer Geron exits field, cuts losses. Nat Biotechnol 30(1):12–13. https://doi.org/10.1038/nbt0112-12
91. Freedman MS, Bar-Or A, Atkins HL, Karussis D, Frassoni F, Lazarus H, Scolding N, Slavin S, Le Blanc K, Uccelli A, Group MS (2010) The therapeutic potential of mesenchymal stem cell transplantation as a treatment for multiple sclerosis: consensus report of the International MSCT Study Group. Mult Scler 16(4):503–510. https://doi.org/10.1177/1352458509359727
92. Friedenstein AJ, Chailakhyan RK, Latsinik NV, Panasyuk AF, Keiliss-Borok IV (1974) Stromal cells responsible for transferring the microenvironment of the hemopoietic tissues. Cloning in vitro and retransplantation in vivo. Transplantation 17(4):331–340
93. Friedenstein AJ, Petrakova KV, Kurolesova AI, Frolova GP (1968) Heterotopic of bone marrow. Analysis of precursor cells for osteogenic and hematopoietic tissues. Transplantation 6(2):230–247
94. Gage FH (2000) Mammalian neural stem cells. Science 287(5457):1433–1438
95. Gage FH, Ray J, Fisher LJ (1995) Isolation, characterization, and use of stem cells from the CNS. Annu Rev Neurosci 18:159–192. https://doi.org/10.1146/annurev.ne.18.030195.001111
96. Galipeau J (2013) The mesenchymal stromal cells dilemma – does a negative phase III trial of random donor mesenchymal stromal cells in steroid-resistant graft-versus-host disease represent a death knell or a bump in the road? Cytotherapy 15(1):2–8. https://doi.org/10.1016/j.jcyt.2012.10.002
97. Garber K (2015) RIKEN suspends first clinical trial involving induced pluripotent stem cells. Nat Biotechnol 33(9):890–891. https://doi.org/10.1038/nbt0915-890
98. Gerdoni E, Gallo B, Casazza S, Musio S, Bonanni I, Pedemonte E, Mantegazza R, Frassoni F, Mancardi G, Pedotti R, Uccelli A (2007) Mesenchymal stem cells effectively modulate pathogenic immune response in experimental autoimmune encephalomyelitis. Ann Neurol 61(3):219–227. https://doi.org/10.1002/ana.21076
99. Glenn JD, Smith MD, Kirby LA, Baxi EG, Whartenby KA (2015) Disparate effects of mesenchymal stem cells in experimental autoimmune encephalomyelitis and cuprizone-induced demyelination. PLoS One 10(9):e0139008. https://doi.org/10.1371/journal.pone.0139008
100. Glennie S, Soeiro I, Dyson PJ, Lam EW, Dazzi F (2005) Bone marrow mesenchymal stem cells induce division arrest anergy of activated T cells. Blood 105(7):2821–2827. https://doi.org/10.1182/blood-2004-09-3696
101. Goldman SA, Nottebohm F (1983) Neuronal production, migration, and differentiation in a vocal control nucleus of the adult female canary brain. Proc Natl Acad Sci U S A 80(8):2390–2394

102. Gomez-Nicola D, Valle-Argos B, Pallas-Bazarra N, Nieto-Sampedro M (2011) Interleukin-15 regulates proliferation and self-renewal of adult neural stem cells. Mol Biol Cell 22(12):1960–1970. https://doi.org/10.1091/mbc.E11-01-0053
103. Goodman AD (2016) Stem cell therapy for MS. Mult Scler J 22:28
104. Gordon D, Pavlovska G, Uney JB, Wraith DC, Scolding NJ (2010) Human mesenchymal stem cells infiltrate the spinal cord, reduce demyelination, and localize to white matter lesions in experimental autoimmune encephalomyelitis. J Neuropathol Exp Neurol 69(11):1087–1095. https://doi.org/10.1097/NEN.0b013e3181f97392
105. Gore A, Li Z, Fung HL, Young JE, Agarwal S, Antosiewicz-Bourget J, Canto I, Giorgetti A, Israel MA, Kiskinis E, Lee JH, Loh YH, Manos PD, Montserrat N, Panopoulos AD, Ruiz S, Wilbert ML, Yu J, Kirkness EF, Izpisua Belmonte JC, Rossi DJ, Thomson JA, Eggan K, Daley GQ, Goldstein LS, Zhang K (2011) Somatic coding mutations in human induced pluripotent stem cells. Nature 471(7336):63–67. https://doi.org/10.1038/nature09805
106. Grigoriadis N, Lourbopoulos A, Lagoudaki R, Frischer JM, Polyzoidou E, Touloumi O, Simeonidou C, Deretzi G, Kountouras J, Spandou E, Kotta K, Karkavelas G, Tascos N, Lassmann H (2011) Variable behavior and complications of autologous bone marrow mesenchymal stem cells transplanted in experimental autoimmune encephalomyelitis. Exp Neurol 230(1):78–89. https://doi.org/10.1016/j.expneurol.2011.02.021
107. Guan Y, Jiang Z, Ciric B, Rostami AM, Zhang GX (2008) Upregulation of chemokine receptor expression by IL-10/IL-4 in adult neural stem cells. Exp Mol Pathol 85(3):232–236. https://doi.org/10.1016/j.yexmp.2008.07.003
108. Guo J, Li H, Yu C, Liu F, Meng Y, Gong W, Yang H, Shen X, Ju G, Li Z, Wang J (2010) Decreased neural stem/progenitor cell proliferation in mice with chronic/nonremitting experimental autoimmune encephalomyelitis. Neurosignals 18(1):1–8. https://doi.org/10.1159/000242424
109. Guo Y, Chan KH, Lai WH, Siu CW, Kwan SC, Tse HF, Wing-Lok Ho P, Wing-Man Ho J (2013) Human mesenchymal stem cells upregulate CD1dCD5(+) regulatory B cells in experimental autoimmune encephalomyelitis. Neuroimmunomodulation 20(5):294–303. https://doi.org/10.1159/000351450
110. Halfpenny CA, Scolding NJ. Immune-modifying agents do not impair the survival, migration or proliferation of oligodendrocyte progenitors (CG-4) in vitro. J Neuroimmunol. 2003 Jun;139(1-2):9-16
111. Hammang JP, Archer DR, Duncan ID (1997) Myelination following transplantation of EGF-responsive neural stem cells into a myelin-deficient environment. Exp Neurol 147(1):84–95. https://doi.org/10.1006/exnr.1997.6592
112. Hedayatpour A, Ragerdi I, Pasbakhsh P, Kafami L, Atlasi N, Pirhajati Mahabadi V, Ghasemi S, Reza M (2013) Promotion of remyelination by adipose mesenchymal stem cell transplantation in a cuprizone model of multiple sclerosis. Cell J 15(2):142–151
113. Hentze H, Graichen R, Colman A (2007) Cell therapy and the safety of embryonic stem cell-derived grafts. Trends Biotechnol 25(1):24–32. https://doi.org/10.1016/j.tibtech.2006.10.010
114. Hoch AI, Leach JK (2014) Concise review: optimizing expansion of bone marrow mesenchymal stem/stromal cells for clinical applications. Stem Cells Transl Med 3(5):643–652. https://doi.org/10.5966/sctm.2013-0196
115. Honmou O, Felts PA, Waxman SG, Kocsis JD (1996) Restoration of normal conduction properties in demyelinated spinal cord axons in the adult rat by transplantation of exogenous Schwann cells. J Neurosci 16(10):3199–3208
116. Horwitz EM, Le Blanc K, Dominici M, Mueller I, Slaper-Cortenbach I, Marini FC, Deans RJ, Krause DS, Keating A, International Society for Cellular T (2005) Clarification of the nomenclature for MSC: the International Society for Cellular Therapy position statement. Cytotherapy 7(5):393–395. https://doi.org/10.1080/14653240500319234
117. Howell OW, Reeves CA, Nicholas R, Carassiti D, Radotra B, Gentleman SM, Serafini B, Aloisi F, Roncaroli F, Magliozzi R, Reynolds R (2011) Meningeal inflammation is widespread and linked to cortical pathology in multiple sclerosis. Brain J Neurol 134(Pt 9):2755–2771. https://doi.org/10.1093/brain/awr182

118. Hsieh J, Aimone JB, Kaspar BK, Kuwabara T, Nakashima K, Gage FH (2004) IGF-I instructs multipotent adult neural progenitor cells to become oligodendrocytes. J Cell Biol 164(1):111–122. https://doi.org/10.1083/jcb.200308101
119. Huang JK, Fancy SP, Zhao C, Rowitch DH, Ffrench-Constant C, Franklin RJ (2011) Myelin regeneration in multiple sclerosis: targeting endogenous stem cells. Neurotherapeutics 8(4):650–658. https://doi.org/10.1007/s13311-011-0065-x
120. Hussein SM, Batada NN, Vuoristo S, Ching RW, Autio R, Narva E, Ng S, Sourour M, Hamalainen R, Olsson C, Lundin K, Mikkola M, Trokovic R, Peitz M, Brustle O, Bazett-Jones DP, Alitalo K, Lahesmaa R, Nagy A, Otonkoski T (2011) Copy number variation and selection during reprogramming to pluripotency. Nature 471(7336):58–62. https://doi.org/10.1038/nature09871
121. Irvine KA, Blakemore WF (2008) Remyelination protects axons from demyelination-associated axon degeneration. Brain 131(Pt 6):1464–1477. https://doi.org/10.1093/brain/awn080
122. Ivanova NB, Dimos JT, Schaniel C, Hackney JA, Moore KA, Lemischka IR (2002) A stem cell molecular signature. Science 298(5593):601–604. https://doi.org/10.1126/science.1073823
123. Jaramillo-Merchan J, Jones J, Ivorra JL, Pastor D, Viso-Leon MC, Armengol JA, Molto MD, Geijo-Barrientos E, Martinez S (2013) Mesenchymal stromal-cell transplants induce oligodendrocyte progenitor migration and remyelination in a chronic demyelination model. Cell Death Dis 4:e779. https://doi.org/10.1038/cddis.2013.304
124. Jeffery ND, Crang AJ, O'Leary MT, Hodge SJ, Blakemore WF (1999) Behavioural consequences of oligodendrocyte progenitor cell transplantation into experimental demyelinating lesions in the rat spinal cord. Eur J Neurosci 11(5):1508–1514
125. Jiang XX, Zhang Y, Liu B, Zhang SX, Wu Y, Yu XD, Mao N (2005) Human mesenchymal stem cells inhibit differentiation and function of monocyte-derived dendritic cells. Blood 105(10):4120–4126. https://doi.org/10.1182/blood-2004-02-0586
126. Jordan JD, Ma DK, Ming GL, Song H (2007) Cellular niches for endogenous neural stem cells in the adult brain. CNS Neurol Disord Drug Targets 6(5):336–341
127. Jose AM (2002) Multiple sclerosis: can Schwann cells wrap it up? Yale J Biol Med 75(2):113–116
128. Kan I, Barhum Y, Melamed E, Offen D (2011) Mesenchymal stem cells stimulate endogenous neurogenesis in the subventricular zone of adult mice. Stem Cell Rev 7(2):404–412. https://doi.org/10.1007/s12015-010-9190-x
129. Karussis D, Karageorgiou C, Vaknin-Dembinsky A, Gowda-Kurkalli B, Gomori JM, Kassis I, Bulte JW, Petrou P, Ben-Hur T, Abramsky O, Slavin S (2010) Safety and immunological effects of mesenchymal stem cell transplantation in patients with multiple sclerosis and amyotrophic lateral sclerosis. Arch Neurol 67(10):1187–1194. https://doi.org/10.1001/archneurol.2010.248
130. Kassis I, Grigoriadis N, Gowda-Kurkalli B, Mizrachi-Kol R, Ben-Hur T, Slavin S, Abramsky O, Karussis D (2008) Neuroprotection and immunomodulation with mesenchymal stem cells in chronic experimental autoimmune encephalomyelitis. Arch Neurol 65(6):753–761. https://doi.org/10.1001/archneur.65.6.753
131. Kassis I, Petrou P, Halimi M, Karussis D (2013) Mesenchymal stem cells (MSC) derived from mice with experimental autoimmune encephalomyelitis (EAE) suppress EAE and have similar biological properties with MSC from healthy donors. Immunol Lett 154(1–2):70–76. https://doi.org/10.1016/j.imlet.2013.06.002
132. Keirstead HS, Ben-Hur T, Rogister B, O'Leary MT, Dubois-Dalcq M, Blakemore WF (1999) Polysialylated neural cell adhesion molecule-positive CNS precursors generate both oligodendrocytes and Schwann cells to remyelinate the CNS after transplantation. J Neurosci Off J Soc Neurosci 19(17):7529–7536
133. Kemp K, Gordon D, Wraith DC, Mallam E, Hartfield E, Uney J, Wilkins A, Scolding N (2011) Fusion between human mesenchymal stem cells and rodent cerebellar Purkinje cells. Neuropathol Appl Neurobiol 37(2):166–178. https://doi.org/10.1111/j.1365-2990.2010.01122.x

134. Kim J, Efe JA, Zhu S, Talantova M, Yuan X, Wang S, Lipton SA, Zhang K, Ding S (2011) Direct reprogramming of mouse fibroblasts to neural progenitors. Proc Natl Acad Sci U S A 108(19):7838–7843. https://doi.org/10.1073/pnas.1103113108
135. Kim J, Hematti P (2009) Mesenchymal stem cell-educated macrophages: a novel type of alternatively activated macrophages. Exp Hematol 37(12):1445–1453. https://doi.org/10.1016/j.exphem.2009.09.004
136. Kim SY, Cho HS, Yang SH, Shin JY, Kim JS, Lee ST, Chu K, Roh JK, Kim SU, Park CG (2009) Soluble mediators from human neural stem cells play a critical role in suppression of T-cell activation and proliferation. J Neurosci Res 87(10):2264–2272. https://doi.org/10.1002/jnr.22050
137. Kirschenbaum B, Nedergaard M, Preuss A, Barami K, Fraser RA, Goldman SA (1994) In vitro neuronal production and differentiation by precursor cells derived from the adult human forebrain. Cereb Cortex 4(6):576–589
138. Kokaia Z, Martino G, Schwartz M, Lindvall O (2012) Cross-talk between neural stem cells and immune cells: the key to better brain repair? Nat Neurosci 15(8):1078–1087. https://doi.org/10.1038/nn.3163
139. Komada Y, Yamane T, Kadota D, Isono K, Takakura N, Hayashi S, Yamazaki H (2012) Origins and properties of dental, thymic, and bone marrow mesenchymal cells and their stem cells. PLoS One 7(11):e46436. https://doi.org/10.1371/journal.pone.0046436
140. Kondo Y, Wenger DA, Gallo V, Duncan ID (2005) Galactocerebrosidase-deficient oligodendrocytes maintain stable central myelin by exogenous replacement of the missing enzyme in mice. Proc Natl Acad Sci U S A 102(51):18670–18675. https://doi.org/10.1073/pnas.0506473102
141. Koo JW, Duman RS (2008) IL-1beta is an essential mediator of the antineurogenic and anhedonic effects of stress. Proc Natl Acad Sci U S A 105(2):751–756. https://doi.org/10.1073/pnas.0708092105
142. Kornek B, Storch MK, Weissert R, Wallstroem E, Stefferl A, Olsson T, Linington C, Schmidbauer M, Lassmann H (2000) Multiple sclerosis and chronic autoimmune encephalomyelitis: a comparative quantitative study of axonal injury in active, inactive, and remyelinated lesions. Am J Pathol 157(1):267–276. https://doi.org/10.1016/S0002-9440(10)64537-3
143. Krampera M, Galipeau J, Shi Y, Tarte K, Sensebe L, Therapy MSCCotISfC (2013) Immunological characterization of multipotent mesenchymal stromal cells – the International Society for Cellular Therapy (ISCT) working proposal. Cytotherapy 15(9):1054–1061. https://doi.org/10.1016/j.jcyt.2013.02.010
144. Krampera M, Glennie S, Dyson J, Scott D, Laylor R, Simpson E, Dazzi F (2003) Bone marrow mesenchymal stem cells inhibit the response of naive and memory antigen-specific T cells to their cognate peptide. Blood 101(9):3722–3729. https://doi.org/10.1182/blood-2002-07-2104
145. Kuhlmann T, Miron V, Cui Q, Wegner C, Antel J, Bruck W (2008) Differentiation block of oligodendroglial progenitor cells as a cause for remyelination failure in chronic multiple sclerosis. Brain 131(Pt 7):1749–1758. https://doi.org/10.1093/brain/awn096
146. Kuriyan AE, Albini TA, Townsend JH, Rodriguez M, Pandya HK, Leonard RE 2nd, Parrott MB, Rosenfeld PJ, Flynn HW Jr, Goldberg JL (2017) Vision loss after intravitreal injection of autologous "stem cells" for AMD. N Engl J Med 376(11):1047–1053. https://doi.org/10.1056/NEJMoa1609583
147. Kurte M, Bravo-Alegria J, Torres A, Carrasco V, Ibanez C, Vega-Letter AM, Fernandez-O'Ryan C, Irarrazabal CE, Figueroa FE, Fuentealba RA, Riedel C, Carrion F (2015) Intravenous administration of bone marrow-derived mesenchymal stem cells induces a switch from classical to atypical symptoms in experimental autoimmune encephalomyelitis. Stem Cells Int 2015:140170. https://doi.org/10.1155/2015/140170
148. Lachapelle F, Gumpel M, Baulac M, Jacque C, Duc P, Baumann N (1983) Transplantation of CNS fragments into the brain of shiverer mutant mice: extensive myelination by implanted oligodendrocytes. I. Immunohistochemical studies. Dev Neurosci 6(6):325–334

149. Lang J, Maeda Y, Bannerman P, Xu J, Horiuchi M, Pleasure D, Guo F (2013) Adenomatous polyposis coli regulates oligodendroglial development. J Neurosci Off J Soc Neurosci 33(7):3113–3130. https://doi.org/10.1523/JNEUROSCI.3467-12.2013
150. Lanza C, Morando S, Voci A, Canesi L, Principato MC, Serpero LD, Mancardi G, Uccelli A, Vergani L (2009) Neuroprotective mesenchymal stem cells are endowed with a potent antioxidant effect in vivo. J Neurochem 110(5):1674–1684. https://doi.org/10.1111/j.1471-4159.2009.06268.x
151. Lassmann H (2008) Models of multiple sclerosis: new insights into pathophysiology and repair. Curr Opin Neurol 21(3):242–247. https://doi.org/10.1097/WCO.0b013e3282fee94a
152. Lassmann H, van Horssen J, Mahad D (2012) Progressive multiple sclerosis: pathology and pathogenesis. Nat Rev Neurol 8(11):647–656. https://doi.org/10.1038/nrneurol.2012.168
153. Le Blanc K, Tammik L, Sundberg B, Haynesworth SE, Ringden O (2003) Mesenchymal stem cells inhibit and stimulate mixed lymphocyte cultures and mitogenic responses independently of the major histocompatibility complex. Scand J Immunol 57(1):11–20
154. Lee RH, Pulin AA, Seo MJ, Kota DJ, Ylostalo J, Larson BL, Semprun-Prieto L, Delafontaine P, Prockop DJ (2009) Intravenous hMSCs improve myocardial infarction in mice because cells embolized in lung are activated to secrete the anti-inflammatory protein TSG-6. Cell Stem Cell 5(1):54–63. https://doi.org/10.1016/j.stem.2009.05.003
155. Lendahl U, Zimmerman LB, McKay RD (1990) CNS stem cells express a new class of intermediate filament protein. Cell 60(4):585–595
156. Lepelletier Y, Lecourt S, Renand A, Arnulf B, Vanneaux V, Fermand JP, Menasche P, Domet T, Marolleau JP, Hermine O, Larghero J (2010) Galectin-1 and semaphorin-3A are two soluble factors conferring T-cell immunosuppression to bone marrow mesenchymal stem cell. Stem Cells Dev 19(7):1075–1079. https://doi.org/10.1089/scd.2009.0212
157. Li YP, Paczesny S, Lauret E, Poirault S, Bordigoni P, Mekhloufi F, Hequet O, Bertrand Y, Ou-Yang JP, Stoltz JF, Miossec P, Eljaafari A (2008) Human mesenchymal stem cells license adult CD34+ hemopoietic progenitor cells to differentiate into regulatory dendritic cells through activation of the Notch pathway. J Immunol 180(3):1598–1608
158. Liang G, Zhang Y (2013) Genetic and epigenetic variations in iPSCs: potential causes and implications for application. Cell Stem Cell 13(2):149–159. https://doi.org/10.1016/j.stem.2013.07.001
159. Lie DC, Song H, Colamarino SA, Ming GL, Gage FH (2004) Neurogenesis in the adult brain: new strategies for central nervous system diseases. Annu Rev Pharmacol Toxicol 44:399–421. https://doi.org/10.1146/annurev.pharmtox.44.101802.121631
160. Lister R, Pelizzola M, Kida YS, Hawkins RD, Nery JR, Hon G, Antosiewicz-Bourget J, O'Malley R, Castanon R, Klugman S, Downes M, Yu R, Stewart R, Ren B, Thomson JA, Evans RM, Ecker JR (2011) Hotspots of aberrant epigenomic reprogramming in human induced pluripotent stem cells. Nature 471(7336):68–73. https://doi.org/10.1038/nature09798
161. Lucchinetti C, Bruck W, Parisi J, Scheithauer B, Rodriguez M, Lassmann H (2000) Heterogeneity of multiple sclerosis lesions: implications for the pathogenesis of demyelination. Ann Neurol 47(6):707–717
162. Luskin MB, Zigova T, Soteres BJ, Stewart RR (1997) Neuronal progenitor cells derived from the anterior subventricular zone of the neonatal rat forebrain continue to proliferate in vitro and express a neuronal phenotype. Mol Cell Neurosci 8(5):351–366. https://doi.org/10.1006/mcne.1996.0592
163. Luz-Crawford P, Kurte M, Bravo-Alegria J, Contreras R, Nova-Lamperti E, Tejedor G, Noel D, Jorgensen C, Figueroa F, Djouad F, Carrion F (2013) Mesenchymal stem cells generate a CD4+CD25+Foxp3+ regulatory T cell population during the differentiation process of Th1 and Th17 cells. Stem Cell Res Ther 4(3):65. https://doi.org/10.1186/scrt216
164. Mahad DH, Trapp BD, Lassmann H (2015) Pathological mechanisms in progressive multiple sclerosis. Lancet Neurol 14(2):183–193. https://doi.org/10.1016/S1474-4422(14)70256-X
165. Mallam E, Kemp K, Wilkins A, Rice C, Scolding N (2010) Characterization of in vitro expanded bone marrow-derived mesenchymal stem cells from patients with multiple sclerosis. Mult Scler 16(8):909–918. https://doi.org/10.1177/1352458510371959

166. Mallucci G, Peruzzotti-Jametti L, Bernstock JD, Pluchino S (2015) The role of immune cells, glia and neurons in white and gray matter pathology in multiple sclerosis. Prog Neurobiol 127-128:1–22. https://doi.org/10.1016/j.pneurobio.2015.02.003
167. Mantovani A, Biswas SK, Galdiero MR, Sica A, Locati M (2013) Macrophage plasticity and polarization in tissue repair and remodelling. J Pathol 229(2):176–185. https://doi.org/10.1002/path.4133
168. Marin-Banasco C, Suardiaz Garcia M, Hurtado Guerrero I, Maldonado Sanchez R, Estivill-Torrus G, Leyva Fernandez L, Fernandez Fernandez O (2014) Mesenchymal properties of SJL mice-stem cells and their efficacy as autologous therapy in a relapsing-remitting multiple sclerosis model. Stem Cell Res Ther 5(6):134. https://doi.org/10.1186/scrt524
169. Martino G (2004) How the brain repairs itself: new therapeutic strategies in inflammatory and degenerative CNS disorders. Lancet Neurol 3(6):372–378. https://doi.org/10.1016/S1474-4422(04)00771-9
170. Martino G, Pluchino S (2006) The therapeutic potential of neural stem cells. Nat Rev Neurosci 7(5):395–406. https://doi.org/10.1038/nrn1908
171. Martino G, Pluchino S, Bonfanti L, Schwartz M (2011) Brain regeneration in physiology and pathology: the immune signature driving therapeutic plasticity of neural stem cells. Physiol Rev 91(4):1281–1304. https://doi.org/10.1152/physrev.00032.2010
172. Maslov AY, Barone TA, Plunkett RJ, Pruitt SC (2004) Neural stem cell detection, characterization, and age-related changes in the subventricular zone of mice. J Neurosci Off J Soc Neurosci 24(7):1726–1733. https://doi.org/10.1523/JNEUROSCI.4608-03.2004
173. Mazzanti B, Aldinucci A, Biagioli T, Barilaro A, Urbani S, Dal Pozzo S, Amato MP, Siracusa G, Crescioli C, Manuelli C, Bosi A, Saccardi R, Massacesi L, Ballerini C (2008) Differences in mesenchymal stem cell cytokine profiles between MS patients and healthy donors: implication for assessment of disease activity and treatment. J Neuroimmunol 199(1–2):142–150. https://doi.org/10.1016/j.jneuroim.2008.05.006
174. Meinl E, Krumbholz M, Derfuss T, Junker A, Hohlfeld R (2008) Compartmentalization of inflammation in the CNS: a major mechanism driving progressive multiple sclerosis. J Neurol Sci 274(1–2):42–44. https://doi.org/10.1016/j.jns.2008.06.032
175. Melief SM, Geutskens SB, Fibbe WE, Roelofs H (2013) Multipotent stromal cells skew monocytes towards an anti-inflammatory interleukin-10-producing phenotype by production of interleukin-6. Haematologica 98(6):888–895. https://doi.org/10.3324/haematol.2012.078055
176. Mendes SC, Robin C, Dzierzak E (2005) Mesenchymal progenitor cells localize within hematopoietic sites throughout ontogeny. Development 132(5):1127–1136. https://doi.org/10.1242/dev.01615
177. Menn B, Garcia-Verdugo JM, Yaschine C, Gonzalez-Perez O, Rowitch D, Alvarez-Buylla A (2006) Origin of oligodendrocytes in the subventricular zone of the adult brain. J Neurosci 26(30):7907–7918. https://doi.org/10.1523/JNEUROSCI.1299-06.2006
178. Meyer S, Worsdorfer P, Gunther K, Thier M, Edenhofer F (2015) Derivation of adult human fibroblasts and their direct conversion into expandable neural progenitor cells. J Vis Exp 101:e52831. https://doi.org/10.3791/52831
179. Mi R, Luo Y, Cai J, Limke TL, Rao MS, Hoke A (2005) Immortalized neural stem cells differ from nonimmortalized cortical neurospheres and cerebellar granule cell progenitors. Exp Neurol 194(2):301–319. https://doi.org/10.1016/j.expneurol.2004.07.011
180. Ming GL, Song H (2005) Adult neurogenesis in the mammalian central nervous system. Annu Rev Neurosci 28:223–250. https://doi.org/10.1146/annurev.neuro.28.051804.101459
181. Miron VE, Boyd A, Zhao JW, Yuen TJ, Ruckh JM, Shadrach JL, van Wijngaarden P, Wagers AJ, Williams A, Franklin RJ, ffrench-Constant C (2013) M2 microglia and macrophages drive oligodendrocyte differentiation during CNS remyelination. Nat Neurosci 16(9):1211–1218. https://doi.org/10.1038/nn.3469
182. Mitra NK, Bindal U, Eng Hwa W, Chua CL, Tan CY (2015) Evaluation of locomotor function and microscopic structure of the spinal cord in a mouse model of experimental autoimmune encephalomyelitis following treatment with syngeneic mesenchymal stem cells. Int J Clin Exp Pathol 8(10):12041–12052

183. Moalem G, Leibowitz-Amit R, Yoles E, Mor F, Cohen IR, Schwartz M (1999) Autoimmune T cells protect neurons from secondary degeneration after central nervous system axotomy. Nat Med 5(1):49–55. https://doi.org/10.1038/4734
184. Monje ML, Toda H, Palmer TD (2003) Inflammatory blockade restores adult hippocampal neurogenesis. Science 302(5651):1760–1765. https://doi.org/10.1126/science.1088417
185. Mosna F, Sensebe L, Krampera M (2010) Human bone marrow and adipose tissue mesenchymal stem cells: a user's guide. Stem Cells Dev 19(10):1449–1470. https://doi.org/10.1089/scd.2010.0140
186. Mosser DM, Edwards JP (2008) Exploring the full spectrum of macrophage activation. Nat Rev Immunol 8(12):958–969. https://doi.org/10.1038/nri2448
187. Mothe AJ, Tator CH (2008) Transplanted neural stem/progenitor cells generate myelinating oligodendrocytes and Schwann cells in spinal cord demyelination and dysmyelination. Exp Neurol 213(1):176–190. https://doi.org/10.1016/j.expneurol.2008.05.024
188. Moyon S, Dubessy AL, Aigrot MS, Trotter M, Huang JK, Dauphinot L, Potier MC, Kerninon C, Melik Parsadaniantz S, Franklin RJ, Lubetzki C (2015) Demyelination causes adult CNS progenitors to revert to an immature state and express immune cues that support their migration. J Neurosci Off J Soc Neurosci 35(1):4–20. https://doi.org/10.1523/JNEUROSCI.0849-14.2015
189. Mozafari S, Laterza C, Roussel D, Bachelin C, Marteyn A, Deboux C, Martino G, Baron-Van Evercooren A (2015) Skin-derived neural precursors competitively generate functional myelin in adult demyelinated mice. J Clin Invest 125(9):3642–3656. https://doi.org/10.1172/JCI80437
190. Mueller FJ, Serobyan N, Schraufstatter IU, DiScipio R, Wakeman D, Loring JF, Snyder EY, Khaldoyanidi SK (2006) Adhesive interactions between human neural stem cells and inflamed human vascular endothelium are mediated by integrins. Stem Cells 24(11):2367–2372. https://doi.org/10.1634/stemcells.2005-0568
191. Muller FJ, Snyder EY, Loring JF (2006) Gene therapy: can neural stem cells deliver? Nat Rev Neurosci 7(1):75–84. https://doi.org/10.1038/nrn1829
192. Munoz JR, Stoutenger BR, Robinson AP, Spees JL, Prockop DJ (2005) Human stem/progenitor cells from bone marrow promote neurogenesis of endogenous neural stem cells in the hippocampus of mice. Proc Natl Acad Sci U S A 102(50):18171–18176. https://doi.org/10.1073/pnas.0508945102
193. Muraro PA, Abrahamsson SV (2010) Resetting autoimmunity in the nervous system: the role of hematopoietic stem cell transplantation. Curr Opin Investig Drugs 11(11):1265–1275
194. Murphy MB, Moncivais K, Caplan AI (2013) Mesenchymal stem cells: environmentally responsive therapeutics for regenerative medicine. Exp Mol Med 45:e54. https://doi.org/10.1038/emm.2013.94
195. Najm FJ, Lager AM, Zaremba A, Wyatt K, Caprariello AV, Factor DC, Karl RT, Maeda T, Miller RH, Tesar PJ (2013) Transcription factor-mediated reprogramming of fibroblasts to expandable, myelinogenic oligodendrocyte progenitor cells. Nat Biotechnol 31(5):426–433. https://doi.org/10.1038/nbt.2561
196. Nash B, Ioannidou K, Barnett SC (2011) Astrocyte phenotypes and their relationship to myelination. J Anat 219(1):44–52. https://doi.org/10.1111/j.1469-7580.2010.01330.x
197. Nauta AJ, Kruisselbrink AB, Lurvink E, Willemze R, Fibbe WE (2006) Mesenchymal stem cells inhibit generation and function of both CD34+-derived and monocyte-derived dendritic cells. J Immunol 177(4):2080–2087
198. Nave KA (2010) Myelination and the trophic support of long axons. Nat Rev Neurosci 11(4):275–283. https://doi.org/10.1038/nrn2797
199. Nemeth K, Leelahavanichkul A, Yuen PS, Mayer B, Parmelee A, Doi K, Robey PG, Leelahavanichkul K, Koller BH, Brown JM, Hu X, Jelinek I, Star RA, Mezey E (2009) Bone marrow stromal cells attenuate sepsis via prostaglandin E(2)-dependent reprogramming of host macrophages to increase their interleukin-10 production. Nat Med 15(1):42–49. https://doi.org/10.1038/nm.1905

200. Neri M, Maderna C, Ferrari D, Cavazzin C, Vescovi AL, Gritti A (2010) Robust generation of oligodendrocyte progenitors from human neural stem cells and engraftment in experimental demyelination models in mice. PLoS One 5(4):e10145. https://doi.org/10.1371/journal.pone.0010145
201. Nishiyama A, Lin XH, Giese N, Heldin CH, Stallcup WB (1996) Interaction between NG2 proteoglycan and PDGF alpha-receptor on O2A progenitor cells is required for optimal response to PDGF. J Neurosci Res 43(3):315–330. https://doi.org/10.1002/(SICI)1097-4547(19960201)43:3<315::AID-JNR6>3.0.CO;2-M
202. Nottebohm F (1985) Neuronal replacement in adulthood. Ann N Y Acad Sci 457:143–161
203. O'Leary MT, Blakemore WF (1997) Oligodendrocyte precursors survive poorly and do not migrate following transplantation into the normal adult central nervous system. J Neurosci Res 48(2):159–167
204. Odoardi F, Sie C, Streyl K, Ulaganathan VK, Schlager C, Lodygin D, Heckelsmiller K, Nietfeld W, Ellwart J, Klinkert WE, Lottaz C, Nosov M, Brinkmann V, Spang R, Lehrach H, Vingron M, Wekerle H, Flugel-Koch C, Flugel A (2012) T cells become licensed in the lung to enter the central nervous system. Nature 488(7413):675–679. https://doi.org/10.1038/nature11337
205. Okita K, Ichisaka T, Yamanaka S (2007) Generation of germline-competent induced pluripotent stem cells. Nature 448(7151):313–317. https://doi.org/10.1038/nature05934
206. Ontaneda D, Fox RJ, Chataway J (2015) Clinical trials in progressive multiple sclerosis: lessons learned and future perspectives. Lancet Neurol 14(2):208–223. https://doi.org/10.1016/S1474-4422(14)70264-9
207. Papadaki HA, Tsagournisakis M, Mastorodemos V, Pontikoglou C, Damianaki A, Pyrovolaki K, Stamatopoulos K, Fassas A, Plaitakis A, Eliopoulos GD (2005) Normal bone marrow hematopoietic stem cell reserves and normal stromal cell function support the use of autologous stem cell transplantation in patients with multiple sclerosis. Bone Marrow Transplant 36(12):1053–1063. https://doi.org/10.1038/sj.bmt.1705179
208. Payne NL, Sun G, Herszfeld D, Tat-Goh PA, Verma PJ, Parkington HC, Coleman HA, Tonta MA, Siatskas C, Bernard CC (2012) Comparative study on the therapeutic potential of neurally differentiated stem cells in a mouse model of multiple sclerosis. PLoS One 7(4):e35093. https://doi.org/10.1371/journal.pone.0035093
209. Payne NL, Sun G, McDonald C, Layton D, Moussa L, Emerson-Webber A, Veron N, Siatskas C, Herszfeld D, Price J, Bernard CC (2013) Distinct immunomodulatory and migratory mechanisms underpin the therapeutic potential of human mesenchymal stem cells in autoimmune demyelination. Cell Transplant 22(8):1409–1425. https://doi.org/10.3727/096368912X657620
210. Pearl JI, Kean LS, Davis MM, Wu JC (2012) Pluripotent stem cells: immune to the immune system? Sci Trans Med 4(164):164ps125. https://doi.org/10.1126/scitranslmed.3005090
211. Pincus DW, Keyoung HM, Harrison-Restelli C, Goodman RR, Fraser RA, Edgar M, Sakakibara S, Okano H, Nedergaard M, Goldman SA (1998) Fibroblast growth factor-2/brain-derived neurotrophic factor-associated maturation of new neurons generated from adult human subependymal cells. Ann Neurol 43(5):576–585. https://doi.org/10.1002/ana.410430505
212. Pluchino S, Cossetti C (2013) How stem cells speak with host immune cells in inflammatory brain diseases. Glia 61(9):1379–1401. https://doi.org/10.1002/glia.22500
213. Pluchino S, Gritti A, Blezer E, Amadio S, Brambilla E, Borsellino G, Cossetti C, Del Carro U, Comi G, t Hart B, Vescovi A, Martino G (2009) Human neural stem cells ameliorate autoimmune encephalomyelitis in non-human primates. Ann Neurol 66(3):343–354. https://doi.org/10.1002/ana.21745
214. Pluchino S, Martino G (2008) The therapeutic plasticity of neural stem/precursor cells in multiple sclerosis. J Neurol Sci 265(1–2):105–110. https://doi.org/10.1016/j.jns.2007.07.020
215. Pluchino S, Quattrini A, Brambilla E, Gritti A, Salani G, Dina G, Galli R, Del Carro U, Amadio S, Bergami A, Furlan R, Comi G, Vescovi AL, Martino G (2003) Injection of

adult neurospheres induces recovery in a chronic model of multiple sclerosis. Nature 422(6933):688–694. https://doi.org/10.1038/nature01552

216. Pluchino S, Zanotti L, Brambilla E, Rovere-Querini P, Capobianco A, Alfaro-Cervello C, Salani G, Cossetti C, Borsellino G, Battistini L, Ponzoni M, Doglioni C, Garcia-Verdugo JM, Comi G, Manfredi AA, Martino G (2009) Immune regulatory neural stem/precursor cells protect from central nervous system autoimmunity by restraining dendritic cell function. PLoS One 4(6):e5959. https://doi.org/10.1371/journal.pone.0005959
217. Pluchino S, Zanotti L, Rossi B, Brambilla E, Ottoboni L, Salani G, Martinello M, Cattalini A, Bergami A, Furlan R, Comi G, Constantin G, Martino G (2005) Neurosphere-derived multipotent precursors promote neuroprotection by an immunomodulatory mechanism. Nature 436(7048):266–271. https://doi.org/10.1038/nature03889
218. Politi LS, Bacigaluppi M, Brambilla E, Cadioli M, Falini A, Comi G, Scotti G, Martino G, Pluchino S (2007) Magnetic-resonance-based tracking and quantification of intravenously injected neural stem cell accumulation in the brains of mice with experimental multiple sclerosis. Stem Cells 25(10):2583–2592. https://doi.org/10.1634/stemcells.2007-0037
219. Prevosto C, Zancolli M, Canevali P, Zocchi MR, Poggi A (2007) Generation of CD4+ or CD8+ regulatory T cells upon mesenchymal stem cell-lymphocyte interaction. Haematologica 92(7):881–888
220. Rafei M, Campeau PM, Aguilar-Mahecha A, Buchanan M, Williams P, Birman E, Yuan S, Young YK, Boivin MN, Forner K, Basik M, Galipeau J (2009) Mesenchymal stromal cells ameliorate experimental autoimmune encephalomyelitis by inhibiting CD4 Th17 T cells in a CC chemokine ligand 2-dependent manner. J Immunol 182(10):5994–6002. https://doi.org/10.4049/jimmunol.0803962
221. Raff MC (1989) Glial cell diversification in the rat optic nerve. Science 243(4897):1450–1455
222. Raff MC, Miller RH, Noble M (1983) A glial progenitor cell that develops in vitro into an astrocyte or an oligodendrocyte depending on culture medium. Nature 303(5916):390–396
223. Raff MC, Mirsky R, Fields KL, Lisak RP, Dorfman SH, Silberberg DH, Gregson NA, Leibowitz S, Kennedy MC (1978) Galactocerebroside is a specific cell-surface antigenic marker for oligodendrocytes in culture. Nature 274(5673):813–816
224. Ramos-Zuniga R, Gonzalez-Perez O, Macias-Ornelas A, Capilla-Gonzalez V, Quinones-Hinojosa A (2012) Ethical implications in the use of embryonic and adult neural stem cells. Stem Cells Int 2012:470949. https://doi.org/10.1155/2012/470949
225. Ransohoff RM (2012) Animal models of multiple sclerosis: the good, the bad and the bottom line. Nat Neurosci 15(8):1074–1077. https://doi.org/10.1038/nn.3168
226. Rasmusson I, Le Blanc K, Sundberg B, Ringden O (2007) Mesenchymal stem cells stimulate antibody secretion in human B cells. Scand J Immunol 65(4):336–343. https://doi.org/10.1111/j.1365-3083.2007.01905.x
227. Rasmusson I, Ringden O, Sundberg B, Le Blanc K (2003) Mesenchymal stem cells inhibit the formation of cytotoxic T lymphocytes, but not activated cytotoxic T lymphocytes or natural killer cells. Transplantation 76(8):1208–1213. https://doi.org/10.1097/01.TP.0000082540.43730.80
228. Ren G, Zhao X, Zhang L, Zhang J, L'Huillier A, Ling W, Roberts AI, Le AD, Shi S, Shao C, Shi Y (2010) Inflammatory cytokine-induced intercellular adhesion molecule-1 and vascular cell adhesion molecule-1 in mesenchymal stem cells are critical for immunosuppression. J Immunol 184(5):2321–2328. https://doi.org/10.4049/jimmunol.0902023
229. Reynolds BA, Tetzlaff W, Weiss S (1992) A multipotent EGF-responsive striatal embryonic progenitor cell produces neurons and astrocytes. J Neurosci 12(11):4565–4574
230. Reynolds BA, Weiss S (1992) Generation of neurons and astrocytes from isolated cells of the adult mammalian central nervous system. Science 255(5052):1707–1710
231. Ricci-Vitiani L, Lombardi DG, Signore M, Biffoni M, Pallini R, Parati E, Peschle C, De Maria R (2007) Human neural progenitor cells display limited cytotoxicity and increased oligodendrogenesis during inflammation. Cell Death Differ 14(4):876–878. https://doi.org/10.1038/sj.cdd.4402078

232. Rice CM, Mallam EA, Whone AL, Walsh P, Brooks DJ, Kane N, Butler SR, Marks DI, Scolding NJ (2010) Safety and feasibility of autologous bone marrow cellular therapy in relapsing-progressive multiple sclerosis. Clin Pharmacol Ther 87(6):679–685. https://doi.org/10.1038/clpt.2010.44
233. Rice CM, Marks DI, Ben-Shlomo Y, Evangelou N, Morgan PS, Metcalfe C, Walsh P, Kane NM, Guttridge MG, Miflin G, Blackmore S, Sarkar P, Redondo J, Owen D, Cottrell DA, Wilkins A, Scolding NJ (2015) Assessment of bone marrow-derived Cellular Therapy in progressive Multiple Sclerosis (ACTiMuS): study protocol for a randomised controlled trial. Trials 16:463. https://doi.org/10.1186/s13063-015-0953-1
234. Rice CM, Marks DI, Walsh P, Kane NM, Guttridge MG, Redondo J, Sarkar P, Owen D, Wilkins A, Scolding NJ (2015) Repeat infusion of autologous bone marrow cells in multiple sclerosis: protocol for a phase I extension study (SIAMMS-II). BMJ Open 5(9):e009090. https://doi.org/10.1136/bmjopen-2015-009090
235. Richardson WD, Pringle N, Mosley MJ, Westermark B, Dubois-Dalcq M (1988) A role for platelet-derived growth factor in normal gliogenesis in the central nervous system. Cell 53(2):309–319
236. Rosenbluth J, Liu Z, Guo D, Schiff R (1993) Myelin formation by mouse glia in myelin-deficient rats treated with cyclosporine. J Neurocytol 22(11):967–977
237. Roy NS, Cleren C, Singh SK, Yang L, Beal MF, Goldman SA (2006) Functional engraftment of human ES cell-derived dopaminergic neurons enriched by coculture with telomerase-immortalized midbrain astrocytes. Nat Med 12(11):1259–1268. https://doi.org/10.1038/nm1495
238. Roy NS, Wang S, Harrison-Restelli C, Benraiss A, Fraser RA, Gravel M, Braun PE, Goldman SA (1999) Identification, isolation, and promoter-defined separation of mitotic oligodendrocyte progenitor cells from the adult human subcortical white matter. J Neurosci Off J Soc Neurosci 19(22):9986–9995
239. Rozenberg A, Rezk A, Boivin MN, Darlington PJ, Nyirenda M, Li R, Jalili F, Winer R, Artsy EA, Uccelli A, Reese JS, Planchon SM, Cohen JA, Bar-Or A (2016) Human mesenchymal stem cells impact Th17 and Th1 responses through a prostaglandin E2 and myeloid-dependent mechanism. Stem Cells Transl Med. https://doi.org/10.5966/sctm.2015-0243
240. Saccardi R, Freedman MS, Sormani MP, Atkins H, Farge D, Griffith LM, Kraft G, Mancardi GL, Nash R, Pasquini M, Martin R, Muraro PA, European B, Marrow Transplantation G, Center for International B, Marrow R, Group HiMIS (2012) A prospective, randomized, controlled trial of autologous haematopoietic stem cell transplantation for aggressive multiple sclerosis: a position paper. Mult Scler 18(6):825–834. https://doi.org/10.1177/1352458512438454
241. Sacchetti B, Funari A, Michienzi S, Di Cesare S, Piersanti S, Saggio I, Tagliafico E, Ferrari S, Robey PG, Riminucci M, Bianco P (2007) Self-renewing osteoprogenitors in bone marrow sinusoids can organize a hematopoietic microenvironment. Cell 131(2):324–336. https://doi.org/10.1016/j.cell.2007.08.025
242. Sacchetti B, Funari A, Remoli C, Giannicola G, Kogler G, Liedtke S, Cossu G, Serafini M, Sampaolesi M, Tagliafico E, Tenedini E, Saggio I, Robey PG, Riminucci M, Bianco P (2016) No identical "mesenchymal stem cells" at different times and sites: human committed progenitors of distinct origin and differentiation potential are incorporated as adventitial cells in microvessels. Stem Cell Reports 6(6):897–913. https://doi.org/10.1016/j.stemcr.2016.05.011
243. Sakakibara S, Imai T, Hamaguchi K, Okabe M, Aruga J, Nakajima K, Yasutomi D, Nagata T, Kurihara Y, Uesugi S, Miyata T, Ogawa M, Mikoshiba K, Okano H (1996) Mouse-Musashi-1, a neural RNA-binding protein highly enriched in the mammalian CNS stem cell. Dev Biol 176(2):230–242
244. Sato K, Ozaki K, Oh I, Meguro A, Hatanaka K, Nagai T, Muroi K, Ozawa K (2007) Nitric oxide plays a critical role in suppression of T-cell proliferation by mesenchymal stem cells. Blood 109(1):228–234. https://doi.org/10.1182/blood-2006-02-002246
245. Scolding N, Barnes D, Cader S, Chataway J, Chaudhuri A, Coles A, Giovannoni G, Miller D, Rashid W, Schmierer K, Shehu A, Silber E, Young C, Zajicek J (2015) Association of British

Neurologists: revised (2015) guidelines for prescribing disease-modifying treatments in multiple sclerosis. Pract Neurol 15(4):273–279. https://doi.org/10.1136/practneurol-2015-001139
246. Scolding NJ, Pasquini M, Reingold SC, Cohen JA (2017) Cell-based therapeutic strategies for multiple sclerosis. Brain J Neurol. https://doi.org/10.1093/brain/awx154
247. Selai CE, Trimble MR (1997) Patient's assessments of disability in multiple sclerosis. Most patients have difficulty in rating themselves on visual analogue scales. BMJ 315(7118):1305–1306
248. Severino V, Alessio N, Farina A, Sandomenico A, Cipollaro M, Peluso G, Galderisi U, Chambery A (2013) Insulin-like growth factor binding proteins 4 and 7 released by senescent cells promote premature senescence in mesenchymal stem cells. Cell Death Dis 4:e911. https://doi.org/10.1038/cddis.2013.445
249. Shechter R, London A, Varol C, Raposo C, Cusimano M, Yovel G, Rolls A, Mack M, Pluchino S, Martino G, Jung S, Schwartz M (2009) Infiltrating blood-derived macrophages are vital cells playing an anti-inflammatory role in recovery from spinal cord injury in mice. PLoS Med 6(7):e1000113. https://doi.org/10.1371/journal.pmed.1000113
250. Shi J, Marinovich A, Barres BA (1998) Purification and characterization of adult oligodendrocyte precursor cells from the rat optic nerve. J Neurosci Off J Soc Neurosci 18(12):4627–4636
251. Sim FJ, McClain CR, Schanz SJ, Protack TL, Windrem MS, Goldman SA (2011) CD140a identifies a population of highly myelinogenic, migration-competent and efficiently engrafting human oligodendrocyte progenitor cells. Nat Biotechnol 29(10):934–941. https://doi.org/10.1038/nbt.1972
252. Skihar V, Silva C, Chojnacki A, Doring A, Stallcup WB, Weiss S, Yong VW (2009) Promoting oligodendrogenesis and myelin repair using the multiple sclerosis medication glatiramer acetate. Proc Natl Acad Sci U S A 106(42):17992–17997. https://doi.org/10.1073/pnas.0909607106
253. Smith KJ, Blakemore WF, McDonald WI (1979) Central remyelination restores secure conduction. Nature 280(5721):395–396
254. Soares S, Sotelo C (2004) Adult neural stem cells from the mouse subventricular zone are limited in migratory ability compared to progenitor cells of similar origin. Neuroscience 128(4):807–817. https://doi.org/10.1016/j.neuroscience.2004.07.031
255. Spaggiari GM, Capobianco A, Becchetti S, Mingari MC, Moretta L (2006) Mesenchymal stem cell-natural killer cell interactions: evidence that activated NK cells are capable of killing MSCs, whereas MSCs can inhibit IL-2-induced NK-cell proliferation. Blood 107(4):1484–1490. https://doi.org/10.1182/blood-2005-07-2775
256. Sun Y, Deng W, Geng L, Zhang L, Liu R, Chen W, Yao G, Zhang H, Feng X, Gao X, Sun L (2015) Mesenchymal stem cells from patients with rheumatoid arthritis display impaired function in inhibiting Th17 cells. J Immunol Res 2015:284215. https://doi.org/10.1155/2015/284215
257. Takahashi K, Tanabe K, Ohnuki M, Narita M, Ichisaka T, Tomoda K, Yamanaka S (2007) Induction of pluripotent stem cells from adult human fibroblasts by defined factors. Cell 131(5):861–872. https://doi.org/10.1016/j.cell.2007.11.019
258. Takahashi K, Yamanaka S (2006) Induction of pluripotent stem cells from mouse embryonic and adult fibroblast cultures by defined factors. Cell 126(4):663–676. https://doi.org/10.1016/j.cell.2006.07.024
259. Teng YD, Lavik EB, Qu X, Park KI, Ourednik J, Zurakowski D, Langer R, Snyder EY (2002) Functional recovery following traumatic spinal cord injury mediated by a unique polymer scaffold seeded with neural stem cells. Proc Natl Acad Sci U S A 99(5):3024–3029. https://doi.org/10.1073/pnas.052678899
260. Thier M, Worsdorfer P, Lakes YB, Gorris R, Herms S, Opitz T, Seiferling D, Quandel T, Hoffmann P, Nothen MM, Brustle O, Edenhofer F (2012) Direct conversion of fibroblasts into stably expandable neural stem cells. Cell Stem Cell 10(4):473–479. https://doi.org/10.1016/j.stem.2012.03.003

261. Thiruvalluvan A, Czepiel M, Kap YA, Mantingh-Otter I, Vainchtein I, Kuipers J, Bijlard M, Baron W, Giepmans B, Bruck W, t Hart BA, Boddeke E, Copray S (2016) Survival and functionality of human induced pluripotent stem cell-derived oligodendrocytes in a nonhuman primate model for multiple sclerosis. Stem Cells Transl Med 5(11):1550–1561. https://doi.org/10.5966/sctm.2016-0024
262. Thomas JL, Spassky N, Perez Villegas EM, Olivier C, Cobos I, Goujet-Zalc C, Martinez S, Zalc B (2000) Spatiotemporal development of oligodendrocytes in the embryonic brain. J Neurosci Res 59(4):471–476. https://doi.org/10.1002/(SICI)1097-4547(20000215)59:4<471::AID-JNR1>3.0.CO;2-3
263. Tse WT, Pendleton JD, Beyer WM, Egalka MC, Guinan EC (2003) Suppression of allogeneic T-cell proliferation by human marrow stromal cells: implications in transplantation. Transplantation 75(3):389–397. https://doi.org/10.1097/01.TP.0000045055.63901.A9
264. Uccelli A, Moretta L, Pistoia V (2008) Mesenchymal stem cells in health and disease. Nat Rev Immunol 8(9):726–736. https://doi.org/10.1038/nri2395
265. Van Harmelen V, Rohrig K, Hauner H (2004) Comparison of proliferation and differentiation capacity of human adipocyte precursor cells from the omental and subcutaneous adipose tissue depot of obese subjects. Metabolism 53(5):632–637
266. Volpe G, Bernstock JD, Peruzzotti-Jametti L, Pluchino S (2016) Modulation of host immune responses following non-hematopoietic stem cell transplantation: translational implications in progressive multiple sclerosis. J Neuroimmunol. https://doi.org/10.1016/j.jneuroim.2016.12.005
267. von Bahr L, Batsis I, Moll G, Hagg M, Szakos A, Sundberg B, Uzunel M, Ringden O, Le Blanc K (2012) Analysis of tissues following mesenchymal stromal cell therapy in humans indicates limited long-term engraftment and no ectopic tissue formation. Stem Cells 30(7):1575–1578. https://doi.org/10.1002/stem.1118
268. Wang L, Shi J, van Ginkel FW, Lan L, Niemeyer G, Martin DR, Snyder EY, Cox NR (2009) Neural stem/progenitor cells modulate immune responses by suppressing T lymphocytes with nitric oxide and prostaglandin E2. Exp Neurol 216(1):177–183. https://doi.org/10.1016/j.expneurol.2008.11.017
269. Wang S, Bates J, Li X, Schanz S, Chandler-Militello D, Levine C, Maherali N, Studer L, Hochedlinger K, Windrem M, Goldman SA (2013) Human iPSC-derived oligodendrocyte progenitor cells can myelinate and rescue a mouse model of congenital hypomyelination. Cell Stem Cell 12(2):252–264. https://doi.org/10.1016/j.stem.2012.12.002
270. Weiss S, Reynolds BA, Vescovi AL, Morshead C, Craig CG, van der Kooy D (1996) Is there a neural stem cell in the mammalian forebrain? Trends Neurosci 19(9):387–393
271. Windrem MS, Nunes MC, Rashbaum WK, Schwartz TH, Goodman RA, McKhann G 2nd, Roy NS, Goldman SA (2004) Fetal and adult human oligodendrocyte progenitor cell isolates myelinate the congenitally dysmyelinated brain. Nat Med 10(1):93–97. https://doi.org/10.1038/nm974
272. Windrem MS, Roy NS, Wang J, Nunes M, Benraiss A, Goodman R, McKhann GM 2nd, Goldman SA (2002) Progenitor cells derived from the adult human subcortical white matter disperse and differentiate as oligodendrocytes within demyelinated lesions of the rat brain. J Neurosci Res 69(6):966–975. https://doi.org/10.1002/jnr.10397
273. Windrem MS, Schanz SJ, Guo M, Tian GF, Washco V, Stanwood N, Rasband M, Roy NS, Nedergaard M, Havton LA, Wang S, Goldman SA (2008) Neonatal chimerization with human glial progenitor cells can both remyelinate and rescue the otherwise lethally hypomyelinated shiverer mouse. Cell Stem Cell 2(6):553–565. https://doi.org/10.1016/j.stem.2008.03.020
274. Windrem MS, Schanz SJ, Morrow C, Munir J, Chandler-Militello D, Wang S, Goldman SA (2014) A competitive advantage by neonatally engrafted human glial progenitors yields mice whose brains are chimeric for human glia. J Neurosci 34(48):16153–16161. https://doi.org/10.1523/JNEUROSCI.1510-14.2014
275. Wolswijk G (1998) Chronic stage multiple sclerosis lesions contain a relatively quiescent population of oligodendrocyte precursor cells. J Neurosci 18(2):601–609

276. Woodbury D, Schwarz EJ, Prockop DJ, Black IB (2000) Adult rat and human bone marrow stromal cells differentiate into neurons. J Neurosci Res 61(4):364–370. https://doi.org/10.1002/1097-4547(20000815)61:4<364::AID-JNR2>3.0.CO;2-C
277. Yamout B, Hourani R, Salti H, Barada W, El-Hajj T, Al-Kutoubi A, Herlopian A, Baz EK, Mahfouz R, Khalil-Hamdan R, Kreidieh NM, El-Sabban M, Bazarbachi A (2010) Bone marrow mesenchymal stem cell transplantation in patients with multiple sclerosis: a pilot study. J Neuroimmunol 227(1–2):185–189. https://doi.org/10.1016/j.jneuroim.2010.07.013
278. Yaochite JN, de Lima KW, Caliari-Oliveira C, Palma PV, Couri CE, Simoes BP, Covas DT, Voltarelli JC, Oliveira MC, Donadi EA, Malmegrim KC (2016) Multipotent mesenchymal stromal cells from patients with newly diagnosed type 1 diabetes mellitus exhibit preserved in vitro and in vivo immunomodulatory properties. Stem Cell Res Ther 7:14. https://doi.org/10.1186/s13287-015-0261-4
279. Zappia E, Casazza S, Pedemonte E, Benvenuto F, Bonanni I, Gerdoni E, Giunti D, Ceravolo A, Cazzanti F, Frassoni F, Mancardi G, Uccelli A (2005) Mesenchymal stem cells ameliorate experimental autoimmune encephalomyelitis inducing T-cell anergy. Blood 106(5):1755–1761. https://doi.org/10.1182/blood-2005-04-1496
280. Zhang SC, Ge B, Duncan ID (1999) Adult brain retains the potential to generate oligodendroglial progenitors with extensive myelination capacity. Proc Natl Acad Sci U S A 96(7):4089–4094
281. Zhang SC, Wernig M, Duncan ID, Brustle O, Thomson JA (2001) In vitro differentiation of transplantable neural precursors from human embryonic stem cells. Nat Biotechnol 19(12):1129–1133. https://doi.org/10.1038/nbt1201-1129
282. Zheng G, Ge M, Qiu G, Shu Q, Xu J (2015) Mesenchymal stromal cells affect disease outcomes via macrophage polarization. Stem Cells Int 2015:989473. https://doi.org/10.1155/2015/989473
283. Ziv Y, Avidan H, Pluchino S, Martino G, Schwartz M (2006) Synergy between immune cells and adult neural stem/progenitor cells promotes functional recovery from spinal cord injury. Proc Natl Acad Sci U S A 103(35):13174–13179. https://doi.org/10.1073/pnas.0603747103
284. Ziv Y, Ron N, Butovsky O, Landa G, Sudai E, Greenberg N, Cohen H, Kipnis J, Schwartz M (2006) Immune cells contribute to the maintenance of neurogenesis and spatial learning abilities in adulthood. Nat Neurosci 9(2):268–275. https://doi.org/10.1038/nn1629
285. Zuk PA, Zhu M, Mizuno H, Huang J, Futrell JW, Katz AJ, Benhaim P, Lorenz HP, Hedrick MH (2001) Multilineage cells from human adipose tissue: implications for cell-based therapies. Tissue Eng 7(2):211–228. https://doi.org/10.1089/107632701300062859

Part II
Stem Cells for Cardiovascular Diseases

Chapter 7
Cardiac Stem Cells: A Plethora of Potential Therapies for Myocardial Regeneration Within Reach

Enrique Gallego-Colon and Eduardo Oliver

Abstract The limited regenerative capacity of the adult heart hinders cardiac regeneration and still remains an unsolved therapeutic target. In the last decade, the development of cell-based approaches for the treatment of myocardial infarction has made substantial progress. The use of cardiac stem cells (CSC) represents a promising therapeutic tool upon massive loss of cardiac tissue such as after myocardial infarction. Different putative CSC populations, with self-renewing capacity and potential to differentiate into cardiomyocytes, smooth muscle, or endothelial cells, are under evaluation. Preclinical studies have discovered various pools of putative CSC including c-kit cardiac progenitor cells (CPCs), epicardium-derived cells (EPDC), Sca-1, cardiac side population (cSP), islet-1 CPC, cardiosphere-derived cells (CDCs), cardiac colony-forming unit fibroblast (cCFU-F), cardiac atrial appendage stem cells (CASCs), bone marrow-derived CSC (BMdCSC), and modified embryonic or induced pluripotent stem cells. Preclinical studies provided encouraging results on the ability of CSC to restore cardiac contractility, regional and global function, myocyte survival, and the remodeling process. Although CSC therapy is safe and feasible in the clinics, heterogeneous outcomes have cast doubts on CSC efficacy to promote cardiac regeneration. Overall, the poor engraftment of transplanted CSC highlights the possibility that the observed beneficial effects may be attributed to the release of paracrine factors rather than CSC ability to reconstitute functional well-differentiated myocardium. Unfortunately, controversy remains around the benefit of CSC therapy, and big challenges such as ideal dose, identification and selection of the best CSC type, delivery system, and concomitant treatments must be addressed. The success of CSC regenerative therapies implies more basic research, and further clinical trials are needed before to reach the clinical practice and fulfill the long-lasting goal of cardiac regeneration.

E. Gallego-Colon
Department of Biochemistry, Medical University of Silesia, School of Medicine in Katowice, Katowice, Poland
e-mail: egallego@med.sum.edu.pl

E. Oliver (✉)
Centro Nacional de Investigaciones Cardiovasculares Carlos III (CNIC), Madrid, Spain
e-mail: eduardo.oliver@cnic.es

R. Delgado-Morales (ed.), *Stem Cell Genetics for Biomedical Research*,
https://doi.org/10.1007/978-3-319-90695-9_7

Keywords Cardiac regeneration · Myocardial infarction · Ischemic disease · Cardiac stem cells · Cardiac progenitor cell · Cardiac function · Clinical trials · Preclinical studies

Abbreviations

5-aza	5′-Azacytidine
bFGF	Basic fibroblast growth factor
BM	Bone marrow
BMdC	Bone marrow-derived cells
CABG	Coronary artery bypass surgery
CASCs	Cardiac atrial appendage stem cells
cCFU-F	Cardiac colony-forming unit fibroblast
CDCs	Cardiosphere-derived cells
CPCs	Cardiac progenitor cells
CSC	Cardiac stem cell
cSP	Cardiac side population
EF	Ejection fraction
EMT	Epithelial-to-mesenchymal transition
EPDC	Epicardium-derived cells
ESC	Embryonic stem cells
GFP	Green fluorescent protein
HF	Heart failure
IGF-1	Insulin growth factor 1
IHD	Ischemic heart disease
iPSC	Induced pluripotent stem cells
MI	Myocardial infarction
MRI	Magnetic resonance imaging
MSC	Mesenchymal stem cells
Sca-1	Stem cell antigen-1

7.1 Introduction

Cardiovascular diseases (CVD) are one of the major causes of death globally [1]. The World Health Organization has estimated that 17.5 million people die each year from CVD, standing for 31% of all deaths worldwide and 45% of all deaths in Europe. Among the most common types of CVD, stroke and ischemic heart diseases (IHD) represent 80% of total deaths [2]. From an economic standpoint, the European Heart Network estimated that total costs for all CVD are 210 billion € a year [3].

Considered one of the most representative IHD, myocardial infarction (MI), also known as heart attack, is characterized by sudden loss of cardiomyocytes due to prolonged ischemia originated upon occlusion or narrowing of the coronary artery that supplies the heart muscle with oxygen-rich blood [4]. The lack of oxygen supply damages the myocardium and becomes necrotic within 2–4 h after injury. The extent of damage to the myocardium depends both on the collateral circulation of the ischemic zone and the length of time the occlusion is present. The entire process leading to a healed myocardium usually takes 5–6 weeks [5]. Due to the inability of the heart to regenerate, the healing myocardium undergoes remodeling through a succession of molecular, cellular, and extracellular changes that culminate in thinning, dilation, and fibrosis of the ventricular wall [6]. Additionally, tissue ischemia can jeopardize electrical coupling and electrical conduction of the heart leading to lethal arrhythmias and potentially sudden death. To that extent, implantable medical devices are designed to support the damaged heart and prevent cardiac complications [7, 8].

Adverse effects of MI can be reduced by combining interventional and pharmacological strategies. Current interventions for ischemic injury are targeted at restoring blood flow to ischemic tissue. These interventions fall into two main categories: coronary artery bypass graft surgery (CABG) and percutaneous coronary intervention also called coronary angioplasty [9]. A quick restoration of the oxygen supply after removing the obstruction has the potential to rescue myocardium at risk. However, this reperfusion is of little help for the myocardium that is unable to survive the ischemic injury. In fact, reperfusion of the tissue increases radical oxygen species (ROS) production potentially compromising the surviving myocardium in a process known as reperfusion injury [10]. Current pharmacological strategies for the treatment of ischemic disease are facilitated to ameliorate cardiac remodeling and dysfunction in an attempt to prevent heart failure (HF) [11]. In particular, pharmacological approaches target the neurohormonal pathways and inhibition of the renin-angiotensin-aldosterone. The main drugs are angiotensin-converting enzyme inhibitors and β-blockers, which have been demonstrated to prevent left ventricular remodeling and reduce mortality after MI [12, 13]. Additionally, treatment is also managed by antithrombotics, lipid-lowering drugs, nitrates, angiotensin II receptor blockers, and mineralocorticoids [14]. Besides, diuretics are also given, to relieve the symptoms and signs of congestion [11]. All these drugs improve the course of HF after MI, exerting a positive effect on the outcome of the patient as well as reducing mortality and hospitalization times. Unfortunately, interventional and pharmacological strategies only ameliorate MI-derived symptoms and to some extent prevent the worsening of the heart condition. However, such strategies do not promote regeneration of the missing myocardium. Nowadays, the technological improvements in the field of regenerative medicine have provided researchers the opportunity to exploit the use of cell-based [15] and/or gene-based therapies [16–18] to promote repair and cardiomyocyte regeneration. Recent studies indicate that the heart may have hidden the potential to regenerate through a self-renewal, multipotent, and clonogenic cell population: the cardiac stem cells (CSC). In the next sections, the potential therapeutic role of endogenous and non-endogenous CSC will be reviewed.

7.2 Regeneration of the Mammalian Heart

Through evolution, higher mammals have lost the ability to regenerate damaged tissue and/or appendages limiting this capacity to organs such as the lung or liver. Scientists have struggled in understanding tissue regeneration and attempted to answer some fundamental questions for centuries: Why does regeneration occur only in certain organs and not in others? Why was this remarkable ability not preserved or maintained through mammalian natural selection? Is there a regenerative genetic program hidden deep within the genome?

Interestingly, in certain species, the heart still preserves this capacity to some degree. Recent research shows that zebrafish are capable of fully regenerating hearts within 2 months after 20% ventricular resection. In this model, regeneration occurs through robust activation and proliferation of cardiomyocytes localized at the leading epicardial edge of the new myocardium [19, 20]. Genetic ablation studies in zebrafish showed reversal of HF after 60% ablation of total cardiomyocytes. In this model, up to fourfold increase in cardiomyocyte proliferation was observed. Specific ablation of cardiomyocytes and sparing of non-cardiomyocyte tissue optimally engaged regeneration by promoting cardiomyocyte proliferation, epicardial and endocardial production of retinoic acid, epicardial proliferation, de novo vascularization, and macrophage recruitment in the absence of a large clot [21].

Furthermore, the endocardium and epicardium have been proposed to support de novo vascularization during zebrafish regeneration. As early as 3 h after apical heart resection, robust epicardial and endocardial production of Raldh2 was associated with cardiomyocyte proliferation and de novo vascularization [22]. In zebrafish, injury-induced cardiomyocyte proliferation overcomes scar formation, allowing cardiac muscle to regenerate [19–21]. Conversely, another teleost, the medaka fish, responded rather differently to cardiac injury. After heart injury, the medaka heart displayed excessive collagen deposition by activated myofibroblasts and no substantial proliferation of existing cardiomyocytes. Interestingly, Raldh2 gene expression was not found thus suggesting that the endocardium remained in an inactive state [23]. Despite extensive research, fundamental aspects of the mammalian heart's capacity for self-renewal are actively debated. The failure of the adult heart to recover from injury has been largely attributed to the ineffective replacement of affected cardiomyocytes, which withdraw from the cell cycle soon after birth. Instead, cardiac function is maintained through cardiac growth dependent on cellular hypertrophy and proliferation of other cell types acting as a compensatory mechanism.

Recent studies in mammals identified a brief window of time which neonatal mice hearts retain significant regenerative capabilities in response to either resection of the apex or myocardial ischemia by permanent ligation of the left anterior descending artery [24, 25]. Although limited to the first 7 days of postnatal development, mouse hearts were characterized by a robust cardiomyocyte proliferation from preexisting cardiomyocytes and de novo formation of large coronary vessels in the infarcted area. Interestingly, not all newly formed cardiomyocytes were derived from preexisting cardiomyocytes. In fact, neonatal post-MI hearts are characterized by local expression of the progenitor stem cell markers c-kit and Nkx2.5 that contribute to the myogenic and vascular fates [26]. Strikingly, fibrotic scar tissue in neonatal mouse hearts is almost completely resolved within 21 days post-MI [24]. Another study on heart inflammation and repair proved that macrophages are essential for neonatal heart regeneration. In this model, neonatal mice depleted of monocytes/macrophages were unable to regenerate the myocardium or form new vessels. Instead, the damaged area was replaced by fibrotic scars [27]. A completely different but exciting investigation suggests that sympathetic innervation is required for regeneration of the mammalian heart. Particularly, a study suggested that age-related nerve regeneration may be the underlying mechanism of the inadequate adult regeneration after MI [28]. Overall, cardiac regeneration and repair is achieved when the three hallmarks of regeneration, cell replacement and proliferation, de novo vascularization, and quick inflammatory resolution, are met. However, the loss of regenerative potential beyond 7 days after birth is yet not fully understood.

7.3 Cardiac Stem Cells in Myocardial Regeneration and Repair

As discussed in the previous section, HF and MI therapeutic approaches are based on interventional and pharmacological approaches. Consequently, the shortage of heart donors and the limited efficacy of conventional therapies have led to the design of novel regenerative approaches to overcome such limitations.

The prevailing view is that the heart is considered a postmitotic organ without regenerative capacity. At physiological level, cardiomyocyte turnover is an approximate 0.45–1% per year ultimately replacing up to 45% of total cardiomyocytes during adult life [29–32]. After injury, however, the necrotic and/or nonfunctional myocardium is replaced by fibrotic scar tissue compromising cardiac function. The general acceptance is that there are three main cell sources that may contribute to replenish the missing myocardium. Theoretically, cardiomyogenesis can be achieved from (1) division of existing cardiomyocytes [33], (2) from endogenous CSC residing within the heart [34], or (3) from exogenous niches such as bone marrow (BM) [35]. After insult, these cell types, either individually or working synergically, contribute in maintaining homeostasis, genesis of new cardiomyocytes and formation of new vessels. In 2013, Senyo and coworkers determined by genetic mapping techniques that replenishment of the myocardium by new cardiomyocytes occurs at a low rate by the division of preexisting cardiomyocytes during normal aging and in myocardial injury [36]. Indeed, replenishment of heart muscle by division of existing cardiomyocytes is sufficient to replace aging cardiomyocytes from a homeostatic point of view. However, after acute ischemia, homeostatic regenerative strategies are not sufficient to replenish large areas of the myocardium which are affected or lost.

The first indication of the presence of CSC or cardiac progenitor cells (CPCs) in the heart was made in 2003. Beltrami and coworkers discovered the existence of a myogenic stem cell population negative for blood lineage markers (Lin$^-$) but positive for the common stem cells markers c-kit, Sca-1, and MDR-1. The newly discovered CSC were also negative for leukocyte marker and endothelial/hematopoietic progenitor marker. In cell culture, Lin$^-$ c-kit$^+$ cells showed self-renewal potential, clonogenicity, and multipotency. In fact, Beltrami's research marked a breakthrough in the cardiovascular field demonstrating that the adult heart contains a population of self-renewal, multipotent CSC population with potential to generate new cardiomyocytes. Nevertheless, his work at that time did not clarify the endogenous or non-endogenous origin of the Lin$^-$/c-kit$^+$ [34].

7.3.1 Endogenous Cardiac Stem Cell

The origin of newly formed cardiomyocytes in the studies discussed above is still controversial, as research has mainly addressed in understanding the intrinsic cell populations of the adult heart rather than the origin. Compared to non-endogenous CSC, endogenous CSC are considered more suited for myocardial regeneration since these are committed to the cardiomyocytes lineage and cardiac-related structures. Currently, CSC are characterized and isolated based on the expression of selected cell surface markers. Additionally, CSC have been classified according to shared characteristics with stem cell such as clonogenicity, self-renewal, and multipotency. Nowadays, CSC can be isolated from the hearts of multiple species including humans [37–40], mouse [38, 41], rats [34], and dog [42]. The absence of a CSC niche makes the identification of putative CSC a tedious work. Nevertheless, studies have shown a higher number of CSC in the atrial and apical myocardium than in other areas of young and aged heart [41, 43].

Over the past decade, a long list of potential CSC have been identified, isolated, and purified mainly based on the expression of cell surface antigens (Table 7.1).

BM-cSP, bone marrow-derived cardiac side population; BMdCSC, bone marrow-derived cardiac stem cells; CASC, cardiac atrial appendage stem cells; cCFU-F, cardiac colony-forming unit fibroblast; CDCs, cardiosphere-derived cells; CPCs, cardiac progenitor cells; EPDC, epicardium-derived cells; cSP, cardiac side population cells

C-kit$^+$ cardiac progenitor cells (CPCs)

In 2003, c-kit$^+$ cells were also discovered by Beltrami and coworkers in adult rat hearts. C-kit$^+$ cells are distributed in a nonhomogeneous manner throughout the

Table 7.1 Cardiac stem cells (CSC) and flow cytometry immunophenotype

Endogenous CSC	**Cell surface markers**	**References**
C-kit$^+$ CPC	C-Kit$^+$, Sca-1$^+$, MDR$^+$ (abcg2$^+$), CD34$^-$, CD41$^-$, CD133$^-$, Lin$^-$, CD45$^-$, CD20$^-$, CD45RO$^-$, CD8$^-$, TER119$^-$, and cardiogenic factors (GATA4$^+$, NKX2.5$^+$, MEF2C$^+$)	[34, 44–50]
EPDC	WT1$^+$, Tbx18$^+$, Raldh1/2$^+$, Isl1$^+$, CD44$^+$, CD46$^+$, CD90$^+$, CD105$^+$, CD45$^-$, CD11b$^-$, and cardiogenic factors (NKX2.5$^+$)	[51–60]
Endogenous Sca1$^+$ CPC	Sca-1$^+$, CD105$^+$, endothelial cell progenitors (CD34$^-$, FLK1$^-$, Flt$^-$), CD31$^{+/-}$, CD38$^+$, CD45$^-$, c-kit$^-$, and cardiogenic factors (GATA4$^+$, NKX2.5$^+$, MEF2C$^+$)	[61–67]
cSP	Sca-1$^+$, CD31$^{+/-}$, CD29$^+$, ABCG2$^+$, CD45$^-$, CD34$^-$, c-Kit$^-$, and negative for cardiogenic factors (GATA4^{low} and NKX2.5^{low})	[68–80]
Islet-1 CPC	Isl-1$^+$, CD31$^-$, Sca1$^-$, c-kit$^-$, and cardiogenic factors (GATA4$^+$ and NKX2.5$^+$)	[81–86]
CDC	CD31$^+$, CD34$^+$, c-kit$^+$, Sca-1$^+$, CD105$^+$, and cardiogenic factors (NKX2.5$^+$)	[38, 87–96]
cCFU-F	PDGFR-α^+, Sca-1$^+$, CD31$^+$, CD90$^+$, c-kit$^+$, CD45$^-$, FLK1$^+$, CD44$^+$, CD29$^+$, CD105$^+$, and embryonic markers (Oct4$^+$ and cMyc$^+$)	[97, 98]
CASC	ALDH$^+$, CD34$^+$, CD73$^+$, Isl-1$^+$, CD45$^-$, and c-kit$^-$	[99–101]
Non-endogenous CSC	**Cell surface markers**	**References**
BMd CSC	C-kit$^+$, Lin$^-$/c-kit$^+$, Lin$^-$, Sca-1$^+$	[34, 102, 103]
BM-cSP	Sca-1$^+$, CD31$^{+/-}$, CD34$^-$, CD45$^-$, c-kit$^-$	[104–106]

heart and at a density of one c-kit$^+$ every 1×10^4 myocytes [34]. When isolated, Lin$^-$/c-kit$^+$ were found to express genes of the myocyte lineage (Nkx2.5, GATA-4, and MEF2), whereas skeletal muscle transcription factors (MyoD, Myogenin, and Myf5) were not detected. The cells were also negative for hematopoietic lineages CD34 and CD45, indicating an embryonic cardiac origin, thus truly cardiac endogenous cells. Additionally, c-kit$^+$ cell do not display cardiac side population phenotype as will later be discussed in this section. The CSC population are found to be self-renewing, clonogenic, and multipotent. In cell culture, c-kit$^+$ cells can give rise to three main cardiac lineages: endothelial cells, smooth muscle cells, and cardiomyocytes. However, differentiated cardiomyocytes were found to have an immature phenotype with immature sarcomeres. Interestingly, c-kit$^+$/Lin$^-$ cells injected into an ischemic heart resulted in the formation of blood-carrying new vessels. Newly formed cardiomyocytes were also observed in the left ventricle forming a thin layer of well-differentiated sarcomere-containing myocardial cells expressing cardiac myosin heavy chain, α-sarcomeric actin, α-cardiac actinin, N-cadherin, and connexin 43 [34].

In 2015, more new evidences suggested that the endogenous c-kit$^+$ population is not able to differentiate into mature functional cardiomyocytes, and only cardiac endothelial cells express c-kit [44, 45]. Additionally, Keith and collaborators performed perhaps one of the most compelling and extensive comprehensive analyses which reconciled all the discrepant results regarding the cardiomyogenic potential of c-kit$^+$ cells. In the review, Keith concluded that c-kit$^+$ has been indeed misinterpreted and truly defines CSC. C-kit marker is expressed in more than one fetal cardiac progenitor population. Notably, this expression varies temporally and spatially in cardiac development but also in terms of absolute protein levels as evidenced by the expression of c-kit^{low} and c-kit^{high}. C-kit^{low} population give rise to smooth muscle cells and cardiomyocytes during fetal development without persisting in the adult heart. Conversely, c-kit^{high} population are derived from the pro-epicardium with a mesenchymal phenotype which could potentially give rise to fibroblast and endothelial cells. C-kit^{high} can undergo epithelial-to-mesenchymal transition (EMT) migrating to the underlying myocardium in response to injury such as MI [46–48]. Overall, the presence of a c-kit$^+$ epicardium-derived non-cardiomyogenic precursors with mesenchymal phenotype [49] provides a reasonable explanation of the beneficial effects of these cells upon administration into the heart after MI. In another study in rats, investigators focused on the long-term effects of intracoronary delivery of c-kit$^+$ CSC. In the study, c-kit$^+$ CSC from a male donor were infused into a recipient female heart 4 h after reperfusion injury. One year later, rats showed improvement in regional and global left ventricular function. At histological level, the area at risk exhibited increased viable myocardium, sustained proliferative response of endogenous cells leading to the formation of new vessels and cells that express cardiomyocytic proteins (α-SA) but lack of mature cardiomyocyte phenotype. Interestingly, transplanted c-kit$^+$ CSC engrafted and persisted in the heart but failed to produce and/or to differentiate into mature cardiomyocytes. The authors suggested that the underlying mechanism of action behind the improvement in cardiac function persisting over a year period after CSC therapy was mediated by the release of paracrine factors and proliferation of endogenous cells [50]. Despite c-kit$^+$ CSC not potentially differentiating into cardiomyocytes, it is proven that these cells hold potential to ameliorate the adverse effect of myocardial injury.

Epicardium-derived cells (EPDC)

First mentioned in 1998, EPDCs are an embryonic cell population persisting through development and found in the outer layer of the heart known as epicardium. This population is characterized by the expression of WT1, Tbx18, and Raldh1/2 markers. During embryogenesis, EMT of EPDCs give rise to coronary smooth muscle cells [51], endothelial cells, adventitial and interstitial fibroblast, and possibly cardiomyocytes [52–54]. Studies in rodents indicate that myocardial injury activates EPDCs by recapitulating embryonic gene expression programs that control proliferation and migration of EPDC. In the infarcted myocardium, EPDCs contribute to enhance survival and growth of coronary vessels [55, 56]. Besides EPDCs' role in cellular contribution, the epicardium can modulate reparative processes after MI mediated by the release paracrine factors [57]. The EPDC contribution to newly formed cardiomyocytes has not yet been established. Studies in mice show that EPDC retain certain regenerative capacity although the potential is insufficient to replace the extensive loss of myocardium after MI [58]. Another study demonstrates that thymosin b4 can stimulate mobilization and neovascularization of EPDC potentially promoting the formation of new cardiomyocytes [59].

Currently, there is no consensus on the role of EPDC in myocardial repair and regeneration, mainly due to the lack of adequate cell surface markers, and advance experiments with reporter lines to track EPDC after MI are also required. Nonetheless, the potential contribution of EPDC to vasculature and myocardium during heart development, the dormant state in the adult heart, and the possibility of EPDC reactivation by compounds such as thymosin β4 makes EPDC a clinically relevant approach for future ischemic diseases and HF therapies. In fact, the EPDC as potential therapeutic tool is currently being explored. Paulitschek and collaborators generated two induced pluripotent cell (iPSC) lines from EPDC, and the iPSC differentiation potential was evaluated by comparative transcriptome profiler [60]. Further research is underway.

Endogenous Sca1$^+$ CPC

In 2003, an endogenous CPCs expressing stem cell antigen-1 (Sca-1) but negative for c-kit and hematopoietic markers (CD4, CD8, B220, Gr-1, Mac-1, and TER119) was identified in adult mouse heart [61]. In the cardiac muscle, Sca-1$^+$ cells do not express cardiac structural genes (α- and β-MHC, Bmp1r1) nor cardiogenic factors (NKx2.5, MEF2C). However, in vitro culture with 5′-azacytidine (5-aza) induced differentiation into cardiogenic lineages. Furthermore, when infused intravenously after ischemia/reperfusion, Sca-1$^+$ cells homed to injured myocardium. In light of these findings, several authors attempted to isolate Sca-1$^+$ based on the lack of expression of endothelial marker CD31. Sca-1$^+$/CD31$^-$ cells were able to differentiate cardiomyocytes, smooth muscle, and endothelial cells [62–64]. Preclinical studies in rodents have shown the potential of Sca-1$^+$/CD31$^-$ in ameliorating adverse effects of MI, promoting neovascularization and to some extent cardiomyocyte regeneration [65]. The delivery of Sca1$^+$ CPC with adjuvant factors such as insulin growth factor and hepatocyte growth factor further improved cardiac parameters, ejection fraction (EF), and left ventricular volume than standard therapy alone [66].

Sca-1$^+$/CD31$^-$ CPC nature and markers have also been questioned. A recent review by Valente and coworkers pointed out the heterogeneity of the Sca-1$^+$ marker with a fibroblast-like phenotype. In fact, Sca-1 contribution to the formation of new cardiomyocytes is low but continuous during adult life and after injury [64]. Unfortunately, no orthologous counterpart of Sca-1 is found in humans thereby laborious isolation and expansion limit future therapeutic applications. Interestingly, using anti-mouse Sca-1 antibody, human Sca-1-like cells have been isolated. Further characterization revealed that putative Sca1$^+$ CPC displayed some cardiac transcription factors such as Nkx-2.5, MEF2c, Islet-1, and GATA-4. Furthermore, human Sca-1-like cells were also able to differentiate into cardiomyocytes upon stimulation of with 5-aza [67].

Cardiac Side Population cells (cSP)

In 2002, Hierlihy and collaborators described for the first time a CSC population phenotypically characterized by the ability to efflux Hoechst dye through an

ATP-binding cassette transporter, Abcg2, hence appearing to the "side" on FACS analysis [68, 69]. The cardiac side population (cSP) besides expressing Abcg2 also express Sca-1$^+$ and CD29 [70]. The cSP also differ from c-kit$^+$ population, previously discussed, as having phenotypically different gene expression profiles [71]. The cSP cells represent from 0.8% to 2% of non-cardiomyocytes [72] with percentages varying with time both in rat [73] and human [74, 75]. Lineage tracing analysis, using Cre recombinase and a green fluorescent protein (GFP) double transgene, indicated that neural crest cells may contribute to the cSP pool [76]. Interestingly, Liang and collaborators isolated and characterized two distinct Sca-1$^+$ cSP: Sca-1$^+$/CD31$^-$ and Sca-1$^+$/CD31$^+$. Sca-1$^+$/CD31$^-$ cell population is involved in cardiac regeneration [77], whereas the Sca-1$^+$/CD31$^+$ cell population acts as endothelial progenitor cells population [78]. In vitro studies have shown that Sca-1$^+$/CD31$^-$ cSP have cardiomyogenic potential and mature into cardiomyocytes through a process mediated by cellular coupling with adult cardiomyocytes. In co-culture with cardiomyocytes, Sca-1$^+$/CD31$^-$ cSP express cardiac structural proteins (troponin I and α-actinin) and show contraction and intracellular calcium flux [79]. Notably, such features are only observed when co-cultured with cardiomyocytes. Otherwise, neither sarcomeric organization nor contractile activity is observed. Murine models of MI have shown that Sca-1$^+$/CD31$^-$ cSP have the potential to migrate into the infarcted heart after ischemic injury and differentiate into cardiomyocytes, smooth muscle cells, endothelial cells, and fibroblasts [73, 77].

Noseda and collaborators attempted to identify and clarify the origin and phenotype of cSP as well as their contribution in tissue repair and regeneration. Single-cell qRT-PCR and genetic lineage-tracing studies clarified some common features between CSC and Sca-1$^+$. The research also unmasked PDGFRα as potential new marker for cSP cells which better demarcate the cSP population than does the SP phenotype alone [80]. The lack of markers and difficulties of in vivo differentiation manifest the fact that more basic research is warrant to understand the biology of these cells in tissue repair and regeneration [70].

Islet-1 CPC

In 2005, Islet-1 CPC was described as part of the second heart field progenitors during heart development. CPCs play a crucial role in the formation of the right ventricle, atria, and outflow tract. Islet-1 phenotype is based on the expression of Islet-1 and lack Sca-1, c-kit, CD31, Gata-4, and Nkx2.5 expression [81]. Islet-1 cells display cardiomyogenic potential and capacity to differentiate into smooth muscle and endothelial cells [82]. So far, these cells can be found in and isolated from human and rodent hearts [83–85]. However, in the human adult heart, Islet-1 cells are found in reduced numbers hence limiting its therapeutic clinical applicability [86].

Cardiosphere-derived cells (CDCs)

In 2004, another type of CSC was discovered, and the potential clinical use in cardiac regenerative therapies is being evaluated. Messina and collaborators first isolated CDC from atrial and ventricular human biopsy specimens and also from murine hearts. Upon culture, these heterogeneous stem cell populations grow as cell clusters expressing stem and endothelial progenitor cell markers [38]. Interestingly, spatial distribution within the clusters is observed with c-kit$^+$ expression at the core, and differentiating cells expressing vascular, endothelial, and cardiac cell markers are located in the periphery [87–89]. Additionally, several studies have indicated that CDCs show cardiomyogenic potential in vitro [39, 90, 91]. Preclinical studies indicated that mouse-derived CDC can ameliorate MI-derived adverse effects maintaining long-term functional improvement of the injured heart [92]. Importantly, CDC beneficial effects are based on the release of growth factors such as VEGF, HGF, IGF-1, and endoglin in a paracrine fashion [90–93]. In parallel, in a porcine model of MI, porcine-derived CDC infused intracoronary can adequately engraft, reduce infarct size, attenuate adverse left ventricular remodeling, improve hemodynamic functions and cardiac function to some extent. The improvement, assessed by the LacZ$^+$ reporter gene, was achieved partly due to the formation of new cardiac and vascular tissue [94].

Conversely, some studies disputed the cardiomyogenic potential of CDC. Research on lineage-tracing experiments failed to prove that CDC can differentiate into cardiomyocytes in vivo [95]. In fact, the study reported that the spontaneous beating of cardiomyocytes was due to contaminating myocardial tissue fragments [92]. Furthermore, a recent study in rats indicated that human CDC therapy failed to improve cardiac function parameters such as EF, end-diastolic volume, end-systolic volume, and hemodynamic function between the groups receiving human cells, rat cells, or placebo. Overall, no improvement in cardiac remodeling was observed when using CDC cell therapy [96]. C-kit$^+$ CDC potential was tested in the CADUCEUS trial showing improvement in cardiac function as discussed in the following sections.

Cardiac colony-forming unit fibroblast (cCFU-F)

In 2011, cCFU-Fs were described as multipotent, cardiac-resident mesenchymal stem cell (MSC)-like populations capable of long-term in vitro culture and multilineage differentiation. The cCFU-F population has been identified mainly by the expression of PDGFRα and represent a subset of cardiac Sca-1$^+$ cells in murine [97] and in human heart [98]. After MI, individual GFP cCFU-F labeled cells contributed to the formation of cardiomyocytes, endothelial cells, and smooth muscle [20, 98]. Currently, cCFU-F studies are underway to widen understanding and potential therapeutic applications.

Cardiac atrial appendage stem cells (CASCs)

In 2013, CASCs were isolated from human left atrial appendage based on the high expression of aldehyde dehydrogenase (ALDH). CASCs have been described to be phenotypically closer to cCFU-Fs with higher clonogenicity than c-kit$^+$ and CDC. CASCs also express CD34 and Islet-1, which suggests a second heart field progenitor population origin. In vitro analysis demonstrated that CASCs possess cardiomyogenic differentiation capacity [99]. In a clinically relevant setting, in vivo studies on minipigs revealed that autologous administration of CASC after ischemia/reperfusion resulted in left ventricular function improvement. According to the authors, cell engraftment was 19% on average and 98% of engrafted CASC differentiated into cardiomyocytes [99, 100]. Angiogenesis assays also showed that CASCs stimulate blood vessel formation by paracrine mechanisms [101].

7.3.2 Non-endogenous Cardiac Stem Cells

Adequate repair and regeneration of the damaged muscle require the presence of exogenous and/or endogenous stem cells in order to maintain cellular turnover and homeostasis. Non-endogenous stem cells can be recruited into the damaged heart to contribute in tissue repair upon insult (Table 7.1).

Bone marrow-derived cardiac stem cells (BMdCSC)
BM as potential source of CSC was studied by Dr. Piero Anversa and colleagues which suggested that stem cells found in the BM could potentially be differentiated into functional cardiomyocytes [34]. However, in 2004, Dr. Charles Murry and collaborators used cardiomyocyte-restricted and ubiquitously expressed reporter transgenes to track the fate of hematopoietic stem cells. Experiments were performed using two BM-derived stem cell populations either (a) c-kit$^+$, Lin$^-$ or (b) c-kit$^+$, Lin$^-$, Sca-1$^+$. Evaluation of the infarcted area in both non-injured and injured adult mice hearts indicated that neither transdifferentiation into cardiomyocytes nor increased number of cardiomyocytes had occurred [102]. In 2016, Liu and coworkers demonstrated that BMdCSC can differentiate into cardiomyocytes in vivo [103]. In the study, GFP-labeled BM cells were used to reconstitute lethally irradiated non-transgenic, syngeneic mice. After ischemia/reperfusion injury, engraftment of GFP$^+$ cells into the heart was about 1–2% per month, and around 50% of the CSC pool, either Sca-1$^+$ or c-kit$^+$, was indeed GFP$^+$. Although GFP-CSC transdifferentiated into CMs, no increase of CSC was detected after cardiac injury, which indicated that the rate of cardiac transdifferentiation of BM-derived CSC is low [103]. BM-derived CSC studies were further pursued in several clinical trials as discussed later.

Bone marrow-derived cardiac side population (BM-cSP)
In 2005, Mouquet and collaborators determined that cSP in the heart is replenished by selective homing of BM-derived stem cells following MI [104]. When compared with cSP, BM-cSP do not express cardiomyocyte transcription factors such as MEF2C, GATA4, and CD45. Conversely, c-kit expression is detected when cultured under same conditions as cSP [79]. Mouquet discovered that BM-cSP do not contribute to the physiological cSP turnover. However, in mice, GFP-labeled BM-cSP were capable of contributing up to 25% of total cSP pool by day 7 posttreatment after MI [104]. The BM-cSP was also observed to contribute to the skeletal muscle cell populations [105]. Interestingly, the heart milieu encountered by BM-cSP lead to an immunophenotypic conversion with the loss of the hematologic marker CD45. Unfortunately, the cardiomyogenic potential of BM-cSP to differentiate into cardiomyocytes was not explored in this article. Another study also showed that both rodent and human BM-cSP enhanced cardiac function following MI [106]. Collectively, these studies highlight the superiority of BM-cSP in myocardial

regenerative therapies; however, the field seems halted. The signals triggering the recruitment of BM-cSP into the heart and further transition into cSP remain to be clarified; nonetheless, BM-cSP hold an enormous potential for therapeutic myocardial regeneration.

7.3.3 Embryonic and Induced Pluripotent Stem Cells in Cardiac Repair

Recent advances in the field of cardiac regenerative medicine, especially in stem cell biology, have provided scientists with new tools to pursue the long-standing goal of cardiac regeneration. The advances in cellular reprogramming of embryonic stem cells (ESC) and induced pluripotent stem cells (iPSC) have indeed contributed to that goal. ESC are derived from the inner cell mass (ICM) of blastocyst stage embryos in various species including mouse [107] and human [108, 109]. ESC have the unique capacity to proliferate extensively while maintaining pluripotency constituting an invaluable model for basic investigations in developmental biology, nuclear reprogramming, and cell differentiation processes. At the same time, human ESC are promising donor source for cell transplantation therapies for the treatment of any kind of disease [110]. However, ESC therapy is limited by the ethical concerns of human embryos usage.

The discovery of induced pluripotent stem (iPSC) cells bypasses these concerns due to the fact that iPSC technology is based on extensive adult somatic cell reprogramming of differentiated tissue to an embryonic-like state. In 2006, Yamanaka and colleagues first described a method for iPSC generation by reprogramming differentiated mouse skin fibroblast. Reprogramming of differentiated cells was achieved by retroviral-mediated genetic incorporation of four transcription factors: octamer-binding transcription factor-3/4 (OCT3/4), SRY-related high-mobility-group (HMG) box protein-2 (SOX2), MYC, and Kruppel-like factor-4 (KLF4) [111]. Since then, variations of this original protocol have been applied to other types of mouse cells and to human somatic cells to produce cells of an embryonic-like state [112–116]. Almost 10 years later, embryonic fibroblasts can now be reprogrammed by the combination of chemical compounds hence dismissing the adverse effects caused by virally mediated reprogramming such as genetic mutations, altered gene expression, and incomplete epigenetic reprogramming [117]. Furthermore, transdifferentiation from skin fibroblast to cardiomyocyte by defined factors has also been achieved [118, 119]. Today, iPS technology has three major applications in the field of cardiac regenerative medicine, generation of new replacement therapies, disease modeling, and drug discovery [120, 121]. In fact, there are three well-established pathways to generate new cardiomyocytes from differentiated fibroblasts for cell replacement therapies: (1) full reprogramming of fibroblast into iPS cells and subsequent cardiac differentiation [122], (2) partial reprogramming of fibroblasts to cardiac progenitor cells and subsequent differentiation [120], and (3) direct reprogramming of fibroblasts into cardiomyocytes [123]. Although iPSC-derived cardiomyocyte future therapy is very promising, scientist must solve first several issues concerning iPSC use such as tumorigenicity, variable differentiation, and low efficiency reprogramming. Furthermore, adequate characterization of iPSC-derived cardiomyocytes is required before these cells can be tested in clinical trials. Additionally, iPSC technology can be used to model cardiomyopathies in vitro to investigate pathophysiological process as well as in drug discovery [112].

7.3.4 The Role of Paracrine Factors: Enhancement of Stem Cell Potential in the Injured Heart

The accepted view is that the heart has very limited regenerative capacity; hence, scientists have embarked on a quest to find factors to boost CSC's intrinsic potential to promote cardiac repair. Clinical trials with either BM-derived stem cells or CSC resulted in some degree of improvement in ameliorating the adverse effects of cardiac injury. However, the relatively poor survival and limited engraftment of these cells limits its applicability. As a consequence, new approaches are focused searching for factors that promote CSC proliferation and differentiation into cardiac-derived tissue. In fact, enhancement of the CSC potential can also be achieved by preconditioning, either ex vivo and in vivo, with growth factors and/or pharmacological compounds [28]. Preclinical studies in mice indicated that adjuvant therapy with VEGF improves the MI outcomes by enhancing CSC engraftment [124–126]. In clinical studies, VEGF potential was evaluated in the NOGA angiogenesis revascularization therapy: assessment by RadioNuclide imaging (NORTHERN) trial, that was a double-blind, placebo-controlled study of intramyocardial VEGF gene therapy. Unfortunately, the study showed that VEGF gene therapy caused no benefit in cardiac function in patients with ischemic defects or angina [127].

In animal models, another potential preconditioning factor, insulin growth factor 1 (IGF-1), has shown to improve myocardial repair, neovascularization, and BM mobilization through the expression of specific cytokines after MI [128, 129]. IGF-1 supplementation improved cardiac function, reduced left ventricular remodeling, modulated the inflammatory response by promoting anti-inflammatory microenvironment, and improved myocardial repair partly by activation of CSC in the damaged myocardium [129–131]. In clinical trials, the Evaluation of the Safety and Efficacy of Using Insulin-like Growth Factor-1 in Patients With a Heart Attack (RESUS-AMI) trial is a randomized trial evaluating the safety and efficacy of a single intracoronary dose IGF-1 (mecasermin) following percutaneous coronary intervention in ST-elevated acute MI. Primary outcomes are serum glucose levels and percentage change in global left ventricular EF at baseline and 8 weeks after infusion. The study was completed in December 2016, and results await [132].

Other paracrine factors such as SDF-1, TGF-β, or G-CSF have been shown to accelerate healing by mobilizing BM-derived progenitor cells, promoting adequate remodeling, and improving cardiac function after myocardial ischemic injury [133]. SDF-1 mediated recruitment of CSC and promoted myocardial repair after MI [125, 134]. Interestingly, hypoxia preconditioning has been also shown to improve survival of CSC through the activation of SDF-1–CXCR4 axis [135, 136]. Additionally, preconditioning MSC with transforming growth factor β (TGF-β) has been reported to stimulate VEGF production through p38 MAPK-dependent mechanism, which confers cardioprotection after ischemic injury [137]. Novel delivery methods of paracrine factors are currently being investigated to enhance CSC function and modulate endogenous regeneration using microvesicles, proteins, and microRNA [138, 139]. Pharmacological factors have also been shown to enhance stem cell potential in vivo. Administration of simvastatin, a lipid-lowering drug, increases engraftment and wall thickness and improves cardiac function after MSC therapy [140]. In another study, MSC preconditioning with trimetazidine, an antianginal drug, leads to a significant improvement in cardiac function and decrease of tissue fibrosis [141]. In the infarcted heart tissues, MSC upregulated pAkt signaling pathway and Bcl-2 protein involved in apoptosis [141].

7.4 Evaluation of Cardiac Stem Cells for Clinical Use

Over the past 20 years, scientists have strived with the need to find a stem cell population that can potentially become the source of new cardiomyocytes to replenish the damaged heart after MI. For that goal, scientists have discovered and exploited an array of stem cells skeletal myoblast, BM-derived stem cells, CSC, and iPSC. Indeed, cardiac regenerative therapies must overcome to certain extent the three missing hallmarks of regeneration. The reparative and remodelling processes involve replacing the damaged myocardium with newly formed cardiomyocytes in order to compensate the loss of ventricular mass, besides restoring cardiac contractility and eventually cardiac function.

7.4.1 Clinical Trials Using Endogenous Cardiac Stem Cell-Based Therapies

Several major phase I, II, and III clinical trials have assessed and exploited the potential of CSC to repair the heart (Table 7.2).

ALCADIA trial

Starting in 2009, the AutoLogous Human CArdiac-Derived stem cell to treat Ischemic cArdiomyopathy (ALCADIA) trial is a non-randomized, open-labeled trial aimed to test a combinatorial CSC therapy in order to enhance engraftment of transplanted cells. The ALCADIA trial main aim was to evaluate the safety and efficacy of autologous transplanted human cardiac-derived stem cells (hCSC) with a controlled release formulation of basic fibroblast growth factor (bFGF) using a biodegradable gelatin hydrogel. In this trial, patients with ischemic cardiomyopathy and a left ventricular EF between 15% and 35% were recruited. The researchers previously reported the efficacy of this approach in a preclinical animal model [142].

Autologous CSC were isolated based on ESC and mesenchymal (CD105 and CD90) markers from endomyocardial biopsies [143]. At 4 weeks upon cell isolation, expansion, and characterization, patients received 20 myocardial injections containing 0.5 million cells/kg (patient body weight) of autologous CSC and 200 μg of bFGF in a gelatin-hydrogel format during CABG. Primary and secondary outcomes evaluated and demonstrated safety of CSC infusion in combination with controlled release of bFGF in HF patients with chronic ischemic cardiomyopathy over a 12-month follow-up period. Some adverse effects were observed including acute coronary occlusion in one patient and a HF exacerbation episode in another patient. At month 6, improvement in left ventricular EF, exercise tolerance, and decreased infarct size were also reported [144]. In conclusion, the ALCADIA trial demonstrated the safety of the combinatorial treatment of hCSC with bFGF in a gelatin-hydrogel scaffold in patients with ischemic cardiomyopathy. Although the clinical trial was completed in 2013, no final report has yet been published.

SCIPIO trial

The first clinical trial using CSC was the Stem Cell Infusion in Patients with Ischemic cardiOmyopathy (SCIPIO) trial starting in 2008. SCIPIO is a phase I, randomized, open-labeled trial using CSC in ischemic MI patients with a postinfarction left ventricular EF lower than 40%. Initially, 98 patients were enrolled into the study; of those only 23 patients (16 CSC-treated and 7 control group) passed the final eligibility criteria [145]. In this trial, c-kit$^+$/lin$^-$ CSC were isolated and expanded in vitro from the right atrial appendage in patients undergoing CABG surgery. Autologous CSC were characterized and sorted by flow cytometry based on the expression of c-kit$^+$, Lin$^-$, GATA4$^+$, Nkx2.5$^+$, MEF2C$^+$, α-sarcomeric actin, and connexin 43. Autologous CSC were also sorted by flow cytometry based on negative phenotype for smooth muscle cells (α-smooth muscle actin) and endothelial cells (Von Willebrand factor) markers. Senescence and ability to proliferate were also measured. Patients received, based on the size of the infarct, intracoronary injections containing from 0.5 to 1×10^6 autologous CSC, 4 months after CABG surgery [146]. No CSC-related adverse effects were reported. In this study, the primary endpoint was to assess short-term safety, and the secondary endpoint was to evaluate efficacy by various parameters including magnetic resonance imaging (MRI) measurements at 1 and 4 months. After infusion, patients presented increased left ventricular EF from $27.5 \pm 1.6\%$ before treatment to $35.1 \pm 2.4\%$ at 4 months and $41.2 \pm 4.5\%$ at 12 months after infusion. The authors speculated that the amount of cells injected cannot account for the functional and structural improvement of the CSC treatment; hence, such functional improvement was also attributed to the release of paracrine factors by autologous CSC [146]. The authors also estimated that the decrease in infarct size was partly due to regenerated cardiomyocytes

Table 7.2 Selected clinical trials using CSC

	ALCADIA		**SCIPIO**		**CADUCEUS**		**TICAP**		**PERSEUS**	
Clinical trial	Phase I		Phase I		Phase I		Phase I		Phase II	
Cell type	Autologous CDC with bFGF		Autologous c-kit+ CDC		Autologous CDC		Autologous atrial CDC		Autologous atrial CDC	
Delivery method	Intracoronary		Intracoronary		Intracoronary		Intracoronary		Intracoronary	
Number of CSC	0.5×10^6 cell/kg 200 μg of bFGF		$0.5–1 \times 10^6$		$12.5–25 \times 10^6$		$2–3 \times 10^6$		0.3×10^6 cell/kg	
NCT ID	NCT00981006		NCT00474461		NCT00893360		NCT01273857		NCT01829750	
Condition	CHF, ICM, ventricular dysfunction		Coronary artery disease/ CHF/ICM		Ischemic LV dysfunction/recent MI		HLHS, single ventricle, HF		HLHS, single ventricle, HF	
I.C: LVEF	15–35%		<40%		25–45%		N/A		<60%	
Follow-up	12 months		18 months		12 months		18 months		12 months	
	Patient	Control	Patient	Control	Patient	Control	Patient	Control	Patient	Control
Final recruit	6	N/A	16	7	17	8	7	7	24	17
Status	Completed		Completed		Completed		Completed		Completed	
	APOLLON		**CONCERT-HF**		**ALLSTAR**		**DYNAMIC**		**HOPE-Duchenne**	
Clinical trial	Phase III		Phase II		Phase I/II		Phase I		Phase I/II	
Cell type	Autologous atrial CDC		MSC and c-Kit+		Allogeneic CDC		Allogeneic CDC		Allogeneic CDC	
Delivery method	Single dose, intracoronary		Transendocardial injection		Single dose, intracoronary infusion		Single dose, intracoronary infusion		Single dose, intracoronary infusion	
Number of CSC	3×10^5 cell/kg		5×10^6		25×10^6		N/A		$50–75 \times 10^6$ (25×10^6 each ventricular area)	
NCT ID	NCT02781922		NCT02501811		NCT01458405		NCT02293603		NCT02485938	
Condition	HLHS, single ventricle, HF		Ischemic cardiomyopathy		MI		Dilated cardiomyopathy		Cardiomyopathy secondary to DMD	

(continued)

Table 7.2 (continued)

I.C: LVEF	<55%	<40%		N/A		<35%		<35%	
Follow-up	12 months	12 months		12 months		12 months		12 months	
	Patient	Control	Patient	Control	Patient	Control	Patient	Control	Patient
Final recruit	N/A	36	108	N/A	N/A	N/A	14 + 28[a]	N/A	14 + 28[a]
Status	Recruiting	Recruiting		Ongoing, not recruiting		Ongoing, not recruiting		Ongoing, not recruiting	

CDC, cardiosphere-derived cells; CHF, chronic heart failure; DMD, Duchenne muscle dystrophy; HLHS, hypoplastic left heart syndrome; I.C, inclusion criteria; ICM, ischemic cardiomyopathy; MI, myocardial infarction; MSC, mesenchymal stem cells

[a]Phase Ia (14) and Phase Ib (28)

estimated to be up to 415 × 10^6 after 12 month post-infusion [145]. Despite the promising results of SCIPIO trial, the future clinical trials on c-kit$^+$ CSC are unclear, with no immediate plans to start phase II studies. Concerns regarding the results of two supplementary figures and reproducibility of the initial results have put this clinical trial into a halt until scientists gather more understanding about the biology of the c-kit$^+$ cells [147].

CADUCEUS trial
In 2009, the CArdiosphere-Derived aUtologous stem CElls to reverse ventricUlar dySfunction (CADUCEUS) trial started recruiting. CADUCEUS trial is a prospective, randomized, controlled study evaluating the safety and efficacy of intracoronary delivery of autologous CDC in patients with a MI and left ventricular EF between 25% and 40%. A total of 25 patients were randomized in a 2 : 1 ratio and received either CDC therapy (17 patients) or standard care (8 patients). CDC was isolated and expanded from endomyocardial biopsies and infused after 1.5–3 months following coronary angioplasty in patients with signs of left ventricular dysfunction. Cardiosphere was characterized based on the criteria that 95% of CSC express CD105, and less than 5% express CD45 and cytogenetic integrity. Patients received 12.5 to 25 × 10^6 CDC via intracoronary infusion [148]. After 1-year follow-up, MRI showed improvement in cardiac parameters, smaller scar mass, increase in viable myocardium, and systolic wall thickening as well as improvement in regional function in infarcted area. No major adverse effects, such as MI, or serious arrhythmias were observed. The results of the CADUCEUS trial showed safety and effectiveness of intracoronary administration of autologous CDC for the treatment of MI. However, no improvement in global cardiac function parameters such as end-diastolic and end-systolic volume and left ventricular EF was observed. The study points out the fact that CDC therapy may increase the amount of viable myocardium hence contributing to long-term viability.

TICAP trial
Similarly, the Transcoronary Infusion of Cardiac Progenitor (TICAP) trial was designed to test efficacy and safety of intracoronary administration of autologous cardiosphere-derived cells in children with hypoplastic left heart syndrome. A total of 14 patients were treated, 7 patients with CDC and 7 controls. Autologous CDC was isolated from atrial tissue and expanded in vitro. Characterization and validation of CDC was performed by analysis of the markers signal regulatory protein-α (SIRPα) and mesenchymal stem cell markers such as CD90, CD105, and vimentin in more than 60% of isolated CDCs. Patient-derived CDC also expressed typical cardiac transcription factors, such as GATA4, MEF2C, TBX5, HAND2, myocardin, MESP1, and NKX2.5, as well as a vascular endothelial progenitor marker, FLK1. Less than 1% of the cells expressed discoidin domain receptor 2 (DDR2), CD31, CD45, and the cardiac structural gene tropomyosin. Following 4–5 weeks after reconstructive surgery, patients received 2–3 × 10^6 CDC [149]. TICAP trial demonstrated that intracoronary infusion of autologous CDC is safe and feasible in children with hypoplastic left heart syndrome. Besides, staged palliation was feasible and safe, and the preliminary efficacy was evidenced by improvement in ventricular

function, somatic growth, as well as reduction of HF condition. A phase II clinical trial – the PERSEUS trial – is underway to determine the long-term effects and potential improvement in cardiac function.

PERSEUS trial
The cardiac ProgEnitoR cell infuSion to treat hEart Univentricular heart diseaseS (PERSEUS) trial is a phase II, randomized, and controlled study. Approved in 2013, the main objective was to determine efficacy of intracoronary infusion of CDC in children with hypoplastic left heart syndrome as a following step forward based on the results of the TICAP trial. The trial was divided into primary and secondary outcomes. The first outcome assessed intracoronary infusion of CDC after staged palliation compared with controls receiving standard surgical treatment, after 3-month follow-up (study A). The second study was to offer patients in the control group the option to receive CDC infusion 4 months after staged palliation (study B). The secondary outcome was to demonstrate efficacy of CDC infusion after 12-month follow-up in both studies. A total of 41 patients, 24 patients receiving CDC and 17 controls, with hypoplastic left heart syndrome were randomly assigned to receive either CDC infusion or standard care. The researchers found that cardiac function was significantly improved and left ventricular volumes and fibrosis were reduced. Additionally, a general improvement in HF condition and somatic growth in the CDC-treated group was observed when compared with control group for both outcomes following 3 and 12 months after CDC treatment. Phase III APOLLON clinical trial is warrant to confirm the risk and benefit of this therapy [150].

APOLLON trial
Starting in 2016, the Efficacy and Safety Study of Autologous Cardiac Stem Cells Treated After Reconstructive Surgery in Pediatric Patients With Congenital Heart Disease (APOLLON) trial will become the first clinical trial where CSC will reach clinical trial phase III. APOLLON is an industry-sponsored, multicenter randomized single-blind parallel-group study phase III trial based on the positive results of the TICAP and PERSEUS trials. In this trial, three hospitals are involved to confirm the risk/benefit of intracoronary infusion of CDC in children with hypoplastic left heart syndrome with less than 55% of left ventricular EF. The primary outcome is to evaluate changes in EF by MRI at baseline, 6 and 12 months. Secondary outcome will evaluate EF by echocardiography, and by cardiac catheterization from baseline and at 6 and 12 months. Other measurements included are effective arterial elastase and end-systolic elastase, ventricular stiffness, HF index and quality of life index as well as a number of adverse effects. The APOLLON trial is expected to be completed by July 2018 [151].

CONCERT-HF trial
The Combination of Mesenchymal and c-kit$^+$ Cardiac Stem Cells as Regenerative Therapy for Heart Failure (CONCERT-HF) is a phase II, randomized, placebo-controlled study designed to assess feasibility, safety, and effect of autologous BM-derived MSC and c-kit+ CSC, alone or in combination. Infused cells will be delivered by transendocardial injection in patients with ischemic

cardiomyopathy. For the study, 144 patients will be enrolled and randomized in four groups, (1) combined MSC and CSC therapy, (2) MSC only, (3) CSC, and (4) placebo at a 1 : 1 : 1 : 1 ratio. Patients will receive 15 transendocardial injections of 400 μl of either group administered to the left ventricle. The target dose for group one is 150×10^6 MSC and 5×10^6 of CSC, 150×10^6 MSC for group two, 5×10^6 of CSC for group three, and placebo (Plasma-Lyte A) for group four. Cardiac function will be measured by MRI at baseline and then after injection at day 1, week 2, 1 month, 3 months, 6 months, and 12 months. Inclusion criteria include patients with coronary heart disease with evidence of myocardial dysfunction, left ventricular dysfunction, EF less than 40% and/or evidence of HF by MRI. CONCERT-HF trial started in October 2015, and the completion date is expected to be in February 2019 [152].

ALLSTAR trial

The ALLogeneic Heart STem Cells to Achieve Myocardial Regeneration (ALLSTAR) trial is a randomized, multicenter, double-blind, placebo-controlled phase I/II study to assess the safety and efficacy of intracoronary delivery of allogeneic CDC in patients with a MI. In phase I, 14 patients will be enrolled in an open-label, non-randomized, dose-escalation safety trial. In phase II trial, 120 patients will be randomized with patients receiving CDC and/or placebo in a 2 : 1 ratio, in a double-blind, randomized, placebo-controlled trial. Patients will receive a single intracoronary dose of 25×10^6 CDCs (CAP-1002) or placebo. The primary endpoint for both phases is safety at 1 month as patients may develop cardiac complication (active myocarditis) due to the presence of circulating autoantibodies against donor CDC. Other primary endpoints include death due to ventricular tachycardia, ventricular fibrillation, and sudden death. The primary efficacy endpoint is relative percentage improvement in infarct size after 12 months assessed by MRI in both groups [153]. The study started recruiting in October 2012; unfortunately, in November 2017, the American Heart Association Annual Scientific Sessions in Anaheim announced that the ALLSTAR trial failed to show reduction in infarct size from intracoronary injection of stem cells after MI [154].

DYNAMIC trial

The Dilated cardiomYopathy iNtervention with Allogeneic MyocardIally-regenerative Cells (DYNAMIC) is a randomized, double-blind, placebo-controlled, phase I study. The DYNAMIC trial will assess the safety of intracoronary infusion of allogeneic human CDC in patients with dilated cardiomyopathy. The study is divided in phase Ia and phase Ib. Phase Ia will recruit 14 patients in an open-label, single-arm, study design to dose escalation of allogeneic CDC. Phase Ib will recruit 28 patients and consists of a double-blind, randomized, placebo-controlled study. In the study, the intracoronary delivery of allogeneic CDC is performed into the main coronary arteries supplying the anterior, lateral and inferior/posterior ventricular areas. For phase Ia, follow-up will be performed at week 2 and then at 1, 2, 3, 6, and 12 months after intracoronary infusion. The trial is ongoing and not recruiting participants [155].

HOPE-Duchenne trial
The Halt cardiomyOPathy progrEssion (HOPE) in Duchenne trial is a randomized, open-label study that will assess the safety and efficacy of intracoronary delivery of allogeneic CDC in patients with cardiomyopathy secondary to Duchenne muscular dystrophy. Patients will be randomized in a 1 : 1 ratio and will receive either a single intracoronary infusion of allogeneic CDC into the main coronary arteries supplying the anterior, lateral, and inferior/posterior ventricular areas or standard care. Patients will receive a total dose of 75×10^6 cells delivered as 25×10^6 cells into selected ventricular areas. Follow-up will be performed at "follow-up" telephone call on the third day after infusion and study visits at weeks 2 and 6 and then 3, 6, and 12 months after intracoronary delivery. The trial is ongoing and not recruiting participants [156].

7.4.2 Clinical Trials Using Non-endogenous Cardiac Stem Cell-Based Therapies

One of the first cell-based clinical trials that aimed to achieve improvement of cardiac function was the Myoblast Autologous Grafting in Ischemic Cardiomyopathy (MAGIC) [157]. The MAGIC trial evaluated the use of skeletal myoblasts; although these cells do not have a cardiac origin, the transplantation was reported to improve cardiac functionality after MI. The results of the MAGIC trial showed that skeletal myoblast failed to improve cardiac functionality and ventricular contractile function. Interestingly, skeletal myoblasts engrafted and differentiated into skeletal myocytes instead of becoming cardiomyocytes. Furthermore, the engrafted skeletal myocytes were not electrically coupled with the healthy myocardium [157].

In 2011, BMdCSC began to be considered the cell type of choice for cell-based therapies mainly due to the potential to differentiate into various phenotypes; BMdCSC can be easily procured and do not require extensive culture techniques [158, 159]. Many preclinical and clinical trials carried out after these discoveries also produced interesting, controversial, and conflicting results [160–165]. At clinical level, the use of BMdCSC to treat myocardial injury provided modest benefit in most but not all studies. Unfortunately, the disparity between results has sent researchers back to the bench to gain a clearer understanding on the exact mechanism of action of BMdCSC. Other aspects that are being considered include the type of donor stem cell, the enhancement of homing capacities, dose and route of administration, and improvement of long-term engraftment [166–168]. The most promising clinical trials to date on cell-based therapy for the treatment of MI are the Reinfusion of Enriched Progenitor cells And Infarct Remodeling in Acute Myocardial Infarction (REPAIR-AMI) study and the Clinical Benefit and Long-Term Outcome After Intracoronary Autologous Bone Marrow Cell Transplantation in Patients With Acute Myocardial Infarction (BALANCE) study.

In 2006, the REPAIR-AMI study, the first of this kind, quantified global improvement in left ventricular EF. Two hundred and four patients were randomized to receive an intracoronary injection of 2.36×10^7 BMdC within 3–7 days after MI. After 1 year, BMdC-transplanted patients showed a reduction in death and recurrence of MI [169]. In 2009, the BALANCE study, following a similar approach as the REPAIR-AMI study, investigated the quantitative improvement in ventricular function and contractility of patients treated with autologous BM cells after MI. From 124 patients, 62 underwent intracoronary BMdC infusion within 5–9 days after MI. In this case, $6.1 \pm 3.9 \times 10^7$ BMdC were infused directly into the infarct area via an angioplasty balloon catheter. Three months after BMdC therapy, patients displayed an improvement in cardiac function and stroke volume index. The study also showed a reduction in mortality only in the BMdC-treated group [170]. In 2001, the Transplantation of Progenitor Cells and Regeneration Enhancement in Acute Myocardial Infarction (TOPCARE-AMI) started recruiting patients. The trial evaluated safety and feasibility of autologous progenitor cell transplantation in patients with IHD [171]. In 2011, the results of a 5-year follow-up of TOPCARE-AMI trial concluded that long-term cell therapy with autologous progenitor cells expanded ex vivo is safe and results in coronary blood flow and left ventricular function improvement. Additionally, the authors hypothesized that autologous cells enhanced endogenous repair through the release of paracrine factors [172]. An extensive review on non-CSC in ischemic cardiomyopathy has recently been elaborated by Wang and collaborators [173].

7.4.3 Meta-Analysis on the Use of Cardiac Stem Cell-Based Therapies

Although a multitude of clinical trials have been performed to date, the results remain ambiguous and no single-cell-based therapy for heart disease has conclusively proven to be effective in improving cardiac function and/or in regenerating the missing myocardium. In 2015, Gyongyosi and collaborators published a comprehensive meta-analysis including multinational databases and individual data from patients with IHD treated with stem cell therapy. The meta-analysis was elaborated with individual data from 1252 patients from 12 randomized trials including: ASTAMI, Aalst, BOOST, BONAMI, CADUCEUS, FINCELL, REGENT, REPAIR-AMI, SCAMI, SWISS-AMI, TIME, LATE [164]. The overall conclusion of the meta-analysis revealed that an *"intracoronary cell therapy provided no benefit, in terms of clinical events or changes in left ventricular function"*. In 2016, another meta-analysis by Fisher and collaborators found low-quality evidences that BMdC therapy can effectively reduce mortality and incidence of non-fatal MI beside improvement of cardiac function in patients with chronic IHD, HF secondary to IHD, and refractory angina [165]. The meta-analysis included 38 clinical trials and 1907 patients in total, 1114 patients under BMdC therapy and 793 no cell therapy.

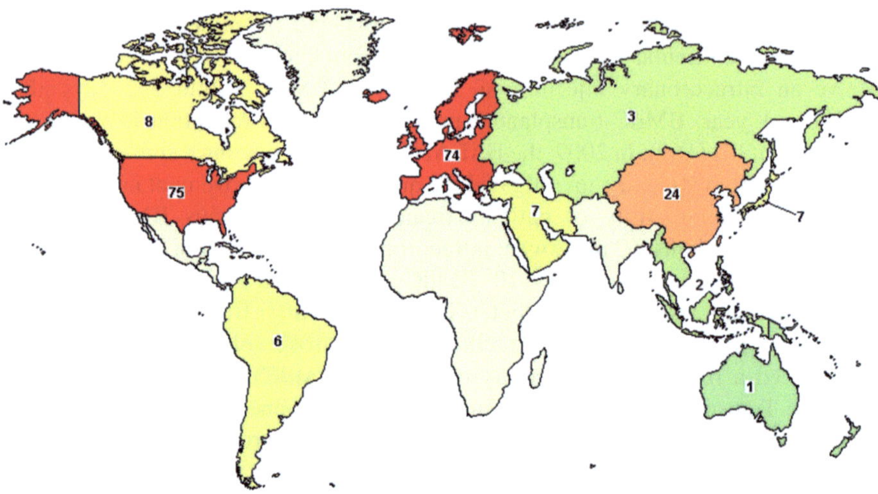

Fig. 7.1 Worldwide distribution of ongoing cardiac stem/progenitor cell trials. (From https://clinicaltrials.gov/ct2/home)

Importantly, the authors indicated that the results were subjected to a reduced number of participants, and hence strong conclusions cannot be drawn.

Overall, in this review, we selected the most interesting past, present, and future clinical trials with CSC. Currently, a total of 203 clinical trials containing the term "cardiac stem cell" and "heart diseases" can be found. The status of such trials is either approved, recruiting, active but not recruiting, or completed (Fig. 7.1).

7.5 Future Directions in Cardiac Regenerative Medicine

Nowadays, except for heart transplantation, current pharmacological and interventional strategies only treat symptoms but are unable to cure the disease. Treatments for ischemic injury are focused in restoring the blood supply and improving heart function in an attempt to ameliorate the adverse effects of myocardial damage. However, these strategies do not potentiate or activate the intrinsic capacity of the heart to regenerate or repair itself. Currently, cardiovascular regenerative medicine is focused on studying and understanding the potential of the mammalian heart to regenerate as well as to design new strategies that will eventually lead to not only complete repair of the heart but also tissue regeneration after cardiac damage. A recent statement from the Transnational AllianCe for regenerative Therapies In Cardiovascular Syndromes (TACTICS) defined "cell implantation" as one of the products for cardiovascular regenerative medicine as well as the "injection of biological or synthetic factors" or "genetic and epigenetic modifications" [174]. Such therapeutic strategies are focused on enhancing stem cell potential either from endogenous or non-endogenous CSC. All these approaches, if successful, should minimize the need for organs, as in general the need of a heart often exceeds availability.

The use of CSC in regenerative medicine has evolved in the last years, but still must face enormous challenges ahead. In the 14th International Symposium on Stem Cell Therapy and Cardiovascular Innovations celebrated in Madrid, experts reached into an agreement and concluded an interesting take-home message: althought cell therapy is promising, it is not ready yet for clinical application. Therefore, more meticulously designed phase III pivotal clinical trials are warranted, and concomitant basic research must continue to guide further clinical evaluations [174, 175]. New therapeutic strategies must focus on several crucial aspects such as repetitive treatments, cell-free products, tissue engineering approaches and in situ cell reprogramming, selection of the optimal dosage, best delivery route, and tracking technology for each condition [174, 175]. The functionality of CSC is influenced by multiple factors, including the initial source, harvesting and isolation techniques, and manufacturing which needs to be standardized. More specific and efficient delivery technologies aiming to achieve the optimal dosage of biological material are also needed. Today, tissue bioengineering offers scientists new tools such as 3D printing, biological scaffolds that can allow us to reinvent therapies to treat diseases of various etiologies. One of the main problems in CSC therapy is poor retention of infused cells; currently the design of biological and non-biological scaffolds may serve for the generation of biological patches to allow CSC to engraft, proliferate, and form new tissues [176–178]. For cardiac cell therapies, these scaffolds are divided into injectable biomaterials – gel, foam, and nanofibrous forms –, decellularized matrices and cardiac patches made of synthetic or natural hydrogels [179–182]. Scientists are currently striving with selecting an appropriate scaffold and cell type which facilitate cell integration onto the damaged myocardium. Ultimately, the strategy is to transplant in vitro generated patches or even an in vitro-built fully grown heart scaffold seeded with autologous CSC [183]. The first clinical application of tissue-engineered technology was performed by Menasche and collaborators in 2015, as part of the proof-of-concept ESCORT trial [184]. In a case report, Menasche embedded ESC-derived CSC into a fibrin scaffold and applied it onto the infarct area of a patient suffering from severe heart failure. The procedure was performed concomitantly to CABG surgery. In fact, ESC were amplified and directed toward a cardiac progenitor phenotype by stimulating the cells with bone morphogenetic protein 2 (BMP-2) and fibroblast growth factor receptor (FGFR)-tyrosine kinase inhibitor. ESC-derived CSC expressed cardiac transcription factor Islet-1 and SSEA-1. At recruitment time, patient presented with NYHA functional class III and left ventricular EF or 26%. After 3 months, the patient presented with NYHA functional class I and left ventricular EF of 36% [185]. Although the authors did not extrapolate strong conclusions, this case report is clearly a promising first outcome regarding the positive combinatorial effects of CSC therapy and a biological tissue-engineered scaffold.

The heart is a remarkable organ, made of various cell types with different functionalities evolving over time. Within such complexity, a single approach to promote cardiac regeneration has proved to be challenging; therefore, the ideal cardiac regenerative therapy would comprise of a combination of key cells working in synergy with paracrine factors to create a permissive environment for the heart to heal or even regenerate [57, 90, 93, 186]. The predominant view is that the positive effects of adult CSC therapy are mediated by the release of paracrine, proangiogenic factors, production of immunomodulatory signals and anti-apoptotic mediators. All these players preserve viability of damaged myocardium, promote viable myocardium to regenerate, induce vascularization and favorable changes in scar tissue. Additionally, hypoxia conditions are used for maintenance of proliferative competency of stem cells; thus, the use of tolerable hypoxic conditions can enhance metabolic reprogramming of adult cardiomyocytes resulting in cell cycle reentry and myocardial regeneration [187, 188]. In the future, one can envisage new therapies with factors that preconditions and protects the heart to potentially harmful myocardial ischemia [131].

7.6 Final Remarks

The last decade of regenerative medicine has led to the discovery, isolation, and characterization of a number of CSC capable of regulating cardiac homeostasis and cellular turnover upon insult. Despite the extensive characterization, there is still controversy and lack of consensus regarding cell surface markers, nature, origin, and function of CSC populations. To some extent, the lack of consensus arises from attempting to identify putative CSC populations by the expression of specific cell surface markers. Furthermore, cell handling and preparation, differences in CSC culture, priming, isolation, purification, and expansion also contribute to discord. Additionally, the genetic background of current animal models may confer different wound healing response patterns after MI which may further overestimate CSC efficacy among different organisms in preclinical studies [189]. Besides, little work has systematically been performed in order to compare the developmental origins and hierarchical structure of isolated CSC. Unlike cells from BM, CSC residing in the heart are predisposed to differentiate into cardiac, endothelial, and smooth muscle cells. Unfortunately, it appears that despite their presence in the heart, CSC functions are limited to maintenance and homeostasis of the cardiac muscle. In fact, even though CSC can differentiate into endothelial, smooth muscle and in a much slower capacity into cardiomyocytes, the CSC population cannot compensate for the sudden loss of cardiomyocytes after MI. Notwithstanding these uncertainties, CSC have been taken into clinical trials with more or less success. Another important aspect yet to be elucidated is how to increase the rate at which CSC replenish the missing myocardium with new cardiomyocytes upon insult [36]. Additionally, if the damaged myocardium is replenished, there is no guarantee that electrical coupling will be achieved; thus, further studies should tackle that aspect as well [190].

In conclusion, despite the fact that some of the results obtained on clinical studies are potentially very exciting, there is still not sufficient understanding to go into clinical practice and debate regarding optimal trial design parameters. Furthermore, the small number of patients, the short period of follow-up, and the putative nature of the results have caused in most cases mild to modest success in clinical trials. Consequently, long-term studies of CSC infusion in patients should be performed to assess whether or not myocardial regeneration occurs following injury. In light of these results, CSC research has to be deconstructed from the bedside to the bench and backward. Future of CSC regenerative therapies will need to combine aspects of stem cell biology, concomitant cardiovascular pharmacological therapies including modulation of inflammatory response, tissue engineering, and the use of specific paracrine factors. The reported progress clearly shows the great potential of CSC in myocardial repair; nevertheless more research is needed to "reach" long-lasting goal of cardiac regeneration.

Acknowledgments E.J.G-C wants to acknowledge the EU FP7-Marie Curie-ITN fellows and lecturers for the scientific input over the years when E.J.G-C was a Marie Curie Fellow. During the preparation of this chapter, E.O. was recipient of a CNIC-IPP Research Fellowship funded by the FP7-PEOPLE-2012-COFUND - Marie Curie Action (600396 / 17214-2016), and is currently supported by the 'Talent Attraction program' of Comunidad de Madrid (2017-T1/BMD-5185). The CNIC is supported by the Ministerio de Economía, Industria y Competitividad (MEIC) and the Pro CNIC Foundation and is a Severo Ochoa Center of Excellence (SEV-2015-0505).

Disclosures E.J.G-C. and E.O. declare no competing financial interests or conflict of interest.

References

1. Benjamin EJ, Blaha MJ, Chiuve SE et al (2017) Heart disease and stroke statistics-2017 update: a report from the American Heart Association. Circulation 135:e146–e603. https://doi.org/10.1161/CIR.0000000000000485
2. http://www.who.int/cardiovascular_diseases/en/
3. http://www.ehnheart.org
4. Thygesen K, Alpert JS, Jaffe AS et al (2012) Third universal definition of myocardial infarction. Circulation 126:2020–2203. https://doi.org/10.1161/CIR.0b013e31826e1058
5. Yancy CW, Jessup M, Bozkurt B et al (2013) 2013 ACCF/AHA guideline for the management of heart failure: a report of the american college of cardiology foundation/american heart association task force on practice guidelines. Circulation 128:240–327. https://doi.org/10.1161/CIR.0b013e31829e8776
6. Frangogiannis NG (2014) The inflammatory response in myocardial injury, repair, and remodelling. Nat Rev Cardiol 11:255–265. https://doi.org/10.1038/nrcardio.2014.28
7. Frazier OH (1994) First use of an untethered, vented electric left ventricular assist device for long-term support. Circulation 89:2908–2914
8. Sajgalik P, Grupper A, Edwards BS et al (2016) Current status of left ventricular assist device therapy. Mayo Clin Proc 91:927–940. https://doi.org/10.1016/j.mayocp.2016.05.002
9. Deb S, Wijeysundera HC, Ko DT et al (2013) Coronary artery bypass graft surgery vs percutaneous interventions in coronary revascularization. JAMA 310:2086. https://doi.org/10.1001/jama.2013.281718
10. Zweier JL (1988) Measurement of superoxide-derived free radicals in the reperfused heart. Evidence for a free radical mechanism of reperfusion injury. J Biol Chem 263:1353–1357

11. Minicucci MF, Azevedo PS, Polegato BF et al (2011) Heart failure after myocardial infarction: clinical implications and treatment. Clin Cardiol 34:410–414. https://doi.org/10.1002/clc.20922
12. Pfeffer MA (1998) ACE inhibitors in acute myocardial infarction: patient selection and timing. Circulation 97:2192–2194. https://doi.org/10.1161/01.CIR.97.22.2192
13. Gottlieb SS, McCarter RJ, Vogel RA (1998) Effect of beta-blockade on mortality among high-risk and low-risk patients after myocardial infarction. N Engl J Med 339:489–497. https://doi.org/10.1056/NEJM199808203390801
14. Ibanez B, James S, Agewall S et al (2018) 2017 ESC Guidelines for the management of acute myocardial infarction in patients presenting with ST-segment elevation. Eur Heart J 39:119–177. https://doi.org/10.1093/eurheartj/ehx393
15. Ptaszek LM, Mansour M, Ruskin JN, Chien KR (2012) Towards regenerative therapy for cardiac disease. Lancet 379:933–942. https://doi.org/10.1016/S0140-6736(12)60075-0
16. Fang H, Lai NC, Gao MH et al (2013) Comparison of adeno-associated virus serotypes and delivery for cardiac gene transfer. Hum Gene Ther Methods 23:234–241. https://doi.org/10.1089/hgtb.2012.105.Comparison
17. Kaspar BK, Roth DM, Lai NC et al (2005) Myocardial gene transfer and long-term expression following intracoronary delivery of adeno-associated virus. J Gene Med 7:316–324. https://doi.org/10.1002/jgm.665
18. Müller OJ, Leuchs B, Pleger ST et al (2006) Improved cardiac gene transfer by transcriptional and transductional targeting of adeno-associated viral vectors. Cardiovasc Res 70:70–78. https://doi.org/10.1016/j.cardiores.2005.12.017
19. Poss KD, Wilson LG, Keating MT (2002) Heart regeneration in zebrafish. Science 298:2188–2190. https://doi.org/10.1126/science.1077857
20. Kikuchi K, Poss KD (2012) Cardiac regenerative capacity and mechanisms. Annu Rev Cell Dev Biol 28:719–741. https://doi.org/10.1146/annurev-cellbio-101011-155739
21. Wang J, Panáková D, Kikuchi K et al (2011) The regenerative capacity of zebrafish reverses cardiac failure caused by genetic cardiomyocyte depletion. Development 138:3421–3430. https://doi.org/10.1242/dev.068601
22. Kikuchi K, Holdway JE, Major RJ et al (2011) Retinoic acid production by endocardium and epicardium is an injury response essential for zebrafish heart regeneration. Dev Cell 20:397–404. https://doi.org/10.1016/j.devcel.2011.01.010
23. Ito K, Morioka M, Kimura S et al (2014) Differential reparative phenotypes between zebrafish and medaka after cardiac injury. Dev Dyn 243:1106–1115. https://doi.org/10.1002/dvdy.24154
24. Porrello ER, Mahmoud AI, Simpson E et al (2011) Transient regenerative potential of the neonatal mouse heart. Science 331:1078–1080. https://doi.org/10.1126/science.1200708
25. Porrello ER, Mahmoud AI, Simpson E et al (2013) Regulation of neonatal and adult mammalian heart regeneration by the miR-15 family. Proc Natl Acad Sci 110:187–192. https://doi.org/10.1073/pnas.1208863110
26. Jesty SA, Steffey MA, Lee FK et al (2012) c-Kit^{+} precursors support postinfarction myogenesis in the neonatal, but not adult, heart. Proc Natl Acad Sci 109:13380–13385. https://doi.org/10.1073/pnas.1208114109
27. Aurora AB, Porrello ER, Tan W et al (2014) Macrophages are required for neonatal heart regeneration. J Clin Invest 124:1382–1392. https://doi.org/10.1172/JCI72181DS1
28. Broughton KM, Sussman MA (2016) Empowering adult stem cells for myocardial regeneration V2.0. Circ Res 118:867–880. https://doi.org/10.1161/CIRCRESAHA.115.305227
29. Soonpaa MH, Field LJ (1997) Assessment of cardiomyocyte DNA synthesis in normal and injured adult mouse hearts. Am J Physiol 272:H220–H226
30. Bergmann O, Bhardwaj RD, Bernard S et al (2009) Evidence for cardiomyocyte renewal in humans. Science 324:98–102. https://doi.org/10.1126/science.1164680
31. Bergmann O et al (2009) Evidence for cardiomyocyte renewal in humans. Science 98–102([PubMed: 19342590]):324

32. Walsh S, Pontén A, Fleischmann BK, Jovinge S (2010) Cardiomyocyte cell cycle control and growth estimation in vivo—an analysis based on cardiomyocyte nuclei. Cardiovasc Res 86:365–373. https://doi.org/10.1093/cvr/cvq005
33. Kikuchi K, Holdway JE, A a W et al (2010) Primary contribution to zebrafish heart regeneration by gata4+ cardiomyocytes. Nature 464:601–605. https://doi.org/10.1038/nature08804
34. Beltrami AP, Barlucchi L, Torella D et al (2003) Adult cardiac stem cells are multipotent and support myocardial regeneration. Cell 114:763–776
35. Orlic D, Kajstura J, Chimenti S et al (2001) Bone marrow cells regenerate infarcted myocardium. Nature 410:701–705. https://doi.org/10.1038/35070587
36. Senyo SE, Steinhauser ML, Pizzimenti CL et al (2012) Mammalian heart renewal by pre-existing cardiomyocytes. Nature 493:433–436. https://doi.org/10.1038/nature11682
37. Dixit P, Katare R (2015) Challenges in identifying the best source of stem cells for cardiac regeneration therapy. Stem Cell Res Ther 6:26. https://doi.org/10.1186/s13287-015-0010-8
38. Messina E (2004) Isolation and expansion of adult cardiac stem cells from human and murine heart. Circ Res 95:911–921. https://doi.org/10.1161/01.RES.0000147315.71699.51
39. Smith RR, Barile L, Cho HC et al (2007) Regenerative potential of cardiosphere-derived cells expanded from percutaneous endomyocardial biopsy specimens. Circulation 115:896–908. https://doi.org/10.1161/CIRCULATIONAHA.106.655209
40. Bearzi C, Rota M, Hosoda T et al (2007) Human cardiac stem cells. Proc Natl Acad Sci 104:14068–14073. https://doi.org/10.1073/pnas.0706760104
41. Urbanek K, Cesselli D, Rota M et al (2006) Stem cell niches in the adult mouse heart. Proc Natl Acad Sci 103:9226–9231. https://doi.org/10.1073/pnas.0600635103
42. Linke A, Muller P, Nurzynska D et al (2005) Stem cells in the dog heart are self-renewing, clonogenic, and multipotent and regenerate infarcted myocardium, improving cardiac function. Proc Natl Acad Sci 102:8966–8971. https://doi.org/10.1073/pnas.0502678102
43. Sanada F, Kim J, Czarna A et al (2014) c-Kit–positive cardiac stem cells nested in hypoxic niches are activated by stem cell factor reversing the aging myopathy novelty and significance. Circ Res 114:41–55. https://doi.org/10.1161/CIRCRESAHA.114.302500
44. Khan M, Koch WJ (2016) c-Kit^{+} cardiac stem cells. Circ Res 118:783–785. https://doi.org/10.1161/CIRCRESAHA.115.308103
45. Sultana N, Zhang L, Yan J et al (2015) Resident c-kit^{+} cells in the heart are not cardiac stem cells. Nat Commun 6:8701. https://doi.org/10.1038/ncomms9701
46. Di Meglio F, Castaldo C, Nurzynska D et al (2010) Localization and origin of cardiac CD117-positive cells: identification of a population of epicardially-derived cells in adult human heart. Ital J Anat Embryol 115:71–78
47. Di Meglio F, Castaldo C, Nurzynska D et al (2010) Epithelial–mesenchymal transition of epicardial mesothelium is a source of cardiac CD117-positive stem cells in adult human heart. J Mol Cell Cardiol 49:719–727. https://doi.org/10.1016/j.yjmcc.2010.05.013
48. Limana F, Bertolami C, Mangoni A et al (2010) Myocardial infarction induces embryonic reprogramming of epicardial c-kit+ cells: role of the pericardial fluid. J Mol Cell Cardiol 48:609–618. https://doi.org/10.1016/j.yjmcc.2009.11.008
49. Keith MCL, Bolli R (2015) "String theory" of c-kitpos cardiac cells: a new paradigm regarding the nature of these cells that may reconcile apparently discrepant results. Circ Res 116:1216–1230. https://doi.org/10.1161/CIRCRESAHA.116.305557
50. Tang XL, Li Q, Rokosh G et al (2016) Long-term outcome of administration of c-kit^{+} cardiac progenitor cells after acute myocardial infarction: transplanted cells do not become cardiomyocytes, but structural and functional improvement and proliferation of endogenous cells persist for at least one year. Circ Res 118:1091–1105. https://doi.org/10.1161/CIRCRESAHA.115.307647
51. Dettman RW, Denetclaw W, Ordahl CP, Bristow J (1998) common epicardial origin of coronary vascular smooth muscle, perivascular fibroblasts, and intermyocardial fibroblasts in the avian heart. Dev Biol 193:169–181. https://doi.org/10.1006/dbio.1997.8801

52. Christoffels VM, Grieskamp T, Norden J et al (2009) Tbx18 and the fate of epicardial progenitors. Nature 458:E8–E9. https://doi.org/10.1038/nature07916
53. Ding Z, Temme S, Quast C, Friebe D, Jacoby C, Zanger K et al (2016 May) Epicardium-derived cells formed after myocardial injury display phagocytic activity permitting in vivo labeling and tracking. Stem Cells Transl Med [Internet] 5(5):639–650
54. Cai C-L, Martin JC, Sun Y et al (2008) A myocardial lineage derives from Tbx18 epicardial cells. Nature 454:104–108. https://doi.org/10.1038/nature06969
55. Smits A, Riley P (2014) Epicardium-derived heart repair. J Dev Biol 2:84–100. https://doi.org/10.3390/jdb2020084
56. Smart N, Dubé KN, Riley PR (2013) Epicardial progenitor cells in cardiac regeneration and neovascularisation. Vascul Pharmacol 58:164–173. https://doi.org/10.1016/j.vph.2012.08.001
57. Zhou B, Honor LB, He H et al (2011) Adult mouse epicardium modulates myocardial injury by secreting paracrine factors. J Clin Invest 121:1894–1904. https://doi.org/10.1172/JCI45529
58. van Wijk B, Gunst QD, Moorman AFM, van den MJB H (2012) Cardiac regeneration from activated epicardium. PLoS One 7:e44692. https://doi.org/10.1371/journal.pone.0044692
59. Smart N, Bollini S, Dubé KN et al (2011) De novo cardiomyocytes from within the activated adult heart after injury. Nature 474:640–644. https://doi.org/10.1038/nature10188
60. Paulitschek C, Schulze-Matz P, Hesse J et al (2017) Generation and characterization of two iPSC lines from human epicardium-derived cells. Stem Cell Res 20:50–53. https://doi.org/10.1016/j.scr.2017.02.007
61. Oh H, Bradfute SB, Gallardo TD et al (2003) Cardiac progenitor cells from adult myocardium: homing, differentiation, and fusion after infarction. Proc Natl Acad Sci U S A 100:12313–12318. https://doi.org/10.1073/pnas.2132126100
62. Freire AG, Nascimento DS, Forte G et al (2014) Stable phenotype and function of immortalized Lin^- $Sca\text{-}1^+$ cardiac progenitor cells in long-term culture: a step closer to standardization. Stem Cells Dev 23:1012–1026. https://doi.org/10.1089/scd.2013.0305
63. Uchida S, De Gaspari P, Kostin S et al (2013) Sca1-derived cells are a source of myocardial renewal in the murine adult heart. Stem Cell Reports 1:397–410. https://doi.org/10.1016/j.stemcr.2013.09.004
64. Valente M, Nascimento DS, Cumano A, Pinto-do-Ó P (2014) $Sca\text{-}1^+$ cardiac progenitor cells and heart-making: a critical synopsis. Stem Cells Dev 23:2263–2273. https://doi.org/10.1089/scd.2014.0197
65. Wang X, Hu Q, Nakamura Y et al (2006) The role of the $Sca\text{-}1^+/CD31^-$ cardiac progenitor cell population in postinfarction left ventricular remodeling. Stem Cells 24:1779–1788. https://doi.org/10.1634/stemcells.2005-0386
66. Wang X, Li Q, Hu Q et al (2014) Intra-myocardial injection of both growth factors and heart derived $Sca\text{-}1^+/CD31^-$ cells attenuates post-MI LV remodeling more than does cell transplantation alone: neither intervention enhances functionally significant cardiomyocyte regeneration. PLoS One 9:e95247. https://doi.org/10.1371/journal.pone.0095247
67. Smits AM, van Vliet P, Metz CH et al (2009) Human cardiomyocyte progenitor cells differentiate into functional mature cardiomyocytes: an in vitro model for studying human cardiac physiology and pathophysiology. Nat Protoc 4:232–243. https://doi.org/10.1038/nprot.2008.229
68. Hierlihy AM, Seale P, Lobe CG et al (2002) The post-natal heart contains a myocardial stem cell population. FEBS Lett 530:239–243
69. Martin CM, Meeson AP, Robertson SM et al (2004) Persistent expression of the ATP-binding cassette transporter, Abcg2, identifies cardiac SP cells in the developing and adult heart. Dev Biol 265:262–275
70. Unno K, Jain M, Liao R (2012) Cardiac side population cells: moving toward the center stage in cardiac regeneration. Circ Res 110:1355–1363. https://doi.org/10.1161/CIRCRESAHA.111.243014

71. Dey D, Han L, Bauer M et al (2013) Dissecting the molecular relationship among various cardiogenic progenitor cells. Circ Res 112:1253–1262. https://doi.org/10.1161/CIRCRESAHA.112.300779
72. Yellamilli A, van Berlo JH (2016) The role of cardiac side population cells in cardiac regeneration. Front Cell Dev Biol 4:1–9. https://doi.org/10.3389/fcell.2016.00102
73. Oyama T, Nagai T, Wada H et al (2007) Cardiac side population cells have a potential to migrate and differentiate into cardiomyocytes in vitro and in vivo. J Cell Biol 176:329–341. https://doi.org/10.1083/jcb.200603014
74. Alfakir M, Dawe N, Eyre R et al (2012) The temporal and spatial expression patterns of ABCG2 in the developing human heart. Int J Cardiol 156:133–138. https://doi.org/10.1016/j.ijcard.2010.10.025
75. Sandstedt J, Jonsson M, Kajic K et al (2012) Left atrium of the human adult heart contains a population of side population cells. Basic Res Cardiol 107:255. https://doi.org/10.1007/s00395-012-0255-7
76. Tomita Y, Matsumura K, Wakamatsu Y et al (2005) Cardiac neural crest cells contribute to the dormant multipotent stem cell in the mammalian heart. J Cell Biol 170:1135–1146. https://doi.org/10.1083/jcb.200504061
77. Liang SX, Tan TYL, Gaudry L, Chong B (2010) Differentiation and migration of $Sca1^{+}$/$CD31^{-}$ cardiac side population cells in a murine myocardial ischemic model. Int J Cardiol 138:40–49. https://doi.org/10.1016/j.ijcard.2008.08.032
78. Liang SX, Khachigian LM, Ahmadi Z et al (2011) In vitro and in vivo proliferation, differentiation and migration of cardiac endothelial progenitor cells ($SCA1^{+}$/$CD31^{+}$ side-population cells). J Thromb Haemost 9:1628–1637. https://doi.org/10.1111/j.1538-7836.2011.04375.x
79. Pfister O (2005) $CD31^{-}$ but not $CD31^{+}$ cardiac side population cells exhibit functional cardiomyogenic differentiation. Circ Res 97:52–61. https://doi.org/10.1161/01.RES.0000173297.53793.fa
80. Noseda M, Harada M, McSweeney S et al (2015) PDGFR? demarcates the cardiogenic clonogenic $Sca1^{+}$ stem/progenitor cell in adult murine myocardium. Nat Commun 6:6930. https://doi.org/10.1038/ncomms7930
81. Laugwitz K-L, Moretti A, Lam J et al (2005) Postnatal $isl1^{+}$ cardioblasts enter fully differentiated cardiomyocyte lineages. Nature 433:647–653. https://doi.org/10.1038/nature03215
82. Lui KO, Zangi L, Silva EA et al (2013) Driving vascular endothelial cell fate of human multipotent Isl1+ heart progenitors with VEGF modified mRNA. Cell Res 23:1172–1186. https://doi.org/10.1038/cr.2013.112
83. Moretti A, Lam J, Evans SM, Laugwitz K-L (2007) Cardiovascular development: towards biomedical applicability. Cell Mol Life Sci 64:674–682. https://doi.org/10.1007/s00018-007-6520-5
84. Bu L, Jiang X, Martin-Puig S et al (2009) Human ISL1 heart progenitors generate diverse multipotent cardiovascular cell lineages. Nature 460:113–117. https://doi.org/10.1038/nature08191
85. Moretti A, Caron L, Nakano A et al (2006) Multipotent embryonic $Isl1^{+}$ progenitor cells lead to cardiac, smooth muscle, and endothelial cell diversification. Cell 127:1151–1165. https://doi.org/10.1016/j.cell.2006.10.029
86. Simpson DL, Mishra R, Sharma S et al (2012) A strong regenerative ability of cardiac stem cells derived from neonatal hearts. Circulation 126:S46–S53. https://doi.org/10.1161/CIRCULATIONAHA.111.084699
87. Barile L, Chimenti I, Gaetani R et al (2007) Cardiac stem cells: isolation, expansion and experimental use for myocardial regeneration. Nat Clin Pract Cardiovasc Med 4:S9–S14. https://doi.org/10.1038/ncpcardio0738
88. Barile L, Messina E, Giacomello A, Marbán E (2007) Endogenous cardiac stem cells. Prog Cardiovasc Dis 50:31–48. https://doi.org/10.1016/j.pcad.2007.03.005

89. Chimenti I, Gaetani R, Barile L et al (2012) Isolation and expansion of adult cardiac stem/progenitor cells in the form of cardiospheres from human cardiac biopsies and murine hearts. Methods Mol Biol 879:327–338
90. Chimenti I, Smith RR, Li T-S et al (2010) Relative roles of direct regeneration versus paracrine effects of human cardiosphere-derived cells transplanted into infarcted mice. Circ Res 106:971–980. https://doi.org/10.1161/CIRCRESAHA.109.210682
91. Davis DR, Kizana E, Terrovitis J et al (2010) Isolation and expansion of functionally-competent cardiac progenitor cells directly from heart biopsies. J Mol Cell Cardiol 49:312–321. https://doi.org/10.1016/j.yjmcc.2010.02.019
92. Redgrave RE, Tual-Chalot S, Davison BJ et al (2017) Cardiosphere-derived cells require endoglin for paracrine-mediated angiogenesis. Stem Cell Reports 8:1287–1298. https://doi.org/10.1016/j.stemcr.2017.04.015
93. Li T-S, Cheng K, Malliaras K et al (2012) Direct comparison of different stem cell types and subpopulations reveals superior paracrine potency and myocardial repair efficacy with cardiosphere-derived cells. J Am Coll Cardiol 59:942–953. https://doi.org/10.1016/j.jacc.2011.11.029
94. Johnston PV, Sasano T, Mills K et al (2009) Engraftment, differentiation, and functional benefits of autologous cardiosphere-derived cells in porcine ischemic cardiomyopathy. Circulation 120:1075–1083. https://doi.org/10.1161/CIRCULATIONAHA.108.816058
95. Shenje LT, Field LJ, Pritchard CA et al (2008) Lineage tracing of cardiac explant derived cells. PLoS One 3:e1929. https://doi.org/10.1371/journal.pone.0001929
96. Kasai-Brunswick TH, da Costa AR, Barbosa RAQ et al (2017) Cardiosphere-derived cells do not improve cardiac function in rats with cardiac failure. Stem Cell Res Ther 8:36. https://doi.org/10.1186/s13287-017-0481-x
97. Chong JJH, Chandrakanthan V, Xaymardan M et al (2011) Adult cardiac-resident MSC-like stem cells with a proepicardial origin. Cell Stem Cell 9:527–540. https://doi.org/10.1016/j.stem.2011.10.002
98. Chong JJH, Reinecke H, Iwata M et al (2013) Progenitor cells identified by PDGFR-alpha expression in the developing and diseased human heart. Stem Cells Dev 22:1932–1943. https://doi.org/10.1089/scd.2012.0542
99. Koninckx R, Daniëls A, Windmolders S et al (2013) The cardiac atrial appendage stem cell: a new and promising candidate for myocardial repair. Cardiovasc Res 97:413–423. https://doi.org/10.1093/cvr/cvs427
100. Fanton Y, Robic B, Rummens JL et al (2015) Cardiac atrial appendage stem cells engraft and differentiate into cardiomyocytes in vivo: a new tool for cardiac repair after MI. Int J Cardiol 201:10–19. https://doi.org/10.1016/j.ijcard.2015.07.066
101. Fanton Y, Houbrechts C, Willems L et al (2016) Cardiac atrial appendage stem cells promote angiogenesis in vitro and in vivo. J Mol Cell Cardiol 97:235–244. https://doi.org/10.1016/j.yjmcc.2016.06.005
102. Murry CE, Soonpaa MH, Reinecke H et al (2004) Haematopoietic stem cells do not transdifferentiate into cardiac myocytes in myocardial infarcts. Nature 428:664–668. https://doi.org/10.1038/nature02446
103. Liu N, Qi X, Han Z et al (2016) Bone marrow is a reservoir for cardiac resident stem cells. Sci Rep 6:28739. https://doi.org/10.1038/srep28739
104. Mouquet F (2005) Restoration of cardiac progenitor cells after myocardial infarction by self-proliferation and selective homing of bone marrow-derived stem cells. Circ Res 97:1090–1092. https://doi.org/10.1161/01.RES.0000194330.66545.f5
105. Kawada H, Ogawa M (2001) Bone marrow origin of hematopoietic progenitors and stem cells in murine muscle. Blood 98:2008–2013
106. Sadek HA, Martin CM, Latif SS et al (2009) Bone-marrow-derived side population cells for myocardial regeneration. J Cardiovasc Transl Res 2:173–181. https://doi.org/10.1007/s12265-009-9090-0

107. Evans MJ, Kaufman MH (1981) Establishment in culture of pluripotential cells from mouse embryos. Nature 292:154–156. https://doi.org/10.1038/292154a0
108. Martin MJ, Muotri A, Gage F, Varki A (2005) Human embryonic stem cells express an immunogenic nonhuman sialic acid. Nat Med 11:228–232. https://doi.org/10.1038/nm1181
109. Thomson JA, Itskovitz-Eldor J, Shapiro SS et al (1998) Embryonic stem cell lines derived from human blastocysts. Science 282:1145–1147
110. Mahla RS (2016) Stem cells applications in regenerative medicine and disease therapeutics. Int J Cell Biol 2016:1–24. https://doi.org/10.1155/2016/6940283
111. Takahashi K, Yamanaka S (2006) Induction of pluripotent stem cells from mouse embryonic and adult fibroblast cultures by defined factors. Cell 126:663–676. https://doi.org/10.1016/j.cell.2006.07.024
112. Devalla HD, Schwach V, Ford JW et al (2015) Atrial-like cardiomyocytes from human pluripotent stem cells are a robust preclinical model for assessing atrial-selective pharmacology. EMBO Mol Med 7:394–410. https://doi.org/10.15252/emmm.201404757
113. Takahashi K, Tanabe K, Ohnuki M et al (2007) Induction of pluripotent stem cells from adult human fibroblasts by defined factors. Cell 131:861–872. https://doi.org/10.1016/j.cell.2007.11.019
114. Yu J, Vodyanik MA, Smuga-Otto K et al (2007) Induced pluripotent stem cell lines derived from human somatic cells. Science 318:1917–1920. https://doi.org/10.1126/science.1151526
115. Lowry WE, Richter L, Yachechko R et al (2008) Generation of human induced pluripotent stem cells from dermal fibroblasts. Proc Natl Acad Sci 105:2883–2888. https://doi.org/10.1073/pnas.0711983105
116. Blelloch R, Venere M, Yen J, Ramalho-Santos M (2007) Generation of induced pluripotent stem cells in the absence of drug selection. Cell Stem Cell 1:245–247. https://doi.org/10.1016/j.stem.2007.08.008
117. Hou P, Li Y, Zhang X et al (2013) Pluripotent stem cells induced from mouse somatic cells by small-molecule compounds. Science 341:651–654. https://doi.org/10.1126/science.1239278
118. Ieda M, Fu J-D, Delgado-Olguin P et al (2010) Direct reprogramming of fibroblasts into functional cardiomyocytes by defined factors. Cell 142:375–386. https://doi.org/10.1016/j.cell.2010.07.002
119. Zhao Z, Xu M, Wu M et al (2015) Transdifferentiation of fibroblasts by defined factors. Cell Reprogram 17:151–159. https://doi.org/10.1089/cell.2014.0089
120. Sayed N, Liu C, Wu JC (2016) Translation of human-induced pluripotent stem cells. J Am Coll Cardiol 67:2161–2176. https://doi.org/10.1016/j.jacc.2016.01.083
121. Youssef AA, Ross EG, Bolli R et al (2016) The promise and challenge of induced pluripotent stem cells for cardiovascular applications. JACC Basic Transl Sci 1:510–523. https://doi.org/10.1016/j.jacbts.2016.06.010
122. Zhang J, Wilson G, Soerens A (2009) Functional cardiomyocytes derived from human induced pluripotent stem cells. Circ Res 104:e30–e41. https://doi.org/10.1161/CIRCRESAHA.108.192237.Functional
123. Yamakawa H (2016) Heart regeneration for clinical application update 2016: from induced pluripotent stem cells to direct cardiac reprogramming. Inflamm Regen 36:23. https://doi.org/10.1186/s41232-016-0028-z
124. Deuse T, Peter C, Fedak PWM et al (2009) Hepatocyte growth factor or vascular endothelial growth factor gene transfer maximizes mesenchymal stem cell-based myocardial salvage after acute myocardial infarction. Circulation 120:S247–S254. https://doi.org/10.1161/CIRCULATIONAHA.108.843680
125. Tang J-M, Wang J-N, Zhang L et al (2011) VEGF/SDF-1 promotes cardiac stem cell mobilization and myocardial repair in the infarcted heart. Cardiovasc Res 91:402–411. https://doi.org/10.1093/cvr/cvr053
126. Tang JM, Luo B, Xiao JH et al (2015) VEGF-A promotes cardiac stem cell engraftment and myocardial repair in the infarcted heart. Int J Cardiol 183:221–231. https://doi.org/10.1016/j.ijcard.2015.01.050

127. Stewart DJ, Kutryk MJ, Fitchett D et al (2009) VEGF gene therapy fails to improve perfusion of ischemic myocardium in patients with advanced coronary disease: results of the NORTHERN trial. Mol Ther 17:1109–1115. https://doi.org/10.1038/mt.2009.70
128. Santini MP, Lexow J, Borsellino G et al (2011) IGF-1Ea induces vessel formation after injury and mediates bone marrow and heart cross-talk through the expression of specific cytokines. Biochem Biophys Res Commun 410:201–207. https://doi.org/10.1016/j.bbrc.2011.05.081
129. Gallego-Colon E, Sampson RD, Sattler S et al (2015) Cardiac-restricted IGF-1Ea overexpression reduces the early accumulation of inflammatory myeloid cells and mediates expression of extracellular matrix remodelling genes after myocardial infarction. Mediators Inflamm 2015:1–10. https://doi.org/10.1155/2015/484357
130. Ellison GM, Torella D, Dellegrottaglie S et al (2011) Endogenous cardiac stem cell activation by insulin-like growth factor-1/hepatocyte growth factor intracoronary injection fosters survival and regeneration of the infarcted pig heart. J Am Coll Cardiol 58:977–986. https://doi.org/10.1016/j.jacc.2011.05.013
131. Gallego-Colon E, Villalba M, Tonkin J et al (2016) Intravenous delivery of adeno-associated virus 9-encoded IGF-1Ea propeptide improves post-infarct cardiac remodelling. NPJ Regen Med 1:16001. https://doi.org/10.1038/npjregenmed.2016.1
132. Evaluation of the Safety and Efficacy of Using Insulin-like Growth Factor-1 in Patients With a Heart Attack (RESUS-AMI) (NCT01438086). https://clinicaltrials.gov/.
133. Minatoguchi S (2004) Acceleration of the healing process and myocardial regeneration may be important as a mechanism of improvement of cardiac function and remodeling by postinfarction granulocyte colony-stimulating factor treatment. Circulation 109:2572–2580. https://doi.org/10.1161/01.CIR.0000129770.93985.3E
134. Askari AT, Unzek S, Popovic ZB et al (2003) Effect of stromal-cell-derived factor 1 on stem-cell homing and tissue regeneration in ischaemic cardiomyopathy. Lancet 362:697–703. https://doi.org/10.1016/S0140-6736(03)14232-8
135. Hu X, Yu SP, Fraser JL et al (2008) Transplantation of hypoxia-preconditioned mesenchymal stem cells improves infarcted heart function via enhanced survival of implanted cells and angiogenesis. J Thorac Cardiovasc Surg 135:799–808. https://doi.org/10.1016/j.jtcvs.2007.07.071
136. Yan F, Yao Y, Chen L et al (2012) Hypoxic preconditioning improves survival of cardiac progenitor cells: role of stromal cell derived factor-1α–CXCR4 axis. PLoS One 7:e37948. https://doi.org/10.1371/journal.pone.0037948
137. Herrmann JL, Wang Y, Abarbanell AM et al (2010) Preconditioning mesenchymal stem cells with transforming growth factor-alpha improves mesenchymal stem cell-mediated cardioprotection. Shock 33:24–30. https://doi.org/10.1097/SHK.0b013e3181b7d137
138. Vandergriff AC, de JBM A, Tang J et al (2015) Intravenous cardiac stem cell-derived exosomes ameliorate cardiac dysfunction in doxorubicin induced dilated cardiomyopathy. Stem Cells Int 2015:1–8. https://doi.org/10.1155/2015/960926
139. Nawaz M, Fatima F, Vallabhaneni KC et al (2016) Extracellular vesicles: evolving factors in stem cell biology. Stem Cells Int 2016:1–17. https://doi.org/10.1155/2016/1073140
140. Yang Y-J, Qian H-Y, Huang J et al (2009) Combined therapy with simvastatin and bone marrow-derived mesenchymal stem cells increases benefits in infarcted swine hearts. Arterioscler Thromb Vasc Biol 29:2076–2082. https://doi.org/10.1161/ATVBAHA.109.189662
141. Wisel S, Khan M, Kuppusamy ML et al (2009) Pharmacological preconditioning of mesenchymal stem cells with trimetazidine (1-[2,3,4-trimethoxybenzyl]piperazine) protects hypoxic cells against oxidative stress and enhances recovery of myocardial function in infarcted heart through Bcl-2 expression. J Pharmacol Exp Ther 329:543–550. https://doi.org/10.1124/jpet.109.150839
142. Takehara N, Tsutsumi Y, Tateishi K et al (2008) Controlled delivery of basic fibroblast growth factor promotes human cardiosphere-derived cell engraftment to enhance cardiac repair for chronic myocardial infarction. J Am Coll Cardiol 52:1858–1865. https://doi.org/10.1016/j.jacc.2008.06.052

143. Abstracts LCT, Brighton T, Eikelboom J et al (2015) Late-breaking clinical trial abstracts. Circulation 132:2267–2285. https://doi.org/10.1161/CIR.0000000000000334
144. Yacoub MH, Terrovitis J (2013) CADUCEUS, SCIPIO, ALCADIA: cell therapy trials using cardiac-derived cells for patients with post myocardial infarction LV dysfunction, still evolving. Glob Cardiol Sci Pract 2013:3. https://doi.org/10.5339/gcsp.2013.3
145. Chugh AR, Beache GM, Loughran JH et al (2012) Administration of cardiac stem cells in patients with ischemic cardiomyopathy: the SCIPIO trial: surgical aspects and interim analysis of myocardial function and viability by magnetic resonance. Circulation 126:S54–S64. https://doi.org/10.1161/CIRCULATIONAHA.112.092627
146. Bolli R, Chugh AR, D'Amario D et al (2011) Cardiac stem cells in patients with ischaemic cardiomyopathy (SCIPIO): initial results of a randomised phase 1 trial. Lancet 378:1847–1857. https://doi.org/10.1016/S0140-6736(11)61590-0
147. The Lancet Editors (2014) Expression of concern: the SCIPIO trial. Lancet 383:1279. https://doi.org/10.1016/S0140-6736(14)60608-5
148. Malliaras K, Makkar RR, Smith RR et al (2014) Intracoronary cardiosphere-derived cells after myocardial infarction. J Am Coll Cardiol 63:110–122. https://doi.org/10.1016/j.jacc.2013.08.724
149. Ishigami S, Ohtsuki S, Tarui S et al (2015) Intracoronary autologous cardiac progenitor cell transfer in patients with hypoplastic left heart syndrome: the TICAP prospective phase 1 controlled trial. Circ Res 116:653–664. https://doi.org/10.1161/CIRCRESAHA.116.304671
150. Ishigami S, Ohtsuki S, Eitoku T et al (2017) Intracoronary cardiac progenitor cells in single ventricle physiology novelty and significance. Circ Res 120:1162–1173. https://doi.org/10.1161/CIRCRESAHA.116.310253
151. Cardiac Stem/Progenitor Cell Infusion in Univentricular Physiology (APOLLON Trial) (APOLLON) (NCT02781922). https://clinicaltrials.gov/.
152. Combination of Mesenchymal and C-kit[+] Cardiac Stem Cells as Regenerative Therapy for Heart Failure (CONCERT-HF) (NCT02501811). https://clinicaltrials.gov/.
153. Allogeneic Heart Stem Cells to Achieve Myocardial Regeneration (ALLSTAR) (NCT01458405). https://clinicaltrials.gov/.
154. Presented by Dr. Timothy D. Henry at the American Heart Association Annual Scientific Sessions (AHA 2017), Anaheim, CA, November 15, 2017. http://www.acc.org/latest-in-cardiology/clinical-trials/2017/11/14/15/07/allstar#references-for-article
155. Dilated cardiomYopathy iNtervention with Allogeneic MyocardIally-regenerative Cells (DYNAMIC) (NCT02293603). https://clinicaltrials.gov/.
156. HOPE-Duchenne (Halt cardiomyOPathy progression in Duchenne) (HOPE) (NCT02485938). https://clinicaltrials.gov/.
157. Menasché P, Alfieri O, Janssens S et al (2008) The myoblast autologous grafting in ischemic cardiomyopathy (MAGIC) trial: First randomized placebo-controlled study of myoblast transplantation. Circulation 117:1189–1200. https://doi.org/10.1161/CIRCULATIONAHA.107.734103
158. Chavakis E, Koyanagi M, Dimmeler S (2010) Enhancing the outcome of cell therapy for cardiac repair: progress from bench to bedside and back. Circulation 121:325–335. https://doi.org/10.1161/CIRCULATIONAHA.109.901405
159. Karantalis V, Hare JM (2015) Use of mesenchymal stem cells for therapy of cardiac disease. Circ Res 116:1413–1430. https://doi.org/10.1161/CIRCRESAHA.116.303614
160. Tomita S, Li RK, Weisel RD et al (1999) Autologous transplantation of bone marrow cells improves damaged heart function. Circulation 100:II247–II256
161. Bel A, Messas E, Agbulut O et al (2003) Transplantation of autologous fresh bone marrow into infarcted myocardium: a word of caution. Circulation 108(Suppl):II247–II252. https://doi.org/10.1161/01.cir.0000089040.11131.d4
162. Waksman R, Fournadjiev J, Baffour R et al (2004) Transepicardial autologous bone marrow-derived mononuclear cell therapy in a porcine model of chronically infarcted myocardium. Cardiovasc Radiat Med 5:125–131. https://doi.org/10.1016/j.carrad.2004.10.001

163. Mathieu M, Bartunek J, El Oumeiri B et al (2009) Cell therapy with autologous bone marrow mononuclear stem cells is associated with superior cardiac recovery compared with use of nonmodified mesenchymal stem cells in a canine model of chronic myocardial infarction. J Thorac Cardiovasc Surg 138:646–653. https://doi.org/10.1016/j.jtcvs.2008.12.031
164. Gyongyosi M, Wojakowski W, Lemarchand P, Lunde K, Tendera M, Bartunek J et al (2015) Meta-Analysis of Cell-based CaRdiac stUdiEs (ACCRUE) in patients with acute myocardial infarction based on individual patient data. Circ Res 116:1346–1360. https://doi.org/10.1161/CIRCRESAHA.116.304346
165. Fisher SA, Doree C, Mathur A et al (2016) Stem cell therapy for chronic ischaemic heart disease and congestive heart failure. In: Martin-Rendon E (ed) Cochrane database of systematic reviews. Wiley, Chichester, UK
166. Wei X, Yang X, Han Z et al (2013) Mesenchymal stem cells: a new trend for cell therapy. Acta Pharmacol Sin 34:747–754. https://doi.org/10.1038/aps.2013.50
167. Kean TJ, Lin P, Caplan AI, Dennis JE (2013) MSCs: delivery routes and engraftment, cell-targeting strategies, and immune modulation. Stem Cells Int 2013:1–13. https://doi.org/10.1155/2013/732742
168. Miao C, Lei M, Hu W et al (2017) A brief review: the therapeutic potential of bone marrow mesenchymal stem cells in myocardial infarction. Stem Cell Res Ther 8:242. https://doi.org/10.1186/s13287-017-0697-9
169. Mills JS, Rao SV (2007) REPAIR-AMI: stem cells for acute myocardial infarction. Future Cardiol 3:137–140. https://doi.org/10.2217/14796678.3.2.137
170. Yousef M, Schannwell CM, Köstering M et al (2009) The BALANCE Study: clinical benefit and long-term outcome after intracoronary autologous bone marrow cell transplantation in patients with acute myocardial infarction. J Am Coll Cardiol 53:2262–2269. https://doi.org/10.1016/j.jacc.2009.02.051
171. Assmus B, Schächinger V, Teupe C et al (2002) Transplantation of progenitor cells and regeneration enhancement in acute myocardial infarction (TOPCARE-AMI). Circulation 106:3009–3017
172. Leistner DM, Fischer-Rasokat U, Honold J et al (2011) Transplantation of progenitor cells and regeneration enhancement in acute myocardial infarction (TOPCARE-AMI): final 5-year results suggest long-term safety and efficacy. Clin Res Cardiol 100:925–934. https://doi.org/10.1007/s00392-011-0327-y
173. Wang X, Zhang J, Zhang F et al (2015) The clinical status of stem cell therapy for ischemic cardiomyopathy. Stem Cells Int 2015:1–13. https://doi.org/10.1155/2015/135023
174. Fernández-Avilés F, Sanz-Ruiz R, Climent AM, Badimon L, Bolli R, Charron D et al (2017) Global position paper on cardiovascular regenerative medicine. Eur Heart J 38:2532–2546. https://doi.org/10.1093/eurheartj/ehx248
175. Sanz-Ruiz R, Climent AM, Fernández-Santos ME, Villa Arranz A, Gutiérrez Ibañes E, Vázquez-Álvarez ME et al (2017) General overview of the 14th international symposium on stem cell therapy and cardiovascular innovations. Circ Res 121:1040–1043. https://doi.org/10.1161/CIRCRESAHA.117.311874
176. Lister Z, Rayner KJ, Suuronen EJ (2016) How biomaterials can influence various cell types in the repair and regeneration of the heart after myocardial infarction. Front Bioeng Biotechnol 4:62. https://doi.org/10.3389/fbioe.2016.00062
177. Leor J, Amsalem Y, Cohen S (2005) Cells, scaffolds, and molecules for myocardial tissue engineering. Pharmacol Ther 105:151–163. https://doi.org/10.1016/j.pharmthera.2004.10.003
178. Vunjak-Novakovic G, Tandon N, Godier A et al (2010) Challenges in cardiac tissue engineering. Tissue Eng Part B Rev 16:169–187. https://doi.org/10.1089/ten.teb.2009.0352
179. Taylor DA, Parikh RB, Sampaio LC (2017) Bioengineering hearts: simple yet complex. Curr Stem Cell Reports 3:35–44. https://doi.org/10.1007/s40778-017-0075-7
180. Segers VFM, Lee RT (2011) Biomaterials to enhance stem cell function in the heart. Circ Res 109:910–922. https://doi.org/10.1161/CIRCRESAHA.111.249052

181. Georgiadis V, Knight RA, Jayasinghe SN, Stephanou A (2014) Cardiac tissue engineering: renewing the arsenal for the battle against heart disease. Integr Biol 6:111–126. https://doi.org/10.1039/C3IB40097B
182. Hastings CL, Roche ET, Ruiz-Hernandez E et al (2015) Drug and cell delivery for cardiac regeneration. Adv Drug Deliv Rev 84:85–106. https://doi.org/10.1016/j.addr.2014.08.006
183. Guyette JP, Charest JM, Mills RW et al (2016) Bioengineering human myocardium on native extracellular matrix novelty and significance. Circ Res 118:56–72. https://doi.org/10.1161/CIRCRESAHA.115.306874
184. Transplantation of Human Embryonic Stem Cell-Derived Progenitors in Severe Heart Failure (ESCORT) (NCT02057900). https://clinicaltrials.gov/.
185. Menasché P, Vanneaux V, Hagège A, Bel A, Cholley B, Cacciapuoti I et al (2015) Human embryonic stem cell-derived cardiac progenitors for severe heart failure treatment: first clinical case report: figure 1. Eur Heart J 36:2011–2017. https://doi.org/10.1093/eurheartj/ehv189
186. Lui KO, Zangi L, Chien KR (2014) Cardiovascular regenerative therapeutics via synthetic paracrine factor modified mRNA. Stem Cell Res 13:693–704. https://doi.org/10.1016/j.scr.2014.06.007
187. Nakada Y, Canseco DC, Thet S et al (2016) Hypoxia induces heart regeneration in adult mice. Nature 541:222–227. https://doi.org/10.1038/nature20173
188. Hu X, Xu Y, Zhong Z et al (2016) A large-scale investigation of hypoxia-preconditioned allogeneic mesenchymal stem cells for myocardial repair in nonhuman primates novelty and significance. Circ Res 118:970–983. https://doi.org/10.1161/CIRCRESAHA.115.307516
189. Van Den SWM B, Van De VAM S, Strzelecka AE et al (2009) Mouse strain determines the outcome of wound healing after myocardial infarction. Cardiovasc Res 84:273–282. https://doi.org/10.1093/cvr/cvp207
190. Monteiro LM, Vasques-Nóvoa F, Ferreira L et al (2017) Restoring heart function and electrical integrity: closing the circuit. NPJ Regen Med 2:9. https://doi.org/10.1038/s41536-017-0015-2

Chapter 8
Human-Induced Pluripotent Stem Cell-Derived Cardiomyocytes in the Evaluation of Cardiotoxic Potential of Drugs

Yee Xiang Chang, and Christine L. Mummery

Abstract Drug development process is currently a lengthy and costly process. One contributing factor to the high cost of drugs is the expense of failed products. To address this problem, early detection of potentially harmful drugs is important so that their development does not advance into expensive clinical trials and then fail to get regulatory approval. Drugs that cause heart damage are among those most commonly removed from the market. Current in vitro safety pharmacology screenings only focus on the evaluation of drug-induced QT prolongation or torsade des pointes (TdP) risks by examining individual cardiac ion channel currents or the action potential. These screenings are oversimplified, resulting in inaccurate prediction, high drug attrition rates, and drug withdrawals from the market. Human-induced pluripotent stem cell-derived cardiomyocytes (hiPSC-CMs) are promising models for cardiotoxicity screening as some human cardiomyocyte features are recapitulated and disease- and patient-specific lines can be generated. The new generation of iPSCs has solved the problem of inefficient cardiomyocyte differentiation and provides optimism in the application of hiPSC-CMs in cardiotoxicity screening. In this chapter, an overview is given of safety pharmacological assays evaluating hiPSC-CM function, and the validity of the application of hiPSC-CM models in cardiotoxicity screening will be assessed.

Keywords Human pluripotent stem cells · Cardiomyocyte differentiation · Cardiac disease · Drug discovery · Cardiac arrhythmia

Y. X. Chang (✉)
University of Utrecht, Utrecht, The Netherlands
e-mail: y.x.chang@students.uu.nl

C. L. Mummery
Department of Anatomy and Embryology, Leiden University Medical Centre, Leiden, The Netherlands
e-mail: c.l.mummery@lumc.nl

R. Delgado-Morales (ed.), *Stem Cell Genetics for Biomedical Research*,
https://doi.org/10.1007/978-3-319-90695-9_8

Abbreviations

AP	Action potential
APA	Action potential amplitude
APD	Action potential duration
CM	Cardiomyocyte
CSD	Calcium-sensitive detector
EC50	Half maximal effective concentration
ECAR	Extracellular acidification rate
ECG	Electrocardiogram
EHT	Engineered heart tissue
ESC	Embryonic stem cell
FDA	Food and Drug Administration
FP	Field potential
FPA	Field potential amplitude
FPD	Field potential duration
HEK-293 cells	Human embryonic kidney 293 cells
hERG	Human Ether-à-go-go-Related Gene
hiPSC-CM	Human-induced pluripotent stem cell-derived cardiomyocyte
IC50	Half maximal inhibition concentration
ICH	The International Council on Harmonization of Technical Requirements for Registration of Pharmaceuticals for Human Use
I_{Kr}	Inward-rectifier K^+ current
iPSC	Induced pluripotent stem cell
ISI	Interspike interval
MEA	Multielectrode arrays
NCE	New chemical entities
OCR	Oxygen consumption rate
TdP	Torsade de pointes
V_{max}	Maximal upstroke velocity
VSD	Voltage-sensitive dye
XF	Extracellular flux

8.1 Introduction

The current drug development process is inefficient and lengthy: only 1 on 5000–10,000 compounds that enter preclinical testing receives a Food and Drug Administration (FDA) approval, and the average process takes 10–15 years [1]. Moreover, the pharmaceutical industry has to deal with the costs of approximately $2 billion to take a candidate drug from the beginning of the pipeline to the marketplace; these development costs include those of failed products included [2]. Therefore, decreasing the failure rate of novel compounds could result in reduction of the costs for the pharmaceutical industry and drug prices in healthcare.

In the past decades, cardiac safety has been the leading cause of drug development failure and withdrawal from the market in the USA [3, 4]. Between 1998 and 2008, 17 cardiovascular and nonvascular drugs were withdrawn from the market as a result of (increased risks of) cardiac events such as myocardial infarction, arrhythmia, and QT prolongation [5]. Cardiac dysfunction resulting in QT prolongation on an electrocardiogram (ECG), which is used as an indicator of delayed ventricular repolarization, is associated with an increased risk of ventricular arrhythmias [6]. An example of a ventricular arrhythmia is *torsade de pointes* (TdP, translated as "twisting of the points") which is characterized by a twist of the QRS complex around the baseline of an ECG and can result in ventricular fibrillation and sudden cardiac death [7]. The predominant mechanism of drug-induced TdP and QT prolongation is the inhibition of the human *Ether-à-go-go-Related* Gene (hERG) ion channel that is responsible for the action potential (AP) repolarization via the outward rectifier potassium current (I_{Kr}) [8].

After the withdrawal of multiple drugs, because of their association with QT prolongation and TdP in the 1990s, the International Council on Harmonization of Technical Requirements for Registration of Pharmaceuticals for Human Use (ICH) S7B guidance was issued in 2005. The drug development guidance document concerns preclinical testing strategies to evaluate the potential of new chemical entities (NCE) to induce delayed ventricular repolarization and can be divided in two components: (I) in vitro electrophysiology studies including I_{Kr} assays and (II) in vivo QT assays [9]. The introduction of the ICH 7B guidance has shown to be successful; no drugs were withdrawn from the market due to concerns about TdP risks since 2005 [10]. However, current models still have some limitations (see Sect. 8.2 *Current Models Used for Cardiotoxicity Screening*).

An ideal model system would possess high human physiological relevance, thereby increasing the predictability of drug-induced cardiotoxicity. For instance, the use of human CMs has the advantage of being applicable for the assessment of adverse effects on both AP duration and ion currents, though some practical considerations should be made [11]. The use of primary human CMs is limited as collection of cardiac tissues can only be performed in specific patient groups (i.e., primarily those undergoing cardiac surgery or other interventions) resulting in the limited availability of cells representing the general (healthy) population. In addition, challenging isolation procedures, low cell viability, and the tendency to dedifferentiate during (prolonged) cell culture make the application of primary human CMs challenging for drug screening [12, 13]. A more interesting human model to consider for drug safety assays is based on human pluripotent stem cells. Both human embryonic stem cells (hESCs) and induced pluripotent stem cells (iPSCs) can be differentiated in different somatic cell types, including CMs. However, iPSCs circumvent the complex ethical issues concerning the use of human embryos.

In 2007, human iPSCs were generated by Takashi et al. by introducing pluripotency in human fibroblast cells using retroviral transduction of four transcription factors: OCT3/4, SOX2, KLF4, and c-MYC [14]. Later that year, it was shown that reprogramming to pluripotency is also possible with a slightly different set of transcription factors (OCT3, SOX2, NANOG, LIN28) [15]. This method was applied to

Table 8.1 Summary of principle pros and cons of hiPSC-CMs in cardiotoxic screening

Advantages of hiPSC-CM models	*Reference*
hiPSC-CMs can be generated from healthy individuals and patients with genetic variants/mutations associated with heart conditions	[18, 19]
hiPSC-CMs have cross-striated and aligned sacromeric structures that can be evaluated for disruption in response to toxic compounds	[16, 20]
Short- and long-term culture is possible without dedifferentiation (hours to months) allowing chronic and acute effects to be determined Expression of cardiac ion channels and electrophysiological responses similar to human CMs (ventricular, atrial, and pacemaker)	[4, 21] [19, 21, 22]
Disadvantages of hiPSC-CM models	*Reference*
hiPSC-CMs are immature and do not show adult physiology	[18, 19]
Action of (drug and compound) metabolites cannot be modeled	[17, 23, 24]

generate iPSCs which were successfully differentiated into the first functional CMs using embryoid body formation [16]. Since then, several approaches have been developed to generate hCMs. These can be broadly divided in two other main categories: monolayer differentiation and inductive co-culture [17]. Using these various differentiation approaches, many hiPSC-derived CMs (hiPSC-CMs) have been characterized and evaluated for the use in cardiac safety assessment. Over the last several years, hiPSC-CMs have been shown to possess various advantages as human heart models (Table 8.1). Most noteworthy, patient-specific hiPSCs can be obtained to study the cardiotoxicity in cardiac disease or genetically predisposed patients. This implies that by generating patient-specific hiPSC-CMs, cardiac safety of drugs can be tested in groups of individuals most susceptible to cardiotoxicity [25, 26]. However, there are certain limitations of hiPSC-CMs, not least their fetal-like phenotype, which raises questions on their suitability for drug testing. In this chapter, the following question will be answered:

- To what extent have hiPSC-CM models been shown to be useful in predicting drug-induced cardiotoxicity in the past 5 years?
- Will hiPSC-CM models lead to an improved predictability of cardiotoxicity compared to current in vitro models?

This chapter will discuss current and future models to assess safety pharmacology of NCE and the advantages and limitations of various assays assessing hiPSC-CM function and will evaluate the validity of the application of hiPSC-CM models in cardiotoxicity screening.

8.2 Current Models Used for Cardiotoxicity Screening

Since the release of ICH S7B guidance, I_{Kr} measurements in dissociated primary cardiomyocytes (CMs) or expression systems such as human embryonic kidney (HEK-293) cells expressing human hERG channels are extensively used for preclinical cardiotoxicity evaluation by pharmaceutical companies [11]. In addition, I_{Kr} assays play a prominent role in reducing TdP risk for NCE [9, 11]. Yet, it is important to realize that the cardiac action potential is the net result of multiple channel activities (Fig. 8.1) [6, 26]. Therefore, evaluation of hERG channel inhibition alone may fail to detect drug-induced QT prolongation or TdP risks or otherwise result in the reprioritization (or even discard) of potential therapeutic agents. To illustrate, ranolazine is known to be an I_{Kr} blocker and is associated with QT prolongation, although not with the induction of arrhythmia. The proposed explanation for the minimal proarrhythmic liabilities is related to the term "repolarization reserve" and means that the blockage of one current, in this case the I_{Kr}, can be compensated by another current such as the reduction of the late I_{Na}. The reduction of I_{Na} results in a mitigating reduced inward current [27, 28]. Moreover, earlier studies have demonstrated that evaluation of hERG channel inhibition alone is prone to false negatives and that prediction of delayed repolarization or clinical proarrhythmias can be complex [29, 30]. Thus, I_{Kr} assays alone are inadequate to determine product labeling for QT prolongation or TdP, and inclusion of animal studies is needed [30].

In addition, there is societal pressure to reduce the number of animal experiments. Over 1.1 million citizens from 26 countries have requested that the European Commission eliminate the directive that regulates the use of animal in scientific research [31]. This social demand is supported by evidence of the challenging clinical translation of results from animal experiments to humans due to interspecies differences. Certain animals are extremely ill-suited to evaluate the cardiac toxicity of NCE. For instance, in adult mice and rats, the outward potassium current (I_{to}) is the predominant ion current controlling repolarization, while in humans, the delayed rectifier currents I_{Kr} and I_{Ks} also play a crucial role [11]. Examples of recommended animal species used for ex vivo experiments and in vivo QT assays are dog, rabbit, and guinea pig [11]. Still, it is crucial to realize that interspecies differences in the types of cardiac ion channels and the relative expression level of each channel type result in determination of various toxicity risks determined by testing the same compound on different species [12]. Consequently, the different toxicity risks may complicate the cardiac safety evaluation.

8.3 Safety Pharmacology Screenings Assessing hiPSC-CM Function

Characterization of hiPSC-CMs is essential to determine the success of cardiac differentiation protocols and also key when performing preclinical safety screening for NCE. To enable prediction of cardiotoxicity in human, by which patient safety is improved and drug attrition is reduced, reliable and sensitive assays are needed to evaluate the morphological and functional effects of a drug. In this chapter, the focus will be on functional assays used to evaluate cardiotoxic potential of drugs in hiPSC-CMs (Table 8.2).

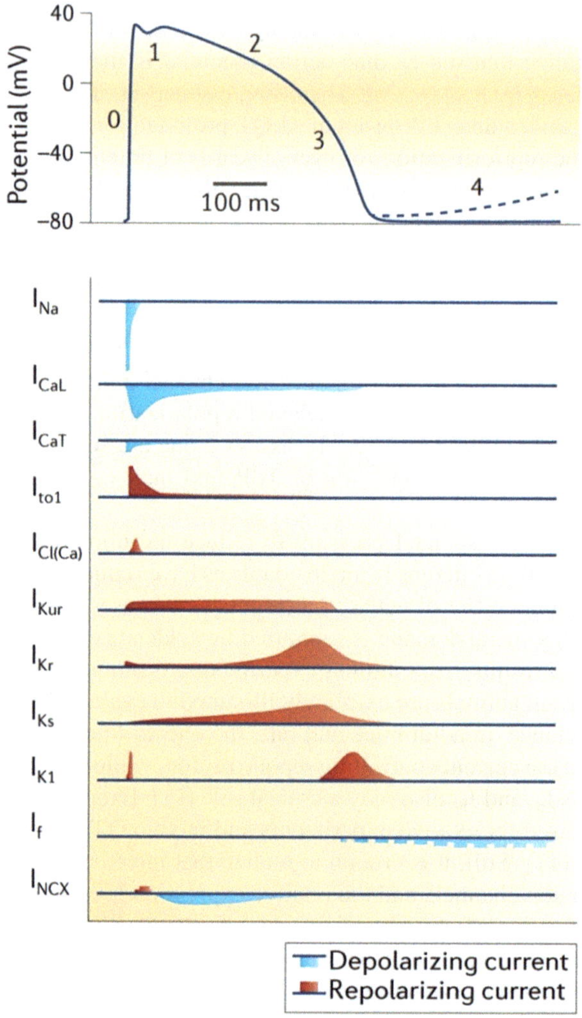

Fig. 8.1 Ventricular action potential is defined by multiple ionic currents. The cardiac action potential is the result of voltage-dependent and time-dependent depolarizing (inward) and repolarizing (outward) currents. Inhibition of the I_{Kr} (also known as the hERG current or $K_v11.1$ current) is the most common mechanism for drug-induced QT prolongation or TdP. The ventricular action potential can be divided in 4 phases: (*phase 0*) rapid depolarization which is dependent of the inward sodium current (I_{Na}), (*phase 1*) early repolarization regulated via the outward potassium current (I_{to}), (*phase 2*) plateau of the action potential which is the consequence of the balance between the inward calcium current (I_{Ca}) and the outward repolarizing potassium currents, (*phase 3*) late repolarization resulting from the delayed rectifier currents (I_{Kr} and I_{Ks}), and (*phase 4*) resting membrane potential (or depolarization, dashed) which is maintained by the inward-rectifier potassium current (I_{K1}). (Figure is adapted from [6]. See also [19])

Table 8.2 Safety pharmacology assays used for hiPSC-CMs

Criteria	Assay(s)	Outcome	Notes
Electrophysiological properties	Patch clamp	Single ion currents, AP (APD, APA, V_{max}), classification of cardiomyocyte subtype (ventricular-, atrial-, or nodal-like)	(+) High accuracy (−) Low-throughput (−) Short-term measurement
Electrophysiological properties	Multielectrode array	FP (FPD, FPA, V_{max}, ISI)	(+) Little inter-assay differences (−) Low- to medium-throughput
Electrophysiological properties	Voltage-sensitive dyes	AP characteristics (as described in patch clamp)	(+) High-throughput (−) Cytotoxicity
Calcium handling	Calcium-sensitive detectors	Calcium transient (calcium transient duration, rise time, decay time and peak amplitude, peak count)	(+) High-throughput (−) Cytotoxicity (−) Underdeveloped calcium regulating organelles in hiPSC-CM
Contraction (and viability)	Impedance assays	Beating frequency, contractility, cell viability	(+) Long-term measurement (+) Medium- to high-throughput
Cardiac energy metabolism	Seahorse assay	OCR and ECAR	(+) Medium- to high-throughput

AP action potential, *APD* action potential duration, *APA* action potential amplitude, V_{max} maximum upstroke velocity, *FP* field potential, *FPD* field potential duration, *FPA* field potential amplitude, *ISI* interspike interval, *OCR* oxygen consumption rate, *ECAR* extracellular acidification rate

8.3.1 Electrophysiology Studies

8.3.1.1 Patch Clamp

A conventional way to characterize the function of hiPSC-CM (disease) models is patch clamp electrophysiology. Using this technique, Zhang et al. were the first to measure the APs in spontaneously beating single cells isolated from hiPSC embryoid bodies (Fig. 8.2a) [16]. After 56–70 days post-differentiation, three major AP types could be distinguished in CMs derived from hiPSC: nodal-, atrial-, and ventricular-like APs, where the ventricular-like phenotype was the most encountered. In addition, similar electrical activity in hiPSC and hESC-CMs was observed. Besides the ability to identify different CM subtypes, patch clamp techniques have also shown their potential to detect drug-induced arrhythmias [34, 35]. iPSC-CMs derived from patients with preexisting heart conditions have demonstrated increased susceptibility to cardiotoxic compounds compared to hiPSC-CM derived from healthy subjects [32]. The increased susceptibility was measured by APD and quantification of early afterdepolarization (EAD) and delayed after depolarizations

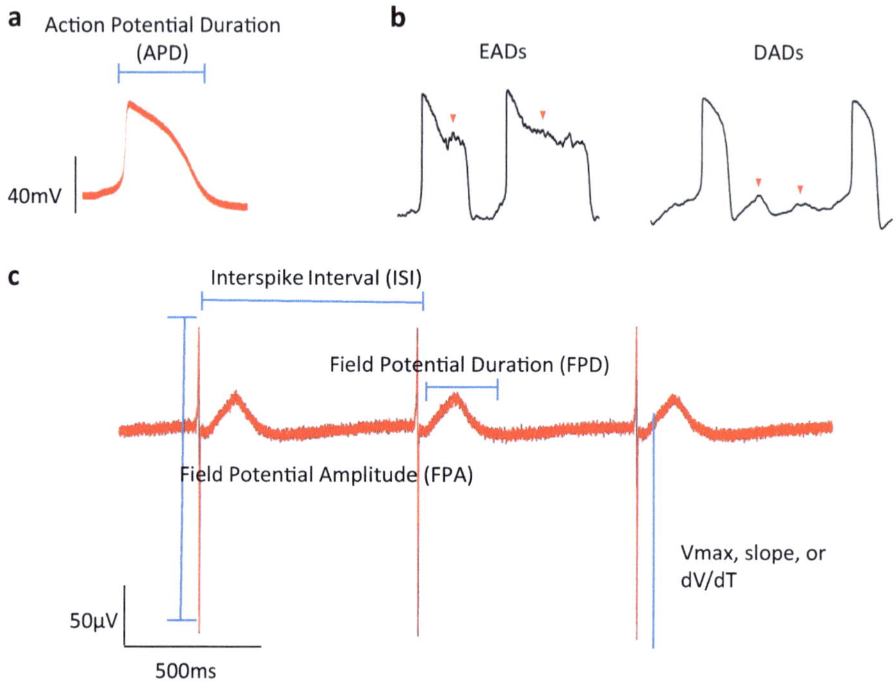

Fig. 8.2 Electrophysiological parameters derived from hiPSC-CMs. (**a**) Using patch clamp, AP recordings can be measured from ventricular-like CMs. The action potential duration (ADP) can be used to describe the wave form. (b) AP recordings from ventricular-like CMs recordings showing early after depolarizations (EAD, *arrow heads*) or delayed after depolarization (DAD, *arrow heads*). (c) Field potential recordings from microelectrode arrays indicating wave form parameters. (Figures are modified from [32, 33])

(DADs), of which the latter two indicate proarrhythmic risk (Fig. 8.2b) [10, 32]. Furthermore, Liang et al. showed that cardiotoxicities can be modeled with a higher accuracy in hiPSC-CMs than in standard hERG assays using HEK293 cells [32].

Although patch clamping has shown to be promising, several limitations of this technique should be considered. Firstly, patch clamp is an invasive method where a tight seal (giga seal) is required for the recording and often the cell membrane has to be broken [36]. Consequently, short-term measurements are only possible. Secondly, it is important to realize that only cells that survive the dissociation process, which are often less mature cells in the population, are measured [37]. Thirdly, patch clamp is an inefficient method involving relatively long time per measurement, in which a single cell can only be measured at once, and high expertise is required to get reliable results [36, 37]. To increase the throughput, Scheel et al. used automated patch clamp to measure the effect of nifedipine, cisapride, and TTX in commercially available Cor.4 U® hiPSC-CMs [34]. Similar drug effects on the APs after drug exposure were observed in hiPSC-CMs compared to manual patch clamp [22, 38].

8.3.1.2 Multielectrode Arrays

Another method to study the electrophysiology of hiPSC-CMs in a higher throughput way is using multielectrode arrays (MEA). MEA may be preferred over patch clamp as less operator skills and specialized instruments are required [23]. In addition, extracellular field potential (FP), corresponding to intracellular AP, and beating frequency can be measured for several hours to days by plating cells on glass slides containing arrays of electrodes [39–41]. The ability to make more extended recordings with MEA may improve the prediction of delayed cardiotoxicity as only short measurements can be made with patch clamp [23]. Another benefit of MEA is the recording of multiple cells, resulting to a decreased vulnerability for the intrinsic differences between cells [4]. Since MEA is a noninvasive and also a nonterminal method, it is often combined with other methods. An example of this is the xCELLigence® RTCA CardioECR instrument which combines MEA technology with impedance recordings (used for the evaluation of contractibility and viability) to detect functional cardiotoxicity (see Sect. 8.3.2 *Contraction and Beating Behavior Studies*).

FP measurements have shown to be capable of detecting the effect of several selective channel blockers and multichannel blockers that are affecting different phases of the APs. Exposure to calcium blockers, verapamil, and nifedipine resulted in shortening of the corrected field potential duration (cFPD), which can be regarded as shortening of the QT interval in ECG recordings, whereas exposure to several I_{Kr} blockers and multichannel blockers resulted in prolonged cFPD [4, 42]. As the effect of multiple channels can be assessed in hiPSC-CMs, it has been shown that MEA recordings in hiPSC-CMs can detect false-negative or false-positives of selected channel/hERG channel assays, which results in a reduction of attrition of drug candidates with therapeutic potential [4]. Furthermore, prediction of drug-induced arrhythmia in hiPSC-CMs was consistent with the results of animal studies [4, 40]. Moreover, most drugs that are clinically associated with TdP also caused EADs in hiPSC-CMs [4, 42]. In contrast to the promising results, a blinded study by Qu et al. concluded that hiPSC-CMs were able to detect drug-induced FPD prolongation and EADs from MEA recordings, however, unable to distinguish the severity of TdP risks, which suggested that MEA have a high false-positive rate [43]. A possible explanation from the authors for the inability to predict the severity of TdP risk was the immature electrophysiological phenotype of the hiPSC-CMs.

8.3.1.3 Fluorescent Imaging

The effects of drug exposure on the function of CMs can also be evaluated with fluorescent imaging methods that detect transmembrane voltage or intracellular calcium transients. Two such fluorescent imaging methods for the characterization of CMs involve the use of voltage-sensitive dyes (VSDs) and calcium-sensitive detectors (CSDs). These dyes can be used on a noninvasive manner to measure the

AP or intracellular ion fluctuation, respectively [36]. *Di*-4ANEPPS, a VSD that intercalates the lipid bilayer of a cell, has been used to compare the electrophysiological characteristics of commercial hiPSC-CMs from Cor.4 U® and iCell® [23, 44]. Differences in the action potential duration (ADP) and cycle length values were found. Nevertheless, the effect of nifedipine, ranolazine, and E4021 on the ADPs was similar for both cell types [44]. Although VSDs are often used to study whole hearts, Hortigon-Vinagre et al. have been one of the few that used VSDs to detect cardiotoxic potential of drugs in hiPSC-CMs. Why this is rarely used for drug safety evaluation in hiPSC-CMs may be due to its cytotoxic nature, associated with the use of voltage dyes on single cells or monolayer cultures, which can affect the outcome of the study.

CSDs can be divided in ratiometric and non-metric which result in absolute or relative measurement of the calcium concentration, respectively [23]. As with some VSDs, the use of CSDs, such as Fluo-4, is also associated with cytotoxicity as the compound (acetoxymethylester) used to introduce CSDs, and UV light, often needed for detection, can be harmful to cells [45]. In addition, when CSDs are excessively loaded, it can affect calcium signaling and interfere with normal functioning of the cell. A number of studies have examined drug effects (> 50 drugs) on calcium transients in hiPSC-CMs as a potential method to detect drug-induced cardiotoxicity [46–48]. Changes in calcium transients after drug exposure were able to recapitulate the expected effect of the drug, known from in vivo models and/or clinical data, for most compounds in the expected concentration range [48, 49]. In contrast, non-inotropics and sodium inhibitors showed mixed results and less accurate predictions. Lu et al. suggested that the effects on sodium channels may not be reflected in the calcium transients since the calcium transient is initiated after depolarization of the membrane by the opening of L-type calcium channels (phase 2 of the AP) [48].

8.3.2 Contraction and Beating Behavior Studies

The beat rate can be determined with electrophysiology-based methods such as AP measurements using patch clamp [22] or kinetic fluorescence imaging with CSDs [47]. Important to note is that some of these methods may involve invasive procedures or the use of fluorescent markers, thereby limiting the performance of multiple assays with the same cells or potentially affecting the functioning of the cell [36]. Automatic label-free beating rate imaging analyzes the actual mechanical beating, reduces the risk of bias, and enables the performance of multiple assays with the same cells [20]. Using this method, Bergström et al. compared the half maximal effective concentration (EC50) or half maximal inhibition concentration (IC50) of verapamil, quinidine, and doxorubicin and compared it to previously reported values determined with other methods [20]. A lower IC50 (or EC50) was found for both verapamil and quinidine by electrophysiological methods, indicating that this method may have a higher sensitivity to detect possible cardiotoxicity. Even so, automatic label-free beating rate imaging could reflect the clinical cardiotoxic effect of the drugs on a relative low labor-intensive manner using a nonspecialized microscope.

Determining the beat rate in cardiotoxicity screening is not only an important parameter to detect heart defects but also the contractility. For example, heart failure is associated with reduced contraction force of the ventricular CMs [50]; therefore, various methods to measure the contractility have been developed, including atomic force microscopy, muscular thin films, and traction force microscopy in which the mechanical displacement or kinetics of contraction force is measured. These techniques are discussed more extensively by Laurila et al. and van Meer et al. [23, 36]. A frequently used medium- to high-throughput platform to evaluate the cardiotoxic risk in hiPSC-CMs is the xCelligence® RTCA instruments which use 48-wells or 96-wells microtiter plates. These instruments combine MEA technology with impedance measurements, using planar gold electrodes, and enable the evaluation of cardiomyocyte electrophysiology, contractility, and cell viability of physiologically relevant interconnected CMs [51]. To further illustrate, impedance assays involve the application of low voltage (< 20 mV) and measure the electrical impedance, which can be influenced by the number of adhered cells, strength of cell adherence, and cell morphology. The impedance signal is displayed as the cell index of which the cell index value reflects the relative contractile force exerted by the cell monolayer.

Cell index monitoring has shown to be useful in cardiotoxicity profiling and even demonstrated to be a more sensitive endpoint than certain structural endpoints [52, 53]. Doherty et al. measured structural (troponin secretion, reactive oxygen species formation, lipid formation, and viability) and functional (beating activity) endpoints in hiPSC-CMs [54]. Later, it was demonstrated that assessment of beat pattern (amplitude, beat rate, or presence of arrhythmia) was able to identify the most drug with known clinical cardiac effects (16 out of 18 drugs), while fewer drugs (maximally 10 out of 18 drugs) with cardiac effects could be identified by assessing structural endpoints. Despite the lower sensitivity of structural endpoint evaluation, these measurements have an added value as they may result in additional insight in the mechanism of cardiotoxicity. Moreover, noninvasive impedance measurements enable long-term recordings (days/weeks) and repeated drug exposure, simulating the approach in most drug therapies, and possibly improve cardiotoxicity prediction. To illustrate, the effect of single and repeated exposure of doxorubicin, a commonly used anticancer therapeutic which has been clinically associated with cardiotoxicity, was studied until day 14 after the initial exposure [55, 56]. In contrast to single exposure, repeated exposure, followed by washout, resulted in a significant decrease in beating amplitude compared to control cells [56]. Since a decreased beating amplitude reflects a reduced contractile force, these hiPSC-CM-derived data are in accordance with findings in rabbit heart which showed a decrease in contractile force after repeated doxorubicin exposure [57].

8.3.3 Bioenergetics and Cell Viability

The heart and liver are highly metabolic active, and simultaneously toxicities associated with these organs are the most common cause of drug withdrawal or "black box warnings" (which is a drug label warning for serious or life-threatening risks) [58]. In addition, most of these drugs that are withdrawn are mitochondrial toxins. Therefore, studying the metabolic state of hiPSC-CMs may also be a valuable endpoint for drug safety screening. Two major metabolic pathways of the cell, oxidative phosphorylation and glycolysis, can be correlated to oxygen consumption rate (OCR) and extracellular acidification rate (ECAR), respectively. A common method used to measure these surrogate markers in hiPSC-CMs is the extracellular flux (XF) assay using Seahorse XF Analyzers [41]. This method is noninvasive and determines the OCR and ECAR based on luminescence intensity [59].

Rana et al. used the Seahorse technology and characterized the mitochondrial function in commercial iCell® hiPSC-CMs which were cultured in medium containing different carbon sources (glucose, galactose, and fatty acids) [60]. In this study by Pfizer, it was shown that the OCR and ECAR of hiPSC-CMs from the same batch could be consistently and reproducibly determined using Seahorse technology. In addition, the relative contribution of oxidative phosphorylation and glycolysis to ATP production was evaluated by calculating OCR/ECAR ratio. The highest OCR/ECAR ratios were observed for the hiPSC-CMs cultured in the presence of galactose, which imply that the CMs in this condition most closely resemble in vivo adult CMs, as adult CMs are primarily dependent on oxidative phosphorylation [60, 61]. hiPSC-CMs cultured in medium containing galactose were used to screen the effect of six anticancer tyrosine kinase inhibitor drugs on OCR and cell viability at day 21 [60]. The cytotoxicity rank order of the compounds was sorafenib > sunitinib > dasatinib > imatinib > lapatinib > nilotinib which corresponded largely with published literature about the clinical cardiovascular toxicity of these tyrosine kinase inhibitors [60, 62]. Moreover, compounds with high cytotoxicity rank showed relatively low OCR compared to other drugs [60]. The agreement in low OCR and cell viability may elucidate the mechanism behind the cardiotoxicity of these drugs, as sorafenib, sunitinib, and imatinib are associated with mitochondria-related cardiotoxicity [63].

More recently, Burridge et al. also showed that assessment of the mitochondrial function may also be a useful method to predict drug-induced cardiotoxicity [18]. A substantial decrease in certain mitochondrial functions (basal and maximal respiration) and a lower mitochondrial content were found in hiPSC-CMs derived from breast cancer patients who experienced clinical cardiotoxicity after doxorubicin treatment, compared to hiPSC-CMs derived from breast cancer patients who did not experienced any cardiotoxicity. In addition, the patient-specific predisposition to doxorubicin-induced cardiotoxicity could not be recapitulated in patient-derived fibroblasts or undifferentiated hiPSCs which confirmed that the effect on the mitochondrial function could not be generalized to other cell types.

To summarize, three classifications emerge from recently developed drug screening assays using hiPSC-CMs: (I) detection of drug-induced arrhythmia by evaluating the electrophysiological parameters resulting from multiple ion channels, (II) detection of cardiotoxicity by assessing the contractility of CMs, or (III) bioenergetics in CMs. In this chapter, it has been demonstrated that all three approaches may lead to an improved safety assessment of (new) drugs. For example, it has been indicated that the use of EADs as a proarrhythmic endpoint might have a greater prognostic value than direct surrogate markers in animal models. This implies that electrophysiological assays could reduce the attrition of new therapeutics that inhibit the I_{Kr} and prolong QT interval yet are not proarrhythmic [64, 65] . Current hERG assays only focus on detection of drug-induced arrhythmia; however this is not enough to predict drug-induced cardiotoxicity. Since the release of the ICH guidance, drugs such as tegaserod, sibutramine, and rosiglitazone have been withdrawn due to late-onset cardiac events and highlight the need of drug screening that are able to detect cardiac events such as myocardial infarct, stroke, and/or heart failure [66]. Contractility or bioenergetic (II and III) assessment in hiPSC-CMs can predict some of these cardiac safety liabilities as [18, 56] cardiomyopathy and heart failure can be caused by loss/damage of intracellular organelles of CMs, or loss of CM viability, resulting in dysfunction in contractility [67, 68]. In addition, the importance of simulating drug therapeutics during drug screening has been shown which may enable detection of late-onset cardiac events after repeated drug exposure and long-term culture of hiPSC-CMs.

8.4 Future Perspectives

In general, it has been shown that hiPSC-CMs are promising with regard to the prediction of the clinically known effects of most drugs. Important to realize is that all of these studies were retrospective studies and that could have biased the outcome. Therefore, a prospective study should be conducted in the near future in which compounds are tested in a blinded way, to confirm whether prediction of cardiotoxicity of a drug is actually possible.

When comparing the IC50 of different pharmacological assays (Table 8.3), large IC50 differences of one or two magnitudes could be observed which has also been observed by Bergström et al. [20]. A possible explanation of the large variations in IC50 can be attributed to the sensitivity of each biophysical parameter, as each parameter is reflecting a different point in the excitation-contraction coupling cascade. The cardiac excitation-contraction coupling consists of action potential formation, calcium-induced calcium release from the sarcoplasmic reticulum, and mechanical contraction in which each parameter has its own time course [52]. Although the processes in the excitation-contraction coupling are dependent on each other, one cannot always predict the other. For example, no effect on the FP recording can be observed after blebbistatin exposure, resulting in inhibition of myosin con-

Table 8.3 Reported IC50 of compounds tested for cardiac safety evaluation

Drug	Method	Effect	IC50/EC50 (μM)	Ref.
I_{Kr} blockers				
Cisapride (+ QT, + clinical TdP)	Patch clamp	↑ ADP	0.035	[32]
	hERG	↓ hERG	0.032	[32]
Ca^{2+} channel blockers				
Nifedipine (- QT, - clinical TdP)	MEA	↓ FPD	0.019	[4]
	hERG	↓ hERG	270	[69]
Multichannel blockers				
Quinidine (+ QT, + clinical TdP)	Beating rate imaging	↓ Beating freq.	40	[20]
	MEA	↑ FPD	0.3*	[4]
	hERG	↓ hERG	0.75	[69]
Verapamil (- QT, - clinical TdP)	Beating rate imaging	↓ Beating freq.	2.25	[20]
	MEA	↓ FPD	0.169	[4]
	CSDs imaging	↓ $[Ca^{2+}]_i$	0.01	[46]
	Patch clamp	↑ APD	5.18	[32]
	Impedance	↓ relative contraction force	0.12	[70]
	hERG assay	↓ hERG	0.2	[69]

*yellow = prolonging effect, green = shortening effect, * is estimated from the dose-response curve*

yellow = prolonging effect, green = shortening effect
[a]*Is estimated from the dose-response curve*

traction, while reduction in impedance signal was observed [71]. Based on this, a multiparameter approach, which reflects each phase in the excitation-contraction cascade, is needed for accurate detection of cardiotoxicity. To detect drug-induced arrhythmia and other cardiac liabilities, it may be useful to extend this multiparameter approach with additional mitochondrial function (and cell viability) assessment.

Besides multiparameter assessment, future studies should also consider the inclusion of isogenic controls while testing for cardiotoxicity. Random selection of wild-type genetic background may not be sufficient as "healthy control." To elaborate, Sala et al. generated isogenic controls via genetic correction in hiPSCs from long QT patients and studied the effect of LUF 7346, a hERG allosteric modulator [72]. No difference was observed when comparing the basal AP of a long QT hiPSC with a non-isogenic healthy control ($LQT2^{N996I}$ vs. $LQT1^{corr}$), while comparing another non-isogenically matched pair ($LQT1^{R190Q}$ vs. LQT2) would lead to inaccurate estimation of LUF7346 concentration needed to correct the AP in long QT-hiPSCs and resulted in shortening of the AP.

To address the main question of this paper, recent studies have provided optimism on the ability of hiPSC models to predict induction of cardiac arrhythmia or even of other functional heart defects. By combining multiple parameter assess-

Table 8.4 Comparison of phenotypical parameters between ventricular hiPSC-CMs and primary ventricular adult CMs

Parameter	Ventricular hiPSC-CM	Primary adult ventricular CMs
Resting membrane potential (mV)	−73.5 [22]	−81.8 [74]
Voltage V_{max} (V/s)	28.9 [32]	215 [74]
Force of contraction (mN/mm^2)	0.43 [75]	51 [76]
Sacromeric organization	Highly organized [32, 77]	Highly organized [23]
Regulation of contraction	Spontaneous	Via SA-node
Size	Relatively small [78]	Relatively large [78]

ment, it might be possible to unravel the mechanism underlying the cardiotoxicity [71]. To date, no in vitro assay has shown 100% accurate prediction of cardiotoxicity, and it is unlikely that hiPSC-CM testing would achieve this as most hiPSC-CM models are a simplified model which are not exposed to drug metabolites and lack other cell types present in the heart such as fibroblasts and endothelial cells. Therefore, not all cardiotoxic events found in the clinic will manifest in the hiPSC-CM model. However, inclusion of hiPSC-CM models in in vitro preclinical tests can still be of added value, as a response to society pressure to reduce the number of animal experiments. Similar responses in hiPSC-CM have also been observed in animal models, implying that hiPSC-CMs can replace certain animal experiments [4, 20, 65]. Although hiPSC-CMs are unlikely to replace animal models entirely due to their simplicity, three-dimensional multicellular organ-on-chip formats may go part way to addressing this [73]. Nevertheless, the use of hiPSC-CMs for replacement of even some animal studies would have financial benefit as, for example, the cost/data point in in vitro studies is usually significantly lower than animal studies. In short, hiPSC-CM models expand the possibilities to detect potential cardiotoxic drugs in humans, including drugs causing heart failure, and could reduce animal use in drug safety assessment most likely resulting in the reduction of costs in preclinical drug safety evaluation.

8.4.1 Limitations in Applicability of hiPSC-CMs to Drug Toxicity Screening

8.4.1.1 Phenotypic Immaturity

A major limitation of the hiPSC-CMs is their phenotypic immaturity (reviewed in [24]). Differences in phenotypic parameters between hiPSC-CMs and adult ventricular CMs (Table 8.4) can lead to inaccurate or limited prediction of toxicity or result in uncharacteristic behavior of CMs [43]. For example, hiPSC-CMs are relatively small and exhibit high pacemaker currents and a reduced I_{Kr}, resulting in a more depolarized resting membrane potential and spontaneously beating of hiPSC-CMs which is not observed in healthy adult ventricular myocytes [78]. Furthermore, maximum contraction forces of an order of magnitude smaller have been observed in hiPSC-CMs compared to adult ventricular CMs. This difference in contraction force may impede the detection between healthy control and damaged cells in hiPSC-CMs [79]. Also, caution is needed when translating hiPSC-CM results to humans, as immature calcium-handling properties have been observed in hiPSCs which is predominately caused by the lack of t-tubules [70, 80].

To increase the accuracy of predictions based on hiPSC-CM models, the maturation state of these cells has to be improved which can be done by prolonged culture [60] or exogenous stimulation, including addition of environmental factors [73], heterotypic cell co-culture [61], 3D environment [81], and mechanical [82] or electrical stimulation [83]. Most likely, a combination of these approaches is needed to improve the maturation state, as mechanical forces, 3D environment, and other cell types contribute to the heart development [79]. Mannhardt et al. have shown promising results regarding the maturation state of hiPSCs [77]. By combining 3D environment and mechanical loading, 3D force-generating engineered heart tissues (EHTs) consisting of classical rod-shaped ventricular CMs showed sarcomeric organization and indications of the presence of an immature t-tubule/SR system. Although the EHTs exhibited underdeveloped t-tubules, Mannhardt et al. found that the EHTs were able to recapitulate the effect of various drug classes with unprecedented precision. Based on this, the question "When are hiPSC-CM mature enough?" could be posed. Since there is no single answer to this question (maturation states can vary depending on the parameter used as readout), the required maturation state to obtain accurate and reproducible toxicity prediction must be carefully assessed and standardized.

8.4.1.2 Phenotypic Variability

Although modest variability in cells derived from one line has been observed, significant line-to-line has been reported and poses a problem to get reproducible results of absolute values [26]. For example, a large variation in values of electrophysiological parameters (e.g., APD, resting membrane potential, and beat rates) has been reported between cell lines and studies [16, 19, 78]. This large variation could be attributed by factors affecting the differential potential and post-differentiation phenotype such as individual backgrounds, DNA epigenetic status, mutation (e.g., caused by reprogramming process, somatic cell type origin), and culture conditions [26, 36, 79, 84].

In one study, DNA methylation and genetic analysis of four isogenic lines from different healthy donors ($n = 4$) demonstrated that iPSC line clustering was mainly donor-dependent [85]. Moreover, analysis of the differently expressed genes across all cell lines showed that most differently expressed genes were genes associated with early development, implying that expression of genes associated with pluripotency and differentiation could be influenced by donor-dependent characteristics. Based on these findings, it was concluded that use of relatively large cohorts of iPSC lines from different donors is important as small cohorts may lead to invalid conclusions. Furthermore, standardization of protocols is key to reduce variability in the results. Promising results have been observed from a multicentered study across ten facilities consisting of pharmaceutical companies and research organizations [65]. By standardizing several testing procedures and the use of commercial iPSC-CM lines, similar variability in absolute FPD and beat rate values was obtained inter- and intra-facility before drug exposure. With minor discrepancies, reliable and consistent MEA results were obtained over facilities that could correctly predict clinical QT prolongation and/or arrhythmic potential.

8.5 Conclusion

Recent studies provided optimism on the ability of hiPSC models to predict drug effects, and hiPSC-CM models could complement or even eventually replace current in vitro preclinical models. Moreover, evaluation of electrophysiology, cellular contraction, Ca transients, and/or cytotoxicity has the potential to predict various cardiac safety liabilities, including arrhythmia, contractile dysfunction, and structural toxicity, while only QT prolongation and TdP can be addressed in current (animal and heterologous overexpression) in vitro models. Consequences of the increasing predictive power of these humanized models will most likely reduce later-stage attrition, the number of animal studies, and risks for participants in clinical studies. A prerequisite is that the challenges discussed here must be overcome, such as the immature phenotype and line-to-line variability; this could be achieved by more closely recapitulating the in vivo environment of CMs and further standardization of experimental protocols.

Acknowledgments We thank Dr. Joost O. Fledderus from the University Medical Center Utrecht for his expertise and comments that improved the manuscript. In addition, we apologize to all those authors whose excellent work could not be cited due to space restrictions.

Disclosures C.L.M. is a cofounder of Pluriomics B.V.

References

1. Pharmaceutical Research and Manufacturers of America (2013) Biopharmaceutical Research Industry Profile. http://phrma-docs.phrma.org/sites/default/files/pdf/PhRMA%20Profile%202013.pdf. Accessed 5 Jan 2017
2. Kola I, Landis J (2004) Can the pharmaceutical industry reduce attrition rates? Nat Rev Drug Discov 3:1–5. https://doi.org/10.1038/nrd1470
3. Piccini JP, Whellan DJ, Berridge BR et al (2009) Current challenges in the evaluation of cardiac safety during drug development: translational medicine meets the critical path initiative. Am Heart J 158:317–326. https://doi.org/10.1016/j.ahj.2009.06.007

4. Navarrete EG, Liang P, Lan F et al (2013) Screening drug-induced arrhythmia events using human induced pluripotent stem cell-derived cardiomyocytes and low-impedance microelectrode arrays. Circulation 128:S3. https://doi.org/10.1161/CIRCULATIONAHA.112.000570
5. MacDonald JS, Robertson RT (2009) Toxicity testing in the 21st century: a view from the pharmaceutical industry. Toxicol Sci 110:40–46. https://doi.org/10.1093/toxsci/kfp088
6. Gintant G, Sager PT, Stockbridge N (2016) Evolution of strategies to improve preclinical cardiac safety testing. Nat Rev Drug Discov 15:457–471. https://doi.org/10.1038/nrd.2015.34
7. Yap YG, Camm AJ (2012) Drug induced QT prolongation and torsades de pointes. Heart 89:1363–1372
8. Spector PS, Curran ME, Keating MT, Sanguinetti MC (1996) Class III antiarrhythmic drugs block HERG, a human cardiac delayed rectifier K+ channel. Open- channel block by methanesulfonanilides. Circ Res 78:499 LP–499503
9. International Conference on Harmonization (2005) Guidance for Industry S7B Nonclinical Evaluation of the Potential for Delayed Ventricular Repolarization (QT Interval Prolongation) by Human Pharmaceuticals. https://www.fda.gov/downloads/Drugs/GuidanceComplianceRegulatoryInformation/Guidances/ucm074963.pdf. Accessed 5 Jan 2017
10. Fermini B, Hancox JC, Abi-Gerges N et al (2016) A new perspective in the field of cardiac safety testing through the comprehensive in vitro proarrhythmia assay paradigm. J Biomol Screen 21:1–11. https://doi.org/10.1177/1087057115594589
11. Friedrichs GS, Patmore L, Bass A (2005) Non-clinical evaluation of ventricular repolarization (ICH S7B): results of an interim survey of international pharmaceutical companies. J Pharmacol Toxicol Methods 52:6–11. https://doi.org/10.1016/j.vascn.2005.05.001
12. Mordwinkin NM, Burridge PW, Wu JC (2013) A review of human pluripotent stem cell-derived cardiomyocytes for high-throughput drug discovery, cardiotoxicity screening, and publication standards. J Cardiovasc Transl Res 6:22–30. https://doi.org/10.1007/s12265-012-9423-2
13. Mitcheson JS, Hancox JC, Levi a J (1998) Cultured adult cardiac myocytes: future applications, culture methods, morphological and electrophysiological properties. Cardiovasc Res 39:280–300. https://doi.org/10.1016/S0008-6363(98)00128-X
14. Takahashi K, Tanabe K, Ohnuki M et al (2007) Induction of pluripotent stem cells from adult human fibroblasts by defined factors. Cell 131:861–872. https://doi.org/10.1016/j.cell.2007.11.019
15. Yu J, Vodyanik MA, Smuga-Otto K et al (2008) Induced pluripotent stem cell lines derived from human somatic cells. Science 318(80):1917–1920. https://doi.org/10.1126/science.1151526
16. Zhang J, Wilson GF, Soerens AG et al (2009) Functional cardiomyocytes derived from human induced pluripotent stem cells. Circ Res 104:e30–e41. https://doi.org/10.1161/CIRCRESAHA.108.192237
17. Mummery CL, Zhang J, Ng ES et al (2012) Differentiation of human embryonic stem cells and induced pluripotent stem cells to cardiomyocytes: a methods overview. Circ Res 111:344–358. https://doi.org/10.1161/CIRCRESAHA.110.227512
18. Burridge PW, Li YF, Matsa E et al (2016) Human induced pluripotent stem cell-derived cardiomyocytes recapitulate the predilection of breast cancer patients to doxorubicin-induced cardiotoxicity. Nat Med 22:547–556. https://doi.org/10.1038/nm.4087
19. Hoekstra M, Mummery CL, Wilde AAM et al (2012) Induced pluripotent stem cell derived cardiomyocytes as models for cardiac arrhythmias. Front Physiol 3:1–14. https://doi.org/10.3389/fphys.2012.00346
20. Bergström G, Christoffersson J, Schwanke K et al (2015) Stem cell derived in vivo-like human cardiac bodies in a microfluidic device for toxicity testing by beating frequency imaging. Lab Chip 15:3242–3249. https://doi.org/10.1039/C5LC00449G
21. Ivashchenko CY, Pipes GC, Lozinskaya IM et al (2013) Human induced pluripotent stem cell derived cardiomyocytes exhibit temporal changes in phenotype. Am J Physiol Heart Circ Physiol 305:H913–H922. https://doi.org/10.1152/ajpheart.00819.2012

22. Ma J, Guo L, Fiene SJ et al (2011) High purity human-induced pluripotent stem cell-derived cardiomyocytes : electrophysiological properties of action potentials and ionic currents. Am J Physiol Heart Circ Physiol 301:2006–2017. https://doi.org/10.1152/ajpheart.00694.2011
23. van Meer BJ, Tertoolen LGJ, Mummery CL (2016) Concise review : measuring physiological responses of human pluripotent stem cell derived cardiomyocytes to drugs and disease. Stem Cells 1863:2008–2015. https://doi.org/10.1002/stem.2403
24. Yang X, Pabon L, Murry CE (2014) Engineering adolescence: maturation of human pluripotent stem cell-derived cardiomyocytes. Circ Res 114:511–523. https://doi.org/10.1161/CIRCRESAHA.114.300558
25. Chi KR (2013) Revolution dawning in cardiotoxicity testing. Nat Rev Drug Discov 12:565–567. https://doi.org/10.1038/nrd4083
26. Sala L, Bellin M, Mummery CL (2016) Integrating cardiomyocytes from human pluripotent stem cells in safety pharmacology: has the time come? Br J Pharmacol 174:1–17. https://doi.org/10.1111/bph.13577
27. Wu L, Rajamani S, Li H et al (2009) Reduction of repolarization reserve unmasks the proarrhythmic role of endogenous late Na(+) current in the heart. Am J Physiol Heart Circ Physiol 297:H1048–H1057. https://doi.org/10.1152/ajpheart.00467.2009
28. Tzeis S, Andrikopoulos G (2012) Antiarrhythmic properties of ranolazine-from bench to bedside. Expert Opin Investig Drugs 21:1733–1741. https://doi.org/10.1517/13543784.2012.716826
29. Cox BF, Gintant GA, Martin RL et al (2004) The utility of hERG and repolarization assays in evaluating delayed cardiac repolarization: influence of multi-channel block. J Cardiovasc Pharmacol 43:369–379. doi:00005344-200403000-00007 [pii]
30. Gintant G (2011) An evaluation of hERG current assay performance: translating preclinical safety studies to clinical QT prolongation. Pharmacol Ther 129:109–119. https://doi.org/10.1016/j.pharmthera.2010.08.008
31. European Commission (2012) Stop vivisection. http://ec.europa.eu/citizens-initiative/public/initiatives/successful/details/2012/000007/en?lg=en. Accessed 19 Dec 2016
32. Liang P, Lan F, Lee AS et al (2013) Drug screening using a library of human induced pluripotent stem cell-derived cardiomyocytes reveals disease-specific patterns of cardiotoxicity. Circulation 127:1677–1691. https://doi.org/10.1161/CIRCULATIONAHA.113.001883
33. Navarrete EG, Liang P, Lan F et al (2013) Screening drug-induced arrhythmia using human induced pluripotent stem cell-derived cardiomyocytes and low-impedance microelectrode arrays. Circulation 128:S3. https://doi.org/10.1161/CIRCULATIONAHA.112.000570
34. Scheel O, Frech S, Amuzescu B et al (2014) Action potential characterization of human induced pluripotent stem cell-derived cardiomyocytes using automated patch-clamp technology. Assay Drug Dev Technol 12:457–469. https://doi.org/10.1089/adt.2014.601
35. Honda M, Kiyokawa J, Tabo M, Inoue T (2011) Electrophysiological characterization of cardiomyocytes derived from human induced pluripotent stem cells. J Pharmacol Sci 117:149–159. https://doi.org/10.1254/jphs.11038FP
36. Laurila E, Ahola A, Hyttinen J, Aalto-Setälä K (2016) Methods for in vitro functional analysis of iPSC derived cardiomyocytes - special focus on analyzing the mechanical beating behavior. Biochim Biophys Acta, Mol Cell Res 1863:1864–1872. https://doi.org/10.1016/j.bbamcr.2015.12.013
37. Bellin M, Mummery CL (2016) Inherited heart disease? What can we expect from the second decade of human iPS cell research? FEBS Lett 590:2482–2493. https://doi.org/10.1002/1873-3468.12285
38. Sheng X, Reppel M, Nguemo F et al (2012) Human pluripotent stem cell-derived cardiomyocytes: response to TTX and Lidocain reveals strong cell to cell variability. PLoS One 7:1–12. https://doi.org/10.1371/journal.pone.0045963
39. Spira ME, Hai A (2013) Multi-electrode array technologies for neuroscience and cardiology. Nat Nanotechnol 8:83–94. https://doi.org/10.1038/nnano.2012.265

40. Gilchrist KH, Lewis GF, Gay EA et al (2015) High-throughput cardiac safety evaluation and multi-parameter arrhythmia profiling of cardiomyocytes using microelectrode arrays. Toxicol Appl Pharmacol 288:249–257. https://doi.org/10.1016/j.taap.2015.07.024
41. Denning C, Borgdorff V, Crutchley J et al (2016) Cardiomyocytes from human pluripotent stem cells: from laboratory curiosity to industrial biomedical platform. Biochim Biophys Acta, Mol Cell Res 1863:1728–1748. https://doi.org/10.1016/j.bbamcr.2015.10.014
42. Nozaki Y, Honda Y, Tsujimoto S, Watanabe H (2014) Availability of human induced pluripotent stem cell-derived cardiomyocytes in assessment of drug potential for QT prolongation. Toxicol Appl Pharmacol 278:72–77. https://doi.org/10.1016/j.taap.2014.04.007
43. Qu Y, Vargas HM (2015) Proarrhythmia risk assessment in human induced pluripotent stem cell-derived cardiomyocytes using the maestro MEA platform. Toxicol Sci 147:286–295. https://doi.org/10.1093/toxsci/kfv128
44. Hortigon-Vinagre MP, Zamora V, Burton FL et al (2016) The use of ratiometric fluorescence measurements of the voltage sensitive dye Di-4-ANEPPS to examine action potential characteristics and drug effects on human induced pluripotent stem cell-derived cardiomyocytes. Toxicol Sci 154:kfw171. https://doi.org/10.1093/toxsci/kfw171
45. Bootman MD, Rietdorf K, Collins T, et al (2014) Ca 2 + -sensitive fluorescent dyes and intracellular Ca 2 + imaging. 25–50. doi:https://doi.org/10.1101/pdb.top066050
46. Pointon A, Harmer AR, Dale IL et al (2015) Assessment of cardiomyocyte contraction in human-induced pluripotent stem cell-derived cardiomyocytes. Toxicol Sci 144:227–237. https://doi.org/10.1093/toxsci/kfu312
47. Sirenko O, Cromwell EF, Crittenden C et al (2013) Assessment of beating parameters in human induced pluripotent stem cells enables quantitative in vitro screening for cardiotoxicity. Toxicol Appl Pharmacol 273:500–507. https://doi.org/10.1016/j.taap.2013.09.017.Assessment
48. Lu HR, Whittaker R, Price JH et al (2015) High throughput measurement of ca++ dynamics in human stem cell-derived cardiomyocytes by kinetic image cytometery: a cardiac risk assessment characterization using a large panel of cardioactive and inactive compounds. Toxicol Sci 148:503–516. https://doi.org/10.1093/toxsci/kfv201
49. Pointon A, Pilling J, Dorval T et al (2016) High-throughput imaging of cardiac microtissues for the assessment of cardiac contraction during drug discovery. Toxicol Sci 155:1–14. https://doi.org/10.1093/toxsci/kfw227
50. Davies CH, Davia K, Bennett JG et al (1995) Reduced contraction and altered frequency response of isolated ventricular myocytes from patients with heart failure. Circulation 92:2540–2549. https://doi.org/10.1161/01.CIR.92.9.2540
51. ACEA Bio (n.d.) xCELLigence RTCA CardioECR. https://aceabio.com/products/rtca-cardio-ecr/. Accessed 3 Feb 2017
52. Peters MF, Lamore SD, Guo L et al (2015) Human stem cell-derived cardiomyocytes in cellular impedance assays: bringing cardiotoxicity screening to the front line. Cardiovasc Toxicol 15:127–139. https://doi.org/10.1007/s12012-014-9268-9
53. Talbert DR, Doherty KR, Trusk PB et al (2015) A multi-parameter in vitro screen in human stem cell-derived cardiomyocytes identifies ponatinib-induced structural and functional cardiac toxicity. Toxicol Sci 143:147–155. https://doi.org/10.1093/toxsci/kfu215
54. Doherty KR, Talbert DR, Trusk PB, et al (2015) Structural and functional screening in human induced-pluripotent stem cell-derived cardiomyocytes accurately identifies cardiotoxicity of multiple drug types. 285:51–60. doi:https://doi.org/10.1016/j.taap.2015.03.008
55. Bellin M, Mummery CL (2016) Stem cells: the cancer's gone, but did chemotherapy damage your heart? Nat Rev Cardiol 13:383. https://doi.org/10.1038/nrcardio.2016.88
56. Chaudhari U, Nemade H, Wagh V et al (2015) Identification of genomic biomarkers for anthracycline-induced cardiotoxicity in human iPSC-derived cardiomyocytes: an in vitro repeated exposure toxicity approach for safety assessment. Arch Toxicol 90:2763–2777. https://doi.org/10.1007/s00204-015-1623-5
57. Boucek RJ, Dodd DA, Atkinson JB et al (1997) Contractile failure in chronic doxorubicin-induced cardiomyopathy. J Moll Cell Cardiol 29:2631–2640. https://doi.org/10.1006/jmcc.1997.0494

58. Hunt CM (2010) Mitochondrial and immunoallergic injury increase risk of positive drug rechallenge after drug-induced liver injury: a systematic review. Hepatology 52:2216–2222. https://doi.org/10.1002/hep.24022
59. Agilent (n.d.) How seahorse XF analyzers work. http://www.agilent.com/en-us/products/cell-analysis-(seahorse)/how-seahorse-xf-analyzers-work. Accessed 27 Jan 2017
60. Rana P, Anson B, Engle S, Will Y (2012) Characterization of human-induced pluripotent stem cell-derived cardiomyocytes: bioenergetics and utilization in safety screening. Toxicol Sci 130:117–131. https://doi.org/10.1093/toxsci/kfs233
61. Robertson C, Tran D, George S (2013) Concise review: maturation phases of human pluripotent stem cell-derived cardiomyocytes. Stem Cells 31:1–17. https://doi.org/10.1002/stem.1331
62. Hasinoff BB (2010) The cardiotoxicity and myocyte damage caused by small molecule anticancer tyrosine kinase inhibitors is correlated with lack of target specificity. Toxicol Appl Pharmacol 244:190–195. https://doi.org/10.1016/j.taap.2009.12.032
63. Varga ZV, Ferdinandy P, Liaudet L, Pacher P (2015) Drug-induced mitochondrial dysfunction and cardiotoxicity. Am J Physiol Hear Circ Physiol 00554:2015. https://doi.org/10.1152/ajpheart.00554.2015
64. Takasuna K, Asakura K, Araki S et al (2017) Comprehensive in vitro cardiac safety assessment using human stem cell technology: overview of CSAHi HEART initiative. J Pharmacol Toxicol Methods 83:42–54. https://doi.org/10.1016/j.vascn.2016.09.004
65. Kitaguchi T, Moriyama Y, Taniguchi T et al (2016) CSAHi study: evaluation of multi-electrode array in combination with human iPS cell-derived cardiomyocytes to predict drug-induced QT prolongation and arrhythmia - effects of 7 reference compounds at 10 facilities. J Pharmacol Toxicol Methods 78:93–102. https://doi.org/10.1016/j.vascn.2015.12.002
66. Coloma PM, Trifirò G, Patadia V, Sturkenboom M (2013) Postmarketing safety surveillance: where does signal detection using electronic healthcare records fit into the big picture? Drug Saf 36:183–197. https://doi.org/10.1007/s40264-013-0018-x
67. Cross MJ, Berridge BR, Clements PJM et al (2015) Physiological, pharmacological and toxicological considerations of drug-induced structural cardiac injury. Br J Pharmacol 172:957–974. https://doi.org/10.1111/bph.12979
68. Pointon A, Abi-gerges N, Cross MJ, Sidaway JE (2013) Phenotypic profiling of structural cardiotoxins in vitro reveals dependency on multiple mechanisms of toxicity. Toxicol Sci 132:317–326. https://doi.org/10.1093/toxsci/kft005
69. Clements M, Thomas N (2014) High-throughput multi-parameter profiling of electrophysiological drug effects in human embryonic stem cell derived cardiomyocytes using multi-electrode arrays. Toxicol Sci 140:445–461. https://doi.org/10.1093/toxsci/kfu084
70. Himmel HM (2013) Drug-induced functional cardiotoxicity screening in stem cell-derives human and mouse cardiomyocytes: effects of reference compounds. J Pharmacol Toxicol Methods 68(1):97–111. https://doi.org/10.1016/j.vascn.2013.05.005
71. Zhang X, Guo L, Zeng H et al (2016) Multi-parametric assessment of cardiomyocyte excitation-contraction coupling using impedance and field potential recording: a tool for cardiac safety assessment. J Pharmacol Toxicol Methods 81:201–216. https://doi.org/10.1016/j.vascn.2016.06.004
72. Sala L, Yu Z, Ward-van Oostwaard D et al (2016) A new hERG allosteric modulator rescues genetic and drug-induced long-QT syndrome phenotypes in cardiomyocytes from isogenic pairs of patient induced pluripotent stem cells. EMBO Mol Med 8:1065–1081. https://doi.org/10.15252/emmm.201606260
73. Passier R, Orlova V, Mummery C (2016) Review complex tissue and disease modeling using hiPSCs. Stem Cell 18:309–321. https://doi.org/10.1016/j.stem.2016.02.011
74. Magyar J, Iost N, Körtvély Á et al (2000) Effects of endothelin-1 on calcium and potassium currents in undiseased human ventricular myocytes. Pflügers Arch 441:144–149. https://doi.org/10.1007/s004240000400
75. Ribeiro MC, Tertoolen LG, Guadix JA et al (2015) Functional maturation of human pluripotent stem cell derived cardiomyocytes in vitro - correlation between contraction force and electrophysiology. Biomaterials 51:138–150. https://doi.org/10.1016/j.biomaterials.2015.01.067

76. van der Velden J, Klein L, van der Bijl M et al (1998) Force production in mechanically isolated cardiac myocytes from human ventricular muscle tissue. Cardiovasc Res 38:414–423. https://doi.org/10.1016/S0008-6363(98)00019-4
77. Mannhardt I, Breckwoldt K, Letuffe-Brenière D et al (2016) Human engineered heart tissue: analysis of contractile force. Stem Cell Reports 7:29–42. https://doi.org/10.1016/j.stemcr.2016.04.011
78. Knollmann BC (2013) Induced pluripotent stem cell – derived cardiomyocytes boutique science or valuable arrhythmia model ? Circ Res 112:969–977. https://doi.org/10.1161/CIRCRESAHA.112.300567
79. Bellin M, Marchetto MC, Gage FH, Mummery CL (2012) Induced pluripotent stem cells: the new patient? Nat Rev Mol Cell Biol 13:713–726. https://doi.org/10.1038/nrm3448
80. Novak A, Barad L, Zeevi-Levin N et al (2012) Cardiomyocytes generated from CPVTD307H patients are arrhythmogenic in response to β-adrenergic stimulation. J Cell Mol Med 16:468–482. https://doi.org/10.1111/j.1582-4934.2011.01476.x
81. Zhang J, Klos M, Wilson GF et al (2012) Extracellular matrix promotes highly efficient cardiac differentiation of human pluripotent stem cells: the matrix sandwich method. Circ Res 111:1125–1136. https://doi.org/10.1161/CIRCRESAHA.112.273144
82. Mathur A, Loskill P, Shao K et al (2015) Human iPSC-based cardiac microphysiological system for drug screening applications. Sci Rep 5:8883. https://doi.org/10.1038/srep08883
83. Chan Y-C, Ting S, Lee Y-K et al (2013) Electrical stimulation promotes maturation of cardiomyocytes derived from human embryonic stem cells. J Cardiovasc Transl Res 6:989–999. https://doi.org/10.1007/s12265-013-9510-z
84. Miyamoto T, Furusawa C, Kaneko K (2015) Pluripotency, differentiation, and reprogramming: a gene expression dynamics model with epigenetic feedback regulation. PLoS Comput Biol 11:1–25. https://doi.org/10.1371/journal.pcbi.1004476
85. Kyttälä A, Moraghebi R, Valensisi C et al (2016) Genetic variability overrides the impact of parental cell type and determines iPSC differentiation potential. Stem Cell Reports 6:200–212. https://doi.org/10.1016/j.stemcr.2015.12.009

Part III
Stem Cells for General Medicine

Chapter 9
Regenerative Medicine for Diabetes Treatment: New β-Cell Sources

Rosa Gasa and Meritxell Rovira

Abstract Diabetes results from an insufficient number of functional insulin-producing pancreatic β cells. Hence, the replenishment of the β cell population, either by induction of endogenous regeneration or by replacement with substitute β cells generated ex vivo, has long been viewed as a promising strategy to definitely cure this disease.One main topic of debate in the field has been the existence of facultative or dedicated progenitors that can give rise to new insulin-producing cells. On the other hand, recent major advances in in vitro directed differentiation protocols have put embryonic and induced pluripotent stem cells on the spotlight as promising cell sources of transplantable β cells. In this chapter, we will discuss current knowledge and future perspectives on both of theseapproaches. The identity of the optimal candidate cell source for β cell replenishment in diabetic patients remains an open and exciting question that will need to be addressed in future investigations.

Keywords Diabetes · Progenitor · Stem · Differentiation · Transdifferentiation · Regeneration

9.1 Introduction

The pancreas is an endoderm-derived gland formed by an endocrine and exocrine compartment. The exocrine pancreas comprises 95% of the gland, while the endocrine compartment represents only 5% [12]. The exocrine compartment contains

R. Gasa
Centro de Investigación Biomédica en Red de Diabetes y Enfermedades Metabólicas Asociadas (CIBERDEM), Barcelona, Spain

Diabetes and Obesity Research Laboratory, August Pi i Sunyer Biomedical Research Institute (IDIBAPS), Barcelona, Spain

M. Rovira (✉)
Center of Regenerative Medicine in Barcelona (CMRB), Hospital Duran i Reynals, Barcelona, Spain
e-mail: mrovira@cmrb.eu

R. Delgado-Morales (ed.), *Stem Cell Genetics for Biomedical Research*,
https://doi.org/10.1007/978-3-319-90695-9_9

acinar cells, organized in acini, which produce and secrete digestive enzymes into the lumen of the ducts. The ductal tree is an intricate network of tubular structures formed by ductal cells. These cells secrete a bicarbonate-containing fluid that together with the acinar secretions are driven into the duodenum; ductal secretions neutralize gastric acid to provide an optimum pH environment for digestive enzymes [61]. The endocrine pancreas is organized in islets of Langerhans that are scattered within the exocrine tissue and contain at least five types of hormone-producing cells. These are insulin-producing β cells, glucagon-producing α cells, somatostatin-producing δ cells, pancreatic polypeptide-producing PP cells, and ghrelin-producing ε cells, from most to least abundant, respectively [75]. These different endocrine cell types are the main regulators of nutrient metabolism and glucose homeostasis.

Diabetes mellitus is a metabolic disorder characterized by loss and/or dysfunction of pancreatic β cells. Over 415 million people worldwide suffer from diabetes, and this number is increasing and is expected to rise up to 642 million in 2040, making diabetes a major public health issue worldwide [50]. The main function of β cells is to produce and secrete insulin, the hormone responsible for the regulation of blood glucose levels. Around 5% of diabetic patients suffer from type 1 diabetes, an autoimmune disease in which the immune system recognizes β cells as foreign and destroys them resulting in almost absolute loss of insulin production and impaired glucose homeostasis. Type 1 diabetes patients are therefore dependent on exogenous administration of insulin for survival. On the other hand, 95% of diabetic patients suffer from type 2 diabetes, which is the result of impaired insulin sensitivity as well as relative insulin deficiency; over time, type 2 diabetic patients can become insulin dependent as well [29]. Therefore, finding ways to expand/regenerate β-cell mass and restore the damaged cells could be an effective therapeutic approach and an important milestone in regenerative medicine for diabetes treatment.

Currently, islet transplantation is the only way to provide new β cells to diabetic patients, but the scarcity of compatible cadaveric donors makes this approach available to only few patients; moreover it requires lifelong immune suppression [14]. Therefore in the past two decades there have been three main lines of research focused in the search of alternative sources of β cells that have already found some success: first, to enhance replication of β cells "in vitro" to expand the limited supply in diabetic patients; second, to identify a pancreatic progenitor that can be isolated and expand in vitro and give rise to new β cells, this progenitors could also be differentiated in vivo under regenerative conditions; third, in vitro differentiation of human embryonic and induced pluripotent stem cells toward insulin-producing cells.

The first of these approaches will be briefly summarized in Sect. 9.3, as it is beyond the scope of this chapter (for a detailed discussion of this approach, you can review the following articles [54, 98]). With respect to the second approach, we will consider the existence of facultative or dedicated pancreatic progenitors in the adult pancreas and discuss how they might be exploited clinically. Finally, we will discuss the great progress made in the field of in vitro differentiation of human embryonic and induced pluripotent stem cells toward insulin-producing cells. The success of both strategies will depend on our understanding of normal pancreas formation, differentiation, and homeostasis; thus, we start this chapter with a section depicting a brief summary of pancreas development.

9.2 Pancreas Development

The pancreas is an endoderm-derived tissue that arises from two buds, one dorsal and one ventral, in the posterior foregut that will later fuse to give rise to the totality of the gland. The pre-pancreatic domain is defined at embryonic day E8.5 in the mouse by the expression of the transcription factors Pdx1 and Ptf1a. Lineage-tracing studies clearly showed that all pancreatic lineages, endocrine, ductal, and acinar cells, derive from this Pdx1/Ptf1a multipotent pancreatic progenitors [78]. Moreover, in mice and humans, knockouts or inactivating mutations in either of these genes result in almost complete pancreatic agenesis [39, 40, 96] (Figs. 9.1, 9.2, and 9.3).

During early pancreatic development, there is growth and branching morphogenesis prior to the differentiation of the pancreatic lineages. In mice, the secondary transition begins around E12.5–E13.5. At this stage, the pancreas contains a branched ductal epithelium that is compartmentalized into tip and trunk domains [121]. The Ptf1a-expressing tip domain will give rise to acinar cells, while trunk cells retain both duct and endocrine differentiation potential. These bi-potent progenitors maintain a Notch active pathway, which restricts endocrine differentiation in the trunk domain. High levels of Notch induce expression of the transcription factor Hes1, which in turn promotes expression of Nkx6.1 and Sox9 while inhibiting expression of the endocrine determinant Ngn3 [1]. Thus, trunk cells that activate Ngn3 will become endocrine cells, whereas *Ngn3*-negative cells will mature as duct cells [2]. After the secondary transition, a massive differentiation of both endocrine and exocrine cells and growth of the pancreas takes place [43]. Hence, primitive duct-like complexes, located at the trunk region, are the source of both adult endocrine and exocrine duct cells [55, 56, 58, 99]. $Ngn3^+$ cells within this progenitor pool delaminate and differentiate into endocrine cell clusters, while the remaining duct-like complexes downregulate the pancreatic progenitor marker *Pdx1* and differentiate into mature pancreatic duct cells with sustained expression of *Sox9*, *Hnf1β*, and *Hnf6* [57, 97].

In the postnatal pancreas, under homeostatic conditions, there is still some growth and final maturation of differentiated cells types, although there is no evidence of the existence of a multipotent pancreatic progenitor cell contributing to either endocrine or exocrine cell proliferation; therefore the growth of which can be accounted solely by division of existing differentiated cells [4]. This last statement has been the focus of an intense debate in the last two decades, as several groups have searched for the existence of dedicated or facultative progenitors in the adult pancreas under regenerative conditions. We will discuss this still open debate in the next section of the chapter.

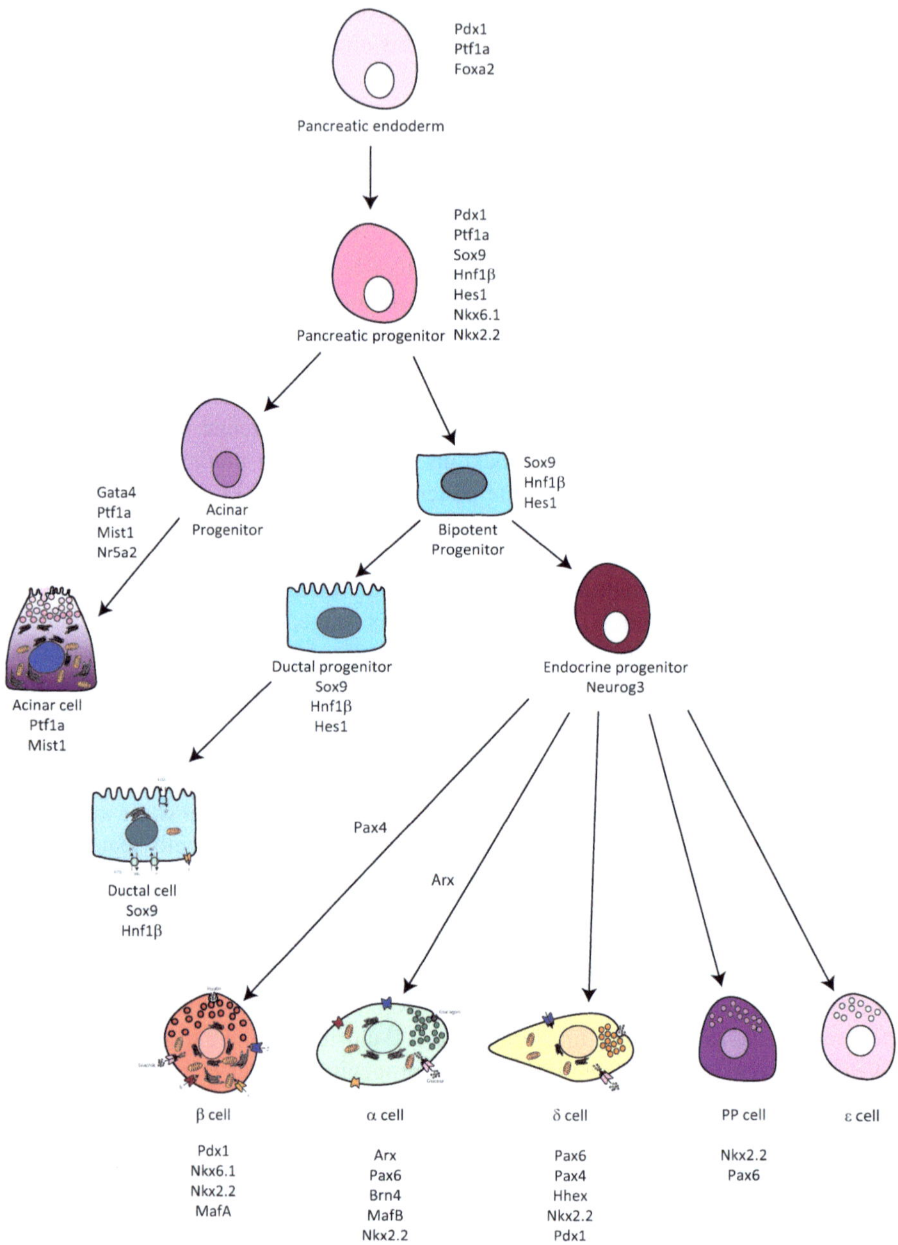

Fig. 9.1 Lineage development in the pancreas. Illustration of transcription factors involved in the key points of cell lineage specification and commitment of the different pancreatic lineages

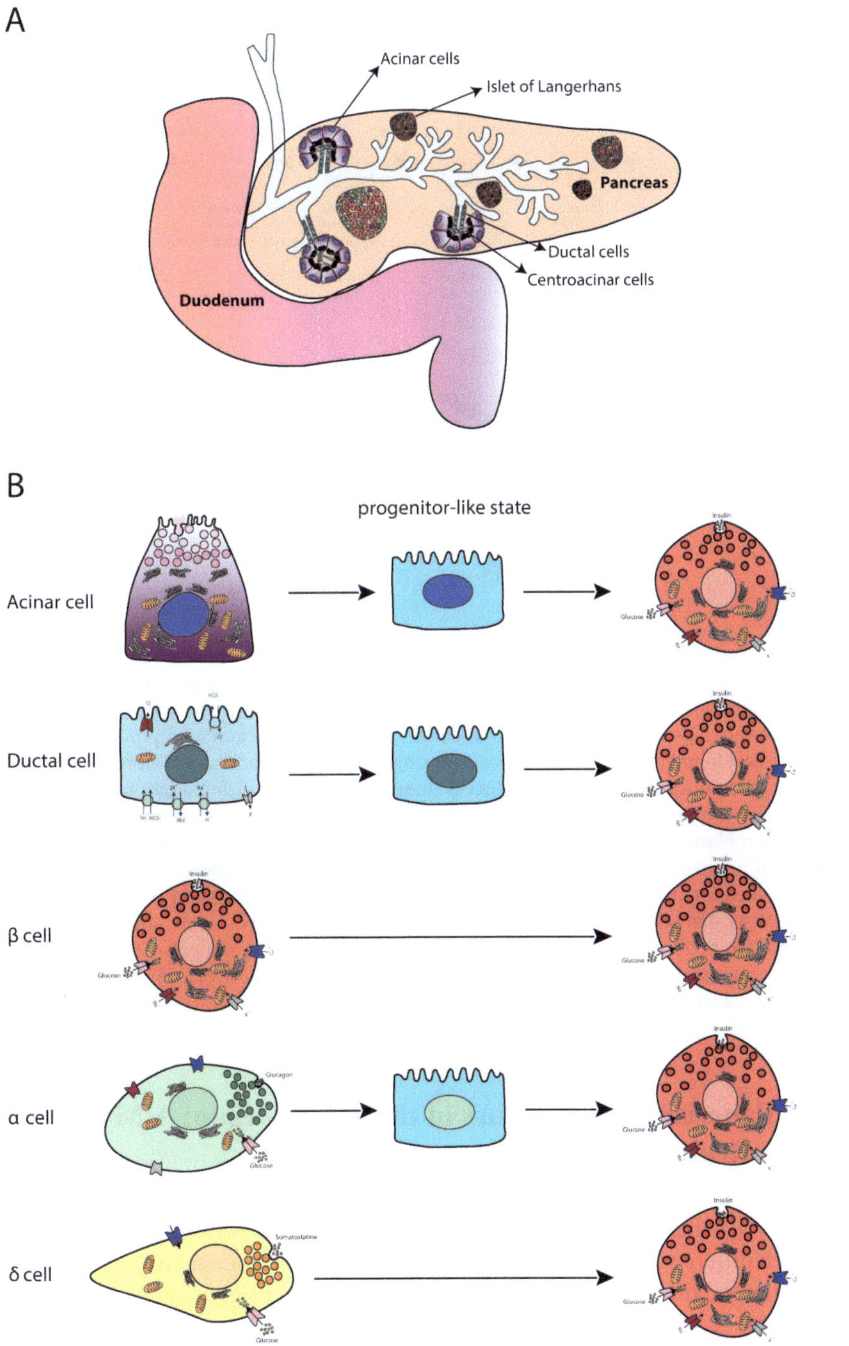

Fig. 9.2 The plastic pancreas. (**a**) Illustration of adult mouse pancreas containing acinar cells, centroacinar cells, and ductal cells and islets containing β cells, α cells, δ cells, PP cells, and ε cells. (**b**) Illustration of cell plasticity between the different pancreatic lineages. Acinar (purple), CAC/ductal (light blue), and α cells (green) can act as facultative progenitors giving rise to β cells thru transdifferentiation to a progenitor-like state (light pink). δ cells (yellow) can give rise to β cells without passing by thru a progenitor-like state. Finally, new β cells (dark pink) can be generated by proliferation of preexisting β cells

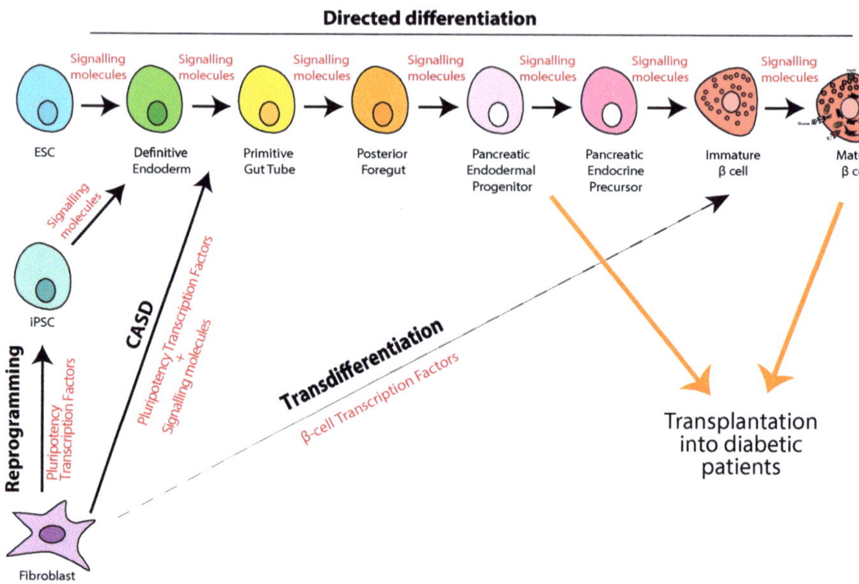

Fig. 9.3 An overview of potential strategies to generate β cells or their precursors for cell replacement therapy in diabetes. Pluripotent stem cells, embryonic (ESC) and induced (iPSC), can be coaxed toward a β-cell fate following a directed multistaged differentiation protocol that recapitulates the development of pancreatic β cells during embryogenesis. The addition of molecules that regulate specific signalling pathways directs stem cells along this differentiation path. An alternate less explored strategy known as CASD (cell activation and signaling directed) entails the conversion of fibroblasts into endodermal progenitors bypassing the pluripotent stage and thus avoiding the risk of tumor formation associated with pluripotency. This paradigm is accomplished through the combined use of pluripotency reprogramming transcription factors and signaling molecules. A third strategy that still needs to be validated in the β-cell field involves the direct reprogramming or transdifferentiation of fibroblasts into β cells using a defined set of β-cell differentiation factors. Whether pancreatic progenitors or mature β cells will be the best option for cell replacement therapy in diabetes remains open

9.3 Pancreatic Progenitors in the Adult Organ: A Myth or a Reality?

There is a striking diversity of regenerative capacities between animals and even between different tissues in a given organism. Often, these remarkable regenerative capacities are due to the activities of pluripotent stem cells or adult progenitor cells residing in a specific tissue [103]. Recent accounts indicate that, in some tissues, certain types of injuries can cause otherwise fully differentiated "unipotent" cells, under homeostatic conditions, to dedifferentiate and adopt a "progenitor-like" state capable of contributing to other cell lineages [124]. These cells have been termed facultative progenitors due to their ability to acquire multipotent qualities during non-homeostatic conditions [81, 116]. There are instances where direct conversion of one specific cell type to another has been observed, this is termed transdifferentiation, and can occur without the acquisition of a "progenitor-like" dedifferentiation state [23, 120].

The presence of adult stem cells in the intestine and skin has been well characterized, as these tissues continuously regenerate to maintain homeostasis. In these tissues, the existence of a niche that maintains such stem cell population through self-renewal and differentiation into terminal fates has been well defined, and its existence determines that of the progenitor cells [46, 52, 93]. Thus, the niche represents a defined anatomical compartment that provides signals to stem cells in the form of secreted and cell surface molecules to control the rate of stem cell proliferation, determine the fate of stem cell daughters, and protect stem cells from exhaustion or death [95]. Such a niche has not yet been identified in the adult pancreas.

In contrast to highly proliferative tissues, such as the intestine, skin, or hematopoietic system, the pancreas, especially the endocrine compartment, has an extremely low turnover rate under physiological conditions. Yet, under high metabolic demands such as pregnancy (physiological condition) or obesity (pathological condition), the pancreas is able to adapt by increasing β-cell mass mainly by self-replication, cellular hypertrophy, as well as increased insulin synthesis and secretion [16, 69]. Although the endocrine and exocrine pancreatic compartments exhibit some capacity for regeneration after injury, this does not lead to full restoration of tissue mass [71, 74, 105]. Fueled, in part, by the search for an unlimited β-cell source for diabetes treatment, there has been an extensive research and debate, as to whether dedicated progenitor cells exist in the pancreas or if instead regeneration comes from facultative progenitors, transdifferentiation, or simply compensatory proliferation of each cell type. Here, we will present detailed data exploring the possibilities of β-cell regeneration from the various lineages of the adult pancreas.

9.3.1 Acinar Cells as a Source of β-Cell Regeneration

Acinar cells are the most abundant cell type in the pancreas; therefore they are an attractive potential source of β cells due to their vast numbers. However, genetic lineage-tracing studies using acinar-specific promoters have so far yielded conflicting results. Lineage-tracing experiments using *Elastase*^{CreER} showed no evidence of acinar-to-β-cell conversion under regenerative settings, such as partial pancreatectomy, pancreatic duct ligation (PDL), or caerulein-induced pancreatitis [33]. Contrary to these results, when a *Ptf1a*^{CreER} line was used to trace acinar cells, Pan et al. showed that Ptf1a^{+} acinar cells after PDL and PDL + streptozotocin (STZ) treatment can act as facultative progenitors, expressing several embryonic multipotency factors, and become competent to produce ductal and endocrine cells, including mature β cells, although with low efficiency [78]. Westphalen et al. recently described a rare population of quiescent acinar cells positive for doublecortin-like kinase-1 (Dclk1). However, this newly defined population under regenerative conditions has a limited capacity to differentiate uniquely into acinar and ductal cells [111]. Thus, these evidences from lineage-tracing studies suggest that, in vivo, the propensity for acinar cells to contribute to new β cells is limited. However, acinar cells possess an intrinsic plasticity to differentiate into endocrine cells in vivo by ectopic expression of three key transcription factors of β-cell differentiation – Pdx1, Ngn3, and MafA [120]. These acinar-reprogrammed β cells persisted for over a year and were shown to have undergone epigenetic, transcriptional, anatomical, and functional changes toward an authentic β-cell phenotype [64]. Subsequent work from the same group also showed that acinar cells can be converted to δ-like and α-like cells by overexpressing Ngn3 and Ngn3 + MafA, respectively. Thus, three major islet endocrine subtypes can be derived from acinar reprogramming. Ngn3 promotes establishment of a generic endocrine state in acinar cells and also promotes δ-specification in the absence of other factors [66]. Recent evidence suggests that hyperglycemia significantly impairs the extent of such acinar-to-β-cell reprogramming in terms of quantity, differentiation status, and function [22].

Interestingly, the plasticity of acinar cells increases in vitro, as the simple fact of culturing them in suspension induces their spontaneous transdifferentiation into a duct-like phenotype resembling that of an embryonic pancreatic progenitor. Moreover, growth factor treatment of acinar cells in culture induces the differentiation of these intermediate duct-like cells into β cells. Knowledge gained by acinar-to-β-cell in vitro differentiation has led to the investigation of the role of such growth factors into the regeneration of β cells coming from acinar cells in vivo. To this end, the role of EGF and CNTF in acinar-to-β-cell transdifferentiation was recently tested in vivo by Baeyens et al., as in vitro experiments showed that these growth factors induced such reprogramming. Baeyens et al. elegantly demonstrated that systemic administration of EGF and CNTF is sufficient to induce acinar reprogramming into β cells in hyperglycemic diabetic mice, generating a functional β-cell mass to restore normoglycemic conditions. Notably, such reprogramming process occurs via *Ngn3* activation and STAT3 signaling [9].

9.3.2 *Contribution of Other Endocrine Cell Types to β-Cell Regeneration*

The common developmental origin as well as function shared between the different endocrine cell types of the islets of Langerhans has aroused a lot of interest into the potential of these closest relatives as the source of new β cells.

Although β-cell replication is not the focus of this section, it is worth noting that several studies based on partial pancreatectomy pointed out at the capacity of β cells to proliferate. However these studies showed controversial results linked to the amount of pancreas removed. Reports in which up to 90% partial pancreatectomy was employed to study β-cell regeneration showed that β-cell regeneration was mediated by both β-cell neogenesis (new β-cell formation from other cell sources) and proliferation of existing β cells [17, 44]. Later studies, which employed 70%

and 50% pancreatectomy, showed that β cells regenerate almost exclusively by self-duplication of existing β cells and not through neogenesis [35, 70, 74, 104]. It is important to note that 90% pancreatectomy induces a hyperglycemic state while 50–70% pancreatectomy does not. It is therefore likely that the hyperglycemic state in subtotal pancreatectomy may provide an additional stimulus for both β-cell neogenesis and self-replication of existing β cells, considering that hyperglycemia is known to stimulate an increase in β-cell number [76, 118].

Since recently β cells have been studied as a homogeneous population: however, recent reports have pointed out the existence of β-cell heterogeneity based on distinct features, such as nuclear size, granularity, polarity, hormone content, glucose responsiveness, and insulin secretion [47, 90]. More recently, cellular and molecular biology techniques and single-cell analysis have led to the discovery of biomarkers of β-cell heterogeneity [8, 10, 36, 73]. Dorrell et al. identified four antigenically distinct subtypes of human β cells, which are distinguished by differential expression of ST8SIA1 and CD9 [36]. Moreover, Bader et al. showed that *Fltp*, a Wnt/planar cell polarity (PCP) effector and reporter gene, acts as a marker that subdivides endocrine cells into two subpopulations and distinguishes proliferation competence from mature β cells with distinct molecular, physiological, and ultrastructural features [8]. Thus, β-cell heterogeneity could be exploited for therapeutic purposes as well.

Although β cells have some capacity for proliferation after pancreatectomy and during pregnancy [16, 69], it is not the case in mouse diabetic models such as streptozotocin, alloxane, and diphtheria toxin-induced β-cell ablation. Thus, several studies have explored the possibility of differentiating other pancreatic endocrine cells such as α, δ, and ε cells into β cells. These studies have revealed that the identity of pancreatic endocrine cell types is not as static or stable as it was once thought to be [5, 23–25, 27, 105].

The endocrine cell type harboring the highest plasticity has been α cells. Seeding work from Herrera's group showed that under conditions of almost complete β-cell loss, following diphtheria toxin-mediated ablation of β cells, α cells switched fate and differentiated into β cells [105]. In the same line, Collombat et al. described how overexpression of *Pax4* (a master regulator of β-cell specification/differentiation) converts embryonic α cells into β cells, albeit with aberrant physiological responses [24]. Similarly, deletion of *Arx* in α cells allows for the activation of some β-cell markers, accompanied by the loss of some α-cell markers [26, 112]. The same group showed that missexpression of *Pax4* in adult α cells induced their conversion into β-like cells, and an unexpected islet neogenesis, presumably from pancreatic duct cells, resulting in the restoration of β-cell mass even after several serial rounds of β-cell ablation and regeneration [5]. Recent work from Collombat and Kubicek laboratories has turned out to be an impressive boost to the field of β-cell regeneration as these authors have uncovered the role of gamma-aminobutyric acid (GABA) receptor signaling in the conversion of α cells into β cells [11, 62]. Two complementary papers nicely illustrate how a high-throughput screen of small molecules to identify compounds that could either induce *Pax4* or inhibit *Arx* expression in α cell lines in vitro led to the identification of the antimalarial drug artemisinin (e.g., artemether) as hit compounds that induce the conversion of α to β cells through activation of $GABA_A$ receptor signaling in α cells [62]. Indeed, either artemisinin treatment [62] or long-term administration of GABA [11] results in robust conversion of α cells into β cells both in vitro and in vivo, either by inhibition of *Arx* function or expression, which consequently led to induction of *Pax4* expression. Interestingly, artemisinins and GABA treatment induced β-cell neogenesis from α cells in vivo in zebrafish and rodent models of diabetes, as well as in human islets in vitro. GABA treatment acts via $GABA_A$ receptor in α cells, whereas artemisinins stabilize neuronal assembly protein gephyrin in this receptor, thus enhancing $GABA_A$ receptor signaling and consequently inhibiting glucagon secretion. This suggests that a reduced extracellular concentration of glucagon is required for α-to-β-cell conversion mediated by GABA receptor signaling; the role of extracellular glucagon concentration in α-to-β-cell conversion was already described in previous work from Collombat laboratory [5], showing that glucagon administration inhibited α-to-β-cell conversion in vivo in a mouse model with ectopic expression of *Pax4* in α cells [5].

Although α-to-β-cell conversion is maintained through the lifetime of a mouse, from puberty to adulthood, work from Herrera's group indicates that prior to puberty, there is no detectable α-to-β-cell conversion, instead, β-cell reconstitution after injury prior to puberty appears to be even more efficient at this stage, consistently leading to diabetes recovery. Chera et al. elegantly demonstrated that the reconstitution of β-cell mass during prepubescence is fueled by the spontaneous reprogramming of somatostatin-producing δ cells [23]. Finally, Arnes et al. have also demonstrated, using $Ghrl^{Cre}$:eGFP knock-in mice, that ghrelin-expressing ε cells give rise to significant numbers of α and PP cells and rare β cells in the adult islet [6, 13].

9.3.3 Ductal/Centroacinar Cells as Facultative Progenitor Cells for β-Cell Regeneration

Pancreatic ductal/centroacinar cells (CACs) have long been pointed out as the source of stem/progenitor cells or facultative progenitors in the adult pancreas. Already at the beginning of the twentieth century, ductal/centroacinar cells were suggested as putative progenitors as the result of the observation of the existence of small endocrine cell clusters budding from the ducts [13]. Since then, several studies in rodents and human pancreas pointed out that the location of endocrine cells, either in close proximity to or embedded in the ductal tree during normal growth or regeneration, suggested a role of ductal cells in β-cell neogenesis in the adult organ [17, 37, 45, 68, 80, 86, 87, 110, 113, 115, 117]. Moreover, ductal/centroacinar cells share a common bi-potent progenitor with endocrine cells during pancreatic development. This common developmental lineage coupled with the fact that in all the regenerative scenarios presented so far, acinar-to-β, ductal-to-β, and α-to-β (with the exception of δ-to-β), a transitional duct-like/progenitor-like state involving *Ngn3* activation is necessary for the reprogramming to occur has also provided a basis for the hypothesis that ductal/centroacinar cells serve as facultative progenitors for β-cell regeneration.

Ductal/centroacinar cells acquire a progenitor-like state under regenerative conditions, as seen by the reactivation of *Pdx1* and canonical Wnt signaling, which is usually active in embryonic pancreatic progenitor cells [17, 68, 88]. Notwithstanding the above evidence, whether or not these duct cells indeed contribute to the regeneration of multiple pancreatic lineages in vivo has been a source of major controversy, as different genetic lineage-tracing studies have reached different conclusion.

Before duct-specific inducible Cre transgenic mice were available, work from Heimberg's group provided very strong evidence of in vivo β-cell neogenesis [115]. This study reported that pancreatic duct ligation (PDL), an injury model that causes the destruction of the acinar tissue in the tail of the pancreas by the blockage of the main pancreatic duct, induces pancreatic ducts to form β cells in the ligated region. Interestingly, PDL differs from other injury models in that it does not involve loss of preexisting β cells nor even transient hyperglycemia. So it is worth to note that although PDL does not induce "regeneration" per se, as it involves production of extra β cells rather than the replacement of β cells lost to injury, the reexpression of the pro-endocrine transcription factor Ngn3 in duct cells, reminiscent of the developing pancreas, is linked to an increased β-cell mass by two fold [115]. The availability of several duct-specific inducible Cre lines generated an intense debate and controversial results. The first duct-specific Cre line used the promoter of human carbonic anhydrase II ($CAII^{CreER}$ transgenic mice). Analysis of this transgenic mouse supported the observations made by Heimberg's group as they showed that CAII-expressing cells serve as progenitor cells that contribute both to exocrine and endocrine lineages in homeostatic conditions as well as following injury by PDL [49]. Criscimanna et al. also reported that when acinar and endocrine tissues are selectively ablated by exposure of $Pdx1^{Cre}$;R26DTR mice to diphtheria toxin, the surviving ductal cells contributed to regeneration of both endocrine and acinar cells [28]. Contrary to the conclusion of the above studies, a number of other genetic lineage-tracing studies using several adult pancreatic duct-specific promoters, including *Hnf1β*, *Mucin1*, *Sox9*, and *Hes1*, in transgenic mice have not observed any contribution of ductal cells to β-cell regeneration at postnatal stages [41, 55, 56, 58, 99] even under injury-induced regenerating conditions, including PDL [55, 58, 99]. In these studies, the contribution of duct cells to the endocrine lineage was only limited to embryonic pancreas and not the adult. Although the induction of *Ngn3* by PDL was noted in some of these subsequent studies, as previously reported, these Ngn3+ duct cells were not found to contribute to β-cell neogenesis [58, 60].

The discrepancies in the results and conclusions of the aforementioned studies might be consequence of the specific ductal gene used in every lineage-tracing mouse transgenic line and/or the injury models analyzed. It is important to note that the ductal compartment is an intricate network of centroacinar, intercalated ducts, intralobular ducts, interlobular ducts, and the main duct, indicating a clear structural heterogeneity. Moreover, most, if not all, of the ductal cells express common markers, such as *Hnf1β* or *Sox9*, but differences in the expression levels of such markers in ductal cells have been described as well as the presence of some ductal cells expressing *Sox9* but not *Hnf1β* [18, 58]. These observations reveal molecular heterogeneity between the ductal tree that could explain the controversial results of the studies mentioned above, as some ductal cells may be labeled in one model and not the others. Another possible explanation of these discrepancies is that Inada et al. used the human promoter of CAII [49] while the other studies used the mouse regulatory regions of *Sox9*, *Hnf1β*, *Hes1*, and *Mucin1* [41, 55, 56, 58, 99]. Furthermore, none of the lineage-tracing models studied so far labels the entire population of duct cells as the efficiency of recombination of all the models is rather low; therefore, it remains possible that a cryptic subpopulation of duct cells that exhibit multipotency is not marked effectively by the genetic tools applied in some of the studies [55, 56, 58, 99].

Adding to the controversy, centroacinar cells (CACs) have recently been shown, by two independent laboratories, to be a source of β-cell neogenesis in the adult zebrafish pancreas, mirroring their progenitor function during development [32, 42]. Interestingly, adult zebrafish CACs retain expression of Nkx6.1 and Nkx2.2a, which are not found in mammalian CACs but are expressed in mammalian islet cells and are master regulators of the endocrine differentiation program [15, 79, 92, 94, 100]. Thus, the enhanced regenerative capacity of CACs in zebrafish might be explained by the retained expression of Nkx6.1 and Nkx2.2.

While these discrepancies remain to be resolved, there is evidence to suggest that pancreatic duct cells can be reprogrammed into β cells in vivo by manipulating a limited number of genetic factors or signaling pathways. Collombat group has recently described how $Ngn3^+$ cells lining the ducts contribute to β-cell expansion induced by overexpressing *Pax4* in α cells or GABA treatment [5, 11]. Moreover, ductal cells have been shown to be capable of converting into α cells, δ cells, and β cells following the inactivation of SCF-type E3 ubiquitin ligase substrate component Fbw7. Deletion of *Fbw7* stabilizes Ngn3 and seems to reactivate an endocrine developmental differentiation program in adult pancreatic ductal cells [91]. El-Gohary et al. have also observed that β cells can arise from specialized ductal structures called intra-islet ducts following partial pancreatectomy in adult TGF-β receptor mutant mice [38]. Recently, lineage-tracing studies using *Sox9*^{CreER} transgenic mice have shown that adult pancreatic ductal cells can differentiate into β cells under specific conditions, which include medium hyperglycemia and long-term administration of low dose of gastrin and epidermal growth factor [119].

9.3.4 In Vivo Regeneration from Facultative Pancreatic Progenitors: An Open Door

While in vivo reprogramming of terminally differentiated pancreatic lineages (acinar, ductal, and endocrine) into β cells has been possible thru genetic manipulation of key transcription factors, this approach is useful for experimentation but cannot easily be translated to therapy in humans. To translate in vivo regenerative studies into therapy, it is of paramount importance to identify signaling pathways or small molecules that can specifically target key β-cell reprogramming genes to induce β-cell neogenesis. Therefore, recent papers already named in the previous sections [11, 62] represent a boost into the field of in vivo pancreatic regeneration as these papers focus in the identification of specific signaling pathways or small molecules that induce β-cell regeneration from differentiated pancreatic lineages in vivo *[3]*. Interestingly, the signaling pathways involved in such plasticity change depending on the cell source of the reprogramming (acinar, ductal, or α cells). Although there are common requirements for all of them, such as the acquisition of a transitional progenitor-like state with the activation of *Ngn3*, extracellular glucose and glucagon levels might also play an important role. Thus, these recent findings open the door to potential new therapies to restore β-cell mass in diabetic patients.

Moreover, the knowledge gained from these studies could be easily translated to the growing and promising field of in vitro differentiation of human embryonic or induced pluripotent stem cells toward insulin-producing cells. Advances in this area will be discussed in detailed in the following section.

9.4 Generation of β Cells from Stem Cells for Cell Replacement Therapy

The transplantation of exogenous β cells to replace lost or dysfunctional endogenous β cells is a potential strategy to adequately control glycemia and prevent long-term secondary complications in diabetic patients. Transplantation of cadaveric islets has been proven to work, but limited islet supply, among other obstacles, hampers its broad application to treat diabetes. Human pluripotent stem cells, including both embryonic stem cells (hESC) and induced pluripotent stem cells (hiPSC), provide the basis for scalable and potentially unlimited numbers of β cells for cell replacement therapies. It has been 12 years since the first publications on the generation of insulin-secreting cells from hESC [30, 31]. Since then, many laboratories including commercial entities have undertaken major efforts to generate transplantable β cells from pluripotent cell sources.

9.4.1 Embryonic Stem Cells as a Source of Surrogate β Cells

The possibility to produce β-like cells from mouse and human ESC was first recognized in the early 2000s. Yet, the field took a significant leap forward in 2005–2006 when Kroon et al. published the first directed differentiation protocol based on defined cultured conditions that, in a stepwise manner, recapitulated the multistaged process that progenitor cells undergo to generate β cells during embryonic development in vivo [30, 31]. Thus, using cocktails of growth factors and small molecules, hESC were sequentially coaxed toward definitive endoderm, pancreatic progenitor, endocrine progenitor, and β-cell fate (four-stage protocol). The success of the protocol was evaluated by *(i)* expression of key transcription factors/markers of these stages and by *(ii)* generation of cells producing insulin and positive for C-peptide. The cells generated using these first protocols secreted insulin in response to membrane depolarization agents, while they exhibited absent or very poor glucose responsiveness. Whole-genome expression profiling studies supported that these cells resembled fetal rather than adult β cells [48].

One key demonstration that it is possible to generate stem cell-derived functional insulin-producing cells came from later reports showing that transplanted hESC-derived pancreatic progenitors gave rise, after several months in vivo, to β-like cells that secreted insulin in response to a glucose challenge and ameliorated hyperglycemia in streptozotocin-treated mice [21, 59]. Thus, functional maturation of stem cell-derived β cells became a plausible possibility. Investigations in animal models have indicated that the duration and extent of progenitor cell maturation may be influenced by indeterminate factors in the recipient organism [19], which implies the likely existence of variability among human hosts. Nonetheless, the findings in animal models paved the way for a Phase I/II clinical trial (NCT02239354) aiming at testing progenitor cell transplantation that was initiated in 2014 by Viacyte and has been recently completed.

The positive results obtained using transplanted ESC-derived progenitor cells fueled the search for strategies that could recapitulate the maturation process in vitro. Researchers in the field speculated that engrafted β cells, once vascularized, could be functional within days of transplantation, as opposed to progenitor cells that would require an unpredictable time in vivo before attaining functional maturity. An important development in the quest for stem cell-derived functional β cells was the seven-stage protocol developed by the Kieffer group. One of the main advances of this protocol was targeting the generation of endoderm progenitors that exhibited high co-expression of the transcription factors PDX1 and NKX6.1. In fact, stem cell-derived pancreatic progenitors are a mixed population of endoderm-like nonhormone-producing cells that co-express PDX1 and NKX6.1 and poly-hormonal cells (insulin+glucagon+somatostatin) that lack the expression of these transcription factors. While the first subpopulation distinguishes endocrine progenitors that arise during the secondary transition and give rise to mature β cells during embryonic development [85], the second subpopulation was shown to mainly generate α cells after transplantation in vivo [53, 83, 84], hence mimicking the waves of endocrine differentiation after the primary and secondary transition during pancreas development. Three additional significant improvements included *(i)* the suppression of expression of the pro-endocrine transcription factor NGN3 during early stages of differentiation using vitamin C, *(ii)* moving from a planar cell culture to an air-liquid interface after "stage 4" and using an ALK5 (TGF-β receptor) inhibitor which resulted in upregulation of the genes encoding NGN3 and the pancreatic hormones, and *(iii)* the enhancement of the expression of the β-cell transcription factor MAFA, which is crucial for β-cell functionality, in the last stage of differentiation using an inhibitor of the tyrosine kinase subfamily

AXL [82]. "Stage 7" cells exhibited calcium and insulin secretion responses to glucose in vitro and reversed diabetes in an animal model. In this same line, the Hebrok laboratory published a differentiation protocol with refined culture conditions to precisely control the timing of *NGN3* expression. Specifically, omission of BMP inhibitors (used in other published protocols) during the pancreas specification stage prevented precocious endocrine formation and enhanced the production of glucose-responsive β cells [89].

Coinciding in time, the Melton laboratory devised an alternate protocol that also included fine-tuning of culture conditions to favor the production of NKX6.1+/insulin+ cells [77]. An important novelty of the protocol was the use of a potentially scalable suspension-based 3D culture that would enable the production of billions of insulin-secreting cells. Remarkably, the generated cells (SC-β) were capable of repeatedly performing glucose-induced insulin secretion, a key functional feature of a β cell. Interestingly, global gene expression analysis revealed that SC-β cells resembled adult more than fetal β cells, although gene expression discrepancies still persisted. Importantly, a 2016 follow-up paper demonstrated that encapsulated SC-β cells reversed hyperglycemia in streptozotocin-treated mice [107]. This study signifies the first demonstration of long-term glycemia correction in a diabetic immunocompetent animal model using human stem cell-derived β cells generated in vitro.

9.4.2 Induced Pluripotent Stem Cells as a Source of Surrogate β Cells

iPSC are derived from skin fibroblasts that have been reprogrammed back into an embryonic-like pluripotent state via introduction of a defined set of transcription factors [101, 102]. Since the advent of iPSC technology, many laboratories have applied the directed differentiation protocols originally designed for hESC cells to hiPSC obtaining essentially similar results [77, 107]. In addition to the obvious advantage of iPSC, that is, that they are not derived from human embryos, which is the main bioethical concern in the stem cell field, iPSC technology also permits the creation of cells that are genetically tailored to the patient, thus eliminating the concern of immune rejection. In this regard, the first report on the generation of hiPSC-derived SC-β from patients with type 1 diabetes (T1D) was published in 2016 [72]. The authors found no significant differences in SC-β cells generated from diabetic as compared to nondiabetic patients. These findings are relevant because they demonstrate that T1D-derived β cells could potentially be used as an autologous source for cell replacement therapy in diabetes. Importantly, the potential future applications of iPSC-derived β cells extend beyond cell therapy as they could also be used for drug screening, toxicology testing, and as select patient cellular models to better study β-cell biology and diabetes.

9.4.3 Current Challenges in the Generation of Stem Cell-Derived β Cells

Despite significant progress in the field, one major challenge that remains in our quest toward a cell-based therapy to treat diabetes is the production of functionally competent β cells, that is, cells that produce and secrete sufficient amounts of correctly processed insulin in response to physiological stimuli including but not restricted to glucose. Even with the latest encouraging improvements in directed differentiation protocols, generated β cells are still not comparable to primary β cells. For instance, glucose-induced insulin secretion both in vitro and in vivo is modest, and the dynamic kinetics is not the same than that of primary β cells. Reversal of hyperglycemia in animal models, which is the current gold standard for assaying β-cell functionality, may not be sufficient to prove proper function of the transplanted cells as, similar to insulin pellets, basal insulin release is sufficient to reverse diabetes in mice that receive transplants [20, 51]. Furthermore, adequate shutdown of insulin production when the glucose stimulus is gone (needed for preventing dangerous hypoglycemia) remains to be established. While more research is needed to improve in vitro or in vivo maturation of generated β cells, additional important challenges in the field are the need for a consensus regarding the standards of how a substitute human β cell should be and how one should assess its function in the laboratory. In this scenario, the availability of reliable pure primary β-cell preparations for comparison also becomes an important roadblock [51].

Replacement β cells need to be protected from the stress derived from nutrient deprivation and hypoxia associated with transplantation. It has been suggested that progenitor or less differentiated β cells might be less fragile than functionally mature β cells; however, we currently lack definite experimental evidence for such claim. Irrespective of whether progenitor or more differentiated β cells are the optimal choice for cell therapy, β cells need to be protected from recurrent autoimmunity in people with T1D. Hence, in parallel with optimization of the production of cellular substitutes of human primary β cells, scientists are actively working on devising strategies to protect these cells from the host immune system. While comprehensive discussion on this topic is beyond the scope of this chapter, there are a number of immune isolation strategies being considered which include macro-encapsulation devices, microencapsulation using materials such as alginate, and the identification of immune-privileged transplant sites [20, 34]. In this regard, it should be noted that the clinical trial initiated by Viacyte (see above) involves not only the use of pancreatic progenitor cells (PEC-01) but also of a durable macro-encapsulation device (Encaptra).

9.5 Generation of Surrogate β Cells Bypassing Pluripotency

In the context of cell transplantation, direct lineage conversion emerges as an attractive alternative to the use of stem cells as it avoids the risk of tumorigenesis associated with the pluripotent state. Previously in this chapter, we have discussed several examples of direct lineage conversions or transdifferentiation toward a β-cell fate of cells that share a similar embryological origin with β cells such as the other islet endocrine lineages and ductal and acinar cells. However, most of these cell types are not readily accessible, making it difficult to envision translational utility of these findings. Skin fibroblasts are an easily accessible cell type that has been the cell source of choice for the production of many hiPSC cell lines. Notably, fibroblasts can also be directly reprogrammed into diverse cell types such as cardiomyocytes or neurons using specific sets of transcription factors [7, 106, 108, 109, 114]. Hence theoretically, if we could sort out the right combination of genes, it might be possible to convert fibroblasts into β cells. However, to this date, the feasibility of this strategy has not been proven.

Halfway between reprogramming to pluripotency and direct lineage conversion, there's a transdifferentiation paradigm known as CASD (cell activation and signaling directed) which uses transient expression of iPSC-reprogramming factors in conjunction with lineage-specific soluble signals to convert fibroblasts into highly proliferative lineage progenitors without first establishing a pluripotent state [65, 67, 123]. Once obtained, these lineage progenitors can be expanded to generate significant amount of precursor cells suitable for differentiation into the desired cell types in vitro or in vivo. Following this approach, the Ding lab has shown that mouse and human fibroblasts can be converted into definitive endodermal progenitors and subsequently expanded and stepwise differentiated toward β-like cells that can protect mice from streptozotocin-induced diabetes [63, 122]. Despite that the efficiency of conversion is modest and there remain many open questions about the underlying mechanisms, these findings demonstrate that functional β cells can be obtained directly from fibroblasts and provide a key starting point for a new avenue to produce substitute β cells for cell replacement therapy and drug testing in diabetes.

9.6 Conclusions

The search for a definite treatment for diabetes that is based on the replacement of lost or dysfunctional β cells has driven much of the research in the fields of pancreas biology and pancreas development during these past two decades. The knowledge gained in our understanding of the mechanisms that maintain adult organ homeostasis as well as those that direct differentiation of progenitors into distinct pancreatic cell lineages during embryogenesis has set the foundation for many of the advances seen in regenerative medicine in diabetes today. As discussed in this chapter, the existence of dedicated progenitors in the adult pancreas remains elusive. Experimental evidence suggests that it is more plausible the existence of facultative progenitors that become activated upon specific non-homeostatic conditions; however additional work is needed to devise effective and safe means to promote their activation in situ. The generation of replacement β cells (or their precursors) from ESC and iPSC in vitro is nowadays one of the most promising strategies, but the field is still in its infancy, and further resources and work are needed for it to become a successful therapy with clinical translation. Finally, the concept that a "terminally differentiated" cell can be coaxed to become another cell type through transdifferentiation paradigms is an idea that has gained momentum in other areas of biomedical research and should be tested as an alternate option in the β-cell field. The prospect of finding a cure for diabetes is exciting, and we are certain that we will see great advances toward this objective in the coming years.

References

1. Afelik S, Jensen J (2013) Notch signaling in the pancreas: patterning and cell fate specification. Wiley Interdiscip Rev Dev Biol 2(4):531–544
2. Afelik S, Qu X, Hasrouni E, Bukys MA, Deering T, Nieuwoudt S, Rogers W, Macdonald RJ, Jensen J (2012) Notch-mediated patterning and cell fate allocation of pancreatic progenitor cells. Development 139(10):1744–1753
3. Afelik S, Rovira M (2017a) Pancreatic β-cell regeneration: advances in understanding the genes and signaling pathways involved. Genome Med 9(1):42
4. Afelik S, Rovira M (2017b) Pancreatic β-cell regeneration: facultative or dedicated progenitors? Mol Cell Endocrinol 445:85–94
5. Al-Hasani K, Pfeifer A, Courtney M, Ben-Othman N, Gjernes E, Vieira A, Druelle N, Avolio F, Ravassard P, Leuckx G, Lacas-Gervais S, Ambrosetti D, Benizri E, Hecksher-Sorensen J, Gounon P, Ferrer J, Gradwohl G, Heimberg H, Mansouri A, Collombat P (2013) Adult duct-lining cells can reprogram into β-like cells able to counter repeated cycles of toxin-induced diabetes. Dev Cell 26(1):86–100
6. Arnes L, Hill JT, Gross S, Magnuson MA, Sussel L (2012) Ghrelin expression in the mouse pancreas defines a unique multipotent progenitor population. PLoS One 7(12):e52026
7. Babos K, Ichida JK (2015) Small molecules take a big step by converting fibroblasts into neurons. Cell Stem Cell 17(2):127–129
8. Bader E, Migliorini A, Gegg M, Moruzzi N, Gerdes J, Roscioni SS, Bakhti M, Brandl E, Irmler M, Beckers J, Aichler M, Feuchtinger A, Leitzinger C, Zischka H, Wang-Sattler R, Jastroch M, Tschöp M, Machicao F, Staiger H, Häring HU, Chmelova H, Chouinard JA, Oskolkov N, Korsgren O, Speier S, Lickert H (2016) Identification of proliferative and mature β-cells in the islets of Langerhans. Nature 535(7612):430–434
9. Baeyens L, Lemper M, Leuckx G, De Groef S, Bonfanti P, Stangé G, Shemer R, Nord C, Scheel DW, Pan FC, Ahlgren U, Gu G, Stoffers DA, Dor Y, Ferrer J, Gradwohl G, Wright CV, Van de Casteele M, German MS, Bouwens L, Heimberg H (2014) Transient cytokine treatment induces acinar cell reprogramming and regenerates functional beta cell mass in diabetic mice. Nat Biotechnol 32(1):76–83
10. Baron M, Veres A, Wolock SL, Faust AL, Gaujoux R, Vetere A, Ryu JH, Wagner BK, Shen-Orr SS, Klein AM, Melton DA, Yanai I (2016) A single-cell transcriptomic map of the human and mouse pancreas reveals inter- and intra-cell population structure. Cell Syst 3(4):346–360.e4
11. Ben-Othman N, Vieira A, Courtney M, Record F, Gjernes E, Avolio F, Hadzic B, Druelle N, Napolitano T, Navarro-Sanz S, Silvano S, Al-Hasani K, Pfeifer A, Lacas-Gervais S, Leuckx G, Marroquí L, Thévenet J, Madsen OD, Eizirik DL, Heimberg H, Kerr-Conte J, Pattou F, Mansouri A, Collombat P (2017) Long-term GABA sdministration induces alpha cell-mediated beta-like cell neogenesis. Cell 168(1–2):73–85.e11
12. Benitez CM, Goodyer WR, Kim SK (2012) Deconstructing pancreas developmental biology. Cold Spring Harb Perspect Biol 4(6)
13. Bensley RR (1911) Studies on the pancreas of the guinea pig. Am J Anat 12:297–388
14. Benthuysen JR, Carrano AC, Sander M (2016) Advances in β cell replacement and regeneration strategies for treating diabetes. J Clin Invest 126(10):3651–3660

15. Binot AC, Manfroid I, Flasse L, Winandy M, Motte P, Martial JA, Peers B, Voz ML (2010) Nkx6.1 and nkx6.2 regulate alpha- and beta-cell formation in zebrafish by acting on pancreatic endocrine progenitor cells. Dev Biol 340(2):397–407
16. Bonner-Weir S (2000) Islet growth and development in the adult. J Mol Endocrinol 24(3):297–302
17. Bonner-Weir S, Baxter LA, Schuppin GT, Smith FE (1993) A second pathway for regeneration of adult exocrine and endocrine pancreas. A possible recapitulation of embryonic development. Diabetes 42(12):1715–1720
18. Bonner-Weir S, Guo L, Li WC, Ouziel-Yahalom L, Lysy PA, Weir GC, Sharma A (2012) Islet neogenesis: a possible pathway for beta-cell replenishment. Rev Diabet Stud 9(4):407–416
19. Bruin JE, Asadi A, Fox JK, Erener S, Rezania A, Kieffer TJ (2015a) Accelerated maturation of human stem cell-derived pancreatic progenitor cells into insulin-secreting cells in Immunodeficient rats relative to mice. Stem Cell Reports 5(6):1081–1096
20. Bruin JE, Rezania A, Kieffer TJ (2015b) Replacing and safeguarding pancreatic beta cells for diabetes. Sci Transl Med 7(316):316ps23
21. Bruin JE, Rezania A, Xu J, Narayan K, Fox JK, O'Neil JJ, Kieffer TJ (2013) Maturation and function of human embryonic stem cell-derived pancreatic progenitors in macroencapsulation devices following transplant into mice. Diabetologia 56(9):1987–1998
22. Cavelti-Weder C, Li W, Zumsteg A, Stemann-Andersen M, Zhang Y, Yamada T, Wang M, Lu J, Jermendy A, Bee YM, Bonner-Weir S, Weir GC, Zhou Q (2016) Hyperglycaemia attenuates in vivo reprogramming of pancreatic exocrine cells to beta cells in mice. Diabetologia 59(3):522–532
23. Chera S, Baronnier D, Ghila L, Cigliola V, Jensen JN, Gu G, Furuyama K, Thorel F, Gribble FM, Reimann F, Herrera PL (2014) Diabetes recovery by age-dependent conversion of pancreatic δ-cells into insulin producers. Nature 514(7523):503–507
24. Collombat P, Mansouri A (2009) Pax4 transdifferentiates glucagon-secreting alpha cells to insulin-secreting beta endocrine pancreatic cells. Med Sci (Paris) 25(8–9):763–765
25. Collombat P, Xu X, Ravassard P, Sosa-Pineda B, Dussaud S, Billestrup N, Madsen OD, Serup P, Heimberg H, Mansouri A (2009) The ectopic expression of Pax4 in the mouse pancreas converts progenitor cells into alpha and subsequently beta cells. Cell 138(3):449–462
26. Courtney M, Gjernes E, Druelle N, Ravaud C, Vieira A, Ben-Othman N, Pfeifer A, Avolio F, Leuckx G, Lacas-Gervais S, Burel-Vandenbos F, Ambrosetti D, Hecksher-Sorensen J, Ravassard P, Heimberg H, Mansouri A, Collombat P (2013) The inactivation of Arx in pancreatic α-cells triggers their neogenesis and conversion into functional β-like cells. PLoS Genet 9(10):e1003934
27. Courtney M, Pfeifer A, Al-Hasani K, Gjernes E, Vieira A, Ben-Othman N, Collombat P (2011) In vivo conversion of adult α-cells into β-like cells: a new research avenue in the context of type 1 diabetes. Diabetes Obes Metab 13(Suppl 1):47–52
28. Criscimanna A, Speicher JA, Houshmand G, Shiota C, Prasadan K, Ji B, Logsdon CD, Gittes GK, Esni F (2011) Duct cells contribute to regeneration of endocrine and acinar cells following pancreatic damage in adult mice. Gastroenterology 141(4):1451–1462, 1462.e1–6
29. Cusi K (2016) Treatment of patients with type 2 diabetes and non-alcoholic fatty liver disease: current approaches and future directions. Diabetologia 59:1112
30. D'Amour KA, Agulnick AD, Eliazer S, Kelly OG, Kroon E, Baetge EE (2005) Efficient differentiation of human embryonic stem cells to definitive endoderm. Nat Biotechnol 23(12):1534–1541
31. D'Amour KA, Bang AG, Eliazer S, Kelly OG, Agulnick AD, Smart NG, Moorman MA, Kroon E, Carpenter MK, Baetge EE (2006) Production of pancreatic hormone-expressing endocrine cells from human embryonic stem cells. Nat Biotechnol 24(11):1392–1401
32. Delaspre F, Beer RL, Rovira M, Huang W, Wang G, Gee S, Vitery M e C, Wheelan SJ, Parsons MJ (2015) Centroacinar cells are progenitors that contribute to endocrine pancreas regeneration. Diabetes 64(10):3499–3509
33. Desai BM, Oliver-Krasinski J, De Leon DD, Farzad C, Hong N, Leach SD, Stoffers DA (2007) Preexisting pancreatic acinar cells contribute to acinar cell, but not islet beta cell, regeneration. J Clin Invest 117(4):971–977

34. Desai T, Shea LD (2017) Advances in islet encapsulation technologies. Nat Rev Drug Discov 16(5):338–350
35. Dor Y, Brown J, Martinez OI, Melton DA (2004) Adult pancreatic beta-cells are formed by self-duplication rather than stem-cell differentiation. Nature 429(6987):41–46
36. Dorrell C, Schug J, Canaday PS, Russ HA, Tarlow BD, Grompe MT, Horton T, Hebrok M, Streeter PR, Kaestner KH, Grompe M (2016) Human islets contain four distinct subtypes of β cells. Nat Commun 7:11756
37. El-Gohary Y, Tulachan S, Branca M, Sims-Lucas S, Guo P, Prasadan K, Shiota C, Gittes GK (2012) Whole-mount imaging demonstrates hypervascularity of the pancreatic ducts and other pancreatic structures. Anat Rec (Hoboken) 295(3):465–473
38. El-Gohary Y, Wiersch J, Tulachan S, Xiao X, Guo P, Rymer C, Fischbach S, Prasadan K, Shiota C, Gaffar I, Song Z, Galambos C, Esni F, Gittes GK (2016) Intraislet pancreatic ducts can give rise to insulin-positive cells. Endocrinology 157(1):166–175
39. Fujitani Y, Fujitani S, Boyer DF, Gannon M, Kawaguchi Y, Ray M, Shiota M, Stein RW, Magnuson MA, Wright CV (2006) Targeted deletion of a cis-regulatory region reveals differential gene dosage requirements for Pdx1 in foregut organ differentiation and pancreas formation. Genes Dev 20(2):253–266
40. Fukuda A, Kawaguchi Y, Furuyama K, Kodama S, Horiguchi M, Kuhara T, Kawaguchi M, Terao M, Doi R, Wright CV, Hoshino M, Chiba T, Uemoto S (2008) Reduction of Ptf1a gene dosage causes pancreatic hypoplasia and diabetes in mice. Diabetes 57(9):2421–2431
41. Furuyama K, Kawaguchi Y, Akiyama H, Horiguchi M, Kodama S, Kuhara T, Hosokawa S, Elbahrawy A, Soeda T, Koizumi M, Masui T, Kawaguchi M, Takaori K, Doi R, Nishi E, Kakinoki R, Deng JM, Behringer RR, Nakamura T, Uemoto S (2011) Continuous cell supply from a Sox9-expressing progenitor zone in adult liver, exocrine pancreas and intestine. Nat Genet 43(1):34–41
42. Ghaye AP, Bergemann D, Tarifeño-Saldivia E, Flasse LC, Von Berg V, Peers B, Voz ML, Manfroid I (2015) Progenitor potential of nkx6.1-expressing cells throughout zebrafish life and during beta cell regeneration. BMC Biol 13(1):70
43. Gittes GK (2009) Developmental biology of the pancreas: a comprehensive review. Dev Biol 326(1):4–35
44. Gu D, Arnush M, Sawyer SP, Sarvetnick N (1995) Transgenic mice expressing IFN-gamma in pancreatic beta-cells are resistant to streptozotocin-induced diabetes. Am J Phys 269(6 Pt 1):E1089–E1094
45. Gu D, Sarvetnick N (1993) Epithelial cell proliferation and islet neogenesis in IFN-g transgenic mice. Development 118(1):33–46
46. Haegebarth A, Clevers H (2009) Wnt signaling, lgr5, and stem cells in the intestine and skin. Am J Pathol 174(3):715–721
47. Heimberg H, De Vos A, Vandercammen A, Van Schaftingen E, Pipeleers D, Schuit F (1993) Heterogeneity in glucose sensitivity among pancreatic beta-cells is correlated to differences in glucose phosphorylation rather than glucose transport. EMBO J 12(7):2873–2879
48. Hrvatin S, O'Donnell CW, Deng F, Millman JR, Pagliuca FW, DiIorio P, Rezania A, Gifford DK, Melton DA (2014) Differentiated human stem cells resemble fetal, not adult, beta cells. Proc Natl Acad Sci U S A 111(8):3038–3043
49. Inada A, Nienaber C, Katsuta H, Fujitani Y, Levine J, Morita R, Sharma A, Bonner-Weir S (2008) Carbonic anhydrase II-positive pancreatic cells are progenitors for both endocrine and exocrine pancreas after birth. Proc Natl Acad Sci U S A 105(50):19915–19919
50. International Diabetes Federation (2015) IDF Diabetes, 7 ed. Brussels, Belgium
51. Johnson JD (2016) The quest to make fully functional human pancreatic beta cells from embryonic stem cells: climbing a mountain in the clouds. Diabetologia 59(10):2047–2057
52. Jones DL, Wagers AJ (2008) No place like home: anatomy and function of the stem cell niche. Nat Rev Mol Cell Biol 9(1):11–21
53. Kelly OG, Chan MY, Martinson LA, Kadoya K, Ostertag TM, Ross KG, Richardson M, Carpenter MK, D'Amour KA, Kroon E, Moorman M, Baetge EE, Bang AG (2011) Cell-surface markers for the isolation of pancreatic cell types derived from human embryonic stem cells. Nat Biotechnol 29(8):750–756

54. Klochendler A, Caspi I, Corem N, Moran M, Friedlich O, Elgavish S, Nevo Y, Helman A, Glaser B, Eden A, Itzkovitz S, Dor Y (2016) The genetic program of pancreatic beta-cell replication in vivo. Diabetes 65(7):2081–2093
55. Kopinke D, Brailsford M, Shea JE, Leavitt R, Scaife CL, Murtaugh LC (2011) Lineage tracing reveals the dynamic contribution of Hes1+ cells to the developing and adult pancreas. Development 138(3):431–441
56. Kopinke D, Murtaugh LC (2010) Exocrine-to-endocrine differentiation is detectable only prior to birth in the uninjured mouse pancreas. BMC Dev Biol 10:38
57. Kopp JL, Dubois CL, Hao E, Thorel F, Herrera PL, Sander M (2011a) Progenitor cell domains in the developing and adult pancreas. Cell Cycle 10(12):1921–1927
58. Kopp JL, Dubois CL, Schaffer AE, Hao E, Shih HP, Seymour PA, Ma J, Sander M (2011b) Sox9+ ductal cells are multipotent progenitors throughout development but do not produce new endocrine cells in the normal or injured adult pancreas. Development 138(4):653–665
59. Kroon E, Martinson LA, Kadoya K, Bang AG, Kelly OG, Eliazer S, Young H, Richardson M, Smart NG, Cunningham J, Agulnick AD, D'Amour KA, Carpenter MK, Baetge EE (2008) Pancreatic endoderm derived from human embryonic stem cells generates glucose-responsive insulin-secreting cells in vivo. Nat Biotechnol 26(4):443–452
60. Kushner JA, Weir GC, Bonner-Weir S (2010) Ductal origin hypothesis of pancreatic regeneration under attack. Cell Metab 11(1):2–3
61. Lee MG, Ohana E, Park HW, Yang D, Muallem S (2012) Molecular mechanism of pancreatic and salivary gland fluid and HCO3 secretion. Physiol Rev 92(1):39–74
62. Li J, Casteels T, Frogne T, Ingvorsen C, Honoré C, Courtney M, Huber KV, Schmitner N, Kimmel RA, Romanov RA, Sturtzel C, Lardeau CH, Klughammer J, Farlik M, Sdelci S, Vieira A, Avolio F, Briand F, Baburin I, Májek P, Pauler FM, Penz T, Stukalov A, Gridling M, Parapatics K, Barbieux C, Berishvili E, Spittler A, Colinge J, Bennett KL, Hering S, Sulpice T, Bock C, Distel M, Harkany T, Meyer D, Superti-Furga G, Collombat P, Hecksher-Sørensen J, Kubicek S (2017) Artemisinins Target GABAA Receptor Signaling and Impair α Cell Identity. Cell 168(1–2):86–100.e15
63. Li K, Zhu S, Russ HA, Xu S, Xu T, Zhang Y, Ma T, Hebrok M, Ding S (2014a) Small molecules facilitate the reprogramming of mouse fibroblasts into pancreatic lineages. Cell Stem Cell 14(2):228–236
64. Li W, Cavelti-Weder C, Zhang Y, Clement K, Donovan S, Gonzalez G, Zhu J, Stemann M, Xu K, Hashimoto T, Yamada T, Nakanishi M, Zeng S, Gifford D, Meissner A, Weir G, Zhou Q (2014b) Long-term persistence and development of induced pancreatic beta cells generated by lineage conversion of acinar cells. Nat Biotechnol 32(12):1223–1230
65. Li W, Ding S (2010) Generation of novel rat and human pluripotent stem cells by reprogramming and chemical approaches. Methods Mol Biol 636:293–300
66. Li W, Nakanishi M, Zumsteg A, Shear M, Wright C, Melton DA, Zhou Q (2014c) In vivo reprogramming of pancreatic acinar cells to three islet endocrine subtypes. elife 3:e01846
67. Li W, Wei W, Zhu S, Zhu J, Shi Y, Lin T, Hao E, Hayek A, Deng H, Ding S (2009) Generation of rat and human induced pluripotent stem cells by combining genetic reprogramming and chemical inhibitors. Cell Stem Cell 4(1):16–19
68. Li WC, Rukstalis JM, Nishimura W, Tchipashvili V, Habener JF, Sharma A, Bonner-Weir S (2010) Activation of pancreatic-duct-derived progenitor cells during pancreas regeneration in adult rats. J Cell Sci 123(Pt 16):2792–2802
69. Lingohr MK, Buettner R, Rhodes CJ (2002) Pancreatic beta-cell growth and survival--a role in obesity-linked type 2 diabetes? Trends Mol Med 8(8):375–384
70. Meier JJ, Butler AE, Saisho Y, Monchamp T, Galasso R, Bhushan A, Rizza RA, Butler PC (2008) Beta-cell replication is the primary mechanism subserving the postnatal expansion of beta-cell mass in humans. Diabetes 57(6):1584–1594
71. Menge BA, Breuer TG, Ritter PR, Uhl W, Schmidt WE, Meier JJ (2012) Long-term recovery of β-cell function after partial pancreatectomy in humans. Metabolism 61(5):620–624
72. Millman JR, Xie C, Van Dervort A, Gurtler M, Pagliuca FW, Melton DA (2016) Generation of stem cell-derived beta-cells from patients with type 1 diabetes. Nat Commun 7:11463

73. Muraro, M. J., Dharmadhikari, G., Grün, D., Groen, N., Dielen, T., Jansen, E., van Gurp, L., Engelse, M. A., Carlotti, F., de Koning, E. J. and van Oudenaarden, A. (2016) 'A single-cell transcriptome atlas of the human pancreas', Cell Syst 3(4):385-394.e3
74. Nir T, Melton DA, Dor Y (2007) Recovery from diabetes in mice by beta cell regeneration. J Clin Invest 117(9):2553–2561
75. Oliver-Krasinski JM, Stoffers DA (2008) On the origin of the beta cell. Genes Dev 22(15):1998–2021
76. Otonkoski T, Beattie GM, Lopez AD, Hayek A (1994) Use of hepatocyte growth factor/scatter factor to increase transplantable human fetal islet cell mass. Transplant Proc 26(6):3334
77. Pagliuca FW, Millman JR, Gurtler M, Segel M, Van Dervort A, Ryu JH, Peterson QP, Greiner D, Melton DA (2014) Generation of functional human pancreatic beta cells in vitro. Cell 159(2):428–439
78. Pan FC, Bankaitis ED, Boyer D, Xu X, Van de Casteele M, Magnuson MA, Heimberg H, Wright CV (2013) Spatiotemporal patterns of multipotentiality in Ptf1a-expressing cells during pancreas organogenesis and injury-induced facultative restoration. Development 140(4):751–764
79. Papizan JB, Singer RA, Tschen SI, Dhawan S, Friel JM, Hipkens SB, Magnuson MA, Bhushan A, Sussel L (2011) Nkx2.2 repressor complex regulates islet beta-cell specification and prevents beta-to-alpha-cell reprogramming. Genes Dev 25(21):2291–2305
80. Phillips JM, O'Reilly L, Bland C, Foulis AK, Cooke A (2007) Patients with chronic pancreatitis have islet progenitor cells in their ducts, but reversal of overt diabetes in NOD mice by anti-CD3 shows no evidence for islet regeneration. Diabetes 56(3):634–640
81. Puri S, Folias AE, Hebrok M (2015) Plasticity and dedifferentiation within the pancreas: development, homeostasis, and disease. Cell Stem Cell 16(1):18–31
82. Rezania A, Bruin JE, Arora P, Rubin A, Batushansky I, Asadi A, O'Dwyer S, Quiskamp N, Mojibian M, Albrecht T, Yang YH, Johnson JD, Kieffer TJ (2014) Reversal of diabetes with insulin-producing cells derived in vitro from human pluripotent stem cells. Nat Biotechnol 32(11):1121–1133
83. Rezania A, Bruin JE, Xu J, Narayan K, Fox JK, O'Neil JJ, Kieffer TJ (2013) Enrichment of human embryonic stem cell-derived NKX6.1-expressing pancreatic progenitor cells accelerates the maturation of insulin-secreting cells in vivo. Stem Cells 31(11):2432–2442
84. Rezania A, Riedel MJ, Wideman RD, Karanu F, Ao Z, Warnock GL, Kieffer TJ (2011) Production of functional glucagon-secreting alpha-cells from human embryonic stem cells. Diabetes 60(1):239–247
85. Rieck S, Bankaitis ED, Wright CV (2012) Lineage determinants in early endocrine development. Semin Cell Dev Biol 23(6):673–684
86. Rooman I, Bouwens L (2004) Combined gastrin and epidermal growth factor treatment induces islet regeneration and restores normoglycaemia in C57Bl6/J mice treated with alloxan. Diabetologia 47(2):259–265
87. Rosenberg L (1995) In vivo cell transformation: neogenesis of beta cells from pancreatic ductal cells. Cell Transplant 4(4):371–383
88. Rovira M, Scott SG, Liss AS, Jensen J, Thayer SP, Leach SD (2010) Isolation and characterization of centroacinar/terminal ductal progenitor cells in adult mouse pancreas. Proc Natl Acad Sci U S A 107(1):75–80
89. Russ HA, Parent AV, Ringler JJ, Hennings TG, Nair GG, Shveygert M, Guo T, Puri S, Haataja L, Cirulli V, Blelloch R, Szot GL, Arvan P, Hebrok M (2015) Controlled induction of human pancreatic progenitors produces functional beta-like cells in vitro. EMBO J 34(13):1759–1772
90. Salomon D, Meda P (1986) Heterogeneity and contact-dependent regulation of hormone secretion by individual B cells. Exp Cell Res 162(2):507–520
91. Sancho R, Gruber R, Gu G, Behrens A (2014) Loss of Fbw7 reprograms adult pancreatic ductal cells into alpha, delta, and beta cells. Cell Stem Cell 15(2):139–153
92. Sander M, Sussel L, Conners J, Scheel D, Kalamaras J, Dela Cruz F, Schwitzgebel V, Hayes-Jordan A, German M (2000) Homeobox gene Nkx6.1 lies downstream of Nkx2.2 in the major pathway of beta-cell formation in the pancreas. Development 127(24):5533–5540

93. Sato T, Clevers H (2015) SnapShot: growing organoids from stem cells. Cell 161(7):1700–1700.e1
94. Schaffer AE, Taylor BL, Benthuysen JR, Liu J, Thorel F, Yuan W, Jiao Y, Kaestner KH, Herrera PL, Magnuson MA, May CL, Sander M (2013) Nkx6.1 controls a gene regulatory network required for establishing and maintaining pancreatic Beta cell identity. PLoS Genet 9(1):e1003274
95. Schofield PK, Treherne JE (1978) Kinetics of sodium and lithium movements across the blood-brain barrier of an insect. J Exp Biol 74:239–251
96. Sellick GS, Barker KT, Stolte-Dijkstra I, Fleischmann C, Coleman RJ, Garrett C, Gloyn AL, Edghill EL, Hattersley AT, Wellauer PK, Goodwin G, Houlston RS (2004) Mutations in PTF1A cause pancreatic and cerebellar agenesis. Nat Genet 36(12):1301–1305
97. Seymour PA (2014) Sox9: a master regulator of the pancreatic program. Rev Diabet Stud 11(1):51–83
98. Shirakawa J, Kulkarni RN (2016) Novel factors modulating human beta-cell proliferation. Diabetes Obes Metab 18 Suppl 1:71–77
99. Solar M, Cardalda C, Houbracken I, Martín M, Maestro MA, De Medts N, Xu X, Grau V, Heimberg H, Bouwens L, Ferrer J (2009) Pancreatic exocrine duct cells give rise to insulin-producing beta cells during embryogenesis but not after birth. Dev Cell 17(6):849–860
100. Sussel L, Kalamaras J, Hartigan-O'Connor DJ, Meneses JJ, Pedersen RA, Rubenstein JL, German MS (1998) Mice lacking the homeodomain transcription factor Nkx2.2 have diabetes due to arrested differentiation of pancreatic beta cells. Development 125(12):2213–2221
101. Takahashi K, Tanabe K, Ohnuki M, Narita M, Ichisaka T, Tomoda K, Yamanaka S (2007) Induction of pluripotent stem cells from adult human fibroblasts by defined factors. Cell 131(5):861–872
102. Takahashi K, Yamanaka S (2006) Induction of pluripotent stem cells from mouse embryonic and adult fibroblast cultures by defined factors. Cell 126(4):663–676
103. Tanaka EM, Reddien PW (2011) The cellular basis for animal regeneration. Dev Cell 21(1):172–185
104. Teta M, Rankin MM, Long SY, Stein GM, Kushner JA (2007) Growth and regeneration of adult beta cells does not involve specialized progenitors. Dev Cell 12(5):817–826
105. Thorel F, Népote V, Avril I, Kohno K, Desgraz R, Chera S, Herrera PL (2010) Conversion of adult pancreatic alpha-cells to beta-cells after extreme beta-cell loss. Nature 464(7292):1149–1154
106. Tsunemoto RK, Eade KT, Blanchard JW, Baldwin KK (2015) Forward engineering neuronal diversity using direct reprogramming. EMBO J 34(11):1445–1455
107. Vegas AJ, Veiseh O, Gurtler M, Millman JR, Pagliuca FW, Bader AR, Doloff JC, Li J, Chen M, Olejnik K, Tam HH, Jhunjhunwala S, Langan E, Aresta-Dasilva S, Gandham S, McGarrigle JJ, Bochenek MA, Hollister-Lock J, Oberholzer J, Greiner DL, Weir GC, Melton DA, Langer R, Anderson DG (2016) Long-term glycemic control using polymer-encapsulated human stem cell-derived beta cells in immune-competent mice. Nat Med 22(3):306–311
108. Vierbuchen T, Wernig M (2011) Direct lineage conversions: unnatural but useful? Nat Biotechnol 29(10):892–907
109. Vierbuchen T, Wernig M (2012) Molecular roadblocks for cellular reprogramming. Mol Cell 47(6):827–838
110. Weaver CV, Sorenson RL, Kaung HC (1985) Immunocytochemical localization of insulin-immunoreactive cells in the pancreatic ducts of rats treated with trypsin inhibitor. Diabetologia 28(10):781–785
111. Westphalen CB, Takemoto Y, Tanaka T, Macchini M, Jiang Z, Renz BW, Chen X, Ormanns S, Nagar K, Tailor Y, May R, Cho Y, Asfaha S, Worthley DL, Hayakawa Y, Urbanska AM, Quante M, Reichert M, Broyde J, Subramaniam PS, Remotti H, Su GH, Rustgi AK, Friedman RA, Honig B, Califano A, Houchen CW, Olive KP, Wang TC (2016) Dclk1 defines quiescent pancreatic progenitors that promote injury-induced regeneration and tumorigenesis. Cell Stem Cell 18(4):441–455
112. Wilcox CL, Terry NA, Walp ER, Lee RA, May CL (2013) Pancreatic α-cell specific deletion of mouse Arx leads to α-cell identity loss. PLoS One 8(6):e66214

113. Xu G, Stoffers DA, Habener JF, Bonner-Weir S (1999) Exendin-4 stimulates both beta-cell replication and neogenesis, resulting in increased beta-cell mass and improved glucose tolerance in diabetic rats. Diabetes 48(12):2270–2276
114. Xu J, Du Y, Deng H (2015) Direct lineage reprogramming: strategies, mechanisms, and applications. Cell Stem Cell 16(2):119–134
115. Xu, X., D'Hoker, J., Stangé, G., Bonné, S., De Leu, N., Xiao, X., Van de Casteele, M., Mellitzer, G., Ling, Z., Pipeleers, D., Bouwens, L., Scharfmann, R., Gradwohl, G. and Heimberg, H. (2008) 'Beta cells can be generated from endogenous progenitors in injured adult mouse pancreas', Cell, 132(2), pp. 197–207
116. Yanger K, Stanger BZ (2011) Facultative stem cells in liver and pancreas: fact and fancy. Dev Dyn 240(3):521–529
117. Yoneda S, Uno S, Iwahashi H, Fujita Y, Yoshikawa A, Kozawa J, Okita K, Takiuchi D, Eguchi H, Nagano H, Imagawa A, Shimomura I (2013) Predominance of beta-cell neogenesis rather than replication in humans with an impaired glucose tolerance and newly diagnosed diabetes. J Clin Endocrinol Metab 98(5):2053–2061
118. Zhang M, Lin Q, Qi T, Wang T, Chen CC, Riggs AD, Zeng D (2016a) Growth factors and medium hyperglycemia induce Sox9+ ductal cell differentiation into beta cells in mice with reversal of diabetes. Proc Natl Acad Sci U S A 113(3):650–655
119. Zhang M, Lin Q, Qi T, Wang T, Chen CC, Riggs AD, Zeng D (2016b) Growth factors and medium hyperglycemia induce Sox9+ ductal cell differentiation into β cells in mice with reversal of diabetes. Proc Natl Acad Sci U S A 113(3):650–655
120. Zhou Q, Brown J, Kanarek A, Rajagopal J, Melton DA (2008) In vivo reprogramming of adult pancreatic exocrine cells to beta-cells. Nature 455(7213):627–632
121. Zhou Q, Law AC, Rajagopal J, Anderson WJ, Gray PA, Melton DA (2007) A multipotent progenitor domain guides pancreatic organogenesis. Dev Cell 13(1):103–114
122. Zhu S, Russ HA, Wang X, Zhang M, Ma T, Xu T, Tang S, Hebrok M, Ding S (2016) Human pancreatic beta-like cells converted from fibroblasts. Nat Commun 7:10080
123. Zhu S, Wang H, Ding S (2015) Reprogramming fibroblasts toward cardiomyocytes, neural stem cells and hepatocytes by cell activation and signaling-directed lineage conversion. Nat Protoc 10(7):959–973
124. Zipori D (2004) The nature of stem cells: state rather than entity. Nat Rev Genet 5(11):873–878

Chapter 10
Dental Pulp Stem Cells Promote Wound Healing and Muscle Regeneration

E. Martínez-Sarrà, S. Montori, C. Gil-Recio, R. Núñez-Toldrà, N. Carrio Bertran, A. Al Madhoun, M. Sampaolesi, and M. Atari

Abstract Stem cells (SC) are unspecialized cells that can self-renew and generate specialized progeny through differentiation. These cells are found in almost all multicellular organisms and are capable of renewing themselves through cell division. Moreover, under certain stimuli, they can differentiate into tissue-specific cells. Therefore, stem cells serve as a reservoir and repair system capable of replacing differentiated cells lost either naturally through apoptosis or as a result of trauma or disease. Consequently, the potential of stem cells to renew and differentiate makes them attractive candidates for regenerative medicine. These basic stem cell properties differ among various sources of stem cells, and they can be classified based on their origin and/or differentiation potential. The capacity to differentiate into specialized cell types and be able to originate a mature cell type is referred to as potency. Stem cells can be classified depending on their differentiation potential. Totipotent stem cells can differentiate into embryonic and extra-embryonic cell types; such cells can construct a complete, viable organism. Pluripotent stem cells produce all cells of an organism and have the capacity to

E. Martínez-Sarrà
Regenerative Medicine Research Institute, Universitat Internacional de Catalunya, Barcelona, Spain

Translational Cardiomyology Laboratory, Stem Cell Biology and Embryology Unit, Department of Development and Regeneration, Leuven, Belgium

S. Montori · C. Gil-Recio · R. Núñez-Toldrà · N. Carrio Bertran · M. Atari (✉)
Regenerative Medicine Research Institute, Universitat Internacional de Catalunya, Barcelona, Spain
e-mail: matari@biointelligentsl.com

A. Al Madhoun
Research Division, Dasman Diabetes Institute, Dasman, Kuwait

M. Sampaolesi
Translational Cardiomyology Laboratory, Stem Cell Biology and Embryology Unit, Department of Development and Regeneration, Leuven, Belgium

Human Anatomy Unit, Department of Public Health, Experimental and Forensic Medicine, University of Pavia, Pavia, Italy

R. Delgado-Morales (ed.), *Stem Cell Genetics for Biomedical Research*,
https://doi.org/10.1007/978-3-319-90695-9_10

form representative tissues of all three germ layers of the developing embryo: ectoderm, mesoderm and endoderm. Multipotent stem cells can self-renew and differentiate only in a closely related family of cells from the same germ layer tissues, while unipotent stem cells exhibit limited development potential, giving rise to only a single cell type. Stem cells can be also classified according to the type of cells. Embryonic Stem Cells (ESC) are cells derived from blastocyst, Adult Stem Cells (ASC) refer to any cell found in a developed organism that has the ability to divide and create another cell like itself or even to create a cell more differentiated than itself, and Induced Pluripotent Stem Cells (iPSC) are reprogrammed somatic cells with pluripotent capabilities. Several types of adult stem cells have been isolated from teeth, including Stem Cells from Human Exfoliated Deciduous Teeth (SHED), Periodontal Ligament Stem Cells (PDLSC), Dental Follicle Precursor Cells (DFPC), Stem Cells from Apical Papilla (SCAP) and Dental Pulp Stem Cells (DPSC). These post-natal populations have mesenquimal-like qualities such as the capacity for self-renewal and the potential to differentiate into multiple tissues including adipose, bone, endothelial and neural-like tissue. Dental Pulp Pluripotent-like Stem Cells (DPPSC) are also isolated from the dental pulp of the third molars, express pluripotency markers, and show embryonic-like behaviour differentiating into tissues of the three embryonic layers. Mesoderm-derived cell types are osteogenic cells, chondrogenic cells, adipogenic cells, skeletal muscle cells, smooth muscle cells, cardiac muscle cells and endothelial cells. To date, there exist two commonly used methods to induce vascular cell differentiation from human pluripotent stem cells: embryoid body (EB) formation and monolayer-directed differentiation. The two major cellular components of blood vessels are Endothelial Cells (EC) and Vascular Smooth Muscle Cells (VSMC). A better understanding of the cellular and molecular mechanisms that control VSMC differentiation is essential to help develop new approaches to both prevent and treat several related diseases. Another important mesoderm-derived tissue for regenerative medicine is skeletal muscle, which is responsible for the voluntary movement of the body. Many diseases that affect the musculature belong to the group of muscular dystrophies (MD). Development of reliable and reproducible in vitro cellular models to study these tissues is needed, yet it has been problematic due to intrinsic peculiarities of them.

Keywords DPPSC · Stem cells · Dental pulp · Tissue engineering · Mesoderm tissue

Abbreviations

αSMA	Alpha smooth muscle actin
γc	Common gamma chain/interleukin-2 receptor subunit gamma
AP/ALP	Alkaline phosphatase
ASC	Adult stem cells
BM	Bone marrow
BM-MSC	Bone marrow mesenchymal stromal cells
BMP-4	Bone morphogenetic protein 4
CD31/PECAM-1	Cluster of differentiation 31/platelet endothelial cell adhesion molecule 1
DFPC	Dental follicle precursor cells
DMD	Duchenne muscular dystrophy
DMEM	Dulbecco's Modified Eagle's Medium
DNMT3B	DNA methyltransferase 3 beta
DPMSC	Dental pulp mesenchymal stem cells
DPSC	Dental pulp stem cells
DPPSC	Dental pulp pluripotent-like stem cells
EC	Endothelial cells
EGF	Epidermal growth factor
EGM-2	Endothelial growth medium 2
ESC	Embryonic stem cells
FBS	Fetal bovine serum
FGF	Fibroblast growth factor
GAPDH	Glyceraldehyde 3-phosphate dehydrogenase
GMP	Good manufacturing practices
hESC	Human embryonic stem cells
HLA-DR	Human leukocyte antigen-antigen D related
HS	Human serum
HUVEC	Human umbilical endothelial cells
iPS/iPSC	Induced pluripotent stem (cells)
KLF4	Kruppel-like factor 4
LA-BSA	Linoleic acid-bovine serum albumin
LMNA	Lamin A/C
LEFTY2/EBAF	Left-right determination factor 2
LIF	Leukemia inhibitory factor
DGC	Dystrophin glycoprotein complex
DYS	Dystrophin
MAPC	Multipotent adult progenitor cells
MD	Muscular dystrophy
MyHC	Myosin heavy chain
MIAMI	Marrow-isolated adult multilineage inducible
MPC	Mesenchymal progenitor cells
MSC	Mesenchymal stromal cells

OCT4/POU5F1	Octamer-binding transcription factor 4
PBS	Phosphate-buffered saline
PCR	Polymerase chain reaction
PDGF	Human platelet-derived growth factor
PDLSC	Periodontal ligament stem cells
Rag2	Recombination activating gene 2
REX1	ZFP42 zinc finger protein
RT-PCR	Retrotranscriptase PCR
SC	Stem cells
SCAP	Stem cells from apical papilla
sCGH	Short-comparative genomic hybridization
Scid	Severe combined immunodeficiency
Sgcb	Beta-sarcoglycan
SHED	Stem cells from human exfoliated deciduous teeth
SMC	Smooth muscle cells
SOX2	Sex-determining region Y-box2
SSEA	Stage-specific embryonic antigen
TDGF1	Teratocarcinoma-derived growth factor 1
tGFP	Turbo green fluorescent protein
TGF-β	Transforming growth factor beta
TRA-1-60	Tumor rejection antigen-1-60
TRA-1-81	Tumor rejection antigen-1-81
VE-CAD	Vascular cadherin
VEGFR2/FLK1/KDR	Endothelial growth factor receptor 2
VSEL	Very small embryonic-like
VSMC	Vascular smooth muscle cells
vWF	von Willebrand factor endothelial vascular

10.1 Stem Cells

Stem cells (SC) are undifferentiated cells that have the ability to self-renew and to generate mature, differentiated cells with specialized functions [1]. In humans, stem cells have been identified in the inner cell mass of the early embryo, some fetal tissues, the umbilical cord, the placenta, and in several adult organs. For the last decades, research on stem cells has provided important information on developmental, morphological, and physiological processes that govern tissue and organ formation, maintenance, regeneration, and repair after injuries [2]. More recently, significant advancements in our understanding of stem cell biology have provoked great interest and hold high therapeutic promise based on the possibility of stimulating their ex vivo and in vivo expansion and differentiation into functional progeny that could regenerate injured tissues/organs in humans [3].

So far, different types of stem cells have been used in regenerative medicine studies: (i) embryonic stem cells (ESC), which are pluripotent stem cells that can differentiate into any adult tissue, but they however present donor-recipient immunocompatibility problems and ethical controversy [4]; (ii) adult stem cells (ASC), which are multipotent and can only repair and regenerate the tissue in which they reside, limiting their therapeutic use; and (iii) induced pluripotent stem cells (iPSC), somatic cells reprogrammed to a pluripotent state [5], which however present safety issues that impair their use in clinical applications [6].

10.1.1 Embryonic Stem Cells

ESC is derived from the blastocyst, an early-stage embryo. Fertilization of an oocyte by a sperm cell results in a single totipotent cell called zygote, the earliest embryonic stage. This zygote, during the first hours after fertilization, divides into identical totipotent cells, which can develop into any of the three germ layers of a human organism (ectoderm, mesoderm, or endoderm), into cells of the syncytiotrophoblast layer of the placenta, and into the cytotrophoblast layer. When they reach a 16-cell stage, called morula, the totipotent cells can undergo differentiation. They can turn into cells that will eventually become either the blastocyst's inner cell mass or the outer trophoblasts. Approximately 5 days after fertilization, and after several more cycles of cell division, the morula cells begin to commit, forming a hollow sphere of cells, called a blastocyst. The outer layer of the blastocyst is called the trophoblast, and the cluster of cells inside the sphere is called the inner cell mass. At this stage, there are about 70 trophoblast cells and about 30 cells in the inner cell mass. The cells of the inner cell mass, the source of ESC, have become pluripotent stem cells that give rise to all cell types of the major tissue layers (ectoderm, mesoderm, and endoderm) of the cell mass results in the destruction of the fertilized human embryo, which raises ethical issues [7].

Human embryonic stem cell (hESC) lines express many markers that are common to pluripotent and undifferentiated cells, such as octamer-binding transcription factor 4 (OCT4), NANOG, sex-determining region Y-box2 (SOX2), alkaline phosphatase (ALP), LIN28, ZFP42 zinc finger protein (REX1), teratocarcinoma-derived growth factor 1 (TDGF1/Cripto), DNA methyltransferase 3 beta (DNMT3B), CD9, CD24, left-right determination factor 2 (LEFTY2/EBAF), and Thy-1, as well as stage-specific embryonic antigen-3 and antigen-4 (SSEA-3 and SSEA-4) and tumor rejection antigen-1-60 and 1–81 (TRA-1-60 and TRA-1-81) [8].

10.1.2 Induced Pluripotent Stem Cells

In 2006, Takahashi et al. [5] showed that the introduction of four specific genes encoding transcription factors (OCT4, SOX2, c-MYC, and KLF4) could convert mouse adult cells (fibroblasts) into pluripotent stem cells known as iPSC. These cells have the ability to propagate indefinitely and to give rise to every cell in the adult body, and they are similar to ESC in morphology, proliferation, and teratoma capacity formation. In 2007, iPSC were obtained from human adult stem cells by two different groups and gene introduction: Takahashi et al. [5] using the same four genes (OCT4, SOX2, c-MYC, and KLF4) and Yu et al. using OCT4, NANOG, SOX2, and LIN28 [9].

10.1.3 Adult Stem Cells

ASC are undifferentiated cells that reside in most of adult tissues/organs, including the bone marrow, heart, brain, lungs, liver, skin, and eyes [10–15]. They can renew themselves in the body, making identical copies of themselves, or become specialized to yield the cell types of the tissue of origin [16]. The use of ASC in research and therapy is not controversial, unlike ESC, as they are extracted from adult tissue samples rather than destroyed human embryos.

One type of ASC, mesenchymal stromal cells (MSC), has generated great interest in the fields of regenerative medicine and immunotherapy due to their unique biologic properties. MSC were first discovered in 1968 by Friedenstein and colleagues [17] as adherent fibroblast-like cells in the bone marrow (BM) capable of differentiating into the bone. It was subsequently shown that MSC could be isolated from various tissues such as BM, adipose tissue [18], and umbilical cord blood [19]. These cells can be expanded in vitro, which allows them to rapidly reach the desired cell counts for use in vivo. Using different strategies, several laboratories have identified, isolated, and cultured MSC with specific properties [20].

10.1.3.1 Dental Pulp Stem Cells

SC can be classified based on their capacity for differentiation and also on their tissue of origin, such as bone marrow mesenchymal/stromal cells (BM-MSC), hematopoietic SC, umbilical cord blood SC, epithelial SC, and SC from the dental pulp. The dental pulp is a soft connective tissue within the dental crown thought to be derived from migratory neural crest cells during development. It has been shown to harbor various populations of multipotent stem/progenitor cells. Since their very first isolation in 2000 by Gronthos et al. [21], several types of adult stem cells have been isolated from teeth, including dental pulp stem cells (DPSC), stem cells from human exfoliated deciduous teeth (SHED), periodontal ligament stem cells (PDLSC), dental follicle precursor cells (DFPC), and stem cells from apical papilla (SCAP) [21–25].

These populations have MSC-like qualities, namely, the capacity for self-renewal, the potential to differentiate into multiple lineages including osteoblasts and chondroblasts, and a potential for in vitro differentiation into cell types from various embryonic layers, including adipose, bone, endothelial, and neural-like tissue. They are often compared to BM-MSC and share many similarities with them, but they differ in that dental stem cells seem to be committed to an odontogenic fate, more so than to an osteogenic one. Since they have a neural crest origin, they have stronger neurogenesis capabilities but weaker adipogenesis and chondrogenesis [26].

DPSC are isolated by enzymatic digestion of pulp tissue after separating the crown from the roots. These cells are morphologically similar to fibroblasts, very proliferative and clonogenic. DPSC are multipotent cells that proliferate extensively, can be safely cryopreserved, possess immunosuppressive properties, and express markers such as CD13, CD29, CD44, CD59, CD73, CD90, CD105, CD146, and STRO-1, but do not express CD14, CD24, CD34, CD45, CD19, and HLA-DR (human leukocyte antigen-antigen D related). They have the ability to differentiate into odontoblast-like cells, osteoblasts, adipocytes, neural cells, cardiomyocytes, myocytes, and chondrocytes in vitro. DPSC represent less than 1% of the total cell population present in the dental pulp [27].

10.1.3.2 Dental Pulp Pluripotent-like Stem Cells

In previous studies, our group has described a new population of adult stem cells called dental pulp pluripotent-like stem cells (DPPSC) [28–30]. These cells are isolated from the dental pulp of the third molars, express pluripotency markers such as OCT4, NANOG, and SOX2, and show embryonic-like behavior differentiating into tissues of the three embryonic layers.

These cells are not the first SC population isolated from adult tissues with pluripotency-like capacities. Indeed, several populations have been identified in recent years: very small embryonic-like (VSEL) cells [31], multipotent adult progenitor cells (MAPC) [32], mesodermal progenitor cells (MPC) [33], and marrow-isolated adult multilineage inducible (MIAMI) cells [34].

Compared to other populations, DPPSC have as the main advantage the fact that the third molars are a very accessible source of cells, because wisdom tooth extraction is widely performed and the teeth are usually considered to be medical waste. Since the third molar is the last tooth to develop in humans, it is normally in an early stage of development and is capable of yielding an optimum quantity of dental pulp tissue for the isolation of stem cells. Although the percentage of DPPSC decreases with age, a population of these cells is always present [28].

DPPSC and DPSC share the same isolation protocol, as well as some characteristics. However, they differ in the expression levels of embryonic markers as well as some membrane proteins such as CD73. The culture conditions between them are also different; DPPSC need low density and a specific medium that contains growth factors such as epidermal growth factor (EGF), platelet-derived growth factor (PDGF) and leukemia inhibitory factor (LIF) to allow maintenance of the pluripotent state of DPPSC. The characteristics unique to these cells are still under investigation, but the current evidence gain insights for future comparative studies of the regenerative potency of DPPSC and SC from other sources. It has been demonstrated, for instance, that DPPSC have a greater capacity for generating bone-like cells in comparison with DPSC [30].

In addition, DPPSC have pluripotent-like properties that have not been found in cells of any other adult source: the ability to form EB-like and teratoma-like structures (28), which had been thought to be exclusive to ESC and iPSC [35, 36].

10.1.4 Stem Cells in Regenerative Medicine

Regenerative medicine is an emerging interdisciplinary field of research and clinical applications focused on the repair, replacement, or regeneration of cells, tissues, or organs to restore impaired function resulting from any cause, including congenital defects, disease, trauma, and aging [37]. Organ regeneration is distinct from organ repair as an endpoint of a healing process following injury. Repair is an adaptation to loss of normal organ mass and leads to restoration of the interrupted continuity by synthesis of scar tissue without restoration of the normal tissue. By contrast, regeneration restores the interrupted continuity by synthesis of the missing organ mass at the original anatomical site, yielding a regenerate [38]. Therefore, the ultimate aim of regenerative medicine is to regenerate cells, tissues, or organs.

This field holds the promise of providing therapeutic treatment for conditions where current therapies are inadequate by stimulating the body's own repair mechanisms to functionally heal previously irreparable tissues or organs or by growing tissues and organs in the laboratory and implanting them when the body cannot heal itself [39]. It uses a combination of several technological approaches that moves it beyond traditional transplantation and replacement therapies. These approaches may include, but are not limited to, the use of soluble molecules administered alone or as a secretion by infused cells (immunomodulation therapy), gene therapy, cell transplantation (cell therapy), transplantation of in vitro grown organs and tissues (tissue engineering) and the reprogramming of cell and tissue types [40, 41].

Some of the biomedical approaches within the field of regenerative medicine may involve the use of stem cells [42], named stem-cell therapy. Human body has an endogenous system of regeneration through stem cells, where adult stem cells are found almost in every type of tissue. Their main postnatal function is to repair and regenerate the tissues in which they reside. That is why the idea is that restoration of function is best accomplished by these cells [39].

Stem cells can overcome one of the most limiting aspects of cell therapy and tissue regeneration, which is obtaining enough quantity of cells [43], since they have a high self-renewal potential. Moreover, they can help in the regeneration process via a paracrine effect. If regeneration is achieved using autologous cells, i.e., adult stem cells, it would potentially solve the problem of the shortage of organs available for donation, the problem of organ transplant rejection due to immune complications, and the problem of possible transfer of inherent infections from another donor. Taking into account all these aspects, the use of autologous adult stem cells is highly explored in regenerative medicine.

As said before, other types of stem cells, such as ESC, have a great potential due to their characteristics, but their use is limited by ethical considerations and donor-recipient immunocompatibility [39]. Another type of stem cells, iPSC, presents disadvantages regarding safety issues that impair their use in clinical applications.

A limiting aspect that cell therapy using adult stem cells can present is the difficulty to harvest the cells, since it can require an invasive procedure to the patient that can cause large donor site morbidity [43]. However, dental pulp tissue represents an easily accessible source of adult stem cells with low morbidity. As said before, since the third molar is the last tooth to develop in humans, it is normally in an early stage of development and is capable of yielding an optimum quantity of dental pulp tissue for the isolation of stem cells.

In Rotter et al. [43], it was stated that high numbers of stem cells with an effective and reliable potential for differentiation are needed for clinical applications. Thus, the identification of new stem cell sources and the establishment of optimized cell culture conditions that allow for the amplification of stem cells are of utmost relevance. In addition, the isolation procedure should ideally be minimally invasive and possibly be performed under local anesthesia [43]. As said before, DPPSC, adult stem cells from the dental pulp, present a minimally invasive isolation protocol, and their effective and reliable potential for differentiation is currently under investigation, with promising results already obtained [28, 30]. Nevertheless, more studies to test the differentiation capacity of DPPSC to multiple cell types are still needed.

10.2 Mesoderm-Derived Cells and Tissues

Mesoderm-derived cell types are osteogenic cells, chondrogenic cells, adipogenic cells, skeletal muscle cells, smooth muscle cells, cardiac muscle cells, and endothelial cells. Although the osteogenic differentiation potential of DPPSC is quite established, little is known about their endothelial and smooth and skeletal muscle cell differentiation capability.

A key tissue regenerative process related to any stem cell-based application is the angiogenesis, the generation of new blood vessels. Indeed, blood vessels deliver oxygen and nutrients to all of the tissues and organs in the body. The two major cellular components of blood vessels are endothelial cells (EC) and vascular smooth muscle cells (VSMC). Both EC and VSMC are required for vascular function, including blood pressure control, interactions with immune cells, and the uptake of nutrients [44]. As pointed out before, these two cellular types derive from the mesodermal germ layer of the very early embryo, although some smooth muscle cells (SMC) originate from the neural crest [45].

Blood vessels can be divided into three major types: arteries (and arterioles), veins (and venules), and capillaries. Arteries carry the blood away from the heart, veins carry it to the heart, and capillaries are the ones in which gas, nutrients, and

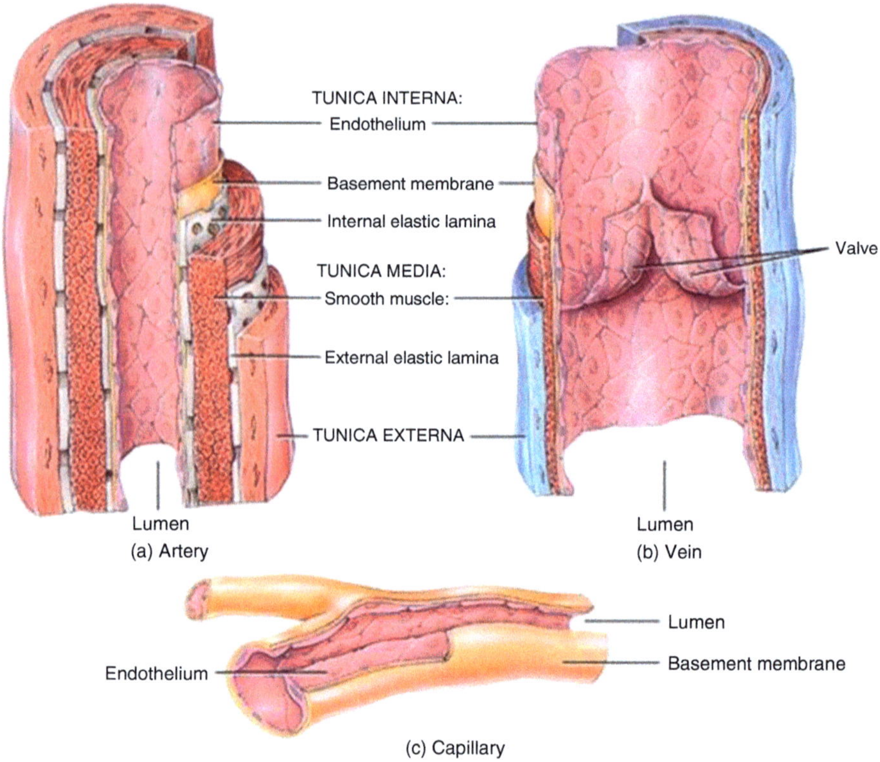

Fig. 10.1 Structure of the three major types of blood vessels. (Extracted from Tortora et al. [46])

wastes exchange with the tissues actually takes place. Regarding their structure, arteries and veins have the same three layers of tissue in their walls, while capillaries, which are much thinner, only present the most internal one of them. These layers are (i) the tunica interna or intima, which is the thinnest of the three and is formed by EC; (ii) the tunica media, which is rich in VSMC; and (iii) the tunica externa or adventitia, made of connective tissue (Fig. 10.1).

Both EC and VSMC play various major physiological and pathological roles in the blood vessels. EC are responsible for subendothelial matrix protein synthesis, homeostasis, thrombolysis, vasomotor properties, antigen presentation, and synthesis of growth factors such as PDGF, insulin-like growth factor 1 (IGF-1), and FGF, which promote cell growth of VSMC [47, 48]. VSMC also play physiological and pathological major roles, since they must continually repair arterial injuries and maintain functional mass in response to changing demands upon the vessel wall. Vascular smooth muscle is composed of multifunctional cells that exhibit spontaneous and agonist-induced contractile properties, secrete and assemble a wide variety of extracellular matrix proteins, display migratory and proliferative responses to tissue injury, and both produce and respond to a variety of paracrine-acting growth factors [49].

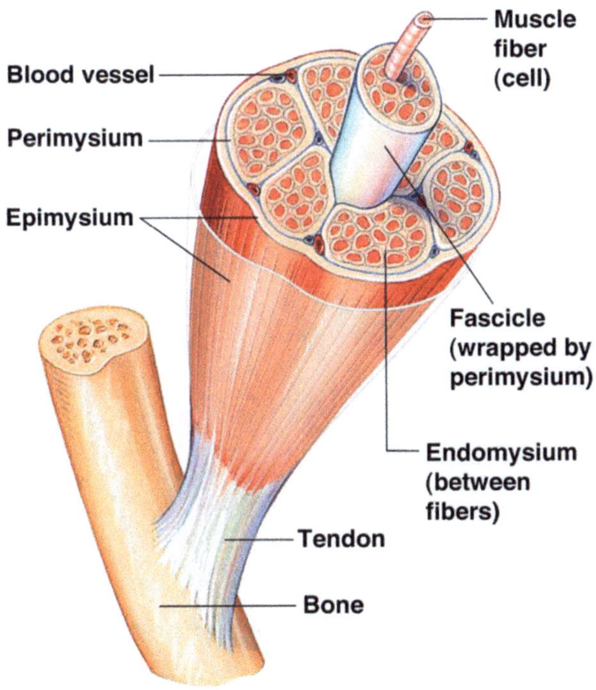

Fig. 10.2 Structure of a skeletal muscle. (Copyright © 2006 Pearson Education, Inc., publishing as Benjamin Cummings)

Currently, there is a growing list of diseases with high incidences with an etiology that is related to blood vessels [50]. More than 500 million people worldwide may benefit from pro- or anti-angiogenesis treatments [51]. Tissue regeneration and stimulation of angiogenesis are important therapeutic considerations in the treatment of myocardial infarction, peripheral vascular disease, and stroke [52]. In addition, angiogenesis is a key process in tissue engineering. If blood supply cannot be established rapidly, there is insufficient oxygen and nutrient transport and necrosis of the implanted tissue will occur [53]. For all these reasons, EC and VSMC differentiation are studied in regenerative medicine approaches.

Another important mesoderm-derived tissue for regenerative medicine is skeletal muscle, which is the largest mesoderm-derived organ and is responsible for the voluntary movement of the body. The smallest unit of skeletal muscle is the muscle fiber or myofiber, which is a long cylindrical cell that contains many nuclei, mitochondria, and sarcomeres. The latter made of actin and myosin filaments represent the functional unit of the muscle and are responsible for muscle contraction.

Each muscle fiber is surrounded by a thin layer of connective tissue called the endomysium (Fig. 10.2). Approximately 20–80 of these muscle fibers are grouped together in a parallel arrangement called a muscle fascicle or fiber bundle that is encapsulated by a perimysium, which is thicker than the epimysium, enclosing each

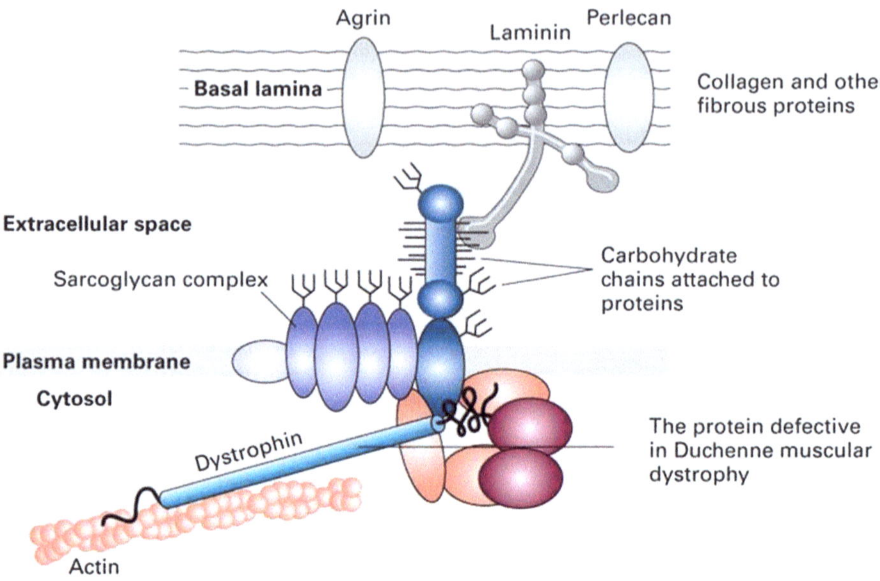

Fig. 10.3 The dystrophin glycoprotein complex (DGC) in skeletal muscle cells. Extracted from Lodish et al. [58]

of the bundled muscle fibers. A distinct muscle is formed by enveloping a large number of muscle fascicles in a thick collagenous external sheath extending from the tendons called the epimysium (Fig. 10.2).

As said before, the smallest units of skeletal muscle are cylindrical, multinucleated muscle fibers. These structures are established during embryogenesis, when mononuclear cells known as myoblasts fuse into immature myofibers (myotubes). The myofiber nuclei (myonuclei) are postmitotic and under normal conditions cannot reenter a proliferative state to contribute additional nuclei. Thus, during postnatal life, myofiber growth, homeostasis, and repair only rely on satellite cells, myogenic stem cells residing between the basal lamina and the muscle fiber membrane [54–57].

Many diseases that affect the musculature belong to the group of muscular dystrophies (MD). MD refers to a group of more than 30 genetic diseases that cause progressive weakness and degeneration of skeletal muscles used during voluntary movement. These disorders vary in age of onset, severity, and pattern of affected muscles. All forms of MD grow worse as muscles progressively degenerate and weaken. At the end, many individuals eventually can lose the ability to walk. For all this, skeletal muscle differentiation is also studied in regenerative medicine approaches.

The proteins that result affected in these diseases called Dystrophin Glycoprotein Complex (DGC) (Fig. 10.3). The most important, dystrophin (DYS), which results mutated in Duchenne muscular dystrophy (DMD), connects the sarcolemmal cytoskeleton to the extracellular matrix. This is accomplished through a series of interactions with a large group of membranous proteins that belong, for instance, to the DGC. Other dystrophies show mutations in genes encoding for sarcoglycans, proteins that are thought to stabilize the DGC and indeed are thought to be part of it (Fig. 10.3).

10.2.1 Endothelial Differentiation from Stem Cells

Currently available vascular grafts cannot reproduce the complex functions of native vessels [59]. In this regard, it has been proved that presence of functional endothelial cells (EC) in the grafts can improve these functions [60]. Cells assigned to cover the tissue-engineered vessels have been already isolated from several human sources such as blood vessels, bone marrow, and adipose tissue. However, these grafts are difficult to prepare due to insufficient autologous sources of EC [61] or the difficulty to obtain them.

To increment the number of obtained cells, stem cell differentiation has been explored, since EC derived from hESC may be used to generate a large supply of transplantable, healthy, functional cells for the repair of ischemic tissues [52]. To date, there exist two commonly used methods to induce vascular cell differentiation from human pluripotent stem cells: (i) embryoid body (EB) formation [62, 63] and (ii) monolayer-directed differentiation [52, 64]. EB formation results in differentiation of human pluripotent stem cells into various cell types, including vascular cells, albeit inefficiently (1%–5%) [62, 65, 66]. Current monolayer differentiation methods offer increased efficiencies (5–20%) but depend on undefined supplements, coculture [64, 67], heterogeneous cell aggregates [68], conditioned medium [52, 69], or lack consistent yields of vascular cells [70].

There have also been studies concerning the differentiation of adult stem cells into EC, mostly using MSC [71, 72]. MSC derived from the bone marrow are able to differentiate into various cell lineages of the bone, cartilage, adipose tissue, and myocardium and to some extent also into EC [73–75]. The disadvantages of BM-MSC are the invasive harvesting procedure and the long-lasting differentiation process. Therefore, other possible sources with less traumatizing intervention are necessary. MSC from adipose tissue represent an abundant and accessible source of such cells, which can as well differentiate into a number of mesodermal lineages, including EC [18, 76, 77]. The capacity of these cells to differentiate into the endothelial cell lineage is however limited [71]. Furthermore, in several studies, endothelial progenitor cells have also been used for autologous transplantation therapy and have been shown to promote the regeneration of ischemic tissues [78–81]. However, difficulties in obtaining sufficient numbers of adult endothelial progenitor cells still limit autologous stem cell therapies [82, 83]. This stresses an urgent need for an easily accessible stem cell source with a high proliferation rate that can also provide sufficient cell numbers for transplantation [53].

In a preliminary study from this group [28], DPPSC showed potential to differentiate to the endothelial lineage. In the project here presented, therefore, we wanted to further investigate DPPSC capacity to undergo endothelial differentiation in vitro, optimizing their differentiation protocol, as well as testing their angiogenic potential in vivo. For the latest, we used a nude mice model to study the wound healing assay process in terms of wound closure, wound matrix organization, and wound revascularization.

10.2.2 Smooth Muscle Differentiation from Stem Cells

The alterations in the differentiated state in smooth muscle cells (SMC) contribute to a variety of major cardiovascular diseases such as atherosclerosis, hypertension, restenosis, and vascular aneurysm [45, 84]. A better understanding of the cellular and molecular mechanisms that control VSMC differentiation is essential to help develop new approaches to both prevent and treat these diseases. Therefore, development of reliable and reproducible in vitro cellular models to study SMC differentiation is needed, yet it has been problematic due to intrinsic peculiarities of SMC [85, 86]. In addition, smooth muscle differentiation is exploited by tissue engineering techniques to design new organ structures consisting of smooth muscle, e.g., vascular constructs or bladder reconstruction substrates [87].

SMC differentiation is a complex and poorly defined process, although much progress has been made in identifying molecular mechanisms controlling the expression of SMC-specific genes [85]. Accumulating evidence has shown that a precisely coordinated molecular network orchestrates the SMC differentiation program involved in a range of signaling pathways, including transforming growth factor beta (TGF-β).

VSMC originate from at least five different sources of progenitors during embryonic development, including neural crest, proepicardium, serosal mesothelium, secondary heart field, and somites [45, 88]; and SMC responses to environmental signals, such as growth factors, have been observed to vary depending on the developmental origins of SMC [45]. In addition, SMC can undergo phenotypic changes, in vitro and in vivo, switching between secretory and contractile phenotypes, thus obscuring our conceptual reference to terminal differentiation in these cells [86].

Several in vitro model systems have been developed to mimic the SMC differentiation in vivo including using embryonic stem cells, embryoid bodies, neural crest cells, pluripotent C3H10T1/2 cells, and multipotent A404 cells. Although these models have significantly contributed to the understanding of SMC differentiation, each of these models has its limitations [85]. Human embryonic stem cell can differentiate to both EC and SMC populations in the same differentiation conditions. Though the cells are excellent for in vivo neoangiogenesis and regeneration of blood vessels, they may not be ideal for precisely dissecting the molecular mechanism governing SMC differentiation because SMC differentiated from embryonic stems cells are heterogenic and thus contain a mixed population. In Shi et al. (2015) [85], human embryonic stem cell-derived mesenchymal cells derived from H9 human embryonic were studied. These stem cells are natural SMC progenitors for mesoderm-derived SMC that account for most of the vascular SMC [45] and could be robustly differentiated to SMC phenotype upon TGF-β stimulation and exhibit a morphology resembling functional SMC.

In this project, we studied the DPPSC ability to undergo smooth muscle differentiation in vitro and in vivo in a mouse model of wound healing and in two mice model of dystrophy. Regarding the different developmental origins of SMC, it would be interesting to evaluate their differentiation from DPPSC, which have an ectomesodermal origin.

10.2.3 Skeletal Muscle Differentiation from Stem Cells

Muscular dystrophies are particularly challenging from a therapeutic point of view, and stem cell-based therapy is considered to be one of the most promising methods for treating muscular dystrophies. Because of the properties of satellite cells, they represented the first candidate and there have been several clinical trials since the early 1980s involving the transplant by intramuscular injections of these cells into several locations of a single muscle or at most a few muscles [89, 90]. Although results in treating DMD patients have been encouraging, this method has been limited by (i) the necessity of a huge number of injections, (ii) immune responses toward injected satellite cells, and (iii) the rapid death of most of the satellite cells in the first 72 hours following injection [91, 92]. In order to overcome these issues, many studies were addressed to find a population of ideal stem cells that could fulfill several criteria needed for the transplantation. Many studies focused on stem cells usually derived from adult bone marrow, hematopoietic SC and MSC. The latter can be obtained from fat, skin, periosteum, synovial membrane, and muscle as well. MSC are multipotent and capable of differentiating into several connective tissue types including osteocytes, chondrocytes, adipocytes, tenocytes, and myoblasts [93]. They can also impose an additional anti-inflammatory and paracrine effect on differentiation and tissue regeneration via cytokine pathways, have anti-apoptotic features [94–96], and can produce extracellular matrix molecules [97]. Compared with pluripotent ESC or iPSC, MSC have a greater biosafety profile and lower risk of tumorigenicity, and perhaps that is why numerous MSC-based therapies have made it to the clinical trial stage [98, 99]. Regarding DPSC, it has been shown that upon infusion into cardiotoxin-induced muscle defects, cloned DPSC engraft and colonize host muscle, as well as express dystrophin and myosin heavy chain more efficaciously than their parent heterogeneous cells. Another study also showed the ability of pre-differentiated DPSC to regenerate dystrophin-expressing myofibers within the injected host muscles, although the number of fibers was very limited [100]. These cells also showed enhancement of angiogenesis in the tissue.

All these data suggested that DPPSC could have therapeutic potential in MD. In this study, we investigated the myogenic potential of DPPSC in vitro and in vivo in an mdx mouse model and in a β-sarcoglycan-null mouse model, which represent two reliable mice models of dystrophies.

References

1. Fuchs E, Segre JA (2000) Stem cells: a new lease on life. Cell 100(1):143–155
2. Bhattacharya N, Stubblefield PG (2012) Regenerative medicine: using non-fetal sources of stem cells. Springer Springer-Verlag London
3. Mimeault M, Hauke R, Batra SK (2007) Stem cells: a revolution in therapeutics-recent advances in stem cell biology and their therapeutic applications in regenerative medicine and cancer therapies. Clin Pharmacol Ther 82(3):252–264
4. Medvedev SP, Shevchenko AI, Zakian SM (2010) Induced pluripotent stem cells: problems and advantages when applying them in regenerative medicine. Acta Nat 2(2):18–28
5. Takahashi K, Tanabe K, Ohnuki M, Narita M, Ichisaka T, Tomoda K et al (2007) Induction of pluripotent stem cells from adult human fibroblasts by defined factors. Cell 131(5):861–872
6. Mimeault M, Batra SK (2006) Concise review: recent advances on the significance of stem cells in tissue regeneration and cancer therapies. Stem Cells 24(11):2319–2345
7. Baldwin T (2009) Morality and human embryo research. Introduction to the talking point on morality and human embryo research. EMBO Rep 10(4):299–300
8. Richards M, Tan SP, Tan JH, Chan WK, Bongso A (2004) The transcriptome profile of human embryonic stem cells as defined by SAGE. Stem Cells 22(1):51–64
9. Yu J, Vodyanik MA, Smuga-Otto K, Antosiewicz-Bourget J, Frane JL, Tian S et al (2007) Induced pluripotent stem cell lines derived from human somatic cells. Science (New York, NY) 318(5858):1917–1920
10. Watts C, McConkey H, Anderson L, Caldwell M (2005) Anatomical perspectives on adult neural stem cells. J Anat 207(3):197–208
11. Kim CF, Jackson EL, Woolfenden AE, Lawrence S, Babar I, Vogel S et al (2005) Identification of bronchioalveolar stem cells in normal lung and lung cancer. Cell 121(6):823–835
12. Leri A, Kajstura J, Anversa P (2005) Cardiac stem cells and mechanisms of myocardial regeneration. Physiol Rev 85(4):1373–1416
13. Schmelzer E, Zhang L, Bruce A, Wauthier E, Ludlow J, Yao HL et al (2007) Human hepatic stem cells from fetal and postnatal donors. J Exp Med 204(8):1973–1987
14. Tumbar T, Guasch G, Greco V, Blanpain C, Lowry WE, Rendl M et al (2004) Defining the epithelial stem cell niche in skin. Science (New York, NY) 303(5656):359–363
15. Lavker RM, Tseng SC, Sun TT (2004) Corneal epithelial stem cells at the limbus: looking at some old problems from a new angle. Exp Eye Res 78(3):433–446
16. Fuchs E, Tumbar T, Guasch G (2004) Socializing with the neighbors: stem cells and their niche. Cell 116(6):769–778
17. Friedenstein AJ, Petrakova KV, Kurolesova AI, Frolova GP (1968) Heterotopic of bone marrow. Analysis of precursor cells for osteogenic and hematopoietic tissues. Transplantation 6(2):230–247
18. Zuk PA, Zhu M, Ashjian P, De Ugarte DA, Huang JI, Mizuno H et al (2002) Human adipose tissue is a source of multipotent stem cells. Mol Biol Cell 13(12):4279–4295
19. In 't Anker PS, Scherjon SA, Kleijburg-van der Keur C, de Groot-Swings GM, Claas FH, Fibbe WE et al (2004) Isolation of mesenchymal stem cells of fetal or maternal origin from human placenta. Stem Cells 22(7):1338–1345
20. Sharma RR, Pollock K, Hubel A, McKenna D (2014) Mesenchymal stem or stromal cells: a review of clinical applications and manufacturing practices. Transfusion 54(5):1418–1437
21. Gronthos S, Mankani M, Brahim J, Robey PG, Shi S (2000) Postnatal human dental pulp stem cells (DPSCs) in vitro and in vivo. Proc Natl Acad Sci U S A 97(25):13625–13630
22. Miura M, Gronthos S, Zhao M, Lu B, Fisher LW, Robey PG et al (2003) SHED: stem cells from human exfoliated deciduous teeth. Proc Natl Acad Sci U S A 100(10):5807–5812
23. Sonoyama W, Liu Y, Yamaza T, Tuan RS, Wang S, Shi S et al (2008) Characterization of the apical papilla and its residing stem cells from human immature permanent teeth: a pilot study. J Endod 34(2):166–171
24. Seo BM, Miura M, Gronthos S, Bartold PM, Batouli S, Brahim J et al (2004) Investigation of multipotent postnatal stem cells from human periodontal ligament. Lancet 364(9429):149–155

25. Vollner F, Driemel O, Reichert T, Morsczeck C (2007) Isolation and characterization of dental follicle precursor cells (DFPCs). J Stem Cells Regen Med 2(1):130
26. Davies OG, Cooper PR, Shelton RM, Smith AJ, Scheven BA (2015) A comparison of the in vitro mineralisation and dentinogenic potential of mesenchymal stem cells derived from adipose tissue, bone marrow and dental pulp. J Bone Miner Metab 33(4):371–382
27. Sloan AJ, Waddington RJ (2009) Dental pulp stem cells: what, where, how? Int J Paediatr Dent / the British Paedodontic Society [and] the International Association of Dentistry for Children 19(1):61–70
28. Atari M, Gil-Recio C, Fabregat M, Garcia-Fernandez D, Barajas M, Carrasco MA et al (2012) Dental pulp of the third molar: a new source of pluripotent-like stem cells. J Cell Sci 125(Pt 14):3343–3356
29. Atari M, Barajas M, Hernandez-Alfaro F, Gil C, Fabregat M, Ferres Padro E et al (2011) Isolation of pluripotent stem cells from human third molar dental pulp. Histol Histopathol 26(8):1057–1070
30. Atari M, Caballe-Serrano J, Gil-Recio C, Giner-Delgado C, Martinez-Sarra E, Garcia-Fernandez DA et al (2012) The enhancement of osteogenesis through the use of dental pulp pluripotent stem cells in 3D. Bone 50(4):930–941
31. Kucia M, Reca R, Campbell FR, Zuba-Surma E, Majka M, Ratajczak J et al (2006) A population of very small embryonic-like (VSEL) CXCR4(+)SSEA-1(+)Oct-4+ stem cells identified in adult bone marrow. Leukemia 20(5):857–869
32. Jiang Y, Jahagirdar BN, Reinhardt RL, Schwartz RE, Keene CD, Ortiz- Gonzalez XR et al (2002) Pluripotency of mesenchymal stem cells derived from adult marrow. Nature 418(6893):41–49
33. Petrini M, Pacini S, Trombi L, Fazzi R, Montali M, Ikehara S et al (2009) Identification and purification of mesodermal progenitor cells from human adult bone marrow. Stem Cells Dev 18(6):857–866
34. D'Ippolito G, Diabira S, Howard GA, Menei P, Roos BA, Schiller PC (2004) Marrow- isolated adult multilineage inducible (MIAMI) cells, a unique population of postnatal young and old human cells with extensive expansion and differentiation potential. J Cell Sci 117(Pt 14):2971–2981
35. Benton G, George J, Kleinman HK, Arnaoutova IP (2009) Advancing science and technology via 3D culture on basement membrane matrix. J Cell Physiol 221(1):18–25
36. Maltman DJ, Przyborski SA (2010) Developments in three-dimensional cell culture technology aimed at improving the accuracy of in vitro analyses. Biochem Soc Trans 38(4):1072–1075
37. Mason C, Dunnill P (2008) A brief definition of regenerative medicine. Regen Med 3(1):1–5
38. Yannas IV (2007) Tissue and organ regeneration in adults. Springer Springer Science+Business Media, LLC, New York
39. Bajada S, Mazakova I, Ashton BA, Richardson JB, Ashammakhi N. (2008) Stem Cells in Regenerative Medicine. Topics in Tissue Engineering, Vol. 4. Eds. N Ashammakhi, R Reis, & F Chiellini 1–28
40. Greenwood HL, Thorsteinsdottir H, Perry G, Renihan J, Singer PA, Daar AS (2006) Regenerative medicine: new opportunities for developing countries. J Biotechnol 8(1–2):60–77
41. Muneoka K, Allan CH, Yang X, Lee J, Han M (2008) Mammalian regeneration and regenerative medicine. Birth Defects Res C Embryo Today 84(4):265–280
42. Riazi AM, Kwon SY, Stanford WL (2009) Stem cell sources for regenerative medicine. Methods Mol Biol 482:55–90
43. Rotter N, Oder J, Schlenke P, Lindner U, Bohrnsen F, Kramer J et al (2008) Isolation and characterization of adult stem cells from human salivary glands. Stem Cells Dev 17(3):509–518
44. Patsch C, Challet-Meylan L, Thoma EC, Urich E, Heckel T, O'Sullivan JF et al (2015) Generation of vascular endothelial and smooth muscle cells from human pluripotent stem cells. Nat Cell Biol 17(8):994–1003
45. Majesky MW (2007) Developmental basis of vascular smooth muscle diversity. Arterioscler Thromb Vasc Biol 27(6):1248–1258

46. Tortora G, Derrickson B (2009) Chapter 21: The cardiovascular system: blood vessels and hemodynamics. In: Principles of anatomy and physiology. Wiley USA
47. Luo Z, Wang G, Wang W, Xiao Q, Xu Q (2011) Signalling pathways that regulate endothelial differentiation from stem cells. Front Biosci (Landmark edition) 16:472–485
48. Grotendorst GR, Seppa HE, Kleinman HK, Martin GR (1981) Attachment of smooth muscle cells to collagen and their migration toward platelet-derived growth factor. Proc Natl Acad Sci U S A 78(6):3669–3672
49. Hirschi KK, Majesky MW (2004) Smooth muscle stem cells. Anat Rec A Discov Mol Cell Evol Biol 276(1):22–33
50. Luttun A, Ross JJ, Verfaillie C, Aranguren XL, Prosper F Differentiation of multipotent adult progenitor cells into functional endothelial and smooth muscle cells. Curr Protoc Immunol / edited by John E Coligan [et al] 2006;Chapter 22:Unit 22F:9
51. Carmeliet P (2005) Angiogenesis in life, disease and medicine. Nature 438(7070):932–936
52. Kane NM, Meloni M, Spencer HL, Craig MA, Strehl R, Milligan G et al (2010) Derivation of endothelial cells from human embryonic stem cells by directed differentiation: analysis of microRNA and angiogenesis in vitro and in vivo. Arterioscler Thromb Vasc Biol 30(7):1389–1397
53. Bronckaers A, Hilkens P, Fanton Y, Struys T, Gervois P, Politis C et al (2013) Angiogenic properties of human dental pulp stem cells. PLoS One 8(8):e71104
54. Katz B (1961) The terminations of the afferent nerve fibre in the muscle spindle of the frog. Philos Trans Royal Soc Lond [Biol] 243:221–240
55. Mauro A (1961) Satellite cell of skeletal muscle fibers. J Biophys Biochem Cytol 9:493–495
56. Hawke TJ, Garry DJ (2001) Myogenic satellite cells: physiology to molecular biology. J Appl Physiol (Bethesda, Md: 1985) 91(2):534–551
57. Yablonka-Reuveni Z, Day K (2011) Chapter 11: Skeletal muscle stem cells in the spotlight: the satellite cell. In: Regenerating the heart: stem cells and the cardiovascular system springer. Humana Press, pp 173–200 Springer Science+Business Media, LLC
58. Lodish H, Berk A, Zipursky SL, Matsudaira P, Baltimore D, Darnell J (2013) Molecular cell biology. W. H. Freeman New York
59. Watanabe M, Shin'oka T, Tohyama S, Hibino N, Konuma T, Matsumura G et al (2001) Tissue-engineered vascular autograft: inferior vena cava replacement in a dog model. Tissue Eng 7(4):429–439
60. Meinhart JG, Deutsch M, Fischlein T, Howanietz N, Froschl A, Zilla P (2001) Clinical autologous in vitro endothelialization of 153 infrainguinal ePTFE grafts. Ann Thorac Surg 71(5 Suppl):S327–S331
61. Seifalian AM, Tiwari A, Hamilton G, Salacinski HJ (2002) Improving the clinical patency of prosthetic vascular and coronary bypass grafts: the role of seeding and tissue engineering. Artif Organs 26(4):307–320
62. James D, Nam HS, Seandel M, Nolan D, Janovitz T, Tomishima M et al (2010) Expansion and maintenance of human embryonic stem cell-derived endothelial cells by TGFbeta inhibition is Id1 dependent. Nat Biotechnol 28(2):161–166
63. Levenberg S, Golub JS, Amit M, Itskovitz-Eldor J, Langer R (2002) Endothelial cells derived from human embryonic stem cells. Proc Natl Acad Sci U S A 99(7):4391–4396
64. Vodyanik MA, Bork JA, Thomson JA, Slukvin II (2005) Human embryonic stem cell- derived CD34+ cells: efficient production in the coculture with OP9 stromal cells and analysis of lymphohematopoietic potential. Blood 105(2):617–626
65. Li Z, Suzuki Y, Huang M, Cao F, Xie X, Connolly AJ et al (2008) Comparison of reporter gene and iron particle labeling for tracking fate of human embryonic stem cells and differentiated endothelial cells in living subjects. Stem Cells 26(4):864–873
66. Wang H, Charles PC, Wu Y, Ren R, Pi X, Moser M et al (2006) Gene expression profile signatures indicate a role for Wnt signaling in endothelial commitment from embryonic stem cells. Circ Res 98(10):1331–1339
67. Wang ZZ, Au P, Chen T, Shao Y, Daheron LM, Bai H et al (2007) Endothelial cells derived from human embryonic stem cells form durable blood vessels in vivo. Nat Biotechnol 25(3):317–318

68. Sumi T, Tsuneyoshi N, Nakatsuji N (2008) Suemori H. Defining early lineage specification of human embryonic stem cells by the orchestrated balance of canonical Wnt/beta-catenin, Activin/nodal and BMP signaling. Development (Cambridge, England) 135(17):2969–2979
69. Inman GJ, Nicolas FJ, Callahan JF, Harling JD, Gaster LM, Reith AD et al (2002) SB-431542 is a potent and specific inhibitor of transforming growth factor-beta superfamily type I activin receptor-like kinase (ALK) receptors ALK4, ALK5, and ALK7. Mol Pharmacol 62(1):65–74
70. Orlova VV, Drabsch Y, Freund C, Petrus-Reurer S, van den Hil FE, Muenthaisong S et al (2014) Functionality of endothelial cells and pericytes from human pluripotent stem cells demonstrated in cultured vascular plexus and zebrafish xenografts. Arterioscler Thromb Vasc Biol 34(1):177–186
71. Colazzo F, Chester AH, Taylor PM, Yacoub MH (2010) Induction of mesenchymal to endothelial transformation of adipose-derived stem cells. J Heart Valve Dis 19(6):736–744
72. Konno M, Hamazaki TS, Fukuda S, Tokuhara M, Uchiyama H, Okazawa H et al (2010) Efficiently differentiating vascular endothelial cells from adipose tissue-derived mesenchymal stem cells in serum-free culture. Biochem Biophys Res Commun 400(4):461–465
73. Caplan AI (2005) Review: mesenchymal stem cells: cell-based reconstructive therapy in orthopedics. Tissue Eng 11(7–8):1198–1211
74. Oswald J, Boxberger S, Jorgensen B, Feldmann S, Ehninger G, Bornhauser M et al (2004) Mesenchymal stem cells can be differentiated into endothelial cells in vitro. Stem Cells 22(3):377–384
75. Pittenger MF, Mackay AM, Beck SC, Jaiswal RK, Douglas R, Mosca JD et al (1999) Multilineage potential of adult human mesenchymal stem cells. Science (New York, NY) 284(5411):143–147
76. Fischer LJ, McIlhenny S, Tulenko T, Golesorkhi N, Zhang P, Larson R et al (2009) Endothelial differentiation of adipose-derived stem cells: effects of endothelial cell growth supplement and shear force. J Surg Res 152(1):157–166
77. Culmes M, Eckstein HH, Burgkart R, Nussler AK, Guenther M, Wagner E et al (2013) Endothelial differentiation of adipose-derived mesenchymal stem cells is improved by epigenetic modifying drug BIX-01294. Eur J Cell Biol 92(2):70–79
78. Heeschen C, Chang E, Aicher A, Cooke JP (2006) Endothelial progenitor cells participate in nicotine-mediated angiogenesis. J Am Coll Cardiol 48(12):2553–2560
79. Huang PP, Yang XF, Li SZ, Wen JC, Zhang Y, Han ZC (2007) Randomised comparison of G-CSF-mobilized peripheral blood mononuclear cells versus bone marrow-mononuclear cells for the treatment of patients with lower limb arteriosclerosis obliterans. Thromb Haemost 98(6):1335–1342
80. Li Z, Han Z, Wu JC (2009) Transplantation of human embryonic stem cell-derived endothelial cells for vascular diseases. J Cell Biochem 106(2):194–199
81. Yang C, Zhang ZH, Li ZJ, Yang RC, Qian GQ, Han ZC (2004) Enhancement of neovascularization with cord blood CD133+ cell-derived endothelial progenitor cell transplantation. Thromb Haemost 91(6):1202–1212
82. Li Z, Wilson KD, Smith B, Kraft DL, Jia F, Huang M et al (2009) Functional and transcriptional characterization of human embryonic stem cell-derived endothelial cells for treatment of myocardial infarction. PLoS One 4(12):e8443
83. Tepper OM, Galiano RD, Capla JM, Kalka C, Gagne PJ, Jacobowitz GR et al (2002) Human endothelial progenitor cells from type II diabetics exhibit impaired proliferation, adhesion, and incorporation into vascular structures. Circulation 106(22):2781–2786
84. Owens GK, Kumar MS, Wamhoff BR (2004) Molecular regulation of vascular smooth muscle cell differentiation in development and disease. Physiol Rev 84(3):767–801
85. Shi N, Chen SY (2015) From nerve to blood vessel: a new role of Olfm2 in smooth muscle differentiation from human embryonic stem cell-derived mesenchymal cells. J Biomed Res 29(4):261–263
86. Xie C, Ritchie RP, Huang H, Zhang J, Chen YE (2011) Smooth muscle cell differentiation in vitro: models and underlying molecular mechanisms. Arterioscler Thromb Vasc Biol 31(7):1485–1494

87. Mahoney WM, Schwartz SM (2005) Defining smooth muscle cells and smooth muscle injury. J Clin Invest 115(2):221–224
88. Yoshida T, Owens G (2005) Molecular determinants of vascular smooth muscle cell diversity. Circ Res 96(3):280–291
89. Miller RG, Sharma KR, Pavlath GK, Gussoni E, Mynhier M, Lanctot AM et al (1997) Myoblast implantation in Duchenne muscular dystrophy: the San Francisco study. Muscle Nerve 20(4):469–478
90. Skuk D, Goulet M, Tremblay JP (2006) Use of repeating dispensers to increase the efficiency of the intramuscular myogenic cell injection procedure. Cell Transplant 15(7):659–663
91. Fan Y, Maley M, Beilharz M, Grounds M (1996) Rapid death of injected myoblasts in myoblast transfer therapy. Muscle Nerve 19(7):853–860
92. Guerette B, Skuk D, Celestin F, Huard C, Tardif F, Asselin I et al (1997) Prevention by anti-LFA-1 of acute myoblast death following transplantation. J Immunol (Baltimore, Md: 1950) 159(5):2522–2531
93. Bongso A, Lee E (2005) Stem cells: from bench to bedside. Stem cells: their definition, classification and sources. Singapur
94. Meirelles Lda S, Nardi NB (2009) Methodology, biology and clinical applications of mesenchymal stem cells. Front Biosci (Landmark edition). 14:4281–4298
95. Keating A (2012) Mesenchymal stromal cells: new directions. Cell Stem Cell 10(6):709–716
96. Uccelli A, Benvenuto F, Laroni A, Giunti D (2011) Neuroprotective features of mesenchymal stem cells. Best Pract Res Clin Haematol 24(1):59–64
97. Meng J, Adkin CF, Arechavala-Gomeza V, Boldrin L, Muntoni F, Morgan JE (2010) The contribution of human synovial stem cells to skeletal muscle regeneration. Neuromuscular disorders : NMD 20(1):6–15
98. Lin H, Otsu M, Nakauchi H (2013) Stem cell therapy: an exercise in patience and prudence. Philosophical Transactions of the Royal Society B: Biological Sciences. 2013; 368(1609):20110334. doi:10.1098/rstb.2011.0334.
99. Ra JC, Shin IS, Kim SH, Kang SK, Kang BC, Lee HY et al (2011) Safety of intravenous infusion of human adipose tissue-derived mesenchymal stem cells in animals and humans. Stem Cells Dev 20(8):1297–1308
100. Pisciotta A, Riccio M, Carnevale G, Lu A, De Biasi S, Gibellini L et al (2015) Stem cells isolated from human dental pulp and amniotic fluid improve skeletal muscle histopathology in mdx/SCID mice. Stem Cell Res Ther 6:156

Chapter 11
From Bench to Bedside of Mesenchymal Stem Cells Use for Rheumatoid Arthritis Treatment

R. A. Contreras, N. Luque, F. Djouad, Roberto Elizondo-Vega, and Patricia Luz-Crawford

Abstract Mesenchymal stem cells (MSCs) are multipotent stem cells with immunosuppressive properties able to control both the innate and the adaptive immune system. Based on this latter characteristic, MSC-based therapies have become a thriving area for autoimmune diseases such as rheumatoid arthritis (RA). RA is a chronic inflammatory disorder that involves an imbalance of both the innate and adaptive immune systems, which leads to cartilage and bone degradations, causing pain, stiffness, and synovitis. Several studies have reported the beneficial effects of MSCs in experimental models of arthritis. However, the complexity of RA as well as the controversy about the mechanisms and molecules involved in the immunosuppressive effect of MSCs, together with the low efficacy showed on clinical trials, have generated insecurity around its therapeutic use, and for this reason it is necessary to increase the number of investigations. In this context, enhancement of MSC immunosuppressive potential, in order to obtain a stable suppressive phenotype, offers a powerful tool to improve MSC-based therapies. Thus, the aim of this review is to discuss the cross talk between MSCs and the intricate pathogenesis of RA and the possible molecular targets that could lead to improve their therapeutic potential on RA progression.

Keywords Mesenchymal stem cells · Rheumatoid arthritis treatment · Immunosuppressive potential

R. A. Contreras
Laboratorio de Inmunología Celular y Molecular, Centro de Investigación Biomédica, Facultad de Medicina, Universidad de los Andes, Santiago, Chile

Inserm, U 1183, Montpellier, France

N. Luque · R. Elizondo-Vega (✉) · P. Luz-Crawford (✉)
Laboratorio de Inmunología Celular y Molecular, Centro de Investigación Biomédica, Facultad de Medicina, Universidad de los Andes, Santiago, Chile
e-mail: relizondo@uandes.cl; pluz@uandes.cl

F. Djouad
Inserm, U 1183, Montpellier, France

Université de Montpellier, Montpellier, France

R. Delgado-Morales (ed.), *Stem Cell Genetics for Biomedical Research*,
https://doi.org/10.1007/978-3-319-90695-9_11

11.1 Introduction

Rheumatoid arthritis (RA) is a chronic autoimmune disease that mostly degenerates bone and cartilage tissue of small and large articulations; however, it can affect other body systems, such as cardiovascular and pulmonary [1, 2]. Even though the etiology of RA is not completely understood, it has been well described to engage an abnormal activation of diverse immune cells, such as T cells, B cells, natural killer cells (NK), dendritic cells (DC), macrophages, and neutrophils. The exacerbated activation of those cells stimulates the production of proinflammatory mediators like cytokines, proteinases, and growth factors that mediate joint destruction and systemic complications [3, 4]. Presently, there is no cure for RA available; therefore, the therapeutic approaches are focused on managing the symptoms and on reducing the progression of the disease, by stopping joint destruction [5]. Existing treatments are unable to repair damaged joints. Thus, novel strategies must be directed to reduce inflammation and to induce tissue regeneration in order to improve RA progression [5].

Mesenchymal stem cells (MSCs) have been reported as a potential tool in the treatment of RA due to their regenerative and anti-inflammatory properties that could both induce the regeneration of damaged joints and to modulate the pathogenic immune response [6].

11.2 Mesenchymal Stem Cells as Immunosuppressive Agents to Damp Down Arthritis Progression

MSCs are heterogeneous self-renewal multipotent stromal cells that are able to differentiate into cells from adult tissues [7]. The International Society for Cellular Therapy (ISCT) has established the minimal criteria for identifying MSCs as a fibroblastic-like morphology that express mesodermal markers and lack of hematopoietic markers and that are capable of differentiating into adipocytes, chondrocytes, and osteoblast cells [8]. They can be isolated from almost all postnatal and mesodermal tissues including the bone marrow, placenta, adipose tissue, dental tissue, and menstrual blood [9–12]. They can be cultured easily in vitro through plastic adherences under regular culture conditions [8]. MSCs are able to secrete several factors and cytokines including vascular endothelial growth factor (VEGF), transforming growth factor beta (TGFβ), basic fibroblast growth factor (bFGF), insulin growth factor (IGF), and hepatocyte growth factor (HGF) that stimulate angiogenesis, inhibit apoptosis and fibrotic remodeling, and induce proliferation of cells [13–15].

Alongside their role as supportive stromal cell and their ability to differentiate to multiple tissues, MSCs are able to modulate both the innate and the adaptive immune response, by inhibiting proinflammatory populations of DC, proinflammatory macrophages, NK cells, B cells, and T cells and inducing the generation of anti-inflammatory phenotypes [16–19]. MSCs have been tested as treatment for autoimmune and alloimmune diseases, including RA, through the use of specific murine models, like collagen-induced arthritis (CIA), a model widely used to study RA [20–26].

Importantly, some controversial results have been reported with MSCs isolated from different tissues or from different strain of mice, which do not exhibit the same therapeutic and immunosuppressive potential on arthritis [27–29]. This might be due to inherent molecular characteristics of MSCs isolated from each source that might be related to the mechanism by which MSCs exert their immunosuppressive properties. Consequently in order to achieve the correct use of MSCs as a treatment of RA patients, it is fundamental to understand the basis of the interaction between MSCs and the immune cells associated with the pathogenesis of RA.

11.3 Innate Immunity in RA and the Effect of MSCs

Although initially seen as rarely present in synovial fluids from patients and therefore thought as no significant for the diagnostic of RA [30], later on, macrophages were found in synovial membranes of patients with RA, which contained phagocytically active, proinflammatory HLA-DR+ macrophages [31]. Currently, macrophages are recognized to have a central role in the pathogenesis of RA. An increase in the macrophage population is directly associated to active disease, and a diminution of them can be observed after effective treatment [32]. They are abundant in the synovial lining of RA patients, and their numbers are also significantly augmented in the adjacent tissue. These observations led to the hypothesis that macrophages in the synovial membranes of patients with RA may trigger T-cell infiltration and activation through antigen presentation and eventually initiate B-cell infiltration altogether creating a positive inflammatory feedback [31].

Macrophages are highly plastic and capable to rapidly respond to multiple stimuli, generating the classically described proinflammatory (M1) or anti-inflammatory (M2) phenotypes [33, 34]. M1 macrophages are polarized by lipopolysaccharide (LPS) either alone or in association with proinflammatory cytokines such as interferon gamma (IFN-γ). They produce cytokines such as interleukin-1β (IL-1β), IL-6, IL-12, IL-23, and tumor necrosis factor alpha (TNF-α). M2 macrophages are polarized by cytokines such as IL-4 and IL-13 and produce anti-inflammatory cytokines such as IL-10 or TGF-β1 [35].

Several studies have showed that proinflammatory cytokines such as TNF-α and IL-1β are abundant in RA, whereas IL-10 is relatively diminished in patients with RA compared to healthy individuals [36]. When CIA is generated on IL-10 knockout mice, the progression of the disease is more aggressive, and the phenotype of synovial macrophages corresponds mainly to proinflammatory M1 [37].

In healthy individuals, there is a balance between bone formation and reabsorption, led by osteoblasts and osteoclasts, respectively. In RA osteoclast activity is chronically induced, causing severe bone destruction [38, 39]. Synovial fibroblasts are activated in response to TNF-α and IL-1β, secreted by M1 macrophages, and produce receptor activator of nuclear factor-κB (NF-κB) ligand (RANKL) and macrophage colony-stimulating factor 1 (M-CSF) which are essential for the formation of osteoclasts by fusion of myeloid precursors of monocytes/macrophages [40–42].

Additionally, macrophages contribute to disease progression by producing reactive oxygen species (ROS), nitric oxide (NO) intermediates, and matrix-degrading enzymes, as well as chemokines like CXCL8 (also known as IL-8) and monocyte chemoattractant protein 1 (MCP-1, also known as CCL2), which initiate neutrophil and monocyte recruitment in the joint [2].

In this context, MSCs have shown promising results, where adipose-derived MSCs (ASCs) have significantly reduced clinical symptoms in a CIA mouse model [23]. ASCs affect the in vitro formation of bone-resorbing osteoclasts, by inhibiting the RANKL-induced osteoclastogenesis, in both a contact-dependent and contact-independent manner. Furthermore, ASCs reduced osteoclast precursors in the bone marrow, diminishing osteoclastogenesis and, overall, preventing periarticular and systemic bone loss in CIA mice [43].

Aside from inhibiting osteoclast formation, MSCs also induce a phenotypic switch from the proinflammatory M1 toward IL-10-producing M2 macrophages [44]. However, the mechanism by which MSCs induce this M1 to M2 switch on RA has not been fully clarified. Under inflammatory conditions, MSCs increase the secretion of prostaglandin E2 (PGE2) through upregulation of COX2 and other components of the arachidonic acid pathway, which could drive the transition of macrophages to an anti-inflammatory phenotype [45]. Similarly, after being stimulated with TNF-α, MSCs produce TNF-α-stimulated gene 6 protein (TSG-6), an anti-inflammatory molecule that impairs the interaction with Toll-like receptor 2 (TLR2) by interacting with the CD44 receptor on macrophages, which modulates TLR2-mediated NF-κB signaling and decreases the inflammatory response [46]. Similarly, MSCs express IL1-receptor antagonist (IL1RA), which is involved in the modulation of the plasticity of macrophages from proinflammatory M1 into M2 phenotype [17, 47].

Natural killer (NK) cells are one of the first barriers from the innate immune system to microbial agents and tumor cells. Additionally, NK cells can affect adaptive immunity by cytokine production and by directly killing other immune cells, which implies a regulatory role of NK cells in autoimmunity [48, 49]. In RA, NK cells have been described to have both a protective and a detrimental role on disease progression.

NK cells can be abundantly found in the joints of RA patients. They express both RANKL and M-CSF, which are associated with triggering the differentiation of CD14+ monocytes into osteoclasts in the synovial membrane. Depletion of NK cells from mice before the induction of CIA diminishes the severity of subsequent arthritis and almost completely prevents bone erosion. These results suggest that NK cells have a role in bone destruction associated with inflammatory arthritis [49].

Controversial results have shown that upon NK cell depletion, immunized mice presented an early onset of arthritis with much stronger clinical symptoms and joint damage [50]. That was associated with increased B-cell generation, autoantibody production, and a significant increment of IL-17-secreting cells in synovial tissue [50]. Furthermore, activation of NK cells through the administration of an anti-CD94/NKG2A-receptor antibody diminished the pathogenic follicular helper T (Tfh) cells and the CD4+ T-helper cells producing IL-17 (Th17), stopping the disease progression [51]. These data indicate that NK cells may play a rather protective role in the development of experimental arthritis, mostly throughout suppression of Th17 cells.

The effect of MSCs on NK cells depends on the activation state of NK and/or on the cytokines present in the milieu. MSCs can strongly inhibit the proliferation induced by IL-2 of freshly isolated resting NK cells [52]. Moreover, the function of NK cell is regulated by several receptors that can generate either inhibitory or activating signals. Exposure of resting NK cells to activating cytokines, such as IL-2, increases the expression of the activating receptors NKp44, CD69, NKp30, and NKG2D [53]. MSCs can significantly inhibit IL-2-induced NK cell effector functions, such as cytotoxic activity and cytokine production, mostly driven by indoleamine 2,3-dioxygenase (IDO) and PGE2 as crucial mediators of the MSC immunosuppressive effect on NK cells [19]. Moreover, it has been reported that human NK cells secrete NAP-2 (CXCL7), a chemokine that can induce MSC migration. The use of specific antagonists of CXCR2, a receptor that recognizes NAP-2, abolished NK cell-mediated MSC recruitment [54]. However, when NK cells are activated, they can also recognize allogeneic MSCs and induce the apoptosis of the latter cells [55]. This could be reversed by activating MSCs with IFN-γ through high levels of HLA, B-C expression [56]. In the context of arthritis, the specific effect of MSCs on NK cells has not been completely understood.

Dendritic cells (DCs) are potent stimulators of adaptive immunity and also play a critical role in the establishment and maintenance of immunological tolerance. In RA, DCs are the main inducer of inflammation by presenting antigens to autoreactive T cells, which then produce different cytokines associated to T-helper cell differentiation [57]. Since DCs have the capacities to modulate T-cell response, a promising new immunotherapeutic strategy for treating RA is through the generation of tolerogenic DCs (tDCs). tDCs induce peripheral tolerance by promoting regulatory T (Treg) cells and suppressing effector T cells [58]. Following this strategy, the stimulation of DCs with synthetic peroxisome proliferator-activated receptor-gamma (PPAR-γ) agonists suppresses the production of inflammatory cytokines and facilitates the generation of tDCs, ameliorating CIA progression [59].

MSCs are able to interfere with the maturation of DCs, generating tDCs, by impairing the TLR activation, both in vitro and in vivo [16, 60, 61]. Upon exposure to MSCs in vitro, DCs diminish the expression of activation markers and are no longer able to process and present antigen to T cells, resulting in significantly decreased T-cell proliferation [60]. MSCs inhibit the cytokine production of LPS-activated DCs through paracrine mediators acting on the mitogen-activated protein kinase (MAPK) cascade of the NF-κB pathway. This has also been seen in vivo where an intravenous administration of MSCs led to impaired priming of antigen-specific naive T cells, with significantly decreased cell numbers in the draining lymph nodes that are attributable to markedly reduced migration of DCs due, at least in part, to the significantly downregulated expression of CCR7 and CD49dβ1, which are involved in DC homing to lymphoid organs [62]. IL-6 has been proposed as the main mediator of MSCs responsible to abrogate DCs proinflammatory immune response [63].

11.4 Adaptive Immunity in RA and the Effect of MSCs

As previously mentioned, cells from the innate immune system do not only have an important role on the pathogenesis of the disease by directly inducing inflammation and bone erosion but also by recruiting and activating different cells from the adaptive immunity including T and B lymphocytes. A key observation concerning the relationship between adaptive immunity and RA is the production of autoantibodies [4]. Most of the antibody reactivities described occur before the onset of disease, and in most individuals the autoantibody stock is developed already at the onset of disease, with very few individuals becoming autoantibody positive later during the disease course [55]. Consequently, B-cell depletion therapy with anti-CD20 antibody (rituximab) has become an important biologic therapy with positive clinical results [56]. B-cell depletion reduces rheumatoid factor and anti-citrullinated protein antibodies (ACPA), which are prevalent biomarkers of RA [56]. Therefore, removal of autoreactive B cells participating in antigen presentation, costimulation, and cytokine production likely plays an important role, but is not the main mediator of RA, since a decrease in autoantibodies does not necessarily correlate with clinical outcome [64]. In this context, it has been seen that the T-cell population is altered after B-cell depletion, resulting in reduced T-cell activation and cytokine production. In proteoglycan-induced arthritis (PGIA), a murine model of RA, Treg cell percentages were elevated in B-cell-depleted mice, compared to control-treated mice, with a higher proportion of CD4+ T cells expressing Foxp3 and CD25 [56]. Looking further into Treg activity, CD4+CD25+ cells from B-cell-depleted mice expressed higher amounts of Foxp3 and were significantly more suppressive than those from the control group. Interestingly, when Treg cells were removed with an anti-CD25 monoclonal antibody simultaneously with B-cell depletion therapy, the severity of PGIA was restored to levels equal to untreated mice [56]. Thus, B cells have the capacity to regulate inflammatory responses by enhancing effector T cells along with suppressing Treg cells.

It has been described that MSCs can inhibit normal B-cell proliferation, antibody secretion, and differentiation to plasma cells in vitro [17, 65, 66]. However, it has been shown that MSCs induce the survival of B cells together with the capacity to stimulate proliferation and differentiation of various subsets of purified B cells, from both healthy donors and systemic lupus erythematosus (SLE) patients [66]. These inconsistent effects were better understood by showing the existence of a complex interaction between MSCs, B cells, and T cells. T-cell signaling is required for MSCs to exert their immunomodulatory effect on B cells, which appears to be dependent of both soluble factors that are released when the three cell types are present as well as by direct cell-cell contact [67]. This second mechanism has been validated by a study showing that the interaction of programmed cell death 1 (PD-1) protein and its ligand (PDL-1) concerning MSCs and B cells, which increases in the presence of IFN-γ, is required for MSC-mediated inhibition of B lymphocyte activation [68]. Blocking such interactions can restore the molecular pathways associated to B-cell stimulation and partially rescue B-cell proliferation [68]. To confirm that soluble factors also play an important role, Asari et al. demonstrated that MSCs are able to inhibit the mRNA expression levels of the B lymphocyte-induced maturation protein-1 (Blimp-1), master transcriptional regulators required for B-cell terminal differentiation, independent on cell contact [69]. Furthermore, conditioned media derived from MSC are also able to inhibit B-cell differentiation in vitro and in vivo when applied to mice immunized with both T-cell-independent and T-cell-dependent antigens; MSCs significantly suppress the antigen-specific immunoglobulin M and immunoglobulin G1 secretion, whereas culture medium alone had no such effect [69]. The mechanism by which MSCs modulate B-cell differentiation and proliferation is still under investigation. It seems that the chemokine CCL2 exerts an important role on the immunomodulatory capacity of MSCs on B cells since MSCs isolated from lupus-like mice and SLE patients have an impaired inhibition activity on B-cell proliferation and differentiation. This has been confirmed by a reduced expression of the chemokine CCL2 on MSCs derived by SLE patients and lupus-like mice compared to healthy or wild-type MSCs since the overexpression of CCL2 in lupus MSCs restored their immunosuppressive function on B cells [70]. Also it has been reported in an in vitro study that MSCs directly promote the development of CD19+ CD24high CD38high IL-10-secreting regulatory B cells, through the chemokine stromal-derived factor-1α (SDF-1α) and its receptor, CXCR7, which contribute to the generation of an immunosuppressive environment [65, 71]. Finally, on RA it has been demonstrated that MSCs are able to inhibit plasmablast generation in vivo in the murine model of CIA, effect that was associated to the expression of the IL-1 receptor antagonist (IL1RA) on MSCs [17].

As previously mentioned T cells play a key role in the pathogenesis of RA. Already in 1975, it was observed that T cells were present in the synovial membrane of patients with RA [72]. Later on, it was identified that both CD4+ and CD8+ T-cell subsets were present in the joints of RA patients [31, 73]. CD8+ T cells are cytotoxic cells that induce cell death of virus-infected and cancer cells via release of cytolytic granules or induction of Fas-mediated apoptosis [74]. It has been suggested that they play an important role on the pathogenesis of autoimmune disease [75]. In particular, in RA the pathogenic role of CD8+ T cells has not been well established; however, it has been described that there is a high association between HLA class I polymorphisms with a higher probability to develop RA. Also, it has been shown that there is a correlation between the amounts of CD8+ T cells in the joint with RA severity. This information support the hypothesis that CD8+ T cells could play a role in RA where a high frequency of CD8+ T cells in the site of joint inflammation has been found [75]. MSCs are able to suppress CD8+ activation in vitro [76]. Moreover, it has been recently described that MSCs are able to induce the generation of CD8+CD28- Treg cells and also enhance their ability of impede CD4+ T-cell proliferation and activation [77]. In the context of RA, the role of MSCs on CD8+ T cells has not been investigated.

On the other hand, CD4+ T cells, also known as T-helper (Th) cells, exert multiple roles in assisting the immune response including B-cell differentiation and function and CD8 activation [78]. They can be polarized into different T-helper subsets depending on the type of immune response required by the organism. Nowadays, the more recognized CD4+ cell subtypes are T-helper type 1 (Th1), T-helper type 2 (Th2), T-helper type 17 (Th17), and T regulatory cells (Treg) [79]. The different T-helper subpopulations can be discriminated by their specific cytokine profile (IFN-γ for Th1, IL-4 for Th2, IL-17 for Th17), their specific transcription factors (T-bet for Th1, GATA3 for Th2, RORδt and RORα for Th17, and Foxp3 for Treg) [78, 80], and the combinations of chemokine receptors to enrich and even discriminate between them, like CCR6, CCR4, and CD161 for Th17, CXCR3 and CCR5 for Th1, and CCR4 for Th2 [78, 81].

Initially, it was suggested that Th1 cells played a central role in different autoimmune diseases. Early studies documented the presence of IFN-γ-secreting CD4+ T cells in the synovium of patients with RA, which supported the hypothesis that RA was one of the Th1-driven diseases [82, 83]. When IL-12, a Th1-polarizing cytokine, was neutralized using an anti-IL-12p40 antibody (one of the two subunits that compose it) in CIA mice, this attenuated the severity of the disease [84]. However, when CIA was induced on IL-12p40 knockout mice, 20% of mice still developed arthritis [84]. Moreover, the specific genetic ablation of the IL-12p35 subunit (the other subunits that compose IL12) exacerbated arthritis symptoms [85]. Later on this was better understood, with the discovery of the IL-23 cytokine who share the IL-12p40 domain with IL-12. Interestingly, the Th1 neither secretes IL-23 nor cooperates in their differentiation but participates in the generation and maintenance of the proinflammatory Th17 lineage [86]. Interestingly, when IFN-γ or its signaling pathway were inhibited, the developed of CIA was more readily, with an earlier onset and worse clinical score that was associated with an increased on IL17 levels in the serum and joints of CIA mice [87]. These results strongly proposing that Th17 cells are the key factors involved in initiation and perpetuation of chronic destructive arthritis [88].

Th17 cells have been detected in the synovium of patients with RA, and when IL-17 has been blocked using soluble IL-17 receptor-Fc on CIA mice, the disease progression was improved in a dose-dependent manner. Moreover, the overexpression of IL-17 in the knee joint of type CIA mice accelerated the onset and aggravated the synovial inflammation at the site, demonstrated the role of Th17 on joint destruction [89].

On the other hand, it has been established that CD4+ T cells display a high plasticity. Indeed, Th17 cells share exceptional close differentiation pathways as well as molecular signatures with their anti-inflammatory Treg counterparts [90]. Treg cells have been characterized by the high expression of CD25 and the transcription factor Foxp3, which is essential for Treg cell activity [89]. They control inflammation using numerous suppressive mechanisms mediated through both soluble and membrane-bound factors [91]. Moreover the lack of the transcription factor Foxp3 in both human and mouse is responsible for the development of multi-organ autoimmune disease [92]. The plasticity between Th17 and Treg cells might have been acquired initially to enable a flexible immune response to rapidly deal with pathogens. However, this cell plasticity can also lead to deregulations of immune responses and subsequently to the development of autoimmune diseases [93]. In the context of RA, it has been suggested that the increased frequency of Th17 cells is due to either a reduction in the number of Tregs or a qualitative defect in their function [94]. Even though MSCs exert a broad immunosuppressive effect on different cells of the immune system, the immunosuppressive properties on CD4 T cells have remained as the main focus of investigation. Indeed the immunosuppressive properties of MSCs were first described in a mixed lymphocyte reaction where the capacity of MSCs to inhibit T-cell proliferation was demonstrated [95]. Afterward, it was described that MSCs are able to inhibit proinflammatory Th1 and Th17 cells and to induce Treg cells in vitro and in vivo [18, 96, 97]. Indeed MSCs prevent the in vitro differentiation and function of naive CD4+ T cells into Th17 cells and inhibited the secretion of IL-17, IL-22, and TNF-α in memory Th17 cells. The inhibition of memory Th17 cells is associated with an increased on the expression of Foxp3 and the production of the anti-inflammatory cytokine IL-10 and is mediated by cell-cell contact through the PD-1 molecular pathway [98, 99]. MSC inhibition of Th17 cells was described to occur through cell-cell contact mediated by a combined involvement of CCR6 and CD11a/CD18 expressed by T cells and their respective ligands, CCL20 and CD54, present on MSCs exposed to inflammatory cytokines. Also, this effect is mediated by paracrine mechanisms through either production of TGFβ1 or the production of PGE2 binding to its receptor EP4 on T cells [98, 100]. In line with this study, it has been demonstrated that MSCs are able to induce CD4+CD25highFoxp3+ regulatory T cells during the differentiation process of Th1 and Th17 cells that is also associated to the secretion of IL-10 [18]. Similarly, it has been reported a direct conversion of Th17 cells into Treg cells through an IL-17A+FoxP3+ double-positive stage [101]. Moreover the generation of a Th1-producing IL-10 cells has been also described upon MSC coculture in vitro [102]. Moreover the capacity of MSCs to inhibit proinflammatory Th1 and Th17 cells and to induced Treg cells has been demonstrated using several murine models. Indeed umbilical cord-derived MSCs (UC-MSCs) improved sepsis progression by inducing a population of CD4+CD25+Foxp3 classical Treg cells in the lymph nodes of MSC-treated animals compared to the non-treated group [103]. In RA it has been described that the beneficial effect of gingiva-derived MSCs is associated to an increased frequency of CD4+CD39+Foxp3+ regulatory T cells and an inhibition on Th1 and Th17 lineages [24]. In accordance with this study, it was recently described that the therapeutic potential display by murine BM-MSCs in arthritis is associated to the generation of IL-10-producing regulatory Th17 cells in the draining lymph nodes (dLNs) of MSC-treated mice [104].

Osteoprotegerin (OPG) is a protein that inhibits the development of osteoclastogenesis, by sequestering RANKL and inhibiting RANK signaling [105]. MSCs derived from the human palatine tonsil (T-CM) have been reported to produce high levels of OPG in basal conditions, which reverts the differentiation of osteoclasts in a psoriatic arthritis (PsA) model. Psoriasis is a chronic autoimmune disease that mostly affects skin tissue, but psoriasis patients frequently develop systemic inflammation and comorbidities, such as PsA, that can be seen in up to 70% of the patients. PsA is associated with osteoclast-associated bone degeneration, and it has been observed that Th17 cells directly stimulate osteoclast differentiation and maturation via expression of RANKL in vitro. T-CM effectively blocked the RANK-RANKL interaction between the osteoclast precursors and Th17 cells via OPG, inhibiting osteoclast activity in vitro. In a psoriasis mouse model, the frequency of osteoclasts in the bone marrow, identified as tartrate-resistant acid phosphatase (TRAP)-positive cells [106], was significantly diminished after T-CM tail vein injection [107].

Even though T-CM have been reported to express OPG in basal conditions, it is important to highlight that MSCs do not constitutively express suppressive factors such as PGE2 or PDL1. Its induction is mediated by proinflammatory cytokines, suggesting that MSCs in a proinflammatory milieu are able to modify the immune response by modulating the cytokine secretion profile of T-cell subsets in favor of an anti-inflammatory environment.

11.5 Adaptive Memory in RA and the Effect of MSCs

Once the initial response of the adaptive immune system against an antigen ends, between 90 and 95% of the effector cells die. The small amount of cells that survive will eventually become part of the immunological memory: immune cells that are able to respond rapidly to a second round of a specific antigen previously encountered [108, 109]. Memory T cells are a heterogeneous population classically distinguished by the expression of the CD45RO isoform and by the lack of CD45RA isoform (CD45RO+CD45RA-) [108, 110]. Lanzavecchia et al. first identified this heterogeneity in human peripheral blood based on the expression of the lymph node homing CC-chemokine receptor CCR7. They identified two subsets of memory T cells called central memory cells and effector memory cells, respectively [111]. Central memory T (TCM) cells are characterized as CD44high CD45ROhigh CD45RAlow CD127high and CCR7high, are generated in secondary lymphoid tissues, and reside there in the absence of antigen. Effector memory T (TEM) cells are CD44high CD45ROhigh CD45RAlow CD127high L-selectinlow and CCR7low, are generated in secondary lymphoid tissues, and recirculate between blood and nonlymphoid tissues in the absence of antigen [112]. Lanzavecchia's group later showed that, after TCR stimulation, TCM cells secrete higher levels of IL-2 compared to TEM cells, which instead produce more effector cytokines (IFN-γ and IL-4) [113]. The T cell that expresses CD27high CD28high CD45ROlow CD45RAhigh CD95high CD122high CD127high L-selectinhigh and CCR7high and produces low levels of IFNγ and intermediate levels of IL-2 is known as stem cell memory (TSC). T cells that express CD45ROhigh CD45RAlow CD69high CD103high and CD103low subsets, are generated in nonlymphoid tissues, and stably reside in these tissues in the absence of antigen are known as tissue-resident memory (TRM) [112].

Memory T cells have play an important role in RA pathogenesis. It has been shown that RA patients have a larger amount of CD4+ memory T cells in the inflamed synovium [82, 114]. In the synovium of these patients, the accumulation of CD4+ memory T cells appears to be stimulated since the expression of cytokines and activation markers has been observed [115–120].

MSCs have been shown to preserve the naive state of lymphocytes when cocultured with peripheral blood mononuclear cells (PBMC), and their activation is prevented. Moreover, when PBMC are stimulated with phytohemagglutinin (PHA), MSCs induce an increase in the proportion of naive T cells and a significant decrease in memory T cells, specifically in the CD8+ TSC and CD4+ and CD8+ TCM subpopulations [121]. Additionally, Laranjeira et al. showed that when PBMC were not stimulated with PHA, MSCs highly diminished TCM, TEM, and effector CD4+ T cells through IL-2 downregulation [122, 123]. MSCs have also been reported to inhibit antigen-specific naive and memory T cells. MSCs affect cell proliferation, cytotoxicity, and the number of interferon gamma (IFN-γ)-producing antigen-specific T cells, in a cell-contact-dependent and dose-dependent manner [124]. Other publications have shown that inhibition of central (CD45RO+ CD62L+) and effector memory (CD45RO+CD62L-) T cells can be cell-contact-independent, when using conditioned medium derived from the mesenchymal layer of human amniotic membrane [125]. Controversially, it has been reported that human BM-MSCs can stimulate IL-17 production from CD4+ CD45RO+ memory T cells, but not CD4 and CD8 T cells. IL-17-producing CD4+ memory T cells generated by BM-MSCs can effectively enhance phagocytic activity of neutrophils [126].

As mentioned earlier, memory T cells express various chemokine receptors. CCL3, CCL4, and CCL5 that are produced in the synovial tissue, probably by distinct cell types, can attract CCR5-expressing memory cells [127, 128]. CCR5 expression is upregulated in synovial tissue T cells and has been correlated with IFN-γ expression by synovial CD4+ memory T cells in RA patients [129–132]. Furthermore, CCR5 deletion has been significantly associated with diminished clinical manifestations of RA [133]. Additionally, it has been reported that CCR6 is expressed by resting peripheral memory T cells [134], whereas LT-α expression is associated with the presence of lymphocytic aggregates in synovial tissue [117]. The correlation between the expressions of these two markers indicates the possibility that CCR6 might play a role in the development of aggregates of CD4+ T cells that are characteristically found in rheumatoid synovium [130]. Similarly, CCR7+ CD4+ memory T cells express lymph node homing receptors, and even though they lack immediate effector function, they can efficiently stimulate dendritic cells [111]. All these results suggest that inhibition of chemokine receptors might be a potential target for RA therapy.

Very interestingly, in RA patients, it has been reported that tissue-infiltrating CD4+ T cells include a population of cells that has lost CD28 expression and gained expression of memory markers [135–137]. Functionally, CD4+ CD28- T cells resemble effector memory cells. This population produces high levels of IFN-γ and presents an increased cytotoxic activity, compared to CD4+ CD28+ T cells [138–141]. Unlike normal helper T cells, CD4+ CD28− T cells do not support B cells;

however, they express NK-cell receptors, in particular killer activating receptors [142–144]. CD4+ CD28– T cells from patients with RA display a phenotype similar to central and effector memory. They express the lymphoid homing chemokine receptors CCR7, CXCR4, and CCR5. Accordingly, they home to lymph nodes as well as to inflamed synovia [145].

It has been established that in vitro cultured MSCs constitutively secrete a wide variety of different members of the chemokine family, such as CCL2 (MCP-1), CCL3 (MIP-1α), CCL4 (MIP-1β), CCL5 (RANTES), CCL7 (MCP-3), CCL20 (MIP-3α), CCL26 (eotaxin-3), CXCL1 (GROα), CXCL2 (GROβ), CXCL5 (ENA-78), CXCL8 (IL-8), CXCL10 (IP-10), CXCL11 (i-TAC), CXCL12 (SDF-1), and CX3CL1 (fractalkine) [146–148]. By expressing these chemokine ligands, MSCs could potentially interact with the receptors expressed on memory T lymphocytes and thus, alter their homing and facilitate the cell-contact-dependent inhibition, decreasing synovial inflammation.

Murine BM-MSCs have been reported to suppress the differentiation of Th17 from naive and memory T-cell precursors, as well as IL-17A secretion from memory CD4+ T cells and naturally occurring effector-memory Th17, in a model of acute kidney inflammation. This suppression mechanism is mediated initially by MSC/T cell contact and subsequently activated by PGE2 secretion via the EP4 receptor [100]. MSCs by suppressing the secretion of IL17-A could decrease the acute or chronic activation of these cells that have been linked to promoting tissue inflammation and the production of autoantibodies in RA patients [149].

γδ T cells are a T-cell subpopulation with multiple biological functions and associated with a variety of diseases. Patients with RA contain numerous T cells from the effector memory Vγ9Vδ2 T cells (the predominant subtype of γδ T cells in peripheral blood) in peripheral blood and in the synovial fluid of the joints. These cells have a high capacity to secrete inflammatory factors, such as IFNγ and IL-17, and display similar characteristics to specific antigen-presenting cells, including high HLA-DR and CD80/86, and been able to stimulate B lymphocytes, exacerbating immune dysfunction and producing abnormal immune damage [150]. MSCs have been observed to strongly suppress γδ-cell proliferation in vitro, as well as their cytolytic response and cytokine production. This is mediated by the COX-2-dependent production of PGE2 from MSCs via inhibitory receptors expressed in Vγ9Vδ2 T lymphocytes (EP2 and EP4) [151, 152].

The production of autoantibodies against citrullinated cyclic peptides (CCP) has been associated with this RA progression. The production of autoantibodies by B cells depends on the cooperation of the Tfh cells [153]. It has been reported that in RA patients, there is a correlation between an increase in memory Tfh cells with the higher titers of serum autoantibodies and severity of the disease [154, 155]. Refractory RA patients that receive a transplant of allogeneic MSC show a reduction in the serum anti-CCP antibody levels. Also allogeneic UC-MSCs suppress Tfh-cell differentiation in RA patients partly via the production of IDO in response to IFN-γ, and MSC transplantation into CIA mice inhibits both the number and function of Tfh cells in vivo, thus preventing arthritis progression [156, 153]. These results indicate that the MSCs could inhibit the differentiation of Tfh toward the different subsets of memory Tfh1, Tfh2, and Tf17 and thus decrease activity of the autoreactive B cells and therefore the production of autoantibodies.

Meta-analysis of genome-wide association studies has identified as many as 101 RA risk loci [157]. In particular, the HLA-DRB1 genotype was the first identified and by far the strongest genetic risk factor for RA [158, 159]. The shared epitope (SE), a common amino acid sequence at positions 70–74 of HLA-DRB1, is recognized for its association with anti-CCP antibody (ACPA)-positive RA [160]. It is thought that citrullinated autoantigen epitopes bind to HLA-DRB1 that contains the SE and are presented to CD4+ T cells, which contribute to autoimmunity [161]. In the RA synovium, there are ectopic lymphoid follicles as well as clonally expanded T cells and antigen-specific B cells that recognize citrullinated autoantigens [162]. It has been reported that chemokine receptor CXCR4 plays a central role in the homing and retention of CD4+ T cells [163, 164]. Interestingly, RA patients with one or more susceptible HLA-DR haplotypes displayed a significantly higher frequency of memory CXCR4+CD4+ T cells, and not Th1 and Th17 cells, suggesting that synovial migration and retention of memory CXCR4+CD4+ T cells is associated with sustained autoimmunity and local inflammation. Moreover, the frequency of memory CXCR4+CD4+ T cells significantly correlated with the expression level of HLA-DR on B cells, implying that B cells might be important antigen-presenting cells in RA, which was elevated in RA patients with SE. In vitro analysis and transcriptomic pathway analysis suggested that the interaction between HLA-DR and T-cell receptors is an important regulator of memory CXCR4+CD4+ T cells [165]. It is important to note that CXCR4, the specific chemokine receptor of SDF-1a, is expressed in MSCs at early passages and plays a pivotal role in the process of MSC homing [166].

11.6 Clinical Trials: Use of MSCs for the Treatment of RA

The immunomodulatory role of MSCs has been demonstrated through several in vitro studies and animal models, which have made clear their potential as therapeutic tool to restore immune tolerance and reduce the damage caused by rheumatoid arthritis [167]. Unfortunately, preclinical studies show some difficulties in analyzing the results, which may be due to the many variables related to each approach, such as the source of MSCs used, the number of cells per treatment, and the duration of the treatment, among others [168]. For this reason, the potential use of MSCs in RA has been overshadowed by controversial results that have failed to confirm the effect of MSC treatment.

On the other hand, few clinical trials have been conducted in RA patients, difficulting to fully understand the true effects of MSCs during active disease. These studies have been developed to evaluate the therapeutic efficacy and safety of MSCs.

Liang et al. evaluated the safety and efficacy of allogeneic MSC transplantation in refractory RA patients, resistant to disease-modifying antirheumatic drugs. They applied allogeneic bone marrow or umbilical cord MSCs through intravenous infusion into four patients, establishing that the procedure was safe; however, the results only showed a partial and transient clinical improvement [156].

Another study, led by Wang et al., evaluated the safety and efficacy of human umbilical cord MSCs in the treatment of RA. In this study they considered 136 patients who were resistant to disease-modifying antirheumatic drugs, and the patients who receive the treatment with MSCs showed a significant clinical improvement and an increase in the population of regulatory T cells in blood, results that were maintained for 3 to 6 months. Also they show that repeated infusion after this period can enhance the therapeutic efficacy [155].

More recently, a phase Ib/IIa clinical trial by Alvaro Garcia et al. evaluated the safety and tolerability of the intravenous administration of allogeneic expanded ASCs (Cx611) in patients with refractory RA. They treated 53 patients (20 in cohort A, 20 in cohort B, 6 in cohort C, and 7 in placebo group) and showed that intravenous administration of allogeneic ASCs is well tolerated, without evidence of dose-related toxicity. In terms of efficacy, the American College of Rheumatology establishes their response criteria as a standard to measure the effectiveness of various arthritis medications or treatments in clinical trials. Patients treated with Cx61150 showed tendency to an improvement over 50% (or ACR50) [169]. It is important to note that the positive effects do not persist after 3 months, suggesting that might be necessary to increase the number of repetitions. Although most patients tolerated treatment, one of them suffered a case of lacunar infarction in the low-dose treatment group. If the therapies are effective, they are required to increase the number of patients per study to allow the identification of particular groups of RA patients, helping to develop more individualized strategies, increasing the chances of a positive outcome after MSC treatment.

It is important to note that these studies evaluate only the short-term effectiveness of a treatment, ranging from 3 to 8 months, and therefore the effects of MSC therapy over a longer period of time are needed. In addition, the use of MSCs in clinical studies is restricted to patients with severe refractory RA to standard therapies, and it has been proposed that the MSC treatment could be more effective at early stages [170].

On the other hand, the cells used in these clinical trials include autologous bone marrow-derived MSC, umbilical cord-derived MSC, and allogeneic adipose-derived stem cells, and even though the findings support that the use of MSCs is safe, none of them has evaluated any type of conditioning treatments that could potentially stimulate the regenerative properties and immunosuppressive capacity of the MSCs, therefore leaving multiple options to evaluate in the future.

11.7 MSCs Preconditioned to Improve Their Therapeutic Skills for RA Treatment

As mentioned before, MSCs need to be activated by proinflammatory cytokines in order to display a complete immunosuppressive phenotype. The immunosuppressive mediators associated to MSCs were initially described to be promoted through environmental factors, by either exposure to inflammatory cytokines, particularly IFN-γ, or by stimulation of TLR signaling [171, 172]. IFN-γ normally acts as a boost signal for T-cell activation and expansion, but when cocultured with MSCs,

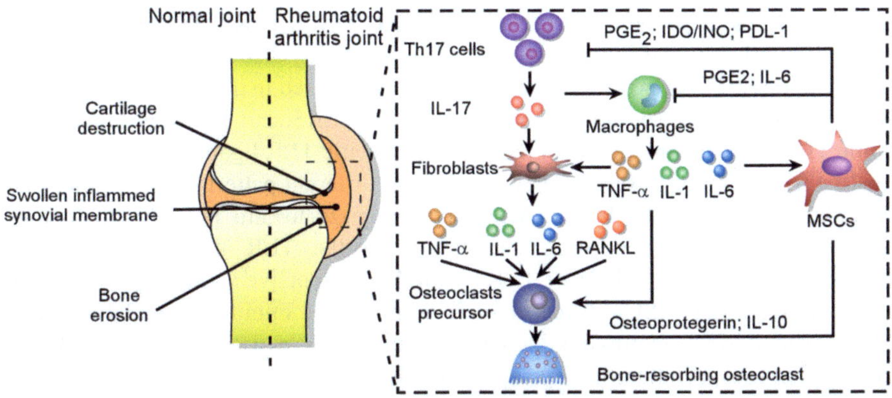

Fig. 11.1 MSCs prevent RA progression through multiple molecular mechanisms that inhibit inflammation and tissue degeneration. Proinflammatory cytokines such as IFN-γ, IL-1β, and TNF-α are able to activate the immunosuppressive potential of MSCs to induce the expression of several suppressive factors that will impact the progression of RA probably through the inhibition of proinflammatory cells of both the innate and the adaptive immune response and impeding the generation of osteoclasts in the joint

this cytokine generates a T-cell immunosuppressive environment, driven by the production of different mediators, some of them specific for human or murine MSCs like IDO on human or NO on murine MSCs and other factors shared among species such as PDL-1, ICAM, PGE2, and IL-6 [20, 173, 174] (Fig. 11.1). Several other proinflammatory cytokines, like TNFα, IL-1α, or IL-1β, have been described to induce a strong anti-inflammatory effect of MSCs, when working together with IFN-γ, through the induction of higher levels of the known suppressive mediators [174]. Moreover MSC stimulation with IFN-γ combined with TNF-α augments the secretion of IL-8, IL-6, hepatocyte growth factor (HGF), and PEG2 production [175]. Furthermore, MSC priming with IFN-γ plus TNF-α has been seen to activate superoxide dismutase 3, an antioxidant and anti-inflammatory enzyme that catalyzes the dismutation of two superoxide radicals into hydrogen peroxide and oxygen. Moreover, the IFN-γ and TNF-α combination has been demonstrated to induce the production of chemokines such as CCR5, CCR10, CXCL9, CXCL10, and CXCR3 that are involved in the chemotaxis and the inhibition of immune effector cells [176].

This IFNγ-induced immunosuppressive effect by MSCs has been seen to inhibit CD4+ and CD8+ T cells, B cells, and NK cells [177]. Importantly, MSCs are not able to suppress proliferation of B lymphocytes without prior IFN-γ stimulation [68].

It has been showed that MSCs are able to sense their microenvironment and identify exogenous and endogenous danger signals, through the expression of different TLRs. Indeed, the expression of TLR1, TLR2, TLR3, TLR4, TLR5, and TLR6 has been reported in human and mice MSCs [178]. Expression and function of TLRs can be modulated in different ways in MSCs. For example, hypoxia can significantly increase the mRNA levels of TLR1, TLR2, TLR5, TLR9, and TLR10 [179]. Remarkably, the inflammatory environment may also modulate the pattern and function of TLRs expressed by MSCs. When cultured in the presence of IFN-α, IFN-γ, TNFα, and IL1β, the expression of TLR2, TLR3, and TLR4 was increased, while TLR6 was downregulated [180]. Using specific TLR ligands has been shown that the triggering of different TLRs resulted in different patterns of cytokines and chemokines secreted by MSCs. MSCs can polarize toward proinflammatory or immunosuppressive phenotypes, depending on the TLR ligand encountered. Following activation of TLR3 or TLR4 with their respective agonists poly (I:C) or LPS, the resulting MSCs differ in their secretion of immune modulators, with TLR3-primed MSCs expressing increased immunosuppressive factors, while TLR4 priming led to a proinflammatory signature [181].

Some contradictory effects have been described for MSCs stimulated with IL-17. As mentioned before, some proinflammatory cytokines stimulate the immunosuppressive phenotype of MSCs, activating multiple molecular mechanisms. Similarly, it has been reported that IL-17 together with IFN-γ and TNF-α can enhance the immunosuppressive effect of MSCs. Interestingly, the effect of IL-17 was dependent on the expression of inducible nitric oxide synthase (iNOS), in MSCs. Moreover, IL-17 was found to modulate the mRNA stability of ARE/poly(U)-binding/degradation factor 1 (AUF1), which is abundant in lymphoid organs, and regulates the mRNA expression of various immune-related molecules including iNOS and IL-6. The role of IL-17 on AUF 1 was further confirmed using auf1-/- MSCs, which after incubation with IFN-γ and TNF-α induced the maximal immunosuppressive effect, both in vitro and in vivo, without the need for IL-17 [182]. Oppositely, very recently it was published that IL-17 also could diminish the immunosuppressive capacity of MSCs. Using olfactory ecto-MSCs (OE-MSCs), which are a population of stem cells that resides in the olfactory lamina propria, IL-17 could significantly diminish the suppressive capacity of OE-MSCs on CD4+ T cells mainly by downregulating the levels of inhibitory factors produced by OE-MSCs, including NO, IL-10, and TGF-β as well as cell surface-expressed inhibitory molecule such as PDL-1. Notably, IL-17 treatment inhibited the capacity of OE-MSCs in inducing Treg expansion and their capacity to inhibit Th1 and Th17 generation both in vitro and in vivo using the CIA model. Furthermore, knockdown of IL-17R in OE-MSCs significantly enhanced their therapeutic effect in ameliorating CIA progression [183]. Even though the studies compared MSCs from different sources and it is well known that MSCs from different sources can exert complete different functions [28], it is clear that further investigation is required to understand the real effect of IL-17 on the immunomodulatory capacity of MSCs (Fig. 11.1).

Notch signaling has recently emerged as an important regulator of immune responses in autoimmune diseases. MSCs have been reported to have an overall increase of CXCR4+ levels, when γ-secretase inhibitor (GSI) interrupts Notch signaling or when generating a knockout of the transcription factor recombination signal-binding protein for immunoglobulin kappa J (RBP-J), which upon activation is a transcriptional activator of the canonical Notch pathway. The GSI-treated or RBP-J deficient MSCs show an increased migration rate in response to stromal cell-derived factor-1α (SDF-1α) (CXCR4 ligand) in vitro, as well as in vivo in a mouse hepatic ischemia/reperfusion injury model, thus decreasing hepatic damage. Therefore, Notch signaling regulates MSC migration and function, at least partially via the modulation of CXCR4 expression [184]. Therefore, interfering with Notch-CXCR4 signaling in MSCs could increase the homing efficiency of therapeutic MSCs in the synovium and potentially inhibiting the activity of memory CXCR4+CD4+ T cells that have been implicated in the pathogenesis of RA, as we previously described.

Finally, a recent publication from our group has suggested the modulation of the activity of peroxisome proliferator-activated receptor (PPAR) β/δ on MSCs, as a way to improve their immunomodulatory effects. PPARβ/δ exhibits multiple biological functions including anti-inflammatory activities through the inhibition of NF-κB signaling and cell adhesion molecule expression [185]. In MSCs it was demonstrated that the inhibition either chemical or genetic significantly increased the immunosuppressive capacities of MSCs in vitro through the activation of NF-κβ signaling that induced the expression of iNOS and adhesive molecules. Moreover using the CIA model, the inhibition of PPARβ/δ significantly improved RA progression compared to control MSCs. Therefore, these results place PPARβ/δ as a master regulator of the immunosuppressive properties of MSCs; it also propose a novel strategy to enhance their therapeutic potential by inhibiting PPARβ/δ function [186].

11.8 Conclusion

The immunoregulatory abilities of MSCs have been studied for several years, demonstrating the wide repertory of mechanisms used to interact with the multiple members of the immune system. Particularly in the context of RA, the complexity of the pathogenesis, involving macrophages, DCs, NKs, T cells, B cells, and other cell types, places MSC-based therapy as the candidate of choice for RA treatment. However, even though the immunosuppressive effects of MSCs have been proved to affect each cell type individually, the plasticity of the phenotype of MSCs, which can be modified depending on the environment, could mislead MSC therapeutic effect. Importantly, the idea of use pretreated MSCs to establish an anti-inflammatory phenotype to harvest the full potential of their immunomodulatory capacities should be considered in the context of refining MSC-based therapeutic approaches.

References

1. Isaacs JD (2010) The changing face of rheumatoid arthritis: sustained remission for all? Nat Rev Immunol 10(8):605–611. https://doi.org/10.1038/nri2804
2. McInnes IB, Schett G (2011) The pathogenesis of rheumatoid arthritis. N Engl J Med 365(23):2205–2219. https://doi.org/10.1056/NEJMra1004965
3. Majithia V, Geraci SA (2007) Rheumatoid arthritis: diagnosis and management. Am J Med 120(11):936–939. https://doi.org/10.1016/j.amjmed.2007.04.005
4. Smolen JS, Aletaha D, Redlich K (2012) The pathogenesis of rheumatoid arthritis: new insights from old clinical data? Nat Rev Rheumatol 8(4):235–243. https://doi.org/10.1038/nrrheum.2012.23
5. Calabro A, Caterino AL, Elefante E, Valentini V, Vitale A, Talarico R, Cantarini L, Frediani B (2016) One year in review 2016: novelties in the treatment of rheumatoid arthritis. Clin Exp Rheumatol 34(3):357–372
6. Tanaka Y (2015) Human mesenchymal stem cells as a tool for joint repair in rheumatoid arthritis. Clin Exp Rheumatol 33(4 Suppl 92):S58–S62
7. Le Blanc K, Pittenger M (2005) Mesenchymal stem cells: progress toward promise. Cytotherapy 7(1):36–45. https://doi.org/10.1080/14653240510018118
8. Dominici M, Le Blanc K, Mueller I, Slaper-Cortenbach I, Marini F, Krause D, Deans R, Keating A, Prockop D, Horwitz E (2006) Minimal criteria for defining multipotent mesenchymal stromal cells. The International Society for Cellular Therapy position statement. Cytotherapy 8(4):315–317. https://doi.org/10.1080/14653240600855905
9. Alcayaga-Miranda F, Cuenca J, Luz-Crawford P, Aguila-Diaz C, Fernandez A, Figueroa FE, Khoury M (2015) Characterization of menstrual stem cells: angiogenic effect, migration and hematopoietic stem cell support in comparison with bone marrow mesenchymal stem cells. Stem Cell Res Ther 6:32. https://doi.org/10.1186/s13287-015-0013-5
10. Gonzalez PL, Carvajal C, Cuenca J, Alcayaga-Miranda F, Figueroa FE, Bartolucci J, Salazar-Aravena L, Khoury M (2015) Chorion mesenchymal stem cells show superior differentiation, immunosuppressive, and angiogenic potentials in comparison with haploidentical maternal placental cells. Stem Cells Transl Med 4(10):1109–1121. https://doi.org/10.5966/sctm.2015-0022
11. Mitrano TI, Grob MS, Carrion F, Nova-Lamperti E, Luz PA, Fierro FS, Quintero A, Chaparro A, Sanz A (2010) Culture and characterization of mesenchymal stem cells from human gingival tissue. J Periodontol 81(6):917–925. https://doi.org/10.1902/jop.2010.090566
12. Romanov YA, Darevskaya AN, Merzlikina NV, Buravkova LB (2005) Mesenchymal stem cells from human bone marrow and adipose tissue: isolation, characterization, and differentiation potentialities. Bull Exp Biol Med 140(1):138–143
13. Crisostomo PR, Wang Y, Markel TA, Wang M, Lahm T, Meldrum DR (2008) Human mesenchymal stem cells stimulated by TNF-alpha, LPS, or hypoxia produce growth factors by an NF kappa B- but not JNK-dependent mechanism. Am J Physiol Cell Physiol 294(3):C675–C682. https://doi.org/10.1152/ajpcell.00437.2007
14. Galderisi U, Giordano A (2014) The gap between the physiological and therapeutic roles of mesenchymal stem cells. Med Res Rev 34(5):1100–1126. https://doi.org/10.1002/med.21322
15. Kyurkchiev D, Bochev I, Ivanova-Todorova E, Mourdjeva M, Oreshkova T, Belemezova K, Kyurkchiev S (2014) Secretion of immunoregulatory cytokines by mesenchymal stem cells. World J Stem Cells 6(5):552–570. https://doi.org/10.4252/wjsc.v6.i5.552
16. Aggarwal S, Pittenger MF (2005) Human mesenchymal stem cells modulate allogeneic immune cell responses. Blood 105(4):1815–1822. https://doi.org/10.1182/blood-2004-04-1559
17. Luz-Crawford P, Djouad F, Toupet K, Bony C, Franquesa M, Hoogduijn MJ, Jorgensen C, Noel D (2016) Mesenchymal stem cell-derived interleukin 1 receptor antagonist promotes macrophage polarization and inhibits B cell differentiation. Stem Cells 34(2):483–492. https://doi.org/10.1002/stem.2254

18. Luz-Crawford P, Kurte M, Bravo-Alegria J, Contreras R, Nova-Lamperti E, Tejedor G, Noel D, Jorgensen C, Figueroa F, Djouad F, Carrion F (2013) Mesenchymal stem cells generate a CD4+CD25+Foxp3+ regulatory T cell population during the differentiation process of Th1 and Th17 cells. Stem Cell Res Ther 4(3):65. https://doi.org/10.1186/scrt216
19. Spaggiari GM, Capobianco A, Abdelrazik H, Becchetti F, Mingari MC, Moretta L (2008) Mesenchymal stem cells inhibit natural killer-cell proliferation, cytotoxicity, and cytokine production: role of indoleamine 2,3-dioxygenase and prostaglandin E2. Blood 111(3):1327–1333. https://doi.org/10.1182/blood-2007-02-074997
20. Bouffi C, Bony C, Courties G, Jorgensen C, Noel D (2010) IL-6-dependent PGE2 secretion by mesenchymal stem cells inhibits local inflammation in experimental arthritis. PLoS One 5(12):e14247. https://doi.org/10.1371/journal.pone.0014247
21. Chen M, Su W, Lin X, Guo Z, Wang J, Zhang Q, Brand D, Ryffel B, Huang J, Liu Z, He X, Le AD, Zheng SG (2013) Adoptive transfer of human gingiva-derived mesenchymal stem cells ameliorates collagen-induced arthritis via suppression of Th1 and Th17 cells and enhancement of regulatory T cell differentiation. Arthritis Rheum 65(5):1181–1193. https://doi.org/10.1002/art.37894
22. Ghannam S, Bouffi C, Djouad F, Jorgensen C, Noel D (2010) Immunosuppression by mesenchymal stem cells: mechanisms and clinical applications. Stem Cell Res Ther 1(1):2. https://doi.org/10.1186/scrt2
23. Gonzalez MA, Gonzalez-Rey E, Rico L, Buscher D, Delgado M (2009) Treatment of experimental arthritis by inducing immune tolerance with human adipose-derived mesenchymal stem cells. Arthritis Rheum 60(4):1006–1019. https://doi.org/10.1002/art.24405
24. Ji S, Guo Q, Han Y, Tan G, Luo Y, Zeng F (2012) Mesenchymal stem cell transplantation inhibits abnormal activation of Akt/GSK3beta signaling pathway in T cells from systemic lupus erythematosus mice. Cell Physiol Biochem 29(5-6):705–712. https://doi.org/10.1159/000178590
25. Ohshima M, Yamahara K, Ishikane S, Harada K, Tsuda H, Otani K, Taguchi A, Miyazato M, Katsuragi S, Yoshimatsu J, Kodama M, Kangawa K, Ikeda T (2012) Systemic transplantation of allogenic fetal membrane-derived mesenchymal stem cells suppresses Th1 and Th17 T cell responses in experimental autoimmune myocarditis. J Mol Cell Cardiol 53(3):420–428. https://doi.org/10.1016/j.yjmcc.2012.06.020
26. Zappia E, Casazza S, Pedemonte E, Benvenuto F, Bonanni I, Gerdoni E, Giunti D, Ceravolo A, Cazzanti F, Frassoni F, Mancardi G, Uccelli A (2005) Mesenchymal stem cells ameliorate experimental autoimmune encephalomyelitis inducing T-cell anergy. Blood 106(5):1755–1761. https://doi.org/10.1182/blood-2005-04-1496
27. Lee JM, Jung J, Lee HJ, Jeong SJ, Cho KJ, Hwang SG, Kim GJ (2012) Comparison of immunomodulatory effects of placenta mesenchymal stem cells with bone marrow and adipose mesenchymal stem cells. Int Immunopharmacol 13(2):219–224. https://doi.org/10.1016/j.intimp.2012.03.024
28. Luz-Crawford P, Torres MJ, Noel D, Fernandez A, Toupet K, Alcayaga-Miranda F, Tejedor G, Jorgensen C, Illanes SE, Figueroa FE, Djouad F, Khoury M (2016) The immunosuppressive signature of menstrual blood mesenchymal stem cells entails opposite effects on experimental arthritis and graft versus host diseases. Stem Cells 34(2):456–469. https://doi.org/10.1002/stem.2244
29. Sullivan C, Murphy JM, Griffin MD, Porter RM, Evans CH, O'Flatharta C, Shaw G, Barry F (2012) Genetic mismatch affects the immunosuppressive properties of mesenchymal stem cells in vitro and their ability to influence the course of collagen-induced arthritis. Arthritis Res Ther 14(4):R167. https://doi.org/10.1186/ar3916
30. Takasugi K, Hollingsworth JW (1967) Morphologic studies of mononuclear cells of human synovial fluid. Arthritis Rheum 10(6):495–501
31. Janossy G, Panayi G, Duke O, Bofill M, Poulter LW, Goldstein G (1981) Rheumatoid arthritis: a disease of T-lymphocyte/macrophage immunoregulation. Lancet 2(8251):839–842

32. Udalova IA, Mantovani A, Feldmann M (2016) Macrophage heterogeneity in the context of rheumatoid arthritis. Nat Rev Rheumatol 12(8):472–485. https://doi.org/10.1038/nrrheum.2016.91
33. Stout RD, Suttles J (2004) Functional plasticity of macrophages: reversible adaptation to changing microenvironments. J Leukoc Biol 76(3):509–513. https://doi.org/10.1189/jlb.0504272
34. Tarique AA, Logan J, Thomas E, Holt PG, Sly PD, Fantino E (2015) Phenotypic, functional, and plasticity features of classical and alternatively activated human macrophages. Am J Respir Cell Mol Biol 53(5):676–688. https://doi.org/10.1165/rcmb.2015-0012OC
35. Moghaddam AS, Mohammadian S, Vazini H, Taghadosi M, Esmaeili SA, Mardani F, Seifi B, Mohammadi A, Afshari JT, Sahebkar A (2018) Macrophage plasticity, polarization and function in health and disease. J Cell Physiol. https://doi.org/10.1002/jcp.26429
36. Kennedy A, Fearon U, Veale DJ, Godson C (2011) Macrophages in synovial inflammation. Front Immunol 2:52. https://doi.org/10.3389/fimmu.2011.00052
37. Ye L, Wen Z, Li Y, Chen B, Yu T, Liu L, Zhang J, Ma Y, Xiao S, Ding L, Li L, Huang Z (2014) Interleukin-10 attenuation of collagen-induced arthritis is associated with suppression of interleukin-17 and retinoid-related orphan receptor gammat production in macrophages and repression of classically activated macrophages. Arthritis Res Ther 16(2):R96. https://doi.org/10.1186/ar4544
38. Kotake S, Yago T, Kawamoto M, Nanke Y (2012) Role of osteoclasts and interleukin-17 in the pathogenesis of rheumatoid arthritis: crucial 'human osteoclastology'. J Bone Miner Metab 30(2):125–135. https://doi.org/10.1007/s00774-011-0321-5
39. Schett G, Teitelbaum SL (2009) Osteoclasts and Arthritis. J Bone Miner Res 24(7):1142–1146. https://doi.org/10.1359/jbmr.090533
40. Hamilton JA, Filonzi EL, Ianches G (1993) Regulation of macrophage colony-stimulating factor (M-CSF) production in cultured human synovial fibroblasts. Growth Factors 9(2):157–165
41. Shigeyama Y, Pap T, Kunzler P, Simmen BR, Gay RE, Gay S (2000) Expression of osteoclast differentiation factor in rheumatoid arthritis. Arthritis Rheum 43(11):2523–2530. https://doi.org/10.1002/1529-0131(200011)43:11<2523::AID-ANR20>3.0.CO;2-Z
42. Teitelbaum SL (2000) Osteoclasts, integrins, and osteoporosis. J Bone Miner Metab 18(6):344–349
43. Garimella MG, Kour S, Piprode V, Mittal M, Kumar A, Rani L, Pote ST, Mishra GC, Chattopadhyay N, Wani MR (2015) Adipose-derived mesenchymal stem cells prevent systemic bone loss in collagen-induced arthritis. J Immunol 195(11):5136–5148. https://doi.org/10.4049/jimmunol.1500332
44. Abumaree MH, Al Jumah MA, Kalionis B, Jawdat D, Al Khaldi A, Abomaray FM, Fatani AS, Chamley LW, Knawy BA (2013) Human placental mesenchymal stem cells (pMSCs) play a role as immune suppressive cells by shifting macrophage differentiation from inflammatory M1 to anti-inflammatory M2 macrophages. Stem Cell Rev 9(5):620–641. https://doi.org/10.1007/s12015-013-9455-2
45. Uccelli A, de Rosbo NK (2015) The immunomodulatory function of mesenchymal stem cells: mode of action and pathways. Ann N Y Acad Sci 1351:114–126. https://doi.org/10.1111/nyas.12815
46. Choi H, Lee RH, Bazhanov N, Oh JY, Prockop DJ (2011) Anti-inflammatory protein TSG-6 secreted by activated MSCs attenuates zymosan-induced mouse peritonitis by decreasing TLR2/NF-kappaB signaling in resident macrophages. Blood 118(2):330–338. https://doi.org/10.1182/blood-2010-12-327353
47. Ortiz LA, Dutreil M, Fattman C, Pandey AC, Torres G, Go K, Phinney DG (2007) Interleukin 1 receptor antagonist mediates the antiinflammatory and antifibrotic effect of mesenchymal stem cells during lung injury. Proc Natl Acad Sci U S A 104(26):11002–11007. https://doi.org/10.1073/pnas.0704421104

48. Fogel LA, Yokoyama WM, French AR (2013) Natural killer cells in human autoimmune disorders. Arthritis Res Ther 15(4):216. https://doi.org/10.1186/ar4232
49. Soderstrom K, Stein E, Colmenero P, Purath U, Muller-Ladner U, de Matos CT, Tarner IH, Robinson WH, Engleman EG (2010) Natural killer cells trigger osteoclastogenesis and bone destruction in arthritis. Proc Natl Acad Sci U S A 107(29):13028–13033. https://doi.org/10.1073/pnas.1000546107
50. Lo CK, Lam QL, Sun L, Wang S, Ko KH, Xu H, Wu CY, Zheng BJ, Lu L (2008) Natural killer cell degeneration exacerbates experimental arthritis in mice via enhanced interleukin-17 production. Arthritis Rheum 58(9):2700–2711. https://doi.org/10.1002/art.23760
51. Leavenworth JW, Wang X, Wenander CS, Spee P, Cantor H (2011) Mobilization of natural killer cells inhibits development of collagen-induced arthritis. Proc Natl Acad Sci U S A 108(35):14584–14589. https://doi.org/10.1073/pnas.1112188108
52. Spaggiari GM, Capobianco A, Becchetti S, Mingari MC, Moretta L (2006) Mesenchymal stem cell-natural killer cell interactions: evidence that activated NK cells are capable of killing MSCs, whereas MSCs can inhibit IL-2-induced NK-cell proliferation. Blood 107(4):1484–1490. https://doi.org/10.1182/blood-2005-07-2775
53. Moretta L, Moretta A (2004) Unravelling natural killer cell function: triggering and inhibitory human NK receptors. EMBO J 23(2):255–259. https://doi.org/10.1038/sj.emboj.7600019
54. Almeida CR, Caires HR, Vasconcelos DP, Barbosa MA (2016) NAP-2 secreted by human NK cells can stimulate mesenchymal stem/stromal cell recruitment. Stem Cell Rep 6(4):466–473. https://doi.org/10.1016/j.stemcr.2016.02.012
55. Klareskog L, Amara K, Malmstrom V (2014) Adaptive immunity in rheumatoid arthritis: anticitrulline and other antibodies in the pathogenesis of rheumatoid arthritis. Curr Opin Rheumatol 26(1):72–79. https://doi.org/10.1097/BOR.0000000000000016
56. Hamel KM, Cao Y, Olalekan SA, Finnegan A (2014) B cell-specific expression of inducible costimulator ligand is necessary for the induction of arthritis in mice. Arthritis Rheum 66(1):60–67. https://doi.org/10.1002/art.38207
57. Abbas AK, Sharpe AH (2005) Dendritic cells giveth and taketh away. Nat Immunol 6(3):227–228. https://doi.org/10.1038/ni0305-227
58. Sennikov SV, Falaleeva SA, Shkaruba NS, Chumasova OA, Obleukhova IA, Sizikov AE, Kurilin VV (2016) Maturation and cytokine production potential of dendritic cells isolated from rheumatoid arthritis patients peripheral blood and induced in vitro. Hum Immunol 77(10):930–936. https://doi.org/10.1016/j.humimm.2016.07.005
59. Byun SH, Lee JH, Jung NC, Choi HJ, Song JY, Seo HG, Choi J, Jung SY, Kang S, Choi YS, Chung JH, Lim DS (2016) Rosiglitazone-mediated dendritic cells ameliorate collagen-induced arthritis in mice. Biochem Pharmacol 115:85–93. https://doi.org/10.1016/j.bcp.2016.05.009
60. Jiang XX, Zhang Y, Liu B, Zhang SX, Wu Y, Yu XD, Mao N (2005) Human mesenchymal stem cells inhibit differentiation and function of monocyte-derived dendritic cells. Blood 105(10):4120–4126. https://doi.org/10.1182/blood-2004-02-0586
61. Li H, Guo Z, Jiang X, Zhu H, Li X, Mao N (2008) Mesenchymal stem cells alter migratory property of T and dendritic cells to delay the development of murine lethal acute graft-versus-host disease. Stem Cells 26(10):2531–2541. https://doi.org/10.1634/stemcells.2008-0146
62. Chiesa S, Morbelli S, Morando S, Massollo M, Marini C, Bertoni A, Frassoni F, Bartolome ST, Sambuceti G, Traggiai E, Uccelli A (2011) Mesenchymal stem cells impair in vivo T-cell priming by dendritic cells. Proc Natl Acad Sci U S A 108(42):17384–17389. https://doi.org/10.1073/pnas.1103650108
63. Djouad F, Charbonnier LM, Bouffi C, Louis-Plence P, Bony C, Apparailly F, Cantos C, Jorgensen C, Noel D (2007) Mesenchymal stem cells inhibit the differentiation of dendritic cells through an interleukin-6-dependent mechanism. Stem Cells 25(8):2025–2032. https://doi.org/10.1634/stemcells.2006-0548
64. Lund FE, Randall TD (2010) Effector and regulatory B cells: modulators of CD4+ T cell immunity. Nat Rev Immunol 10(4):236–247. https://doi.org/10.1038/nri2729

65. Franquesa M, Mensah FK, Huizinga R, Strini T, Boon L, Lombardo E, DelaRosa O, Laman JD, Grinyo JM, Weimar W, Betjes MG, Baan CC, Hoogduijn MJ (2015) Human adipose tissue-derived mesenchymal stem cells abrogate plasmablast formation and induce regulatory B cells independently of T helper cells. Stem Cells 33(3):880–891. https://doi.org/10.1002/stem.1881
66. Traggiai E, Volpi S, Schena F, Gattorno M, Ferlito F, Moretta L, Martini A (2008) Bone marrow-derived mesenchymal stem cells induce both polyclonal expansion and differentiation of B cells isolated from healthy donors and systemic lupus erythematosus patients. Stem Cells 26(2):562–569. https://doi.org/10.1634/stemcells.2007-0528
67. Rosado MM, Bernardo ME, Scarsella M, Conforti A, Giorda E, Biagini S, Cascioli S, Rossi F, Guzzo I, Vivarelli M, Dello Strologo L, Emma F, Locatelli F, Carsetti R (2015) Inhibition of B-cell proliferation and antibody production by mesenchymal stromal cells is mediated by T cells. Stem Cells Dev 24(1):93–103. https://doi.org/10.1089/scd.2014.0155
68. Schena F, Gambini C, Gregorio A, Mosconi M, Reverberi D, Gattorno M, Casazza S, Uccelli A, Moretta L, Martini A, Traggiai E (2010) Interferon-gamma-dependent inhibition of B cell activation by bone marrow-derived mesenchymal stem cells in a murine model of systemic lupus erythematosus. Arthritis Rheum 62(9):2776–2786. https://doi.org/10.1002/art.27560
69. Asari S, Itakura S, Ferreri K, Liu CP, Kuroda Y, Kandeel F, Mullen Y (2009) Mesenchymal stem cells suppress B-cell terminal differentiation. Exp Hematol 37(5):604–615. https://doi.org/10.1016/j.exphem.2009.01.005
70. Che N, Li X, Zhang L, Liu R, Chen H, Gao X, Shi S, Chen W, Sun L (2014) Impaired B cell inhibition by lupus bone marrow mesenchymal stem cells is caused by reduced CCL2 expression. J Immunol 193(10):5306–5314. https://doi.org/10.4049/jimmunol.1400036
71. Qin Y, Zhou Z, Zhang F, Wang Y, Shen B, Liu Y, Guo Y, Fan Y, Qiu J (2015) Induction of regulatory B-cells by mesenchymal stem cells is affected by SDF-1alpha-CXCR7. Cell Physiol Biochem 37(1):117–130. https://doi.org/10.1159/000430338
72. Van Boxel JA, Paget SA (1975) Predominantly T-cell infiltrate in rheumatoid synovial membranes. N Engl J Med 293(11):517–520. https://doi.org/10.1056/NEJM197509112931101
73. Abdel-Nour AN, Elson CJ, Dieppe PA (1986) Proliferative responses of T-cell lines grown from joint fluids of patients with rheumatoid arthritis and other arthritides. Immunol Lett 12(5-6):329–333
74. Gravano DM, Hoyer KK (2013) Promotion and prevention of autoimmune disease by CD8+ T cells. J Autoimmun 45:68–79. https://doi.org/10.1016/j.jaut.2013.06.004
75. Petrelli A, van Wijk F (2016) CD8(+) T cells in human autoimmune arthritis: the unusual suspects. Nat Rev Rheumatol 12(7):421–428. https://doi.org/10.1038/nrrheum.2016.74
76. Engela AU, Baan CC, Litjens NH, Franquesa M, Betjes MG, Weimar W, Hoogduijn MJ (2013) Mesenchymal stem cells control alloreactive CD8(+) CD28(-) T cells. Clin Exp Immunol 174(3):449–458. https://doi.org/10.1111/cei.12199
77. Liu Q, Zheng H, Chen X, Peng Y, Huang W, Li X, Li G, Xia W, Sun Q, Xiang AP (2015) Human mesenchymal stromal cells enhance the immunomodulatory function of CD8(+) CD28(-) regulatory T cells. Cell Mol Immunol 12(6):708–718. https://doi.org/10.1038/cmi.2014.118
78. Homey B (2006) After TH1/TH2 now comes Treg/TH17: significance of T helper cells in immune response organization. Hautarzt 57(8):730–732. https://doi.org/10.1007/s00105-006-1199-3
79. Luckheeram RV, Zhou R, Verma AD, Xia B (2012) CD4(+)T cells: differentiation and functions. Clin Dev Immunol 2012:925135. https://doi.org/10.1155/2012/925135
80. Ivanov II, McKenzie BS, Zhou L, Tadokoro CE, Lepelley A, Lafaille JJ, Cua DJ, Littman DR (2006) The orphan nuclear receptor RORgammat directs the differentiation program of proinflammatory IL-17+ T helper cells. Cell 126(6):1121–1133. https://doi.org/10.1016/j.cell.2006.07.035

81. Annunziato F, Cosmi L, Santarlasci V, Maggi L, Liotta F, Mazzinghi B, Parente E, Fili L, Ferri S, Frosali F, Giudici F, Romagnani P, Parronchi P, Tonelli F, Maggi E, Romagnani S (2007) Phenotypic and functional features of human Th17 cells. J Exp Med 204(8):1849–1861. https://doi.org/10.1084/jem.20070663
82. Morita Y, Yamamura M, Kawashima M, Harada S, Tsuji K, Shibuya K, Maruyama K, Makino H (1998) Flow cytometric single-cell analysis of cytokine production by CD4+ T cells in synovial tissue and peripheral blood from patients with rheumatoid arthritis. Arthritis Rheum 41(9):1669–1676. https://doi.org/10.1002/1529-0131(199809)41:9<1669::AID-ART19>3.0.CO;2-G
83. Simon AK, Seipelt E, Sieper J (1994) Divergent T-cell cytokine patterns in inflammatory arthritis. Proc Natl Acad Sci U S A 91(18):8562–8566
84. Malfait AM, Butler DM, Presky DH, Maini RN, Brennan FM, Feldmann M (1998) Blockade of IL-12 during the induction of collagen-induced arthritis (CIA) markedly attenuates the severity of the arthritis. Clin Exp Immunol 111(2):377–383
85. Murphy CA, Langrish CL, Chen Y, Blumenschein W, McClanahan T, Kastelein RA, Sedgwick JD, Cua DJ (2003) Divergent pro- and antiinflammatory roles for IL-23 and IL-12 in joint autoimmune inflammation. J Exp Med 198(12):1951–1957. https://doi.org/10.1084/jem.20030896
86. Oppmann B, Lesley R, Blom B, Timans JC, Xu Y, Hunte B, Vega F, Yu N, Wang J, Singh K, Zonin F, Vaisberg E, Churakova T, Liu M, Gorman D, Wagner J, Zurawski S, Liu Y, Abrams JS, Moore KW, Rennick D, de Waal-Malefyt R, Hannum C, Bazan JF, Kastelein RA (2000) Novel p19 protein engages IL-12p40 to form a cytokine, IL-23, with biological activities similar as well as distinct from IL-12. Immunity 13(5):715–725
87. Vermeire K, Heremans H, Vandeputte M, Huang S, Billiau A, Matthys P (1997) Accelerated collagen-induced arthritis in IFN-gamma receptor-deficient mice. J Immunol 158(11):5507–5513
88. Sarkar S, Cooney LA, White P, Dunlop DB, Endres J, Jorns JM, Wasco MJ, Fox DA (2009) Regulation of pathogenic IL-17 responses in collagen-induced arthritis: roles of endogenous interferon-gamma and IL-4. Arthritis Res Ther 11(5):R158. https://doi.org/10.1186/ar2838
89. Lubberts E (2015) Role of T lymphocytes in the development of rheumatoid arthritis. Implications for treatment. Curr Pharm Des 21(2):142–146
90. Bettelli E, Carrier Y, Gao W, Korn T, Strom TB, Oukka M, Weiner HL, Kuchroo VK (2006) Reciprocal developmental pathways for the generation of pathogenic effector TH17 and regulatory T cells. Nature 441(7090):235–238. https://doi.org/10.1038/nature04753
91. Shevach EM (2009) Mechanisms of foxp3+ T regulatory cell-mediated suppression. Immunity 30(5):636–645. https://doi.org/10.1016/j.immuni.2009.04.010
92. Buckner JH (2010) Mechanisms of impaired regulation by CD4(+)CD25(+)FOXP3(+) regulatory T cells in human autoimmune diseases. Nat Rev Immunol 10(12):849–859. https://doi.org/10.1038/nri2889
93. Niu Q, Cai B, Huang ZC, Shi YY, Wang LL (2012) Disturbed Th17/Treg balance in patients with rheumatoid arthritis. Rheumatol Int 32(9):2731–2736. https://doi.org/10.1007/s00296-011-1984-x
94. Ehrenstein MR, Evans JG, Singh A, Moore S, Warnes G, Isenberg DA, Mauri C (2004) Compromised function of regulatory T cells in rheumatoid arthritis and reversal by anti-TNFalpha therapy. J Exp Med 200(3):277–285. https://doi.org/10.1084/jem.20040165
95. Di Nicola M, Carlo-Stella C, Magni M, Milanesi M, Longoni PD, Matteucci P, Grisanti S, Gianni AM (2002) Human bone marrow stromal cells suppress T-lymphocyte proliferation induced by cellular or nonspecific mitogenic stimuli. Blood 99(10):3838–3843
96. English K, Ryan JM, Tobin L, Murphy MJ, Barry FP, Mahon BP (2009) Cell contact, prostaglandin E(2) and transforming growth factor beta 1 play non-redundant roles in human mesenchymal stem cell induction of CD4+CD25(High) forkhead box P3+ regulatory T cells. Clin Exp Immunol 156(1):149–160. https://doi.org/10.1111/j.1365-2249.2009.03874.x

97. Tatara R, Ozaki K, Kikuchi Y, Hatanaka K, Oh I, Meguro A, Matsu H, Sato K, Ozawa K (2011) Mesenchymal stromal cells inhibit Th17 but not regulatory T-cell differentiation. Cytotherapy 13(6):686–694. https://doi.org/10.3109/14653249.2010.542456
98. Ghannam S, Pene J, Moquet-Torcy G, Jorgensen C, Yssel H (2010) Mesenchymal stem cells inhibit human Th17 cell differentiation and function and induce a T regulatory cell phenotype. J Immunol 185(1):302–312. https://doi.org/10.4049/jimmunol.0902007
99. Luz-Crawford P, Noel D, Fernandez X, Khoury M, Figueroa F, Carrion F, Jorgensen C, Djouad F (2012) Mesenchymal stem cells repress Th17 molecular program through the PD-1 pathway. PLoS One 7(9):e45272. https://doi.org/10.1371/journal.pone.0045272
100. Duffy MM, Pindjakova J, Hanley SA, McCarthy C, Weidhofer GA, Sweeney EM, English K, Shaw G, Murphy JM, Barry FP, Mahon BP, Belton O, Ceredig R, Griffin MD (2011) Mesenchymal stem cell inhibition of T-helper 17 cell- differentiation is triggered by cell-cell contact and mediated by prostaglandin E2 via the EP4 receptor. Eur J Immunol 41(10):2840–2851. https://doi.org/10.1002/eji.201141499
101. Obermajer N, Popp FC, Soeder Y, Haarer J, Geissler EK, Schlitt HJ, Dahlke MH (2014) Conversion of Th17 into IL-17A(neg) regulatory T cells: a novel mechanism in prolonged allograft survival promoted by mesenchymal stem cell-supported minimized immunosuppressive therapy. J Immunol 193(10):4988–4999. https://doi.org/10.4049/jimmunol.1401776
102. Selleri S, Dieng MM, Nicoletti S, Louis I, Beausejour C, Le Deist F, Haddad E (2013) Cord-blood-derived mesenchymal stromal cells downmodulate CD4+ T-cell activation by inducing IL-10-producing Th1 cells. Stem Cells Dev 22(7):1063–1075. https://doi.org/10.1089/scd.2012.0315
103. Chao YH, Wu HP, Wu KH, Tsai YG, Peng CT, Lin KC, Chao WR, Lee MS, Fu YC (2014) An increase in CD3+CD4+CD25+ regulatory T cells after administration of umbilical cord-derived mesenchymal stem cells during sepsis. PLoS One 9(10):e110338. https://doi.org/10.1371/journal.pone.0110338
104. Luz-Crawford P, Tejedor G, Mausset-Bonnefont AL, Beaulieu E, Morand EF, Jorgensen C, Noel D, Djouad F (2015) Glucocorticoid-induced leucine zipper governs the therapeutic potential of mesenchymal stem cells by inducing a switch from pathogenic to regulatory Th17 cells in a mouse model of collagen-induced arthritis. Arthritis Rheum 67(6):1514–1524. https://doi.org/10.1002/art.39069
105. Perez-Sayans M, Somoza-Martin JM, Barros-Angueira F, Rey JM, Garcia-Garcia A (2010) RANK/RANKL/OPG role in distraction osteogenesis. Oral Surg Oral Med Oral Pathol Oral Radiol Endod 109(5):679–686. https://doi.org/10.1016/j.tripleo.2009.10.042
106. Tsuboi H, Matsui Y, Hayashida K, Yamane S, Maeda-Tanimura M, Nampei A, Hashimoto J, Suzuki R, Yoshikawa H, Ochi T (2003) Tartrate resistant acid phosphatase (TRAP) positive cells in rheumatoid synovium may induce the destruction of articular cartilage. Ann Rheum Dis 62(3):196–203
107. Cho KA, Park M, Kim YH, Ryu KH, Woo SY (2017) Mesenchymal stem cells inhibit RANK-RANKL interactions between osteoclasts and Th17 cells via osteoprotegerin activity. Oncotarget 8(48):83419–83431. https://doi.org/10.18632/oncotarget.21379
108. Bose T (2017) Role of Immunological Memory Cells as a Therapeutic Target in Multiple Sclerosis. Brain Sci 7(11). https://doi.org/10.3390/brainsci7110148
109. Farber DL, Yudanin NA, Restifo NP (2014) Human memory T cells: generation, compartmentalization and homeostasis. Nat Rev Immunol 14(1):24–35. https://doi.org/10.1038/nri3567
110. Sanders ME, Makgoba MW, Sharrow SO, Stephany D, Springer TA, Young HA, Shaw S (1988) Human memory T lymphocytes express increased levels of three cell adhesion molecules (LFA-3, CD2, and LFA-1) and three other molecules (UCHL1, CDw29, and Pgp-1) and have enhanced IFN-gamma production. J Immunol 140(5):1401–1407
111. Sallusto F, Lenig D, Forster R, Lipp M, Lanzavecchia A (1999) Two subsets of memory T lymphocytes with distinct homing potentials and effector functions. Nature 401(6754):708–712. https://doi.org/10.1038/44385

112. Rosenblum MD, Way SS, Abbas AK (2016) Regulatory T cell memory. Nat Rev Immunol 16(2):90–101. https://doi.org/10.1038/nri.2015.1
113. Sallusto F, Geginat J, Lanzavecchia A (2004) Central memory and effector memory T cell subsets: function, generation, and maintenance. Annu Rev Immunol 22:745–763. https://doi.org/10.1146/annurev.immunol.22.012703.104702
114. Kohem CL, Brezinschek RI, Wisbey H, Tortorella C, Lipsky PE, Oppenheimer-Marks N (1996) Enrichment of differentiated CD45RBdim,CD27- memory T cells in the peripheral blood, synovial fluid, and synovial tissue of patients with rheumatoid arthritis. Arthritis Rheum 39(5):844–854
115. Cohen SB, Katsikis PD, Chu CQ, Thomssen H, Webb LM, Maini RN, Londei M, Feldmann M (1995) High level of interleukin-10 production by the activated T cell population within the rheumatoid synovial membrane. Arthritis Rheum 38(7):946–952
116. Dolhain RJ, ter Haar NT, Hoefakker S, Tak PP, de Ley M, Claassen E, Breedveld FC, Miltenburg AM (1996) Increased expression of interferon (IFN)-gamma together with IFN-gamma receptor in the rheumatoid synovial membrane compared with synovium of patients with osteoarthritis. Br J Rheumatol 35(1):24–32
117. Grom AA, Murray KJ, Luyrink L, Emery H, Passo MH, Glass DN, Bowlin T, Edwards C 3rd (1996) Patterns of expression of tumor necrosis factor alpha, tumor necrosis factor beta, and their receptors in synovia of patients with juvenile rheumatoid arthritis and juvenile spondylarthropathy. Arthritis Rheum 39(10):1703–1710
118. MacDonald KP, Nishioka Y, Lipsky PE, Thomas R (1997) Functional CD40 ligand is expressed by T cells in rheumatoid arthritis. J Clin Invest 100(9):2404–2414. https://doi.org/10.1172/JCI119781
119. Mosmann TR, Sad S (1996) The expanding universe of T-cell subsets: Th1, Th2 and more. Immunol Today 17(3):138–146
120. Smeets TJ, Dolhain R, Miltenburg AM, de Kuiper R, Breedveld FC, Tak PP (1998) Poor expression of T cell-derived cytokines and activation and proliferation markers in early rheumatoid synovial tissue. Clin Immunol Immunopathol 88(1):84–90
121. Kapranov NM, Davydova YO, Galtseva IV, Petinati NA, Drize NI, Kuzmina LA, Parovichnikova EN, Savchenko VG (2017) Effect of priming of multipotent mesenchymal stromal cells with interferon gamma on their immunomodulating properties. Biochem Biokhimiia 82(10):1158–1168. https://doi.org/10.1134/S000629791710008X
122. Laranjeira P, Pedrosa M, Pedreiro S, Gomes J, Martinho A, Antunes B, Ribeiro T, Santos F, Trindade H, Paiva A (2015) Effect of human bone marrow mesenchymal stromal cells on cytokine production by peripheral blood naive, memory, and effector T cells. Stem Cell Res Ther 6:3. https://doi.org/10.1186/scrt537
123. Ribeiro A, Laranjeira P, Mendes S, Velada I, Leite C, Andrade P, Santos F, Henriques A, Graos M, Cardoso CM, Martinho A, Pais M, da Silva CL, Cabral J, Trindade H, Paiva A (2013) Mesenchymal stem cells from umbilical cord matrix, adipose tissue and bone marrow exhibit different capability to suppress peripheral blood B, natural killer and T cells. Stem Cell Res Ther 4(5):125. https://doi.org/10.1186/scrt336
124. Krampera M, Glennie S, Dyson J, Scott D, Laylor R, Simpson E, Dazzi F (2003) Bone marrow mesenchymal stem cells inhibit the response of naive and memory antigen-specific T cells to their cognate peptide. Blood 101(9):3722–3729. https://doi.org/10.1182/blood-2002-07-2104
125. Pianta S, Bonassi Signoroni P, Muradore I, Rodrigues MF, Rossi D, Silini A, Parolini O (2015) Amniotic membrane mesenchymal cells-derived factors skew T cell polarization toward Treg and downregulate Th1 and Th17 cells subsets. Stem Cell Rev 11(3):394–407. https://doi.org/10.1007/s12015-014-9558-4
126. Hsu SC, Wang LT, Yao CL, Lai HY, Chan KY, Liu BS, Chong P, Lee OK, Chen HW (2013) Mesenchymal stem cells promote neutrophil activation by inducing IL-17 production in CD4+ CD45RO+ T cells. Immunobiology 218(1):90–95. https://doi.org/10.1016/j.imbio.2012.02.007
127. Akbar AN, Fletcher JM (2005) Memory T cell homeostasis and senescence during aging. Curr Opin Immunol 17(5):480–485. https://doi.org/10.1016/j.coi.2005.07.019

128. Ojdana D, Safiejko K, Lipska A, Radziwon P, Dadan J, Tryniszewska E (2008) Effector and memory CD4+ and CD8+ T cells in the chronic infection process. Folia Histochem Cytobiol 46(4):413–417. https://doi.org/10.2478/v10042-008-0077-5
129. Mack M, Bruhl H, Gruber R, Jaeger C, Cihak J, Eiter V, Plachy J, Stangassinger M, Uhlig K, Schattenkirchner M, Schlondorff D (1999) Predominance of mononuclear cells expressing the chemokine receptor CCR5 in synovial effusions of patients with different forms of arthritis. Arthritis Rheum 42(5):981–988. https://doi.org/10.1002/1529-0131(199905)42:5<981::AID-ANR17>3.0.CO;2-4
130. Nanki T, Lipsky PE (2000) Cytokine, activation marker, and chemokine receptor expression by individual CD4(+) memory T cells in rheumatoid arthritis synovium. Arthritis Res 2(5):415–423. https://doi.org/10.1186/ar120
131. Qin S, Rottman JB, Myers P, Kassam N, Weinblatt M, Loetscher M, Koch AE, Moser B, Mackay CR (1998) The chemokine receptors CXCR3 and CCR5 mark subsets of T cells associated with certain inflammatory reactions. J Clin Invest 101(4):746–754. https://doi.org/10.1172/JCI1422
132. Suzuki N, Nakajima A, Yoshino S, Matsushima K, Yagita H, Okumura K (1999) Selective accumulation of CCR5+ T lymphocytes into inflamed joints of rheumatoid arthritis. Int Immunol 11(4):553–559
133. Garred P, Madsen HO, Petersen J, Marquart H, Hansen TM, Freiesleben Sorensen S, Volck B, Svejgaard A, Andersen V (1998) CC chemokine receptor 5 polymorphism in rheumatoid arthritis. J Rheumatol 25(8):1462–1465
134. Liao F, Rabin RL, Smith CS, Sharma G, Nutman TB, Farber JM (1999) CC-chemokine receptor 6 is expressed on diverse memory subsets of T cells and determines responsiveness to macrophage inflammatory protein 3 alpha. J Immunol 162(1):186–194
135. Appay V, van Lier RA, Sallusto F, Roederer M (2008) Phenotype and function of human T lymphocyte subsets: consensus and issues. Cytometry A 73(11):975–983. https://doi.org/10.1002/cyto.a.20643
136. Goronzy JJ, Weyand CM (2001) Thymic function and peripheral T-cell homeostasis in rheumatoid arthritis. Trends Immunol 22(5):251–255
137. Vallejo AN, Bryl E, Klarskov K, Naylor S, Weyand CM, Goronzy JJ (2002) Molecular basis for the loss of CD28 expression in senescent T cells. J Biol Chem 277(49):46940–46949. https://doi.org/10.1074/jbc.M207352200
138. Appay V, Zaunders JJ, Papagno L, Sutton J, Jaramillo A, Waters A, Easterbrook P, Grey P, Smith D, McMichael AJ, Cooper DA, Rowland-Jones SL, Kelleher AD (2002) Characterization of CD4(+) CTLs ex vivo. J Immunol 168(11):5954–5958
139. Park W, Weyand CM, Schmidt D, Goronzy JJ (1997) Co-stimulatory pathways controlling activation and peripheral tolerance of human CD4+CD28- T cells. Eur J Immunol 27(5):1082–1090. https://doi.org/10.1002/eji.1830270507
140. Teo FH, de Oliveira RT, Mamoni RL, Ferreira MC, Nadruz W Jr, Coelho OR, Fernandes Jde L, Blotta MH (2013) Characterization of CD4+CD28null T cells in patients with coronary artery disease and individuals with risk factors for atherosclerosis. Cell Immunol 281(1):11–19. https://doi.org/10.1016/j.cellimm.2013.01.007
141. Warrington KJ, Takemura S, Goronzy JJ, Weyand CM (2001) CD4+,CD28- T cells in rheumatoid arthritis patients combine features of the innate and adaptive immune systems. Arthritis Rheum 44(1):13–20. https://doi.org/10.1002/1529-0131(200101)44:1<13::AID-ANR3>3.0.CO;2-6
142. Fasth AE, Bjorkstrom NK, Anthoni M, Malmberg KJ, Malmstrom V (2010) Activating NK-cell receptors co-stimulate CD4(+)CD28(-) T cells in patients with rheumatoid arthritis. Eur J Immunol 40(2):378–387. https://doi.org/10.1002/eji.200939399
143. Namekawa T, Snyder MR, Yen JH, Goehring BE, Leibson PJ, Weyand CM, Goronzy JJ (2000) Killer cell activating receptors function as costimulatory molecules on CD4+CD28null T cells clonally expanded in rheumatoid arthritis. J Immunol 165(2):1138–1145
144. Namekawa T, Wagner UG, Goronzy JJ, Weyand CM (1998) Functional subsets of CD4 T cells in rheumatoid synovitis. Arthritis Rheum 41(12):2108–2116. https://doi.org/10.1002/1529-0131(199812)41:12<2108::AID-ART5>3.0.CO;2-Q

145. Zhang X, Nakajima T, Goronzy JJ, Weyand CM (2005) Tissue trafficking patterns of effector memory CD4+ T cells in rheumatoid arthritis. Arthritis Rheum 52(12):3839–3849. https://doi.org/10.1002/art.21482
146. Ivanova-Todorova E, Bochev I, Dimitrov R, Belemezova K, Mourdjeva M, Kyurkchiev S, Kinov P, Altankova I, Kyurkchiev D (2012) Conditioned medium from adipose tissue-derived mesenchymal stem cells induces CD4+FOXP3+ cells and increases IL-10 secretion. J Biomed Biotechnol 2012:295167. https://doi.org/10.1155/2012/295167
147. Kyurkchiev D, Ivanova-Todorova E, Bochev I, Mourdjeva M, Kyurkchiev S (2013) Differences between adipose tissue-derived mesenchymal stem cells and bone marrow-derived mesenchymal stem cells as regulators of the immune response. In: Hayat MA (ed) Stem cells and cancer stem cells, Therapeutic applications in disease and injury, vol 10. Springer, Dordrecht, pp 71–84. https://doi.org/10.1007/978-94-007-6262-6_7
148. Meirelles Lda S, Fontes AM, Covas DT, Caplan AI (2009) Mechanisms involved in the therapeutic properties of mesenchymal stem cells. Cytokine Growth Factor Rev 20(5-6):419–427. https://doi.org/10.1016/j.cytogfr.2009.10.002
149. McGeachy MJ, Singh D, Henkel M, Moreland L (2016) Th17/TfH cells in rheumatoid arthritis: correlations with disease activity and therapy response. J Immunol 196(1 Suppl):51.23
150. Hu C, Qian L, Miao Y, Huang Q, Miao P, Wang P, Yu Q, Nie H, Zhang J, He D, Xu R, Chen X, Liu B, Zhang D (2012) Antigen-presenting effects of effector memory Vgamma9Vdelta2 T cells in rheumatoid arthritis. Cell Mol Immunol 9(3):245–254. https://doi.org/10.1038/cmi.2011.50
151. Liu X, Feng T, Gong T, Shen C, Zhu T, Wu Q, Li Q, Li H (2015) Human umbilical cord mesenchymal stem cells inhibit the function of allogeneic activated Vgamma9Vdelta2 T lymphocytes in vitro. Biomed Res Int 2015:317801. https://doi.org/10.1155/2015/317801
152. Martinet L, Fleury-Cappellesso S, Gadelorge M, Dietrich G, Bourin P, Fournié J-J, Poupot R (2009) A regulatory cross-talk between Vγ9Vδ2 T lymphocytes and mesenchymal stem cells. Eur J Immunol 39(3):752–762. https://doi.org/10.1002/eji.200838812
153. Liu R, Li X, Zhang Z, Zhou M, Sun Y, Su D, Feng X, Gao X, Shi S, Chen W, Sun L (2015) Allogeneic mesenchymal stem cells inhibited T follicular helper cell generation in rheumatoid arthritis. Sci Rep 5:12777. https://doi.org/10.1038/srep12777
154. Ueno H, Banchereau J, Vinuesa CG (2015) Pathophysiology of T follicular helper cells in humans and mice. Nat Immunol 16(2):142–152. https://doi.org/10.1038/ni.3054
155. Wang J, Shan Y, Jiang Z, Feng J, Li C, Ma L, Jiang Y (2013) High frequencies of activated B cells and T follicular helper cells are correlated with disease activity in patients with new-onset rheumatoid arthritis. Clin Exp Immunol 174(2):212–220. https://doi.org/10.1111/cei.12162
156. Liang J, Li X, Zhang H, Wang D, Feng X, Wang H, Hua B, Liu B, Sun L (2012) Allogeneic mesenchymal stem cells transplantation in patients with refractory RA. Clin Rheumatol 31(1):157–161. https://doi.org/10.1007/s10067-011-1816-0
157. Okada Y, Wu D, Trynka G, Raj T, Terao C, Ikari K, Kochi Y, Ohmura K, Suzuki A, Yoshida S, Graham RR, Manoharan A, Ortmann W, Bhangale T, Denny JC, Carroll RJ, Eyler AE, Greenberg JD, Kremer JM, Pappas DA, Jiang L, Yin J, Ye L, Su DF, Yang J, Xie G, Keystone E, Westra HJ, Esko T, Metspalu A, Zhou X, Gupta N, Mirel D, Stahl EA, Diogo D, Cui J, Liao K, Guo MH, Myouzen K, Kawaguchi T, Coenen MJ, van Riel PL, van de Laar MA, Guchelaar HJ, Huizinga TW, Dieude P, Mariette X, Bridges SL Jr, Zhernakova A, Toes RE, Tak PP, Miceli-Richard C, Bang SY, Lee HS, Martin J, Gonzalez-Gay MA, Rodriguez-Rodriguez L, Rantapaa-Dahlqvist S, Arlestig L, Choi HK, Kamatani Y, Galan P, Lathrop M, Consortium R, Consortium G, Eyre S, Bowes J, Barton A, de Vries N, Moreland LW, Criswell LA, Karlson EW, Taniguchi A, Yamada R, Kubo M, Liu JS, Bae SC, Worthington J, Padyukov L, Klareskog L, Gregersen PK, Raychaudhuri S, Stranger BE, De Jager PL, Franke L, Visscher PM, Brown MA, Yamanaka H, Mimori T, Takahashi A, Xu H, Behrens TW, Siminovitch KA, Momohara S, Matsuda F, Yamamoto K, Plenge RM (2014) Genetics of rheumatoid arthritis contributes to biology and drug discovery. Nature 506(7488):376–381. https://doi.org/10.1038/nature12873

158. Diogo D, Okada Y, Plenge RM (2014) Genome-wide association studies to advance our understanding of critical cell types and pathways in rheumatoid arthritis: recent findings and challenges. Curr Opin Rheumatol 26(1):85–92. https://doi.org/10.1097/BOR.0000000000000012
159. Gregersen PK, Silver J, Winchester RJ (1987) The shared epitope hypothesis. An approach to understanding the molecular genetics of susceptibility to rheumatoid arthritis. Arthritis Rheum 30(11):1205–1213
160. Huizinga TW, Amos CI, van der Helm-van Mil AH, Chen W, van Gaalen FA, Jawaheer D, Schreuder GM, Wener M, Breedveld FC, Ahmad N, Lum RF, de Vries RR, Gregersen PK, Toes RE, Criswell LA (2005) Refining the complex rheumatoid arthritis phenotype based on specificity of the HLA-DRB1 shared epitope for antibodies to citrullinated proteins. Arthritis Rheum 52(11):3433–3438. https://doi.org/10.1002/art.21385
161. Scally SW, Petersen J, Law SC, Dudek NL, Nel HJ, Loh KL, Wijeyewickrema LC, Eckle SB, van Heemst J, Pike RN, McCluskey J, Toes RE, La Gruta NL, Purcell AW, Reid HH, Thomas R, Rossjohn J (2013) A molecular basis for the association of the HLA-DRB1 locus, citrullination, and rheumatoid arthritis. J Exp Med 210(12):2569–2582. https://doi.org/10.1084/jem.20131241
162. Amara K, Steen J, Murray F, Morbach H, Fernandez-Rodriguez BM, Joshua V, Engstrom M, Snir O, Israelsson L, Catrina AI, Wardemann H, Corti D, Meffre E, Klareskog L, Malmstrom V (2013) Monoclonal IgG antibodies generated from joint-derived B cells of RA patients have a strong bias toward citrullinated autoantigen recognition. J Exp Med 210(3):445–455. https://doi.org/10.1084/jem.20121486
163. Buckley CD, Amft N, Bradfield PF, Pilling D, Ross E, Arenzana-Seisdedos F, Amara A, Curnow SJ, Lord JM, Scheel-Toellner D, Salmon M (2000) Persistent induction of the chemokine receptor CXCR4 by TGF-beta 1 on synovial T cells contributes to their accumulation within the rheumatoid synovium. J Immunol 165(6):3423–3429
164. Nanki T, Hayashida K, El-Gabalawy HS, Suson S, Shi K, Girschick HJ, Yavuz S, Lipsky PE (2000) Stromal cell-derived factor-1-CXC chemokine receptor 4 interactions play a central role in CD4+ T cell accumulation in rheumatoid arthritis synovium. J Immunol 165(11):6590–6598
165. Nagafuchi Y, Shoda H, Sumitomo S, Nakachi S, Kato R, Tsuchida Y, Tsuchiya H, Sakurai K, Hanata N, Tateishi S, Kanda H, Ishigaki K, Okada Y, Suzuki A, Kochi Y, Fujio K, Yamamoto K (2016) Immunophenotyping of rheumatoid arthritis reveals a linkage between HLA-DRB1 genotype, CXCR4 expression on memory CD4(+) T cells, and disease activity. Sci Rep 6:29338. https://doi.org/10.1038/srep29338
166. Hocking AM (2015) The role of chemokines in mesenchymal stem cell homing to wounds. Adv Wound Care 4(11):623–630. https://doi.org/10.1089/wound.2014.0579
167. MacDonald GI, Augello A, De Bari C (2011) Role of mesenchymal stem cells in reestablishing immunologic tolerance in autoimmune rheumatic diseases. Arthritis Rheum 63(9):2547–2557. https://doi.org/10.1002/art.30474
168. De Bari C (2015) Are mesenchymal stem cells in rheumatoid arthritis the good or bad guys? Arthritis Res Ther 17:113. https://doi.org/10.1186/s13075-015-0634-1
169. Alvaro-Gracia JM, Jover JA, Garcia-Vicuna R, Carreno L, Alonso A, Marsal S, Blanco F, Martinez-Taboada VM, Taylor P, Martin-Martin C, DelaRosa O, Tagarro I, Diaz-Gonzalez F (2017) Intravenous administration of expanded allogeneic adipose-derived mesenchymal stem cells in refractory rheumatoid arthritis (Cx611): results of a multicentre, dose escalation, randomised, single-blind, placebo-controlled phase Ib/IIa clinical trial. Ann Rheum Dis 76(1):196–202. https://doi.org/10.1136/annrheumdis-2015-208918
170. Franceschetti T, De Bari C (2017) The potential role of adult stem cells in the management of the rheumatic diseases. Ther Adv Musculoskelet Dis 9(7):165–179. https://doi.org/10.1177/1759720X17704639
171. Krampera M, Cosmi L, Angeli R, Pasini A, Liotta F, Andreini A, Santarlasci V, Mazzinghi B, Pizzolo G, Vinante F, Romagnani P, Maggi E, Romagnani S, Annunziato F (2006) Role for interferon-gamma in the immunomodulatory activity of human bone marrow mesenchymal stem cells. Stem Cells 24(2):386–398. https://doi.org/10.1634/stemcells.2005-0008

172. Liotta F, Angeli R, Cosmi L, Fili L, Manuelli C, Frosali F, Mazzinghi B, Maggi L, Pasini A, Lisi V, Santarlasci V, Consoloni L, Angelotti ML, Romagnani P, Parronchi P, Krampera M, Maggi E, Romagnani S, Annunziato F (2008) Toll-like receptors 3 and 4 are expressed by human bone marrow-derived mesenchymal stem cells and can inhibit their T-cell modulatory activity by impairing Notch signaling. Stem Cells 26(1):279–289. https://doi.org/10.1634/stemcells.2007-0454
173. Ren G, Zhang L, Zhao X, Xu G, Zhang Y, Roberts AI, Zhao RC, Shi Y (2008) Mesenchymal stem cell-mediated immunosuppression occurs via concerted action of chemokines and nitric oxide. Cell Stem Cell 2(2):141–150. https://doi.org/10.1016/j.stem.2007.11.014
174. Ren G, Zhao X, Zhang L, Zhang J, L'Huillier A, Ling W, Roberts AI, Le AD, Shi S, Shao C, Shi Y (2010) Inflammatory cytokine-induced intercellular adhesion molecule-1 and vascular cell adhesion molecule-1 in mesenchymal stem cells are critical for immunosuppression. J Immunol 184(5):2321–2328. https://doi.org/10.4049/jimmunol.0902023
175. Dorronsoro A, Ferrin I, Salcedo JM, Jakobsson E, Fernandez-Rueda J, Lang V, Sepulveda P, Fechter K, Pennington D, Trigueros C (2014) Human mesenchymal stromal cells modulate T-cell responses through TNF-alpha-mediated activation of NF-kappaB. Eur J Immunol 44(2):480–488. https://doi.org/10.1002/eji.201343668
176. Krampera M (2011) Mesenchymal stromal cell 'licensing': a multistep process. Leukemia 25(9):1408–1414. https://doi.org/10.1038/leu.2011.108
177. Croitoru-Lamoury J, Lamoury FM, Zaunders JJ, Veas LA, Brew BJ (2007) Human mesenchymal stem cells constitutively express chemokines and chemokine receptors that can be upregulated by cytokines, IFN-beta, and Copaxone. J Interf Cytokine Res 27(1):53–64. https://doi.org/10.1089/jir.2006.0037
178. He XX, Bai H, Yang GR, Xue YJ, Su YN (2009) Expression of Toll-like receptors in human bone marrow mesenchymal stem cells. Zhongguo Shi Yan Xue Ye Xue Za Zhi 17(3):695–699
179. DelaRosa O, Lombardo E (2010) Modulation of adult mesenchymal stem cells activity by toll-like receptors: implications on therapeutic potential. Mediat Inflamm 2010:865601. https://doi.org/10.1155/2010/865601
180. Raicevic G, Rouas R, Najar M, Stordeur P, Boufker HI, Bron D, Martiat P, Goldman M, Nevessignsky MT, Lagneaux L (2010) Inflammation modifies the pattern and the function of Toll-like receptors expressed by human mesenchymal stromal cells. Hum Immunol 71(3):235–244. https://doi.org/10.1016/j.humimm.2009.12.005
181. Waterman RS, Tomchuck SL, Henkle SL, Betancourt AM (2010) A new mesenchymal stem cell (MSC) paradigm: polarization into a pro-inflammatory MSC1 or an Immunosuppressive MSC2 phenotype. PLoS One 5(4):e10088. https://doi.org/10.1371/journal.pone.0010088
182. Han X, Yang Q, Lin L, Xu C, Zheng C, Chen X, Han Y, Li M, Cao W, Cao K, Chen Q, Xu G, Zhang Y, Zhang J, Schneider RJ, Qian Y, Wang Y, Brewer G, Shi Y (2014) Interleukin-17 enhances immunosuppression by mesenchymal stem cells. Cell Death Differ 21(11):1758–1768. https://doi.org/10.1038/cdd.2014.85
183. Tian J, Rui K, Tang X, Wang W, Ma J, Tian X, Wang Y, Xu H, Lu L, Wang S (2016) IL-17 down-regulates the immunosuppressive capacity of olfactory ecto-mesenchymal stem cells in murine collagen-induced arthritis. Oncotarget 7(28):42953–42962. https://doi.org/10.18632/oncotarget.10261
184. Xie J, Wang W, Si JW, Miao XY, Li JC, Wang YC, Wang ZR, Ma J, Zhao XC, Li Z, Yi H, Han H (2013) Notch signaling regulates CXCR4 expression and the migration of mesenchymal stem cells. Cell Immunol 281(1):68–75. https://doi.org/10.1016/j.cellimm.2013.02.001
185. Daynes RA, Jones DC (2002) Emerging roles of PPARs in inflammation and immunity. Nat Rev Immunol 2(10):748–759. https://doi.org/10.1038/nri912
186. Luz-Crawford P, Ipseiz N, Espinosa-Carrasco G, Caicedo A, Tejedor G, Toupet K, Loriau J, Scholtysek C, Stoll C, Khoury M, Noel D, Jorgensen C, Kronke G, Djouad F (2016) PPARbeta/delta directs the therapeutic potential of mesenchymal stem cells in arthritis. Ann Rheum Dis 75(12):2166–2174. https://doi.org/10.1136/annrheumdis-2015-208696

Chapter 12
Stem Cells and Cancer

Meredith Davis, Katja Gassner, Ruth Rodriguez-Barrueco, and David Llobet-Navas

Abstract The plasticity of cancer stem cells (CSCs) is one of the greatest challenges in cancer therapeutics. CSCs not only foster tumour development, but they also activate mechanisms of tumour immune surveillance evasion, metabolic reprogramming, and metastatic colonization. The clinical significance of CSCs involves their resistance to chemotherapy and their major role in tumour relapse after treatment. Mechanisms that are essential for the induction, maintenance, and survival of CSCs are ambiguous. A deeper understanding of what triggers CSCs, and helps them survive and spread, will provide insight into new treatment strategies in the field of clinical oncology.

Keyword Cancer stem cells

M. Davis
Institute of Genetic Medicine, Newcastle University, Newcastle upon Tyne, UK

K. Gassner · D. Llobet-Navas (✉)
Institute of Genetic Medicine, Newcastle University, Newcastle upon Tyne, UK

Bellvitge Biomedical Research Institute (IDIBELL), Gran via de l'Hospitalet, 199, L'Hospitalet de Llobregat 08908, Spain
e-mail: dllobet@idibell.cat

R. Rodriguez-Barrueco
Institute of Genetic Medicine, Newcastle University, Newcastle upon Tyne, UK

Bellvitge Biomedical Research Institute (IDIBELL), Gran via de l'Hospitalet, 199, L'Hospitalet de Llobregat 08908, Spain

Universitat de Barcelona, Facultat de Medicina i Ciències de la Salut. Departament d'Anatomia i Terapèutica Experimental. Bellvitge Biomedical Research Institute (IDIBELL), Gran via de l'Hospitalet, 199, Universitat de Barcelona, Facultat de Medicina i Ciències de la Salut, L'Hospitalet de Llobregat 08908, Spain

R. Delgado-Morales (ed.), *Stem Cell Genetics for Biomedical Research*,
https://doi.org/10.1007/978-3-319-90695-9_12

Abbreviations

ABC	ATP-binding cassette
ADC	Antibody-drug conjugate
ALDH1	Aldehyde dehydrogenase 1
ALL	Acute lymphoblastic leukaemia
AML	Acute myeloid leukaemia
BCL-2	B-cell lymphoma-2
CML	Chronic myeloid leukaemia
CSC	Cancer stem cell
DRP1	Dynamin-related protein 1
EMT	Epithelial-mesenchymal transition
ESC	Embryonic stem cell
GSC	Glioma stem cell
HIF	Hypoxia-inducible factor
HNSCC	Head and neck squamous cell carcinoma
HSC	Haematopoietic stem cell
iPSC	Induced pluripotent stem cell
LSC	Leukemic stem cell
MET	Mesenchymal-epithelial transition
miRNA	MicroRNA
mtDNA	Mitochondrial DNA
OXPHOS	Oxidative phosphorylation
PKM	Pyruvate kinase muscle
PDK1	Pyruvate dehydrogenase kinase 1
ROS	Reactive oxygen species
TF	Transcription factor

12.1 Introduction to CSCs

In studies of acute myeloid leukaemia (AML), stem cell biologist John Dick and his team identified a subset of patient-derived AML cells with greater proliferative ability than the remaining populations [1]. This subset of cells displayed the same surface markers ($CD34^{+}CD38^{-}$) as haematopoietic stem cells (HSCs).

By transplanting human AML cells into immune-deficient mice (SCID), the team found that the cells displaying similarities to HSCs were able to generate tumours. The remaining AML populations did not possess tumour-initiating capacity and thus failed to form tumours [2]. Subsequent transplantation studies showed the regenerative power of CSCs in multiple solid tumours including colon [3], colorectal [4], and brain [5]. Injection of just 200 stem-like ($CD44^{+}CD24^{-/low}$) breast cancer cells into SCID mice resulted in tumour formation, whilst injecting thousands of cells with alternative phenotypes did not. Importantly, the tumour consisted of both stem-like ($CD44^{+}$) tumorigenic cells and non-tumorigenic cancer cells with different phenotypes. This indicates that CSCs give rise to heterogeneous popula-

Box 12.1 Definitions

The **cell of origin** acquires the first oncogenic hit or tumorigenic transformation within a tissue.

A **tumour-initiating cell** refers to the cell of origin and arises from a mutated adult stem cell or somatic cell that has undergone dedifferentiation.

Cancer stem cells propagate the growth and regeneration of a tumour. These cells are uniquely tumorigenic in that they give rise to both differentiated cancer cells and self-renewing, indefinitely proliferative cells.

Stem-like cancer cells display a degree of cancer stem cell properties despite incomplete dedifferentiation into a pluripotent state. Transient acquisition of stem-like traits may result from cellular plasticity within a tumour.

Cellular plasticity relates to the flexibility and adaptability of cells to change their identity or specialized function.

Circulating tumour cells represent cells (non-stem, stem-like, and/or cancer stem cells) that have entered the bloodstream after detaching from the primary tumour.

Cells that survive circulation and exit the bloodstream are **metastasis-initiating cells**, which colonize and form tumours within secondary sites.

tions of cells [6]. These studies highlight a key difference between CSCs and mature cancer cells – tumorigenic potential. CSCs are able to self-renew and maintain themselves in an undifferentiated state. They can also produce heterogeneous populations of non-stem cancer cells that form the bulk of a tumour. These CSC properties are analogous with that of normal stem cells. However, normal stem cells function to seed new tissue in a tightly regulated manner, whereas CSCs seed tumours in an abnormal, uncontrollable manner [7].

12.1.1 Isolation of CSCs

Several groups have focused their efforts on the description and characterization of stem cell markers, resulting in the development of multiple techniques to isolate CSCs. The basis for most CSC identification and isolation protocols is either stem cell marker expression or functional characteristics. Methods are not universal for specific types of cancer, and thus it is important to apply a combination of techniques for isolating and validating CSC populations.

Stem cell markers Differential expression and activity of normal stem cell proteins by CSCs have been identified and are commonly used to detect and isolate CSCs (Table 12.1). It has been shown that they are often glycosylated, which is a post-translational modification frequently altered in CSCs [8].

Standardized fluorescence-activated cell sorting (FACS) and immunostaining methods are commonly used to study stem cells as most of the canonical markers

Table 12.1 Markers for identifying CSCs

Cancer	CSC Markers	References
Breast	$CD44^+/CD24^{-/Low}$, $ALDH1^+$	[6, 45]
CNS	$CD44^+$, $CD133^+$	[5, 46]
Colon	$CD44^+$, $CD133^+$, $CD166^+$, $CD24^+$, $EpCAM^+$, $ALDH1^+$	[4, 44, 47–49]
Endometrial	$CD133^+$	[50–52]
Oesophagus	$CD44^+$, $CD24^+$, $CD133^+$, $ABCG2^+$, $CXCR4^+$, $ALDH1^+$	[53]
Ewing's	$CD133^+$	[54]
HNSCC	$CD44^+$, $CD133^+$, $ALDH^+$	[55, 56]
Kidney	$CD105^+$, $CD133^+$, $CD44^+$	[57–61]
Liver	$CD44^+$, $CD133^+$, $CD90^+$, $CD13^+$, $EpCAM^+$	[62]
Lung	$CD44^+$, $CD133^+$, $CD166^+$, $ALDH1^+$	[63]
Melanoma	$CD20^+$, $CD133^+$, $CD271^+$, $ABCB5^+$	[64, 65]
Ovarian	$CD44^+$, $CD133^+$, $CD24^+$, $CD117^+$, $EpCAM^+$, $ALDH1^+$	[66, 67]
Pancreas	$CD44^+/CD24^+$, $CD133^+$, $EpCAM^+$, $ALDH1^+$	[68, 69]
Prostate	$CD44^+/CD24^-$, $CD133^+$, $\alpha 2\beta 1^{high}$, $ALDH1^+$	[70, 71]
Stomach	$CD44^+$, $CD44V8\text{-}10^+$, $CD133^+$, $CD24^+$, $CD54^+$, $CD90^+$, $CD49f^+$, $CD71^+$, $EpCAM^+$, $ALDH1^+$	[72]
Thyroid	$CD133^+$, $CD44^+CD24^-$, $ALDH1^+$, $CD133^+$	[73, 74]
AML	$CD34^{+/-}$, $CD38^{+/-}$, $CD90^{-/+}$, $CD123^+$, $CD45RA^+$, $CD33^+$, $CD13^+$, $CD44^+$, $CD96^+$, $CD47^+$, $CD32^+$, $CD25^+$, $CLL1^+$, $TIM3^+$	[75]
MM	$CD138^-$, $CD19^+$, $CD27^+$, $CD34^-$, $CD138^-$	[76]
B-ALL	$CD34^+$, $CD10^-$, $CD19^{+/-}$, $CD38^-$	[77, 78]
T-ALL	$CD34^+$, $CD4^-$, $CD7^-$	[79]

ALDH aldehyde dehydrogenase, *ALL* acute lymphoblastic leukaemia, *AML* acute myeloid leukaemia, *CNS* central nervous system, *CSC* cancer stem cell, *HNSCC* head and neck squamous cell carcinoma, *MM* multiple myeloma

are surface proteins. Additionally, enzymatic assays to measure the activity of aldehyde dehydrogenase 1 (ALDH1) are used to identify subsets of CSCs. Some of the most widely used CSC markers are the following:

- CD44 (HCAM, PGP1): A transmembrane glycoprotein with roles in adhesion, cell migration, and differentiation [9]. It activates tyrosine kinase receptors, which enhance cell proliferation and survival [10, 11]. CD44 also participates in the intravasation of cells during metastasis [12].
- CD133 (prominin-1): A transmembrane glycoprotein that organizes plasma membrane topology and maintains its lipid structure [13].
- ALDH1: An intracellular enzyme whose activity relates to stem cell proliferation, differentiation, and chemotherapy resistance [14]. The ALDEFLUOR assay and immunostaining with a specific antibody are techniques to measure ALDH1 activity.

The use of stem cell markers is often the chosen method for CSC isolation as the obtained population is homogenous, and the protocol is highly specific. A limitation to this method is the necessity to use a combination of markers for detecting tissue-specific CSCs [15]. Additionally, the complex protocol for dissociation of solid tumours could damage the cells or the surface markers, compromising their detection by specific antibodies [16, 17]. Other disadvantages include the silencing of genes in certain tissues, the presence of splicing variants, and potential differences in detection when using different methods [15].

Side population The side population method is based on the fact that CSCs overexpress several members of the multidrug resistance (MDR) and ATP-binding cassette (ABC) transporters family. The overexpression of these receptors confers the secretion of genotoxic chemicals by CSCs, resulting in chemotherapy resistance [18]. In the same way, CSCs are able to expel Hoechst Dye or Rhodamine 123 with a high efficiency, allowing the identification and isolation of the unstained stem population (side population) by FACS [19–22]. This method has been used to isolate stem-like cells from patients with AML [23], chronic myeloid leukaemia (CML) [24], hepatocellular carcinoma [25], and ovarian cancer [26, 27]. The side population in flow-cytometry has also been used for glioma [28], lung [20, 29], and breast cancer cell lines [30]. The main advantages of this method are its simplicity and usage in several tissues as it does not require a tissue-specific marker. Limitations include low purity and specificity of the side population cells, as well as dye toxicity [17].

Sphere-forming assays CSCs form non-adherent clusters called tumour spheres when grown in suspension and special culture conditions. In such culture conditions, non-malignant or differentiated cells die. Also, it has been shown that tumour cells expressing drug resistance proteins are able to slowly grow and differentiate in sphere-forming media [31, 32]. Purified CSCs from different tumours including prostate [33], brain [5], breast [34], glioblastoma [35], colon [36], stomach [37], melanoma [38], and pancreas [39] have been proven to generate tumour spheres. The sphere-forming assay allows the easy establishment of in vitro CSC cultures to analyse their properties and test therapeutic strategies [34]. However, the obtained spheres are heterogeneous, and the efficiency of the technique can show large differences due to small protocol changes. It is, therefore, important to use the appropriate media and conditions for each cell type [31, 34, 40].

Validation A single CSC should be able to generate a tumour when subcutaneously injected into an immune-compromised mouse. The in vivo limited dilution assay is used to estimate the frequency of purified CSCs in a population [31]. Different doses of cells are implanted into the immune-deficient mice to determine their tumorigenic ability. Subsequently, the cells may be extracted and reinjected to analyse their self-renewal capacity in serial tumour studies [41–44].

12.1.2 CSC-Driven Heterogeneity

Heterogeneity takes place at distinct levels and is marked by changes in cell morphology, molecular profile, surface marker expression, proliferation rate, and long-term replication. Intra-tumour heterogeneity refers to a tumour containing several subpopulations of cells. Each subpopulation carries different combinations of genetic and epigenetic abnormalities, which determine their function, division pattern, and survival. Inter-tumour heterogeneity describes molecular variations occurring between different tumours located in the same organ. Classification of tumours into different subtypes supports the design of novel treatment strategies, many of which target specific genetic or epigenetic abnormalities present within individual tumour types. Additional clinical concerns such as relapse, drug resistance, and metastasis, however, indicate a need to discover more vulnerabilities within tumours and cancer cells [80]. Two main theories aim to describe how heterogeneity arises during tumour initiation and growth: the clonal evolution model and the CSC model [81].

The clonal evolution model In the clonal evolution model (also referred to as the stochastic model), each cancer cell possess equal tumorigenic potential. Here, heterogeneity arises due to stochastic intrinsic and extrinsic signals that result in randomly occurring mutations [82]. In this model, tumorigenesis initially relies on the replicative selection of an abnormal cell that contains a series of mutations. As the mutated cells accumulate, there is an increasing probability of tumour suppressors switching off and oncogenes switching on within the same cell. With each cell cycle, a progressively heterogeneous population of abnormal cells ultimately transition from a benign (non-invasive) to malignant (invasive) phenotype [83]. The clonal model follows rules of Darwinian evolution in that selection of clones with the highest tumorigenic potential would survive and generate progeny with distinct mutations [84, 85]. In this scenario, intrinsic factors such as surface markers expressed by CSCs are not predictors of stemness or tumour-initiating ability [84].

The cancer stem cell model: classical theory (hierarchical model) The CSC model (also referred to as the hierarchical model) provides insights to the origin and heterogeneity of CSCs [85, 86]. In this model, subpopulations of non-stem cancer cells and CSCs are inherently distinct and can vary in terms of tumorigenic potential, longevity, and division [87]. Although it is difficult to determine the exact cellular origin of CSCs within a tumour, similarities between stem cells and CSCs have led to the hypothesis that some CSCs arise from adult stem or progenitor cells that undergo mutational transformation [87, 88]. This classical, unidirectional theory of the CSC model suggests that alterations in stem cell regulatory pathways influence the self-renewal and tumorigenic potential of mutated stem or progenitor cells [87, 89]. In a hierarchically organized tumour, a small population of CSCs can self-replicate to support the long-term growth of a tumour and give rise to progenitors that have diminished clonogenic capacity. Additionally, they can produce a larger population of terminally differentiated cancer cells that form the bulk of tumours. The CSC hierarchy infers that the acquisition of stemness in CSCs is rigid and nonreversible [90]. Under the classical theory, sorting of cancer cells

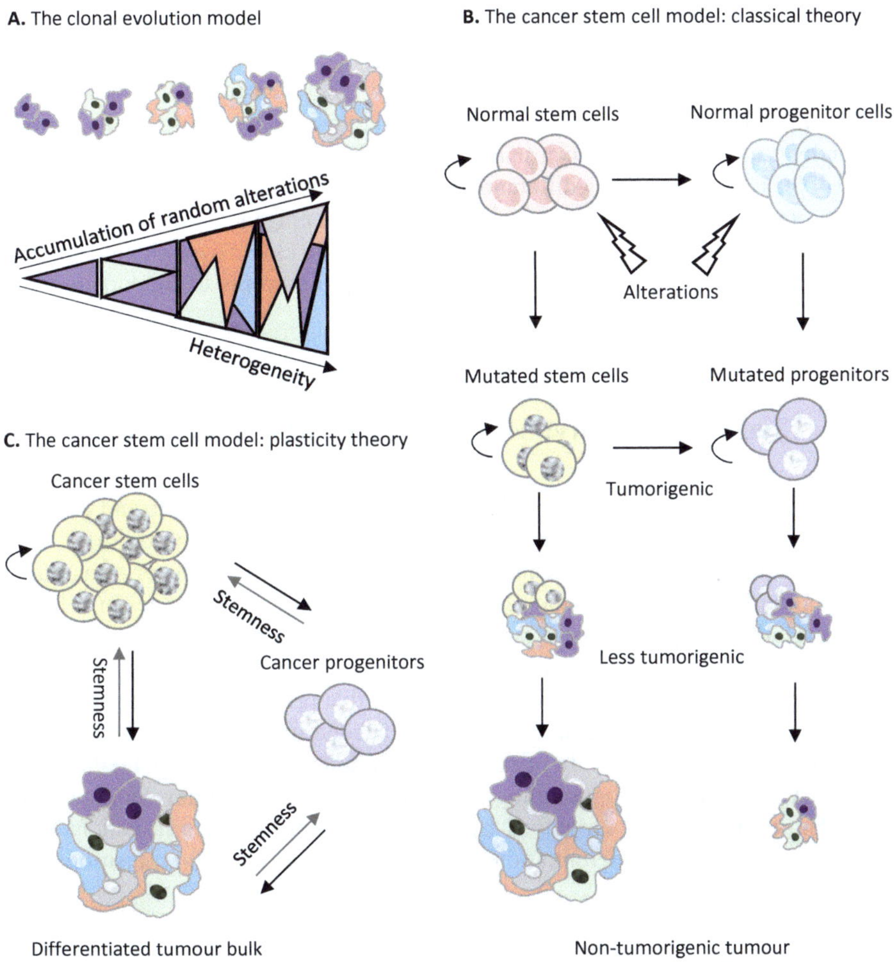

Fig. 12.1 Models of tumourigenesis and plasticity. Models of tumorigenesis and plasticity. (**a**) The clonal evolution model describes a tumour in which heterogeneity results from the accumulation of stochastic mutations. (**b**) In the CSC model, the classical theory refers to a tumour hierarchy in which mutated stem or progenitor cells have the capacity to self-replicate and give rise to differentiated cancer cells. Repopulation of tumour cells is dependent upon the presence of CSCs. (**c**) In the CSC model, the plasticity theory describes a tumour in which differentiated cells reacquire stemness through the process of dedifferentiation. The CSC population is self-supporting, and repopulates the *tumor* bulk during tumour growth. Conversely, *the differentiated compartment* can repopulate a depleted CSC pool via dedifferentiation, which parallels with emerging properties of stemness

according to cell surface markers is possible and may determine tumorigenic potential [81, 85, 86].

The cancer stem cell model: plasticity theory The plasticity theory describes a bidirectional hierarchy in which cancer cells can transition up and down the tumour hierarchy (i.e. transition between states of stemness and differentiation). In this scenario, CSCs may arise from differentiated cells that reacquire properties of stemness through oncogenic signalling or deregulations in epigenetic mechanisms [91]. The plasticity theory combines concepts from both the clonal evolution model and hierarchical CSC model and includes dedifferentiating events to support its basis (Fig. 12.1).

12.2 Changing Identity: Mechanisms of CSC Plasticity

The ability of some cancer cells to interconvert between a CSC state and a differentiated state provides a link between the CSC model and stochastic models. If cells can dedifferentiate to replenish the adult stem cell population or transdifferentiate to replace mature somatic cells in response to injury or changes in cell fate signalling, then why maintain a stem cell population in the first place? Cellular conversions require drastic changes in naturally occurring transcriptional and epigenetic regulation (e.g. chromatic modifications) that predispose cells to malignant transformation [92]. Reprogramming of cellular identity also involves dysregulation of oncogene expression. In principle, maintaining a stem cell pool reduces the frequency of cellular reprogramming events and thus the risk of developing cancer [93]. Three mechanisms employed by CSCs involving changes in cell fate are dedifferentiation, transdifferentiation, and epithelial-mesenchymal transitions.

12.2.1 Dedifferentiation

Dedifferentiation is the process by which somatic, progenitor or multipotent cells reprogram into an unspecialized, stem-like state without changing the underlying DNA sequence [94]. Takahashi and Yamanaka demonstrated that forced expression of four pluripotent transcription factors (TFs) (OCT-4, SOX2, C-MYC, and KLF4) is sufficient to convert somatic cells into induced pluripotent stem cells (iPSCs), in a process called somatic cell reprogramming [95]. These pluripotent cells maintain the ability to self-renew indefinitely and give rise to most cell types present in the three germ layers (endoderm, ectoderm, and mesoderm) [96]. As somatic cell reprogramming presents many similarities to the dedifferentiating events frequently observed in tumours, this process can be used to study multiple steps during tumour progression [97–99]. Noteworthy, OCT-4, Lin28, SOX2, and Nanog are active in embryonic stem cells (ESCs) and also highly expressed in CSCs [100].

12.2.2 Transdifferentiation

Transdifferentiation is the process by which a differentiated somatic cell of a particular lineage (endoderm, mesoderm, or ectoderm) converts into another specialized cell type [101]. This mechanism of cell plasticity differs from dedifferentiation in that the transition does not require reversion into a pluripotent state [93]. Physiological transdifferentiation occurs naturally during regeneration of the retina and in response to liver injury, and multiple studies have demonstrated the importance of cellular plasticity in these processes [102–105]. Artificial induction of transdifferentiation in vitro *or* in vivo requires ectopic expression of certain TFs, chromatin modulators, or protein kinases. A predictive algorithm has been developed to identify key sets of TFs that induce transdifferentiation in different cell types [106]. The myoblast determination (*MYOD*) gene was the first transcriptional activator identified for transdifferentiation and persists as the only TF capable of inducing transdifferentiation without the co-expression of other TFs [107].

Box 12.2 The Defining Properties of Cancer Stem Cells
Cancer stem cells (CSCs) are robust cells endowed with longevity. The most accredited features of CSCs, which are analogous to stem cells, are self-renewal (symmetrical division) and differentiation (asymmetrical division). Differential expression of stem cell markers is a common method for CSC isolation. A key test in declaring that a subpopulation of cells possesses CSC properties is the transplantation of cells into immunocompromised mice. CSCs will seed new tumours and are able to maintain a primary or metastatic tumour bulk. CSCs also have an inherent resistance to conventional therapies. The overall plasticity of CSCs to modify cell identity, metabolic pathways, immune responses, and micro-environmental signals supports their defining properties.

The implications of transdifferentiation in cancer relate to cancer-initiating events and metaplasia [93]. Changes in cell fate require activation of genes specific to the new identity and suppression of genes specific to the old identify. Signalling pathways that mediate changes in cell identity may be oncogenic or dysregulated. For instance, NOTCH signalling and the oncogenic Yes-associated protein (YAP) mediate transdifferentiation of hepatocytes into biliary cells, and this transition is associated with tumour formation in the liver [108, 109]. The contribution of CSCs to tumour lineage plasticity and angiogenesis involves transdifferentiation. Transdifferentiation is also a particularly important event in the regulation of glioblastoma stem cells (GSCs), which are able to transition into endothelial cells that facilitate the formation of tumour blood vessels [110, 111]. A recent study indicates that necrosis stimulates transdifferentiation of GSCs into a migratory phenotype, which is mediated by the enzyme transglutaminase 2 [112].

12.2.3 Epithelial and Mesenchymal Transitions

An issue critical to understanding how CSCs escape the primary tumour and resist therapies whilst the bulk of the tumour cannot is the plasticity of the CSC state. The identification of programmes that transiently endow non-CSCs with stem-like traits, referred to as epithelial-mesenchymal transition (EMT), has presented an additional challenge in eliminating CSCs.

EMT overview The epithelium consists of organized and symmetric layers of cells. Loss of polarity and cellular adhesion enable epithelial cells to shift apart and transdifferentiate towards a mesenchymal phenotype. The weak intracellular adhesion of mesenchymal cells facilitates their increased mobility and invasion into the bloodstream. Enhanced cell motility also requires structural alterations that include degradation of the basal membrane and matrix remodelling. In embryonic morphogenesis, EMT is a crucial programme that enables the dedifferentiation and transdifferentiation of cells throughout development [113, 114]. In the context of cancer, aberrant reactivation of EMT contributes to the initiation and migration of tumour cells [115]. Aside from enhanced metastatic ability, activation of EMT endows epithelial cancer cells with stem-like properties and resistance to chemotherapy [116, 117]. Moreover, EMT is associated with poor clinical outcome [118–120].

Epithelial cancer cells at the invasive front of tumours gain mesenchymal traits in order to travel to secondary sites. However, many metastatic tumours maintain cells in an epithelial state and lack mesenchymal markers such as N-cadherin. This is due to activation of a subsequent programme referred to as the mesenchymal-epithelial transition (MET). MET enables mobile mesenchymal cells to transition back to the epithelial state, during which they reacquire differentiated features that allow colonization within the invaded tissue. Cells that form the bulk of the primary or metastatic tumours are thought to undergo MET [121–123].

The transition between cell identities in cancer is under the influence of multiple mechanisms, including dysregulated transcriptional pathways and aberrant expression of non-coding RNAs. EMT-inducing TFs consist largely of three protein families: SNAIL, TWIST, and ZEB. Considered a master regulator of EMT, SNAIL (*SNAI1* gene) is essential for downregulating E-cadherin during embryonic development and plays a similar role during tumour progression. Cells at the invasive front of tumours express endogenous SNAIL, TWIST, and ZEB, all of which repress the transcription of *E-cadherin* [124, 125]. Moreover, Twist induces stemness by targeting BMI1, a polycomb protein that supports stem cell self-renewal [126, 127]. Various groups have reported that the expression of SNAIL, TWIST, and ZEB positively correlates with features of stemness in multiple cancers [119, 128–131]. For example, in vitro and in vivo upregulation of TWIST and SNAIL paralleled with the expression of pluripotent surface markers and mammosphere formation in breast cancer studies [129, 132]. It is worth noting that EMT does not always equate with stemness as TWIST has been shown to support stemness via mechanisms separate from those which activate EMT [133]. Aside from TFs, non-coding RNAs can also regulate the expression of genes and proteins involved in EMT and MET [134]. MicroRNAs such as miR-200 and miR-205 repress the transcription of *ZEB1* and *ZEB2* mRNA [128, 135, 136]. Loss of miR-205 in aggressive breast cancer subtypes, for example, positively correlated with upregulation of NOTCH2 and ZEB, and the acquisition of stem cell properties [137].

Intermediate EMT induces stemness in cancer The EMT programme is active in non-CSCs, indicating that plasticity in cancer is not exclusive to CSCs. In principle, this would also suggest that preliminary properties of stemness are irrelevant for metastatic growth [121]. Both EMT and MET are permissive for the initiation and progression of cells to metastatic sites [138]. It is less clear, however, the degree to which EMT induces stemness in cancer cells and the stages at which stemness acquisition provides an advantage (i.e. initiation or metastasis). The identification of stem-like cancer cells that exhibit an intermediate EMT has led some groups to propose that transient stemness is acquired during invasion and migration of cancer cells [139]. In an intermediate EMT, cancer cells express both mesenchymal and epithelial genes. This is a feature that has been observed in breast cancer circulating tumour cells (CTCs) [140]. During an intermediate EMT, the shift towards a mesenchymal or epithelial state is in progress, but the conversion into one state or the other is not yet complete. Cancer cells acquire stemness during the transition, but any stem-like properties obtained are lost once the cells commit to either the epithelial or mesenchymal fate [141, 142]. An intermediate EMT, as opposed to adopting a complete mesenchymal identity, facilitates MET and thus metastatic growth [143, 144]. Cancer cells cannot divide or differentiate during migration, which also supports the link between stemness and intermediate EMT.

Transient expression of EMT-TFs facilitates intermediate EMTs. The homeobox factor PRRX1, for example, induces EMT but must be downregulated for cancer cells to acquire stemness, suggesting an antagonistic relationship between EMT and stemness [145]. Similarly, transient expression of TWIST, followed by its deactivation, induces an intermediate EMT that primes cells for stemness. Constitutive expression of TWIST or SNAIL inhibits metastatic growth, suggesting that the TFs are important for inducing EMT but must be switched off for cells to acquire stemness [146–149]. Intermediate EMTs have been identified in ovarian and breast cancer, and are associated with poor clinical outcome in patients [120]. Coordinated transient activation of TWIST and upregulation of miR-424 induced an intermediate EMT in breast cancer cells. Yet, metastases from patients and mouse xenograft tumours presented decreased levels of miR-424 compared to the primary tumours, suggesting that downregulation of the miRNA enables MET [149, 150]. These findings demonstrate that the acquisition of stem-like traits in cancer depends on the expression of transcriptional EMT/MET regulators, and that these regulators are under the control of upstream non-coding RNAs such as miRNAs.

Altogether, these studies support the hypothesis that an intermediate EMT induces stemness in differentiated epithelial cancer cells to facilitate tumour growth and dissemination. An alternative possibility is that tumour growth persists due to epithelial cancer cells that do not require EMT in order to migrate and invade.

Lineage tracing analysis in lung and pancreatic cancer indicated that metastatic spread can occur without EMT activation [116, 117]. In a mouse model of pancreatic cancer, deletion of SNAIL and TWIST did not affect tumour progression, suggesting that EMT may not be an essential effector of metastasis in pancreatic tumours [116]. Non-EMT cells have also been shown to enter the bloodstream and colonize in secondary sites, whilst EMT cells mediated degradation of the basement membrane and facilitated the mobility of non-EMT cells [151].

Some groups have proposed that the relationship between stemness and EMT is not all or nothing. In the context of cancer, EMT may be best viewed as a spectrum in which stemness is acquired when epithelial and mesenchymal genes are co-expressed [141, 152]. Additionally, distinct EMT signatures can originate within a single tumour, and some EMT phenotypes may be exclusive from stemness. In a model of spontaneous EMT, stem-like properties were analysed using phenotypically distinct single cell clones derived from a prostate cancer cell line. Assessment of stem cell features indicated that just one out of the two EMT-active clones displayed stem-like traits. The second EMT-active clone was less stem-like than the EMT-negative clone [153]. They did not determine whether any clones displayed an intermediate EMT phenotype, however.

12.3 Metabolic Reprogramming

To advance, cancer cells must fuel their unrestricted growth and accommodate for the rapid replication of cellular components by readjusting physiological mechanisms of glycolysis and oxidative phosphorylation (OXPHOS). Altered metabolism is a hallmark of cancer, and properties unique to the CSCs pool may lie within their metabolic profiles. Therefore, CSC metabolism is a plausible strategy for targeting CSCs specifically [154].

Glycolytic features of CSCs CSCs that primarily use aerobic glycolysis (i.e. the Warburg effect) to obtain energy characteristically produce and express higher levels of lactate and glycolytic enzymes [155, 156]. Aerobic glycolysis yields two molecules of ATP per glucose, which may seem inefficient in comparison to the production of 36 ATP via OXPHOS (Fig. 12.2). However, non-stem and CSCs benefit from utilizing this pathway because the rate of ATP production is much faster

Fig. 12.2 (continued) to pump hydrogen ions through the mitochondrial membrane via ATP synthase channels. Fig. 12.2 (continued) ATP synthase uses the protons to synthesize the conversion of adenosine diphosphate (ADP) to ATP. Complete oxidation of a glucose molecule via mitochondrial respiration produces around 36 ATP molecules [176, 177]. In the absence of oxygen (anaerobic respiration,) glycolytic pyruvate is reduced to lactate, which can then feed into the Krebs cycle within the mitochondria [178]. HK hexokinase, ADP adenosine diphosphate, ATP adenosine triphosphate, PFK-1 phosphofructosekinase-1. PKM pyruvate kinase muscle, LDH lactate dehydrogenase, PDH pyruvate dehydrogenase, NADH reduced form of nicotinamide adenine dinucleotide, FADH reduced form of flavin adenine dinucleotide, ROS reactive oxygen species, OXPHOS oxidative phosphorylation, ESCs embryonic stem cells, iPSCs induced pluripotent stem cells

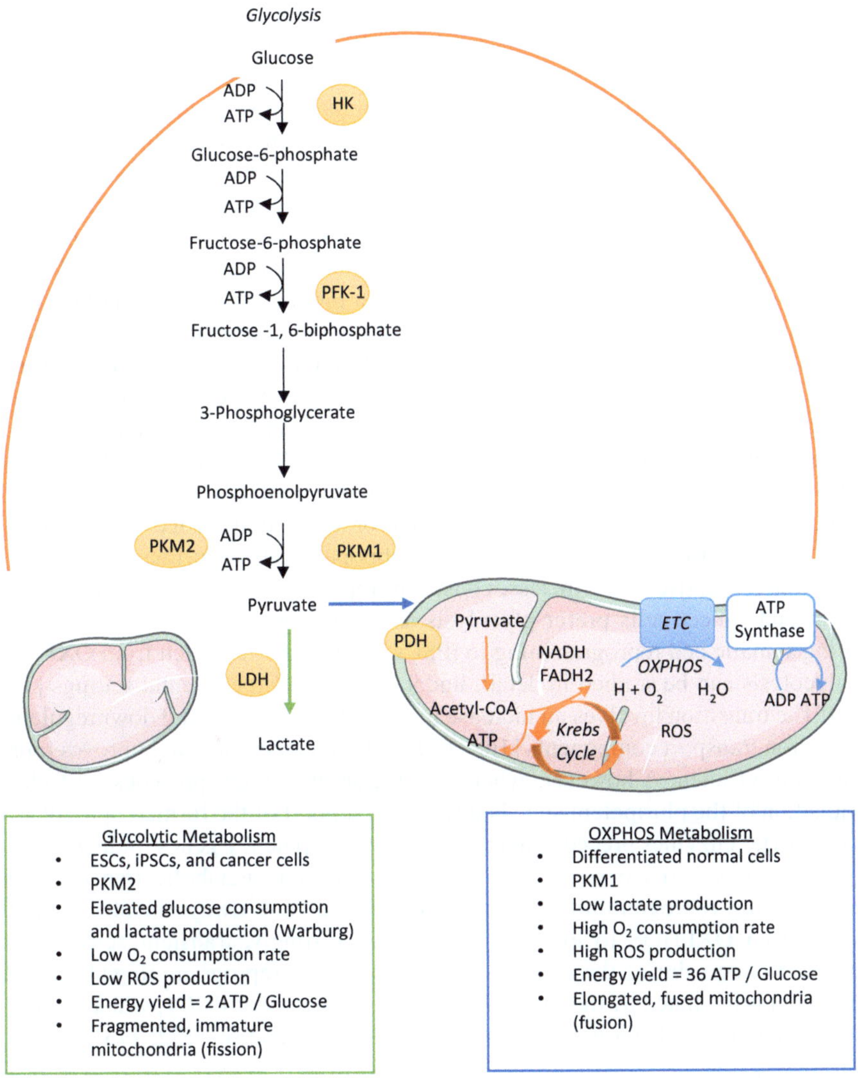

Fig. 12.2 Overview of glycolysis and oxidative phosphorylation. The initial event of cellular respiration is glycolysis – the catalytic breakdown of glucose to pyruvate. This metabolic pathway takes place with or without oxygen (O_2) and yields a net production of two adenosine triphosphates (ATP) and two pyruvate molecules [174]. When O_2 is present, differentiated, slowly proliferating cells obtain their energy by oxidizing glycolytic pyruvate in the mitochondria. Generation of acetyl-CoA triggers a series of reactions in the mitochondrial matrix through a process known as the Krebs cycle (or tricarboxylic acid cycle). Electron carriers (NADH and FADH2) produced during the Krebs cycle help generate the energy needed to power ATP synthesis [175]. Within the inner mitochondrial membrane, energy carried throughout the electron transport chain (ETC) ultimately passes to an O_2 molecule. O_2 reduces to water and carbon dioxide in a process called oxidative phosphorylation (OXPHOS). Coupling of OXPHOS with the ETC creates the energy needed

[157, 158]. Upregulation of glucose transporters leads to a high glycolytic flux and glucose depletion from adjacent tissues [159]. Importantly, glycolysis yields many precursor molecules that serve as building blocks to support division or proliferation. For example, stem cells and cancer cells express the M2 splice variant of pyruvate kinase muscle (PKM). PKM2 prevents lactate from entering the Krebs cycle, allowing cells to use glucose metabolites for generating nucleotides, proteins, and lipids [160, 161]. An additional mechanism of metabolic reprogramming involves the control of pyruvate dehydrogenase kinase-1 (PDK1). Activation of PDK1 supports cancer cells in hypoxic conditions by indirect inhibition of the TCA cycle [162]. Interestingly, PDK isoform 1 is critical for regulating conversion of pyruvate to acetyl-CoA. The kinase enzyme is essential for breast cancer stem cell homeostasis and metastatic colonization of the liver [163, 164]. Other glycolytic enzymes such as lactate dehydrogenase (LDH) and glucose-6-phopshate dehydrogenase are also upregulated in tumours and contribute to maintenance and expansion of CSCs.

Many studies provide support that certain types of CSCs upregulate glycolysis in comparison to their differentiated progeny. However, conflicting reports (due to differences in experimental settings, models, and/or the sheer heterogeneity of cancer) add ambiguity to classifications of CSC subpopulations as primarily glycolytic rather than oxidative. Nonetheless, consistent reports provide evidence that ESCs, iPSCs, and cancer cells prefer glycolysis as a primary energy source [165–167]. During somatic cell reprogramming to iPSCs, the metabolic switch from OXPHOS to glycolysis can be studied in detail, under controlled experimental settings [168, 169]. The transition involves an increase in glycolytic enzymes and downregulation of electron transport components [168, 170]. A high glycolytic rate improves reprogramming efficiency; however, whether a metabolic switch precedes or follows acquisition of the pluripotent state is unclear. In support of the former, upregulation of glycolytic enzymes precedes upregulation of pluripotency genes, indicating that glycolysis may activate stemness [170]. However, the metabolic transition that occurs during reprogramming is complex and does not always involve a single switch from OXPHOS to glycolysis [171]. Interestingly, transient activation of OXPHOS occurs during the early stages of somatic cell reprogramming [171, 172]. This event is mediated by the TF ESRRB and its co-activator PGC1, which are critical for the transient burst in OXPHOS during iPSC generation [172, 173].

Mitochondrial features of CSCs Mitochondrial assets and behaviours relating to CSC metabolism include, alterations in oxygen consumption rate, generation of reactive oxygen species (ROS), survival mechanisms (autophagy and apoptosis), and mitochondrial fission and fusion. Mitochondria are primary producers of ROS, which are signalling byproducts of cellular respiration [177]. Glycolytic CSCs present lower levels of ROS compared to cells that obtain their energy from OXPHOS. Utilizing glycolysis is beneficial for CSCs as low ROS production favours the stem-like phenotype by reducing DNA damage and telomere shortening

[177, 179, 180]. Mitochondria respond to metabolic demands through structural processes of fission (division) and fusion (joining). Fused mitochondria have elongated, tubular morphology (Fig. 12.2). Fusion also enhances OXPHOS and mitochondrial ATP production. Fission, or mitochondrial fragmentation, triggers degradation of defective mitochondria (mitophagy) [181]. Similar to ESCs and iPSCs, CSCs have less mature, fragmented mitochondria compared to differentiated somatic cells. Other mitochondrial-related CSC features include low levels of mitochondrial DNA (mtDNA) copy number, biogenesis, and ATP production from mitochondrial pathways [182, 183]. Additionally, dysregulation of fission genes contributes to the upkeep of CSCs. Dynamin-related protein 1 (DRP1) is a pro-fission protein that plays a role in activating glycolysis and MET during somatic cell reprogramming. Invasive breast cancers present higher levels of DRP1 [184, 185], and phosphorylation of the fission protein also correlates with poor prognosis and growth of tumour-initiating cells in glioma [186]. CSCs may be able to control proportioning of old and young mitochondria during division. In a study of stem-like cells from human breast tumours, stemness was maintained by daughter cells receiving a higher proportion of young mitochondria than old [187].

CSCs have greater resistance to mitochondrial-mediated cell death mechanisms, and apoptotic signals are dysregulated in CSCs. Certain CSC populations, such as pancreatic, generate energy via OXPHOS, which parallels with increases in oxygen consumption rate [188]. Some CSCs prefer to use other mitochondrial pathways, such as fatty acid oxidation (FAO), to obtain their energy. For instance, NANOG reprograms mitochondrial respiration by activating FAO genes and repressing OXPHOS genes in liver CSCs [189]. These studies revealed mitochondrial inhibition as a vulnerability of CSCs, which demonstrates that distinct metabolic profiles of CSCs versus non-CSCs may indicate the effectiveness of certain treatments.

Metabolic flexibility: CSCs that transition between OXPHOS and glycolysis Certain CSCs can switch their metabolic pathways according to changes in the surrounding environment, such as microenvironment acidity and nutrient availability. Metabolic plasticity of CSCs has been demonstrated in GSCs. GSCs from neurospheres relied on OXPHOS, whilst differentiated glioma cells from monolayers were more glycolytic. GSCs had higher levels of ATP compared to progenitor cells but were able to switch to ATP production via OXPHOS upon 2-deoxyglucose treatment, a glycolysis inhibitor. After treatment with oligomycin, an inhibitor of mitochondrial ATP synthesis, GSCs were able to generate ATP from glycolysis [190]. The flexibility of CSCs in glioblastoma highlights the robustness and ability of CSCs to adjust under harsh conditions. A dynamic metabolism may help CSCs survive under stressed conditions or resist treatments such as chemo or radiotherapy (Table 12.2).

Table 12.2 Metabolic phenotypes of CSCs

Cancer	Model	CSC phenotype	CSC identifier	Targeting strategy	Refs.
Breast	Human tissues	Glycolytic	SP^+	Glycolytic inhibition by 2-DG and upregulation of ROS	[191]
	MDA-MB-231 MCF-7	Glycolytic	$CD44^+CD24$, $ALDH^+$, SP^+	PDK1 inhibition Glycolytic inhibition by aspirin	[164]
	MMTV-Wnt-1 mammary tumours	Glycolytic	$CD49f^{high}Epcam^{low}$	PDH activation	[192]
	SUM159PT MDA-MB-436	OXPHOS	$ALDH^+$ or SP^+	Pharmacological inhibition of HiF-1	[193]
GMB	Malignant primary tumours U87MG GBM-176	Adaptable	$CD133^+$, SP^+, low proteasome activity	Co-inhibition of OXPHOS and glycolysis	[190, 194]
	Primary cultures from patients xenotransplantation model	OXPHOS	$CD34^+CD34^+CD38^-$	Inhibition of mitochondrial protein translation	[195]
CML	Leukemic human bone marrow	OXPHOS	ROS low	BCL-2 inhibition	[196]
AML	PDX-derived tumour tissue	OXPHOS	SP^+, $CD133^+$	Metformin Menadione MYC inhibition	[188]
HCC	Xenograft	FAO	$CD133^+CD49f^+$	NANOG inhibition Sorafenib	[189]
Lung	A549	OXPHOS	SP^+	Not determined	[197]
Pancreatic	Patient-derived xenografts	OXPHOS	$CD133^+$, SP^+	Co-inhibition of the KRAS pathway and OXPHOS	[198]

2-DG 2-deoxyglucose, *ALDH* aldehyde dehydrogenase, *CSC* cancer stem cell, *EpCAM* epithelial cell adhesion molecule, *FAO* fatty acid oxidation, *MMTV* mouse mammary virus tumour, *OXPHOS* oxidative phosphorylation, *PDH* pyruvate dehydrogenase, *PDX* patient-derived xenograft, *ROS* reactive oxygen species, *SP* spheroids

12.4 CSC Immunogenicity (or Lack Thereof)

Immunogenicity describes the extent to which an immune response is elicited against pathogens or antigens [199]. Activation of cell death mechanisms in tumours is dependent upon the ability to elicit an immune response. Whilst immunogenic non-CSCs are eliminated by therapies or innate immune responses, CSCs lack immunogenicity and consequently resist immune attacks, therapeutic agents, and radiotherapy.

Immunoediting The immune system recognizes and interacts with cancer cells through a process called immunoediting. A model describing immunoediting in cancer consists of three phases: elimination, equilibrium, and escape [200]. In the elimination (or immune surveillance) phase, the presence of oncogenic mutations in differentiated cells triggers an innate immune response. This antitumour response is dependent upon interferon gamma (IFN-γ), a cytokine that initiates pro-apoptotic, anti-proliferative, and angiostatic mechanisms. In attempt to remove mutated cells and the emerging tumour, natural killer (NK) cells trigger production of IFN-γ. IFN-γ recognizes the transformed cells and induces cytokine production at the tumour site, which subsequently recruits dendritic cells (DCs) to the tumour. DCs are responsible for presenting tumour antigens to lymphocytes, which identify the antigen-positive tumour cells and induce cell death [201]. The elimination response is less likely to recognize oncogenic mutations in poorly immunogenic cells, such as CSCs. Asymmetric division of a CSC gives rise to a proliferative daughter cell that lacks immunogenicity, and a differentiated daughter cell that is immunogenic. Both cells contain the same tumour variants but the differentiated daughter cell is subject to immune surveillance, whilst the proliferative cell remains undetected and maintains the CSC pool. Equilibrium occurs as a balance between the elimination of immunogenic cells and the growth of anti-immunogenic cells. This latency phase may persist over an extended period in which CSCs are ultimately capable of generating a clinically detectable tumour [202].

Immunological molecules expressed by CSCs CSCs lack immunogenicity, which allows them to escape the elimination phase, whilst also employing immune cells to support their own development. Immunogenic properties displayed by CSCs may be dysregulated or absent in comparison to non-CSCs. For example, resistance to immune effectors such as T-cell lymphocytes and natural killer (NK) cells is partly due to the lack of major histocompatibility complex (MHC) molecules expressed by CSCs. MHCs present cell surface antigens for recognition by T cells, which triggers an immune response and destruction of cancer cells [203, 204]. Similar to ESCs, tumours and CSCs exhibit lower levels of MHCs, allowing them to escape immune surveillance. HER2, for example, is an oncogene that supports CSC populations and facilitates immune evasion by downregulating MHC molecules and eliciting anti-apoptotic signals [205–209]. Additional immunological molecules associated with CSC resistance include the B-cell lymphoma-2 (BCL-2) family [210]. BCL-2 are anti-apoptotic proteins that help regulate the balance between cell death and cell survival. BCL-2 elicits pro-survival effects by blocking the release of apoptotic proteins such as cytochrome c from the mitochondria [211]. In CSCs, BCL-2 is elevated and protects them from chemotherapeutic agents by inducing resistance to T cells and NK cells [212]. In addition to aberrant expression of immune molecules, CSCs secrete immunosuppressive molecules, such as TGF-B, that induce stem cell properties including self-renewal and migration [213, 214].

Immune cell regulation of CSC properties CSCs and tumorigenesis are promoted by immune cells such as tumour-associated macrophages (TAMs), DCs, myeloid-derived suppressor cells (MDSCs), and T regulatory cells. The relationship between CSCs and immune cells largely relates to activation of EMT, drug resistance, and the expansion of CSC numbers. Production of TGF-B by TAMs, for instance, has been shown to induce EMT and properties of stemness [215]. TAMs also rely on ZEB1 for their cancer-promoting roles, supporting their link to EMT [216]. CSCs are not only resistant to conventional therapies but can actually benefit from them. Cancer cells targeted by chemotherapy secrete molecules that expand CSC numbers [217]. During chemotherapy, the release of interleukins by cancer cells stimulates proliferation of CSCs. Interleukin-mediated immunosuppression has been shown to increase CSC populations in colon, liver, prostate, and breast cancer [215, 218–220]. The cytokine IL-6 that is crucial for initiating an immune response, for example, has been shown to stimulate CSCs and contribute to chemotherapy resistance [221].

12.5 The Microenvironment and CSC Maintenance

Signals from the tissue microenvironment (TME) influence the activity and state of CSC properties including, EMT, immunogenicity, and altered metabolism. Likewise, CSCs can manipulate the TME to release factors in favour of the CSC phenotype. Within the TME, CSCs create and reside in specialized, protective locations called niches [222]. Similar to normal stem cells, the CSC niches maintain the stem pool by signalling self-renewal and support tissue growth via differentiation. Disseminated prostate cancer cells, for instance, can hijack the haematopoietic stem cell niche to facilitate their expansion [223]. The TME and CSCs corroborate together to maintain a networking system that promotes the survival, migration, and therapeutic resistance of CSCs. Understanding this network and identifying potential windows for its disruption may facilitate new strategies for treating cancer and eliminating CSCs [224].

12.5.1 Components of the Microenvironment

Whilst the core components of the TME are common between different types of CSCs, the proportion of components may vary depending on the type, origin, and stage of cancer. Each component of the TME nevertheless contributes to the survival, maintenance, and metastatic potential of CSCs. The main constituents of the TME fall within three classes: haematopoietic cells, mesenchymal cells, and non-cellular components such as the extracellular matrix (ECM) [19].

Haematopoietic cells Lymphocytes and myelogenous cells originate from haematopoietic stem cells in the bone marrow. Lymphocytes consist of T cells, B cells, and NK cells, whilst those of myeloid lineage include macrophages, neutrophils, and MDSCs [19]. TAMs have an active role in both the initiation and advancement of cancer [225]. TAMs release inflammatory cytokines that mediate the transformation and survival of neoplastic cells. For example, IL-6 is a cytokine that promotes stemness via JAK/STAT3 signalling [226]. TAMs are able to suppress the immune response by secreting factors that recruit cytotoxic T cells. Milk fat globule-EGF-8 (MFG-E8) is a growth factor involved in immune escape, angiogenesis, and phagocytosis. TAMs release high amounts of MFG-E8, which enables CSCs with features such as quiescence and chemoresistance [227]. TAMs also produce stem cell factor (SCF), a cytokine that enhances cell proliferation and self-renewal, and is expressed in human-derived tumours [228, 229]. An underlying mechanism of tumour growth in GMB involves paracrine signalling, which is mediated by TAM-secreted pleiotrophin (PTN). In GSCs, binding of PTN to its receptor, PTPRZ1, was associated with poor prognosis [230].

Mesenchymal cells Mesenchymal stem cells (MSCs), fibroblast, and endothelial cells originate from the mesenchyme [19]. Located in the bone marrow, MSCs mediate the differentiation of surrounding haematopoietic stem cells. They may also differentiate into cell types of different lineages and thus have a functional role in tissue regeneration [231]. Tumour-associated MSCs (TA-MSCs) are MSCs that have migrated to the tumour stroma. TA-MSCs are distinct from MSCs generated from bone marrow or adipose tissue because they release higher levels of factors that help regulate stem cell activity. Among these factors are bone morphogenetic proteins (BMPs). In the presence of IL-1B and TNF, naïve bone marrow MSCs (BM-MSCs) convert into TA-MSCs and promote tumour growth [232]. Interleukins produced from cancer cells promote stem-like phenotypes. IL-1, for example, recruits MSCs and triggers their secretion of prostaglandin E2 (PGE2). Together, PGE2 expression and IL-1 signalling induce activation of B-catenin signalling, thus promoting a CSC phenotype [233]. Additional chemokines secreted by cancer cells and MSCs include CXC-chemokine ligand 12 (CXCL12). CXCL12 receptor, CXCR4, has been used to identify migratory subpopulations of CSCs [234].

Human growth factor (HGF) secreted by surrounding myofibroblasts induced Wnt signalling in colon cancer, which maintained the CSC phenotype and restored stemness in differentiated cells [9]. Cancer-associated fibroblasts (CAFs) are the most abundant cell type of the tumour stroma. CAFs are heterogeneous in that they arise from multiple origins that differ between various cancers. CAFs can promote tumour initiation and metastasis through multiple mechanisms such as degradation of the ECM [235], or by assisting tumour metabolism through secretion of metabolites such as lactate or ketone [236]. It has been reported that CAFs themselves undergo profound metabolic rewiring to support glycolysis in a process involving downregulation of IDH3α by miR-424 [237]. CAFs-specific markers include vimentin, platelet-derived growth factor-α (PDGFα), fibroblast-associated protein (FAP), α-smooth muscle actin (α-SMA), and fibroblast-specific protein 1 (FSP1) [232]. Chemotherapy can activate colorectal CAFs to stimulate secretion of interleukins and thus promote the renewal of CSCs [238]. They may also emerge from transdifferentiating endothelial and epithelial cells or via the EMT process [239].

Endothelial cells can mediate the interaction between CSCs and extracellular components of the tumour. By secreting factors into the surrounding TME, endothelial cells promote stem-like behaviours including transdifferentiation, EMT, and angiogenesis. Vasculature niches are important for the maintenance and promotion of CSCs [240]. Angiogenesis requires the recruitment of endothelial progenitor cells. Glioblastoma stem cells secret VEGF-A, which enhances angiogenic potential of endothelial cells [241]. In squamous cell carcinoma, endothelial growth factor secreted by endothelial cells activated the EMT programme and expression of stem cell markers [242]. Selective targeting of tumour-associated endothelial cells decreased the CSC population in head and neck tumours [243].

The extracellular matrix The ECM is the most prominent noncellular component of the TME, and ECM remodelling is considered a hallmark of cancer [19, 244]. The ECM consists of macromolecule proteins, glycoproteins, and polysaccharides that help regulate cell behaviours, such as differentiation and migration. Hyaluronan, for example, is an abundant polysaccharide within the ECM shown to promote stem cell renewal, and its expression positively correlated with oncogenic signalling and metastasis of CSCs [245, 246]. Tissue architecture affects differentiation and proliferative behaviour of cells. ECM stiffness, cell shape, and collagen abundance act as mechanical cues from the microenvironment that augment the progression and invasion of cancer [247]. For instance, biopsies from aggressive breast cancer subtypes (i.e. Her2 and basal-like) present increased stiffness of the stroma and collagen-enriched ECM compared to luminal subtypes. ECM dynamics are under the control of cellular TME components such as macrophages, which positively correlates with ECM stiffness [248]. For CSCs, the ECM is an important component for retaining and recruiting cells [249].

12.5.2 Hypoxia and Acidic pH

A hypoxic TME results from the rapid proliferation of cancer cells that exceeds the availability of oxygen supplied by blood vessels [250]. Whilst the survival of cancer cells requires oxygen, CSC properties and niches are stimulated by hypoxic environments [251]. Hypoxia and low acidity of the TME correlate with changes in metabolism, expression of pluripotent TFs, induction of EMT, and resistance to anticancer therapies [250]. Hypoxia-inducible factors (HIF) 1 and 2 are TFs that control gene expression in response to hypoxic conditions and/or oncogene signalling [250]. HIF-1a mediates the metabolic shift to glycolysis in cancer cells and iPSCs by upregulating glycolytic enzymes, such as PKM2 and PDK1 [252, 253]. PKM2 is directly targeted and activated by HIF-1a in cancer cells and can also enhance transcriptional activity of HIF-1a via a positive feedback loop [254]. The HIF1a-PKM2 complex induces IL-6 secretion and an inflammatory response in macrophages, and the production of lactate during a metabolic shift to glycolysis [255]. HIF-2 regulates the pluripotency of stem and/or differentiated cells by directly binding to OCT-4 [256]. In GSCs, hypoxia stimulates the release of stem cell marker CD44, which binds to HIF-2a to promote hypoxia-induced stemness [257]. Stimulation of CSC properties by HIF TFs has been demonstrated in ovarian

[258], breast [259, 260], pancreatic [261], and brain cancers [262]. Hypoxic microenvironments also induce EMT and thus stimulate the development of stem-like phenotypes and metastatic potential of cancer [261].

12.5.3 Metastasis of CSCs

Metastasis is the migratory behaviour displayed by both CSCs and non-CSCs, which spread from the primary tumour in phases of intravasation, transportation, and extravasation. CSCs are robust and have a better chance of surviving each phase of dissemination compared to non-stem cancer cells. Therefore, CSCs also have the highest potential to colonize within distant organs and reinitiate tumour growth. Distinct subpopulations of migratory CSCs have been isolated from pancreatic [234], breast [263], and colorectal tumours [130], among others. Circulating tumour cells from primary breast cancer patients are enriched with markers of stemness (EPCAM, CD44, CD47, and MET) and give rise to liver, lung, and bone metastases when injected into immunocompromised mice [264]. Metastases may consist of CSCs originating from the primary tumour or cells originating from the local tumour site that have dedifferentiated into CSCs. In both cases, cues from the microenvironment guide the metastatic behaviours of CSCs. The plasticity of metastatic cells relates to the capability of entering a dormant state, which contributes to the recurrence of high-grade tumours. Such plasticity is characteristic of CSCs and thus raises the possibility that CSCs are the driving force of metastases. In primary colon tumours, metastases formation was traced back to a subpopulation of tumour-initiating cells with self-renewal abilities [265], supporting the idea that stem-like cancer cells initiate metastatic events.

12.6 Therapeutic Targeting of CSCs

Many conventional chemo- and radio-therapies target proliferative cells, within the bulk of a tumour. The plasticity of CSCs partly refers to their ability to adopt transient, quiescent states that enable survival during patient treatment. The long-term survival of stem cells is partly due to their quiescent state [266]. Quiescence refers to the G_0, or non-cycling, reversible phase in which cells reside until signalled to re-enter the cell cycle and resume a proliferative state [267, 268]. Quiescent and slow-cycling subpopulations of CSCs are often responsible for cancer regeneration following treatment [269–272]. One possible explanation for cancer re-emergence involves mechanisms by which chemotherapy reactivates quiescent CSCs [273]. In certain tissues such as the intestine, chemotherapy may eliminate highly proliferative CSCs, whereas differentiated, slow-cycling cancer cells survive. Differentiated cancer cells that survive treatment may subsequently dedifferentiate or generate stem-like progeny that replenish CSC populations [274, 275]. Quiescence, therefore, contributes to the survival and plasticity of both stem-like and non-stem cancer cells and is an important consideration for the design and implementation of therapies aiming to eradicate CSCs. In a similar scenario, selective targeting of CSCs may not be an efficient strategy for cancers that follow the plasticity CSC model opposed to the unidirectional hierarchy theory [90]. According to the plasticity model, differentiated cells or progenitors may repopulate a depleted CSC pool [276, 277] (also see Sect. 12.2 of this book chapter).

An additional challenge relating to CSC-targeted therapeutics is the fact that CSCs share many of their properties with normal stem cells (e.g. self-renewal and potential to differentiate). Signalling pathways (e.g. Hedgehog, NOTCH, and Wnt) that regulate CSCs are also active in normal adult stem cell populations [43, 278, 279]. Moreover, the expression of CSC markers is common to normal stem cells and is not always tissue-specific [280]. Therapies targeting these properties of CSCs may entail potential harmful side effects: damaging the normal adult stem cell populations.

Despite the hurdles imposed by CSC plasticity, a variety of promising strategies aim to eliminate CSCs. Approaches include (1) immunotherapy and antibody-drug conjugates (ADC), (3) inhibition of essential CSC regulatory pathways, (3) epigenetic therapy, and (4) targeting the microenvironment [81, 90, 281]. Clinical approaches for targeting CSCs are summarized in Table 12.3.

Table 12.3 Therapeutic targeting of CSC

Intervention	Mechanism	Cancer	Aim	Clinical trial identifier
Drug: Gemcitabine hydrochloride Drug: Vismodegib	Inhibits hedgehog signalling	Pancreatic	Evaluate number of CSCs in biopsies collected before and after treatment with vismodegib	NCT01195415
Drug: PRI-724	Inhibits Wnt signalling	Myeloid malignancies	Evaluate safety and tolerability of doses	NCT01606579
Novel drug MK-0752 followed by the chemotherapeutic docetaxel	Inhibits Notch signalling	Breast	Evaluate BCSC markers in serial tumour biopsies	NCT00645333
Biological CSC:DC vaccine	Induces immunogenicity of CSC and thus T-cell recognition	Ovarian Glioblastoma Lung Nasopharyngeal Hepatocellular	Determine antibody-binding specificity and immunogenecity in vitro and isolate $ALDH^{high}$ and $ALDH^{low}$ from specimens	NCT02178670 NCT00846456 NCT02084823 NCT02115958 NCT02089919
Dietary supplement: Fursultiamine in conjunction with chemo-radiotherapy	Repression of: OCT-4 SOX-2 NANOG ABCB1 ABCG2	Oesophageal	Evaluate stem cell markers in clinical specimens before and after treatment	NCT02423811

Immunotherapy and antibody-drug conjugates CSCs present gene expression profiles, surface markers, and antigens that are distinct from the bulk of the tumour [282]. Immunological targeting of surface markers or antigens expressed by CSCs is therefore a rational approach for cancer treatments, and represents a basis for the discovery of clinically approved drugs. Strategies for direct targeting of surface markers include immunotherapy, ADCs, and antibody-based immunotoxins [283]. Antibodies blocking cell surface proteins or conjugated to small compounds reportedly diminish the self-renewal capacity of $CD44^{+}$ cells in AML, CML, breast, head and neck, and pancreatic cancer cell lines, and impair tumorigenesis in $CD44^{+}$ xenografts [283]. Translation of monoclonal antibody strategies into the clinic has been challenging due to the lack of known CSC-specific markers. Treatments that eliminate CSCs also target normal stem cells and thus cause severe side effects in patients, and have limited efficacy [281]. Alternative, more specific, strategies may involve the use of bi- or multi-specific antibodies, as well as ADCs, as they target CSC-enriched surface markers. Catumaxomab, for example, is a bispecific, trifunctional antibody approved for the treatment of $CD133^{+}/Epcam^{+}$ malignant ascites in epithelial cancers [283]. Several ADCs are under clinical trial investigation [285], but it is unclear whether any target CSCs specifically [90].

Inhibition of CSC regulatory pathways Wnt/β-catenin, Notch, and Sonic Hedgehog (SHH) signal transduction pathways mediate self-renewal and are dysregulated in different types of cancer. Inhibition of one or more of these stem cell regulatory pathways is an additional strategy for CSC elimination [281].

High-throughput screening (HTS) of compounds that target intracellular components of Wnt, Notch, and SHH pathways has been challenging since CSCs represent a small cell population and their propagation in vitro has been difficult [19]. Nevertheless, some studies using HTS provide proof of principle that selectively killing CSCs is feasible [286] [19, 281]. Several inhibitors targeting Wnt signalling are under clinical trial investigation for the treatment of breast and pancreatic cancers (e.g. LGK974 inhibits porcupine, an enzyme involved in maturation of Wnt proteins) [281]. Inhibition of Notch1 signalling by epigallocatechin gallate (EGCG) in head and neck squamous cell carcinoma decreased the expression of CSC markers and repressed sphere formation. In combination with cisplatin, EGCG also inhibited tumour formation in xenograft models. Improved therapeutic response to cisplatin might be due to EGCG-induced downregulation of ABCC2 and ABCG2 expression, as ABC efflux transporters are highly expressed in CSCs and contribute to chemotherapeutic resistance. Agents reducing drug efflux, in combination with conventional chemotherapy, are thus an attractive treatment strategy [281].

Epigenetic therapy The regulation of CSC plasticity also involves epigenetic modifications, which are heritable changes in gene function and activity that do not alter the underlying DNA sequence [154]. The epigenetic route to targeting CSCs is complex, consisting of changes in transcription, DNA methylation, histone modifications, and miRNA regulation. Epigenetic approaches to CSC elimination include differentiation therapy and pharmacological inhibition of certain epigenetic modifiers [19].

The concept of differentiation therapy began in the 1970s and stems from in vitro studies showing that certain chemicals such as dimethyl sulfoxide [289] and retinoic acid [290] forced differentiation in haematopoietic cells. Inducing CSCs or progenitor cancer cells to differentiate into mature cancer cells increases their proliferation rate and thus sensitivity to anti-proliferative chemotherapies [19]. Treatment with all-*trans* retinoic acid (ATRA) induced the differentiation of leukemic promyelocytes to mature granulocytes and, in combination with cytotoxic chemotherapy, began as a standard treatment for acute promyelocytic leukaemia (APL) in the 1990s [19, 291]. Nearly two decades later, a team at Harvard University identified ATRA's target. The drug directly binds and inhibits prolyl-isomerase PIN1, an epigenetic regulator. In addition to leukaemia, PIN1 mediates oncogenic signalling in multiple solid tumours such as breast [292]. Moreover, PIN1 regulates normal and cancerous mammary stem cell properties such as self-renewal and replication. Mechanistically, PIN1 blocks FBXW7α E3 ubiquitin ligase from binding to NOTCH-1 and NOTCH-4 receptors and thus sustains NOTCH-dependent CSC maintenance [292–294]. The application of ATRA is currently limited to haematological malignancies.

However, deciphering the mechanisms of action of successful and already approved drugs such as ATRA may facilitate drug re-purposing for targeted treatment of CSCs in additional heterogeneous, solid cancers such as breast.

Inhibitors of histone-modifying enzymes such as suberoylanilide hydroxamic acid (SAHA), an inhibitor of a histone deacetylase (HDAC), induced differentiation of human breast [295] and endometrial cancer cells [296]. Methyltransferase inhibitors represent an additional strategy for epigenetic targeting of CSCs through inhibition of transcriptional regulators such as zeste homolog 2 (EZH2), which is overexpressed in CSCs [297]. E7438 is an EZH2 inhibitor currently in phase 1/2 trials for advanced solid tumours and B-cell lymphomas. Acriflavine is an FDA and EPA approved drug that inhibits the epigenetic regulation of HIF-1a, a TF that supports CSC plasticity. In vivo and in vitro studies provide conclusive evidence (i.e. potency, safety, and optimal route of administration) for the use of acriflavine in gliomas, which may provide a basis for jumpstarting a clinical trial [298]. Acriflavine also specifically targets leukaemic stem cells that reside in hypoxic regions of the bone marrow, and represents a novel treatment for CML patients [299].

Altogether, differentiation therapy strategies look promising, but to be clinically relevant, they would need to be combined with conventional chemotherapy to eliminate both CSCs and differentiated and thereby sensitized cell populations within the tumour [19].

Targeting the microenvironment Targeting the TME represents an alternative approach for new cancer therapies. The idea is to disrupt the paracrine signalling in the tumour-associated stroma that sustains CSC identity. Such approaches carry the risk of inhibiting signalling pathways in normal cells. However, CSCs often show a higher dependency on paracrine signalling, suggesting that low-dose treatments might affect CSCs without having an impact on normal tissue homeostasis [19]. For instance, it has been shown that several cancer types depend on extracellular Wnt signalling. Inhibitors of PORCN, an o-acyl transferase of WNT factors, were efficient in blocking renewal of WNT-dependent CSCs in colorectal and pancreatic cancers, and are currently in phase 1 clinical trials. Furthermore, vantictumab, an antibody targeting the WNT Frizzled receptors, is in phase 1 clinical trials for pancreatic cancer, non-small lung cancer, and breast cancer [90]. Antibodies and inhibitors preventing the recruitment of CSC-supporting cells to the tumour niche have also been developed. For example, ARRY-382, a CSF-1 receptor kinase inhibitor that prevents recruitment of CSC-supporting macrophages, has entered phase 2 of clinical trials for the treatment of metastatic cancers, in combination with pembrolizumab [19]. Inhibitors targeting the CXCL2-CXCR4 axis, a complex that supports CSC and tumour stroma communication, are under investigation. Likewise, antibodies targeting the vascular endothelial growth factor (VEGF)/VEGF receptor system may distrupt the CSC niche and a normalized vasculature [300]. Due to the complexity of TME stimuli, along with tumour heterogeneity/plasticity, drug combination strategies may be necessary for certain cancer therapies to be successful [283].

12.7 Future Perspectives

CSCs represent a primary source of tumour heterogeneity. CSC plasticity involves their ability to change cellular identity, reprogram metabolic pathways, evade immune responses, resist therapies, and adapt to their surrounding environments during growth and metastasis. Inherent properties of CSCs such as differential expression of stem cell regulators may provide clinicians with functionally relevant biomarkers that predict therapeutic resistance and/or metastatic potential in patient subsets [301, 302]. However, the plasticity (i.e. heterogeneity) of CSCs presents the greatest challenge in therapeutically eradicating cancer. A complete understanding of the functional and phenotypic differences in CSCs is necessary in order to leverage their plasticity towards the creation and implementation of clinical useful interventions. Additional molecular mechanisms that remain unclear include those that regulate the transient acquisition of stemness in cancer cells [270, 303].

References

1. Bonnet D, Dick JE (1997) Human acute myeloid leukemia is organized as a hierarchy that originates from a primitive hematopoietic cell. Nat Med 3(7):730
2. Lapidot T, Sirard C, Vormoor J, Murdoch B, Hoang T, Caceres-Cortes J et al (1994) A cell initiating human acute myeloid leukaemia after transplantation into SCID mice. Nature 367:645–648

3. O'Brien CA, Pollett A, Gallinger S, Dick JE (2006) A human colon cancer cell capable of initiating tumour growth in immunodeficient mice. Nature 445:106
4. Dalerba P, Dylla SJ, Park I-K, Liu R, Wang X, Cho RW et al (2007) Phenotypic characterization of human colorectal cancer stem cells. Proc Natl Acad Sci U S A 104:10158–10163
5. Singh SK, Hawkins C, Clarke ID, Squire JA, Bayani J, Hide T et al (2004) Identification of human brain tumour initiating cells. Nature 432:396–401
6. Al-Hajj M, Wicha MS, Benito-Hernandez A, Morrison SJ, Clarke MF (2003) Prospective identification of tumorigenic breast cancer cells. Proc Natl Acad Sci 100:3983–3988
7. Damjanov I (1993) Teratocarcinoma: neoplastic lessons about normal embryogenesis. Int J Dev Biol 37:39–46
8. Karsten U, Goletz S (2013) What makes cancer stem cell markers different? Springerplus 2:301
9. Vermeulen L, De Sousa EMF, van der Heijden M, Cameron K, de Jong JH, Borovski T et al (2010) Wnt activity defines colon cancer stem cells and is regulated by the microenvironment. Nat Cell Biol 12:468–476
10. Ghatak S, Misra S, Toole BP (2005) Hyaluronan constitutively regulates ErbB2 phosphorylation and signaling complex formation in carcinoma cells. J Biol Chem 280:8875–8883
11. Misra S, Toole BP, Ghatak S (2006) Hyaluronan constitutively regulates activation of multiple receptor tyrosine kinases in epithelial and carcinoma cells. J Biol Chem 281:34936–34941
12. Aigner S, Ramos CL, Hafezi-Moghadam A, Lawrence MB, Friederichs J, Altevogt P et al (1998) CD24 mediates rolling of breast carcinoma cells on P-selectin. FASEB J 12:1241–1251
13. Mizrak D, Brittan M, Alison M (2008) CD133: molecule of the moment. J Pathol 214:3–9
14. Muzio G, Maggiora M, Paiuzzi E, Oraldi M, Canuto RA (2012) Aldehyde dehydrogenases and cell proliferation. Free Radic Biol Med 52:735–746
15. Medema JP (2013) Cancer stem cells: the challenges ahead. Nat Cell Biol 15:338–344
16. Masters JR, Foley CL, Bisson I, Ahmed A (2003) Cancer stem cells. BJU Int 92:661–662
17. Abbaszadegan MR, Bagheri V, Razavi MS, Momtazi AA, Sahebkar A, Gholamin M (2017) Isolation, identification, and characterization of cancer stem cells: a review. J Cell Physiol 232:2008–2018
18. Moserle L, Ghisi M, Amadori A, Indraccolo S (2010) Side population and cancer stem cells: therapeutic implications. Cancer Lett 288:1–9
19. Pattabiraman DR, Weinberg RA (2014) Tackling the cancer stem cells - what challenges do they pose? Nat Rev Drug Discov 13:497–512
20. Song J, Chang I, Chen Z, Kang M, Wang CY (2010) Characterization of side populations in HNSCC: highly invasive, chemoresistant and abnormal Wnt signaling. PLoS One 5:e11456
21. Goodell MA, Brose K, Paradis G, Conner AS, Mulligan RC (1996) Isolation and functional properties of murine hematopoietic stem cells that are replicating in vivo. J Exp Med 183:1797–1806
22. Goodell MA, McKinney-Freeman S, Camargo FD (2005) Isolation and characterization of side population cells. Methods Mol Biol 290:343–352
23. Feuring-Buske M, Hogge DE (2001) Hoechst 33342 efflux identifies a subpopulation of cytogenetically normal CD34(+)CD38(−) progenitor cells from patients with acute myeloid leukemia. Blood 97:3882–3889
24. Huang FF, Zhang L, Wu DS, Yuan XY, Yu YH, Zhao XL et al (2014) PTEN regulates BCRP/ABCG2 and the side population through the PI3K/Akt pathway in chronic myeloid leukemia. PLoS One 9:e88298
25. Jiang Y, Gao H, Liu M, Mao Q (2016) Sorting and biological characteristics analysis for side population cells in human primary hepatocellular carcinoma. Am J Cancer Res 6:1890–1905
26. Szotek PP, Pieretti-Vanmarcke R, Masiakos PT, Dinulescu DM, Connolly D, Foster R et al (2006) Ovarian cancer side population defines cells with stem cell-like characteristics and Mullerian inhibiting substance responsiveness. Proc Natl Acad Sci U S A 103:11154–11159
27. Yasuda K, Torigoe T, Morita R, Kuroda T, Takahashi A, Matsuzaki J et al (2013) Ovarian cancer stem cells are enriched in side population and aldehyde dehydrogenase bright overlapping population. PLoS One 8:e68187

28. Shen G, Shen F, Shi Z, Liu W, Hu W, Zheng X et al (2008) Identification of cancer stem-like cells in the C6 glioma cell line and the limitation of current identification methods. In Vitro Cell Dev Biol Anim 44:280–289
29. Ho MM, Ng AV, Lam S, Hung JY (2007) Side population in human lung cancer cell lines and tumors is enriched with stem-like cancer cells. Cancer Res 67:4827–4833
30. Nakanishi T, Chumsri S, Khakpour N, Brodie AH, Leyland-Jones B, Hamburger AW et al (2010) Side-population cells in luminal-type breast cancer have tumour-initiating cell properties, and are regulated by HER2 expression and signalling. Br J Cancer 102:815–826
31. Pastrana E, Silva-Vargas V, Doetsch F (2011) Eyes wide open: a critical review of sphere-formation as an assay for stem cells. Cell Stem Cell 8:486–498
32. Cammareri P, Lombardo Y, Francipane MG, Bonventre S, Todaro M, Stassi G (2008) Isolation and culture of colon cancer stem cells. Methods Cell Biol 86:311–324
33. Rajasekhar VK, Studer L, Gerald W, Socci ND, Scher HI (2011) Tumour-initiating stem-like cells in human prostate cancer exhibit increased NF-kappaB signalling. Nat Commun 2:162
34. Ponti D, Costa A, Zaffaroni N, Pratesi G, Petrangolini G, Coradini D et al (2005) Isolation and in vitro propagation of tumorigenic breast cancer cells with stem/progenitor cell properties. Cancer Res 65:5506–5511
35. Vik-Mo EO, Nyakas M, Mikkelsen BV, Moe MC, Due-Tonnesen P, Suso EM et al (2013) Therapeutic vaccination against autologous cancer stem cells with mRNA-transfected dendritic cells in patients with glioblastoma. Cancer Immunol Immunother 62:1499–1509
36. Rao GH, Liu HM, Li BW, Hao JJ, Yang YL, Wang MR et al (2013) Establishment of a human colorectal cancer cell line P6C with stem cell properties and resistance to chemotherapeutic drugs. Acta Pharmacol Sin 34:793–804
37. Chen T, Yang K, Yu J, Meng W, Yuan D, Bi F et al (2012) Identification and expansion of cancer stem cells in tumor tissues and peripheral blood derived from gastric adenocarcinoma patients. Cell Res 22:248–258
38. Fang D, Nguyen TK, Leishear K, Finko R, Kulp AN, Hotz S et al (2005) A tumorigenic subpopulation with stem cell properties in melanomas. Cancer Res 65:9328–9337
39. Gou S, Liu T, Wang C, Yin T, Li K, Yang M et al (2007) Establishment of clonal colony-forming assay for propagation of pancreatic cancer cells with stem cell properties. Pancreas 34:429–435
40. Jensen JB, Parmar M (2006) Strengths and limitations of the neurosphere culture system. Mol Neurobiol 34:153–161
41. Hu Y, Smyth GK (2009) ELDA: extreme limiting dilution analysis for comparing depleted and enriched populations in stem cell and other assays. J Immunol Methods 347:70–78
42. Korkaya H, Paulson A, Charafe-Jauffret E, Ginestier C, Brown M, Dutcher J et al (2009) Regulation of mammary stem/progenitor cells by PTEN/Akt/beta-catenin signaling. PLoS Biol 7:e1000121
43. Liu S, Dontu G, Mantle ID, Patel S, Ahn NS, Jackson KW et al (2006) Hedgehog signaling and Bmi-1 regulate self-renewal of normal and malignant human mammary stem cells. Cancer Res 66:6063–6071
44. O'Brien CA, Kreso A, Jamieson CH (2010) Cancer stem cells and self-renewal. Clin Cancer Res 16:3113–3120
45. Carrasco E, Alvarez PJ, Prados J, Melguizo C, Rama AR, Aranega A et al (2014) Cancer stem cells and their implication in breast cancer. Eur J Clin Investig 44:678–687
46. Jackson M, Hassiotou F, Nowak A (2015) Glioblastoma stem-like cells: at the root of tumor recurrence and a therapeutic target. Carcinogenesis 36:177–185
47. Botchkina G (2013) Colon cancer stem cells--from basic to clinical application. Cancer Lett 338:127–140
48. Ricci-Vitiani L, Lombardi DG, Pilozzi E, Biffoni M, Todaro M, Peschle C et al (2007) Identification and expansion of human colon-cancer-initiating cells. Nature 445:111–115
49. Tseng JY, Yang CY, Yang SH, Lin JK, Lin CH, Jiang JK (2015) Circulating CD133(+)/ESA(+) cells in colorectal cancer patients. J Surg Res 199:362–370
50. Cervello I, Mirantes C, Santamaria X, Dolcet X, Matias-Guiu X, Simon C (2011) Stem cells in human endometrium and endometrial carcinoma. Int J Gynecol Pathol 30:317–327

51. Rutella S, Bonanno G, Procoli A, Mariotti A, Corallo M, Prisco MG et al (2009) Cells with characteristics of cancer stem/progenitor cells express the CD133 antigen in human endometrial tumors. Clin Cancer Res 15:4299–4311
52. Li Z (2013) CD133: a stem cell biomarker and beyond. Exp Hematol Oncol 2:17
53. Qian X, Tan C, Wang F, Yang B, Ge Y, Guan Z et al (2016) Esophageal cancer stem cells and implications for future therapeutics. Onco Targets Ther 9:2247–2254
54. Suva ML, Riggi N, Stehle JC, Baumer K, Tercier S, Joseph JM et al (2009) Identification of cancer stem cells in Ewing's sarcoma. Cancer Res 69:1776–1781
55. Krishnamurthy S, Nor JE (2012) Head and neck cancer stem cells. J Dent Res 91:334–340
56. Prince ME, Sivanandan R, Kaczorowski A, Wolf GT, Kaplan MJ, Dalerba P et al (2007) Identification of a subpopulation of cells with cancer stem cell properties in head and neck squamous cell carcinoma. Proc Natl Acad Sci U S A 104:973–978
57. Bussolati B, Bruno S, Grange C, Ferrando U, Camussi G (2008) Identification of a tumor-initiating stem cell population in human renal carcinomas. FASEB J 22:3696–3705
58. Angelotti ML, Ronconi E, Ballerini L, Peired A, Mazzinghi B, Sagrinati C et al (2012) Characterization of renal progenitors committed toward tubular lineage and their regenerative potential in renal tubular injury. Stem Cells 30:1714–1725
59. Lindgren D, Bostrom AK, Nilsson K, Hansson J, Sjolund J, Moller C et al (2011) Isolation and characterization of progenitor-like cells from human renal proximal tubules. Am J Pathol 178:828–837
60. Bruno S, Bussolati B, Grange C, Collino F, Graziano ME, Ferrando U et al (2006) CD133+ renal progenitor cells contribute to tumor angiogenesis. Am J Pathol 169:2223–2235
61. Lim SD, Young AN, Paner GP, Amin MB (2008) Prognostic role of CD44 cell adhesion molecule expression in primary and metastatic renal cell carcinoma: a clinicopathologic study of 125 cases. Virchows Arch 452:49–55
62. Sun JH, Luo Q, Liu LL, Song GB (2016) Liver cancer stem cell markers: progression and therapeutic implications. World J Gastroenterol 22:3547–3557
63. Lundin A, Driscoll B (2013) Lung cancer stem cells: progress and prospects. Cancer Lett 338:89–93
64. Lang D, Mascarenhas JB, Shea CR (2013) Melanocytes, melanocyte stem cells, and melanoma stem cells. Clin Dermatol 31:166–178
65. Schatton T, Murphy GF, Frank NY, Yamaura K, Waaga-Gasser AM, Gasser M et al (2008) Identification of cells initiating human melanomas. Nature 451:345–349
66. Zhan Q, Wang C, Ngai S (2013) Ovarian cancer stem cells: a new target for cancer therapy. Biomed Res Int 2013:916819
67. Zhang S, Balch C, Chan MW, Lai HC, Matei D, Schilder JM et al (2008) Identification and characterization of ovarian cancer-initiating cells from primary human tumors. Cancer Res 68:4311–4320
68. Li C, Heidt DG, Dalerba P, Burant CF, Zhang L, Adsay V et al (2007) Identification of pancreatic cancer stem cells. Cancer Res 67:1030–1037
69. Zhan HX, Xu JW, Wu D, Zhang TP, Hu SY (2015) Pancreatic cancer stem cells: new insight into a stubborn disease. Cancer Lett 357:429–437
70. Collins AT, Berry PA, Hyde C, Stower MJ, Maitland NJ (2005) Prospective identification of tumorigenic prostate cancer stem cells. Cancer Res 65:10946–10951
71. Sharpe B, Beresford M, Bowen R, Mitchard J, Chalmers AD (2013) Searching for prostate cancer stem cells: markers and methods. Stem Cell Rev 9:721–730
72. Brungs D, Aghmesheh M, Vine KL, Becker TM, Carolan MG, Ranson M (2016) Gastric cancer stem cells: evidence, potential markers, and clinical implications. J Gastroenterol 51:313–326
73. Nagayama Y, Shimamura M, Mitsutake N (2016) Cancer stem cells in the thyroid. Front Endocrinol 7:20
74. Shimamura M, Nagayama Y, Matsuse M, Yamashita S, Mitsutake N (2014) Analysis of multiple markers for cancer stem-like cells in human thyroid carcinoma cell lines. Endocr J 61:481–490

75. Horton SJ, Huntly BJ (2012) Recent advances in acute myeloid leukemia stem cell biology. Haematologica 97:966–974
76. Matsui W, Wang Q, Barber JP, Brennan S, Smith BD, Borrello I et al (2008) Clonogenic multiple myeloma progenitors, stem cell properties, and drug resistance. Cancer Res 68:190–197
77. Cox CV, Evely RS, Oakhill A, Pamphilon DH, Goulden NJ, Blair A (2004) Characterization of acute lymphoblastic leukemia progenitor cells. Blood 104:2919–2925
78. Castor A, Nilsson L, Astrand-Grundstrom I, Buitenhuis M, Ramirez C, Anderson K et al (2005) Distinct patterns of hematopoietic stem cell involvement in acute lymphoblastic leukemia. Nat Med 11:630–637
79. Cox CV, Martin HM, Kearns PR, Virgo P, Evely RS, Blair A (2007) Characterization of a progenitor cell population in childhood T-cell acute lymphoblastic leukemia. Blood 109:674–682
80. Singh D, Minz AP, Sahoo SK (2017) Nanomedicine-mediated drug targeting of cancer stem cells. Drug Discov Today 22(6):952–959. https://doi.org/10.1016/j.drudis.2017.04.005
81. Meacham CE, Morrison SJ (2013) Tumour heterogeneity and cancer cell plasticity. Nature 501:328–337
82. Wicha MS, Liu S, Dontu G (2006) Cancer stem cells: an old idea—a paradigm shift. Cancer Res 66:1883–1890
83. Nowell PC (1976) The clonal evolution of tumor cell populations. Science 194:23–28
84. Dick JE (2008) Stem cell concepts renew cancer research. Blood 112:4793–4807
85. Kreso A, Dick JE (2014) Evolution of the cancer stem cell model. Cell Stem Cell 14:275–291
86. Vermeulen L, de Sousa e Melo F, Richel DJ, Medema JP (2012) The developing cancer stem-cell model: clinical challenges and opportunities. Lancet Oncol 13:e83–e89
87. Hope KJ, Jin L, Dick JE (2004) Acute myeloid leukemia originates from a hierarchy of leukemic stem cell classes that differ in self-renewal capacity. Nat Immunol 5:738
88. Ashkenazi R, Gentry SN, Jackson TL (2008) Pathways to tumorigenesis—modeling mutation acquisition in stem cells and their progeny. Neoplasia 10:1170–1182
89. Jamieson CHM, Ailles LE, Dylla SJ, Muijtjens M, Jones C, Zehnder JL et al (2004) Granulocyte–macrophage progenitors as candidate leukemic stem cells in blast-crisis CML. N Engl J Med 351:657–667
90. Batlle E, Clevers H (2017) Cancer stem cells revisited. Nat Med 23:1124–1134
91. Pardal R, Clarke MF, Morrison SJ (2003) Applying the principles of stem-cell biology to cancer. Nat Rev Cancer 3:895–902
92. Abollo-Jimenez F, Jimenez R, Cobaleda C (2010) Physiological cellular reprogramming and cancer. Semin Cancer Biol 20:98–106
93. Merrell AJ, Stanger BZ (2016) Adult cell plasticity in vivo: de-differentiation and transdifferentiation are back in style. Nat Rev Mol Cell Biol 17:413–425
94. Aguilar-Gallardo C, Simon C (2013) Cells, stem cells, and cancer stem cells. Semin Reprod Med 31:5–13
95. Takahashi K, Yamanaka S (2006) Induction of pluripotent stem cells from mouse embryonic and adult fibroblast cultures by defined factors. Cell 126:663–676
96. Jilkine A, Gutenkunst RN (2014) Effect of dedifferentiation on time to mutation acquisition in stem cell-driven cancers. PLoS Comput Biol 10:e1003481
97. Friedmann-Morvinski D, Verma IM (2014) Dedifferentiation and reprogramming: origins of cancer stem cells. EMBO Rep 15:244–253
98. Iglesias JM, Gumuzio J, Martin AG (2017) Linking pluripotency reprogramming and cancer. Stem Cells Transl Med 6:335–339
99. Yamada Y, Haga H (2014) Concise review: dedifferentiation meets cancer development: proof of concept for epigenetic cancer. Stem Cells Transl Med:1182–1187
100. Ben-Porath I, Thomson MW, Carey VJ, Ge R, Bell GW, Regev A et al (2008) An embryonic stem cell-like gene expression signature in poorly differentiated aggressive human tumors. Nat Genet 40:499–507
101. Wagers AJ, Weissman IL (2004) Plasticity of adult stem cells. Cell 116:639–648

102. Maki N, Martinson J, Nishimura O, Tarui H, Meller J, Tsonis PA et al (2010) Expression profiles during dedifferentiation in newt lens regeneration revealed by expressed sequence tags. Mol Vis 16:72–78
103. Thitoff AR, Call MK, Del Rio-Tsonis K, Tsonis PA. Unique expression patterns of the retinoblastoma (Rb) gene in intact and lens regeneration-undergoing newt eyes. Anat Rec A: Discov Mol Cell Evol Biol 2003; 271A:185–188
104. Tarlow BD, Pelz C, Naugler WE, Wakefield L, Wilson EM, Finegold MJ et al (2014) Bipotential adult liver progenitors are derived from chronically injured mature hepatocytes. Cell Stem Cell 15:605–618
105. Thorel F, Népote V, Avril I, Kohno K, Desgraz R, Chera S et al (2010) Conversion of adult pancreatic [agr]-cells to [bgr]-cells after extreme [bgr]-cell loss. Nature 464:1149–1154
106. Rackham OJL, Firas J, Fang H, Oates ME, Holmes ML, Knaupp AS et al (2016) A predictive computational framework for direct reprogramming between human cell types. Nat Genet 48:331–335
107. Davis RL, Weintraub H, Lassar AB (1987) Expression of a single transfected cDNA converts fibroblasts to myoblasts. Cell 51:987–1000
108. Fan B, Malato Y, Calvisi DF, Naqvi S, Razumilava N, Ribback S et al (2012) Cholangiocarcinomas can originate from hepatocytes in mice. J Clin Invest 122:2911–2915
109. Li H, Wolfe A, Septer S, Edwards G, Zhong X, Abdulkarim AB et al (2012) Deregulation of Hippo kinase signalling in human hepatic malignancies. Liver Int 32:38–47
110. Wang R, Chadalavada K, Wilshire J, Kowalik U, Hovinga KE, Geber A et al (2010) Glioblastoma stem-like cells give rise to tumour endothelium. Nature 468:829–833
111. Ricci-Vitiani L, Pallini R, Biffoni M, Todaro M, Invernici G, Cenci T et al (2010) Tumour vascularization via endothelial differentiation of glioblastoma stem-like cells. Nature 468:824–828
112. Yin J, Oh YT, Kim J-Y, Kim SS, Choi E, Kim TH et al (2017) Transglutaminase 2 inhibition reverses mesenchymal transdifferentiation of glioma stem cells by regulating C/EBPβ signaling. Cancer Res 77:4973–4984
113. Hay ED (1995) An overview of epithelio-mesenchymal transformation. Cells Tissues Organs 154:8–20
114. Hay ED (2005) The mesenchymal cell, its role in the embryo, and the remarkable signaling mechanisms that create it. Dev Dyn 233:706–720
115. Puisieux A, Brabletz T, Caramel J (2014) Oncogenic roles of EMT-inducing transcription factors. Nat Cell Biol 16:488–494
116. Zheng X, Carstens JL, Kim J, Scheible M, Kaye J, Sugimoto H et al (2015) Epithelial-to-mesenchymal transition is dispensable for metastasis but induces chemoresistance in pancreatic cancer. Nature 527:525–530
117. Fischer KR, Durrans A, Lee S, Sheng J, Li F, Wong STC et al (2015) Epithelial-to-mesenchymal transition is not required for lung metastasis but contributes to chemoresistance. Nature 527:472–476
118. Thiery JP, Acloque H, Huang RYJ, Nieto MA (2009) Epithelial-mesenchymal transitions in development and disease. Cell 139:871–890
119. Ye X, Tam WL, Shibue T, Kaygusuz Y, Reinhardt F, Ng Eaton E et al (2015) Distinct EMT programs control normal mammary stem cells and tumour-initiating cells. Nature 525:256–260
120. Huang RYJ, Wong MK, Tan TZ, Kuay KT, Ng AHC, Chung VY et al (2013) An EMT spectrum defines an anoikis-resistant and spheroidogenic intermediate mesenchymal state that is sensitive to e-cadherin restoration by a src-kinase inhibitor, saracatinib (AZD0530). Cell Death Dis 4:e915
121. Beerling E, Seinstra D, de Wit E, Kester L, van der Velden D, Maynard C et al (2016) Plasticity between epithelial and mesenchymal states unlinks EMT from metastasis-enhancing stem cell capacity. Cell Rep 14:2281–2288
122. Celià-Terrassa T, Kang Y (2016) Distinctive properties of metastasis-initiating cells. Genes Dev 30:892–908

123. Nieto MA, Huang Ruby Y-J, Jackson Rebecca A, Thiery Jean P (2016) EMT: 2016. Cell 166:21–45
124. Cano A, Perez-Moreno MA, Rodrigo I, Locascio A, Blanco MJ, del Barrio MG et al (2000) The transcription factor Snail controls epithelial-mesenchymal transitions by repressing E-cadherin expression. Nat Cell Biol 2:76–83
125. Sánchez-Tilló E, Siles L, de Barrios O, Cuatrecasas M, Vaquero EC, Castells A et al (2011) Expanding roles of ZEB factors in tumorigenesis and tumor progression. Am J Cancer Res 1:897–912
126. Martin A, Cano A (2010) Tumorigenesis: Twist1 links EMT to self-renewal. Nat Cell Biol 12:924–925
127. Yang M-H, Hsu DS-S, Wang H-W, Wang H-J, Lan H-Y, Yang W-H et al (2010) Bmi1 is essential in Twist1-induced epithelial-mesenchymal transition. Nat Cell Biol 12:982–992
128. Chaffer Christine L, Marjanovic Nemanja D, Lee T, Bell G, Kleer Celina G, Reinhardt F et al (2013) Poised chromatin at the ZEB1 promoter enables breast cancer cell plasticity and enhances tumorigenicity. Cell 154:61–74
129. Mani SA, Guo W, Liao M-J, Eaton EN, Ayyanan A, Zhou AY et al (2008) The epithelial-mesenchymal transition generates cells with properties of stem cells. Cell 133:704–715
130. Pang R, Law WL, Chu ACY, Poon JT, Lam CSC, Chow AKM et al (2010) A subpopulation of CD26+ cancer stem cells with metastatic capacity in human colorectal cancer. Cell Stem Cell 6:603–615
131. Chen Y-C, Chen Y-W, Hsu H-S, Tseng L-M, Huang P-I, Lu K-H et al (2009) Aldehyde dehydrogenase 1 is a putative marker for cancer stem cells in head and neck squamous cancer. Biochem Biophys Res Commun 385:307–313
132. Hennessy BT, Gonzalez-Angulo AM, Stemke-Hale K, Gilcrease MZ, Krishnamurthy S, Lee JS et al (2009) Characterization of a naturally occurring breast cancer subset enriched in epithelial-to-mesenchymal transition and stem cell characteristics. Cancer Res 69:4116–4124
133. Beck B, Lapouge G, Rorive S, Drogat B, Desaedelaere K, Delafaille S et al (2015) Different levels of Twist1 regulate skin tumor initiation, stemness, and progression. Cell Stem Cell 16:67–79
134. Hao J, Zhang Y, Deng M, Ye R, Zhao S, Wang Y et al (2014) MicroRNA control of epithelial-mesenchymal transition in cancer stem cells. Int J Cancer 135:1019–1027
135. Grelet S, Link LA, Howley B, Obellianne C, Palanisamy V, Gangaraju VK et al (2017) A regulated PNUTS mRNA to lncRNA splice switch mediates EMT and tumour progression. Nat Cell Biol 19:1105–1115
136. Polyak K, Weinberg RA (2009) Transitions between epithelial and mesenchymal states: acquisition of malignant and stem cell traits. Nat Rev Cancer 9:265–273
137. Chao C-H, Chang C-C, Wu M-J, Ko H-W, Wang D, Hung M-C et al (2014) MicroRNA-205 signaling regulates mammary stem cell fate and tumorigenesis. J Clin Invest 124:3093–3106
138. Kalluri R, Weinberg RA (2009) The basics of epithelial-mesenchymal transition. J Clin Invest 119:1420–1428
139. George JT, Jolly MK, Xu J, Somarelli J, Levine H (2017) Survival outcomes in cancer patients predicted by a partial EMT gene expression scoring metric. Cancer Res 77:6415–6428
140. Yu M, Bardia A, Wittner BS, Stott SL, Smas ME, Ting DT et al (2013) Circulating breast tumor cells exhibit dynamic changes in epithelial and mesenchymal composition. Science 339:580–584
141. Jolly MK, Ware KE, Gilja S, Somarelli JA, Levine H (2017) EMT and MET: necessary or permissive for metastasis? Mol Oncol 11:755–769
142. Tian X-J, Zhang H, Xing J (2013) Coupled reversible and irreversible bistable switches underlying TGFβ-induced epithelial to mesenchymal transition. Biophys J 105:1079–1089
143. Celia-Terrassa T, Meca-Cortes O, Mateo F, Martinez de Paz A, Rubio N, Arnal-Estape A et al (2012) Epithelial-mesenchymal transition can suppress major attributes of human epithelial tumor-initiating cells. J Clin Invest 122:1849–1868
144. Micalizzi DS, Farabaugh SM, Ford HL (2010) Epithelial-mesenchymal transition in Cancer: parallels between normal development and tumor progression. J Mammary Gland Biol Neoplasia 15:117–134

145. Ocaña Oscar H, Córcoles R, Fabra Á, Moreno-Bueno G, Acloque H, Vega S et al (2012) Metastatic colonization requires the repression of the epithelial-mesenchymal transition inducer Prrx1. Cancer Cell 22:709–724
146. Schmidt Johanna M, Panzilius E, Bartsch Harald S, Irmler M, Beckers J, Kari V et al (2015) Stem-cell-like properties and epithelial plasticity arise as stable traits after transient Twist1 activation. Cell Rep 10:131–139
147. Tsai Jeff H, Donaher Joana L, Murphy Danielle A, Chau S, Yang J (2012) Spatiotemporal regulation of epithelial-mesenchymal transition is essential for squamous cell carcinoma metastasis. Cancer Cell 22:725–736
148. Tran HD, Luitel K, Kim M, Zhang K, Longmore GD, Tran DD (2014) Transient SNAIL1 expression is necessary for metastatic competence in breast cancer. Cancer Res 74:6330
149. Drasin DJ, Guarnieri AL, Neelakantan D, Kim J, Cabrera JH, Wang CA et al (2015) TWIST1-induced microRNA-424 reversibly drives mesenchymal programming while inhibiting tumor initiation. Cancer Res 75:1908–1921
150. Nandy SB, Orozco A, Lopez-Valdez R, Roberts R, Subramani R, Arumugam A et al (2017) Glucose insult elicits hyperactivation of cancer stem cells through miR-424-cdc42-prdm14 signalling axis. Br J Cancer 117:1665–1675
151. Tsuji T, Ibaragi S, G-f H (2009) Epithelial-mesenchymal transition and cell cooperativity in metastasis. Cancer Res 69:7135–7139
152. Varga J, Greten FR (2017) Cell plasticity in epithelial homeostasis and tumorigenesis. Nat Cell Biol 19:1133–1141
153. Harner-Foreman N, Vadakekolathu J, Laversin SA, Mathieu MG, Reeder S, Pockley AG et al (2017) A novel spontaneous model of epithelial-mesenchymal transition (EMT) using a primary prostate cancer derived cell line demonstrating distinct stem-like characteristics. Sci Rep 7:40633
154. Hanahan D, Weinberg RA (2011) Hallmarks of cancer: the next generation. Cell 144:646–674
155. Warburg O, Wind F, Negelein E (1927) The metabolism of tumors in the body. J Gen Physiol 8:519–530
156. Peiris-Pagès M, Martinez-Outschoorn UE, Pestell RG, Sotgia F, Lisanti MP (2016) Cancer stem cell metabolism. Breast Cancer Res 18:55
157. Guppy M, Greiner E, Brand K (1993) The role of the Crabtree effect and an endogenous fuel in the energy metabolism of resting and proliferating thymocytes. Eur J Biochem 212:95–99
158. Shestov AA, Liu X, Ser Z, Cluntun AA, Hung YP, Huang L et al (2014) Quantitative determinants of aerobic glycolysis identify flux through the enzyme GAPDH as a limiting step. elife 3:e03342
159. Tennant DA, Duran RV, Gottlieb E (2010) Targeting metabolic transformation for cancer therapy. Nat Rev Cancer 10:267–277
160. Christofk HR, Vander Heiden MG, Wu N, Asara JM, Cantley LC (2008) Pyruvate kinase M2 is a phosphotyrosine-binding protein. Nature 452:181–186
161. Christofk HR, Vander Heiden MG, Harris MH, Ramanathan A, Gerszten RE, Wei R et al (2008) The M2 splice isoform of pyruvate kinase is important for cancer metabolism and tumour growth. Nature 452:230–233
162. Kim JW, Tchernyshyov I, Semenza GL, Dang CV (2006) HIF-1-mediated expression of pyruvate dehydrogenase kinase: a metabolic switch required for cellular adaptation to hypoxia. Cell Metab 3:177–185
163. Dupuy F, Tabaries S, Andrzejewski S, Dong Z, Blagih J, Annis MG et al (2015) PDK1-dependent metabolic reprogramming dictates metastatic potential in breast Cancer. Cell Metab 22:577–589
164. Peng F, Wang JH, Fan WJ, Meng YT, Li MM, Li TT et al (2017) Glycolysis gatekeeper PDK1 reprograms breast cancer stem cells under hypoxia. Oncogene 37:1062–1074
165. Pavlova Natalya N, Thompson Craig B (2016) The emerging hallmarks of cancer metabolism. Cell Metab 23:27–47
166. Weber GF (2016) Time and circumstances: cancer cell metabolism at various stages of disease progression. Front Oncol 6

167. Cairns RA, Harris IS, Mak TW (2011) Regulation of cancer cell metabolism. Nat Rev Cancer 11:85–95
168. Panopoulos AD, Yanes O, Ruiz S, Kida YS, Diep D, Tautenhahn R et al (2012) The metabolome of induced pluripotent stem cells reveals metabolic changes occurring in somatic cell reprogramming. Cell Res 22:168–177
169. Wu J, Ocampo A, Belmonte JCI (2016) Cellular metabolism and induced pluripotency. Cell 166:1371–1385
170. Folmes CD, Nelson TJ, Martinez-Fernandez A, Arrell DK, Lindor JZ, Dzeja PP et al (2011) Somatic oxidative bioenergetics transitions into pluripotency-dependent glycolysis to facilitate nuclear reprogramming. Cell Metab 14:264–271
171. Hawkins KE, Joy S, Delhove JM, Kotiadis VN, Fernandez E, Fitzpatrick LM et al (2016) NRF2 orchestrates the metabolic shift during induced pluripotent stem cell reprogramming. Cell Rep 14:1883–1891
172. Kida YS, Kawamura T, Wei Z, Sogo T, Jacinto S, Shigeno A et al (2015) ERRs mediate a metabolic switch required for somatic cell reprogramming to pluripotency. Cell Stem Cell 16:547–555
173. Sone M, Morone N, Nakamura T, Tanaka A, Okita K, Woltjen K et al (2017) Hybrid cellular metabolism coordinated by Zic3 and Esrrb synergistically enhances induction of naive pluripotency. Cell Metab 25:1103–17.e6
174. Berg JM, Tymoczko JL, Stryer L (2002) Biochemistry. 5th edn. Section 16.1, Glycolysis is an energy-conversion pathway in many organisms, W H Freeman, New York. Available from: https://www.ncbi.nlm.nih.gov/books/NBK22593/
175. Stryer JMB, John LT, Lubert S (2002) The citric acid cycle. Biochemistry. W H Freeman, New York
176. Stryer JMB, John LT, Lubert S (2002) Oxidative phosphorylation. Biochemistry. W H Freeman, New York
177. Chen H, Chan DC (2017) Mitochondrial dynamics in regulating the unique phenotypes of cancer and stem cells. Cell Metab 26:39–48
178. San-Millán I, Brooks GA (2017) Reexamining cancer metabolism: lactate production for carcinogenesis could be the purpose and explanation of the Warburg effect. Carcinogenesis 38:119–133
179. Diehn M, Cho RW, Lobo NA, Kalisky T, Dorie MJ, Kulp AN et al (2009) Association of reactive oxygen species levels and radioresistance in cancer stem cells. Nature 458:780–783
180. Yuan S, Lu Y, Yang J, Chen G, Kim S, Feng L et al (2015) Metabolic activation of mitochondria in glioma stem cells promotes cancer development through a reactive oxygen species-mediated mechanism. Stem Cell Res Ther 6:198
181. Paola M, Alvaro M-H, Miriam G-A, Maricruz A-R, Julio R-L, Paulina C-H (2017) Mitochondrial dynamics and cancer. Tumor Biol 39:1010428317698391
182. Prigione A, Fauler B, Lurz R, Lehrach H, Adjaye J (2010) The senescence-related mitochondrial/oxidative stress pathway is repressed in human induced pluripotent stem cells. Stem Cells 28:721–733
183. Prieto J, Torres J (2017) Mitochondrial dynamics: in cell reprogramming as it is in cancer. Stem Cells Int 2017:11
184. Zhang J, Zhang Y, Wu W, Wang F, Liu X, Shui G et al (2017) Guanylate-binding protein 2 regulates Drp1-mediated mitochondrial fission to suppress breast cancer cell invasion. Cell Death Dis 8:e3151
185. Zhao J, Zhang J, Yu M, Xie Y, Huang Y, Wolff DW et al (2013) Mitochondrial dynamics regulates migration and invasion of breast cancer cells. Oncogene 32:4814–4824
186. Xie Q, Wu Q, Horbinski CM, Flavahan WA, Yang K, Zhou W et al (2015) Mitochondrial control by DRP1 in brain tumor initiating cells. Nat Neurosci 18:501
187. Katajisto P, Döhla J, Chaffer CL, Pentinmikko N, Marjanovic N, Iqbal S et al (2015) Asymmetric apportioning of aged mitochondria between daughter cells is required for stemness. Science 348:340–343

188. Sancho P, Burgos-Ramos E, Tavera A, Bou Kheir T, Jagust P, Schoenhals M et al (2015) MYC/PGC-1α balance determines the metabolic phenotype and plasticity of pancreatic cancer stem cells. Cell Metab 22:590–605
189. Chen CL, Uthaya Kumar DB, Punj V, Xu J, Sher L, Tahara SM et al (2016) NANOG metabolically reprograms tumor-initiating stem-like cells through tumorigenic changes in oxidative phosphorylation and fatty acid metabolism. Cell Metab 23:206–219
190. Vlashi E, Lagadec C, Vergnes L, Matsutani T, Masui K, Poulou M et al (2011) Metabolic state of glioma stem cells and nontumorigenic cells. Proc Natl Acad Sci U S A 108:16062–16067
191. Ciavardelli D, Rossi C, Barcaroli D, Volpe S, Consalvo A, Zucchelli M et al (2014) Breast cancer stem cells rely on fermentative glycolysis and are sensitive to 2-deoxyglucose treatment. Cell Death Dis 5:e1336
192. Feng W, Gentles A, Nair RV, Huang M, Lin Y, Lee CY et al (2014) Targeting unique metabolic properties of breast tumor initiating cells. Stem Cells 32:1734–1745
193. Lee K-M, Giltnane JM, Balko JM, Schwarz LJ, Guerrero-Zotano AL, Hutchinson KE et al (2017) MYC and MCL1 cooperatively promote chemotherapy-resistant breast cancer stem cells via regulation of mitochondrial oxidative phosphorylation. Cell Metab 26:633–47.e7
194. Flavahan WA, Wu Q, Hitomi M, Rahim N, Kim Y, Sloan AE et al (2013) Brain tumor initiating cells adapt to restricted nutrition through preferential glucose uptake. Nat Neurosci 16:1373–1382
195. Kuntz EM, Baquero P, Michie AM, Dunn K, Tardito S, Holyoake TL et al (2017) Targeting mitochondrial oxidative phosphorylation eradicates therapy-resistant chronic myeloid leukemia stem cells. Nat Med 23(10):1234–1240
196. Lagadinou ED, Sach A, Callahan K, Rossi RM, Neering SJ, Minhajuddin M et al (2013) BCL-2 inhibition targets oxidative phosphorylation and selectively eradicates quiescent human leukemia stem cells. Cell Stem Cell 12:329–341
197. Ye X-Q, Li Q, Wang G-H, Sun F-F, Huang G-J, Bian X-W et al (2011) Mitochondrial and energy metabolism-related properties as novel indicators of lung cancer stem cells. Int J Cancer 129:820–831
198. Viale A, Pettazzoni P, Lyssiotis CA, Ying H, Sanchez N, Marchesini M et al (2014) Oncogene ablation-resistant pancreatic cancer cells depend on mitochondrial function. Nature 514:628–632
199. Mahanty S, Prigent A, Garraud O (2015) Immunogenicity of infectious pathogens and vaccine antigens. BMC Immunol 16:31
200. Dunn GP, Bruce AT, Ikeda H, Old LJ, Schreiber RD (2002) Cancer immunoediting: from immunosurveillance to tumor escape. Nat Immunol 3:991–998
201. Dunn GP, Old LJ, Schreiber RD (2004) The three Es of cancer immunoediting. Annu Rev Immunol 22:329–360
202. Codony-Servat J, Rosell R (2015) Cancer stem cells and immunoresistance: clinical implications and solutions. Transl Lung Cancer Res 4:689–703
203. Rock KL, Reits E, Neefjes J (2016) Present yourself! By MHC class I and MHC class II molecules. Trends Immunol 37:724–737
204. Abdullah LN, Chow EK-H (2013) Mechanisms of chemoresistance in cancer stem cells. Clin Transl Med 2:3
205. Luen SJ, Salgado R, Fox S, Savas P, Eng-Wong J, Clark E et al (2017) Tumour-infiltrating lymphocytes in advanced HER2-positive breast cancer treated with pertuzumab or placebo in addition to trastuzumab and docetaxel: a retrospective analysis of the CLEOPATRA study. Lancet Oncol 18:52–62
206. Ithimakin S, Day KC, Malik F, Zen Q, Dawsey SJ, Bersano-Begey TF et al (2013) HER2 drives luminal breast cancer stem cells in the absence of HER2 amplification: implications for efficacy of adjuvant trastuzumab. Cancer Res 73:1635–1646
207. Korkaya H, Paulson A, Iovino F, Wicha MS (2008) HER2 regulates the mammary stem/progenitor cell population driving tumorigenesis and invasion. Oncogene 27:6120–6130
208. Seliger B, Kiessling R (2013) The two sides of HER2/neu: immune escape versus surveillance. Trends Mol Med 19:677–684

209. Bianchini G, Gianni L (2014) The immune system and response to HER2-targeted treatment in breast cancer. Lancet Oncol 15:e58-e68
210. Sd CT, Souazé F, Basseville A, Bernard A-C, Pécot J, Lopez J et al (2017) BCL-XL directly modulates RAS signalling to favour cancer cell stemness. Nat Commun 8:1123
211. Kim R, Emi M, Tanabe K (2006) Role of mitochondria as the gardens of cell death. Cancer Chemother Pharmacol 57:545–553
212. Cammareri P, Scopelliti A, Todaro M, Eterno V, Francescangeli F, Moyer MP et al (2010) Aurora-a is essential for the tumorigenic capacity and chemoresistance of colorectal cancer stem cells. Cancer Res 70:4655–4665
213. Worthington John J, Kelly A, Smedley C, Bauché D, Campbell S, Marie Julien C et al (2015) Integrin αvβ8-mediated TGF-β activation by effector regulatory T cells is essential for suppression of T-cell-mediated inflammation. Immunity 42:903–915
214. Scheel C, Eaton Elinor N, Li Sophia H-J, Chaffer Christine L, Reinhardt F, Kah K-J et al (2011) Paracrine and autocrine signals induce and maintain mesenchymal and stem cell states in the breast. Cell 145:926–940
215. Fan Q-M, Jing Y-Y, Yu G-F, Kou X-R, Ye F, Gao L et al (2014) Tumor-associated macrophages promote cancer stem cell-like properties via transforming growth factor-beta1-induced epithelial–mesenchymal transition in hepatocellular carcinoma. Cancer Lett 352:160–168
216. Cortés M, Sanchez-Moral L, de Barrios O, Fernández-Aceñero MJ, Martínez-Campanario MC, Esteve-Codina A et al (2017) Tumor-associated macrophages (TAMs) depend on ZEB1 for their cancer-promoting roles. EMBO J 36:3336–3355
217. Martin OA, Anderson RL, Narayan K, MacManus MP (2017) Does the mobilization of circulating tumour cells during cancer therapy cause metastasis? Nat Rev Clin Oncol 14:32–44
218. Todaro M, Alea MP, Di Stefano AB, Cammareri P, Vermeulen L, Iovino F et al (2007) Colon cancer stem cells dictate tumor growth and resist cell death by production of interleukin-4. Cell Stem Cell 1:389–402
219. Nappo G, Handle F, Santer FR, McNeill RV, Seed RI, Collins AT et al (2017) The immunosuppressive cytokine interleukin-4 increases the clonogenic potential of prostate stem-like cells by activation of STAT6 signalling. Oncogene 6:e342
220. Gaggianesi M, Turdo A, Chinnici A, Lipari E, Apuzzo T, Benfante A et al (2017) IL4 primes the dynamics of breast cancer progression via DUSP4 inhibition. Cancer Res 77:3268–3279
221. Korkaya H, Kim GI, Davis A, Malik F, Henry NL, Ithimakin S et al (2012) Activation of an IL6 inflammatory loop mediates trastuzumab resistance in HER2+ breast cancer by expanding the cancer stem cell population. Mol Cell 47:570–584
222. Plaks V, Kong N, Werb Z (2015) The cancer stem cell niche: how essential is the niche in regulating stemness of tumor cells? Cell Stem Cell 16:225–238
223. Shiozawa Y, Pedersen EA, Havens AM, Jung Y, Mishra A, Joseph J et al (2011) Human prostate cancer metastases target the hematopoietic stem cell niche to establish footholds in mouse bone marrow. J Clin Invest 121:1298–1312
224. Xie T, Li L (2007) Stem cells and their niche: an inseparable relationship. Development 134:2001–2006
225. Song W, Mazzieri R, Yang T, Gobe GC (2017) Translational significance for tumor metastasis of tumor-associated macrophages and epithelial-mesenchymal transition. Front Immunol 8:1106
226. Yu H, Lee H, Herrmann A, Buettner R, Jove R (2014) Revisiting STAT3 signalling in cancer: new and unexpected biological functions. Nat Rev Cancer 14:736–746
227. Jinushi M, Chiba S, Yoshiyama H, Masutomi K, Kinoshita I, Dosaka-Akita H et al (2011) Tumor-associated macrophages regulate tumorigenicity and anticancer drug responses of cancer stem/initiating cells. Proc Natl Acad Sci U S A 108:12425–12430
228. Fatrai S, van Schelven SJ, Ubink I, Govaert KM, Raats D, Koster J et al (2015) Maintenance of clonogenic KIT(+) human colon tumor cells requires secretion of stem cell factor by differentiated tumor cells. Gastroenterology 149:692–704
229. Levina V, Marrangoni A, Wang T, Parikh S, Su Y, Herberman R et al (2010) Elimination of human lung cancer stem cells through targeting of the stem cell factor–c-kit autocrine signaling loop. Cancer Res 70:338–346

230. Shi Y, Ping YF, Zhou W, He ZC, Chen C, Bian BS et al (2017) Tumour-associated macrophages secrete pleiotrophin to promote PTPRZ1 signalling in glioblastoma stem cells for tumour growth. Nat Commun 8:15080
231. Uccelli A, Moretta L, Pistoia V (2006) Immunoregulatory function of mesenchymal stem cells. Eur J Immunol 36:2566–2573
232. Shi Y, Du L, Lin L, Wang Y (2017) Tumour-associated mesenchymal stem/stromal cells: emerging therapeutic targets. Nat Rev Drug Discov 16:35–52
233. Li HJ, Reinhardt F, Herschman HR, Weinberg RA (2012) Cancer-stimulated mesenchymal stem cells create a carcinoma stem cell niche via prostaglandin E2 signaling. Cancer Discov 2:840–855
234. Hermann PC, Huber SL, Herrler T, Aicher A, Ellwart JW, Guba M et al (2007) Distinct populations of cancer stem cells determine tumor growth and metastatic activity in human pancreatic cancer. Cell Stem Cell 1:313–323
235. Boire A, Covic L, Agarwal A, Jacques S, Sherifi S, Kuliopulos A (2005) PAR1 is a matrix metalloprotease-1 receptor that promotes invasion and tumorigenesis of breast cancer cells. Cell 120:303–313
236. Fiaschi T, Marini A, Giannoni E, Taddei ML, Gandellini P, De Donatis A et al (2012) Reciprocal metabolic reprogramming through lactate shuttle coordinately influences tumor-stroma interplay. Cancer Res 72:5130–5140
237. Zhang D, Wang Y, Shi Z, Liu J, Sun P, Hou X et al (2015) Metabolic reprogramming of cancer-associated fibroblasts by IDH3alpha downregulation. Cell Rep 10:1335–1348
238. Lotti F, Jarrar AM, Pai RK, Hitomi M, Lathia J, Mace A et al (2013) Chemotherapy activates cancer-associated fibroblasts to maintain colorectal cancer-initiating cells by IL-17A. J Exp Med 210:2851–2872
239. Cirri P, Chiarugi P (2011) Cancer associated fibroblasts: the dark side of the coin. Am J Cancer Res 1:482–497
240. Calabrese C, Poppleton H, Kocak M, Hogg TL, Fuller C, Hamner B et al (2007) A perivascular niche for brain tumor stem cells. Cancer Cell 11:69–82
241. Treps L, Perret R, Edmond S, Ricard D, Gavard J (2017) Glioblastoma stem-like cells secrete the pro-angiogenic VEGF-A factor in extracellular vesicles. J Extracell Vesicles 6:1359479
242. Zhang Z, Dong Z, Lauxen IS, Filho MSA, Nör JE (2014) Endothelial cell-secreted EGF induces epithelial to mesenchymal transition and endows head and neck cancer cells with stem-like phenotype. Cancer Res 74:2869–2881
243. Krishnamurthy S, Dong Z, Vodopyanov D, Imai A, Helman JI, Prince ME et al (2010) Endothelial cell-initiated signaling promotes the survival and self-renewal of cancer stem cells. Cancer Res 70:9969–9978
244. Lu P, Weaver VM, Werb Z (2012) The extracellular matrix: a dynamic niche in cancer progression. J Cell Biol 196:395–406
245. Shiina M, Bourguignon LYW (2015) Selective activation of cancer stem cells by size-specific hyaluronan in head and neck cancer. Int J Cell Biol 2015:10
246. Okuda H, Kobayashi A, Xia B, Watabe M, Pai SK, Hirota S et al (2012) Hyaluronan synthase HAS2 promotes tumor progression in bone by stimulating the interaction of breast cancer stem-like cells with macrophages and stromal cells. Cancer Res 72:537–547
247. Halder G, Dupont S, Piccolo S (2012) Transduction of mechanical and cytoskeletal cues by YAP and TAZ. Nat Rev Mol Cell Biol 13:591–600
248. Acerbi I, Cassereau L, Dean I, Shi Q, Au A, Park C et al (2015) Human breast cancer invasion and aggression correlates with ECM stiffening and immune cell infiltration. Integr Biol (Camb) 7:1120–1134
249. Fuchs E, Tumbar T, Guasch G (2004) Socializing with the neighbors. Stem Cells Their Niche Cell 116:769–778
250. Nakazawa MS, Keith B, Simon MC (2016) Oxygen availability and metabolic adaptations. Nat Rev Cancer 16:663–673
251. Das B, Tsuchida R, Malkin D, Koren G, Baruchel S, Yeger H (2008) Hypoxia enhances tumor stemness by increasing the invasive and tumorigenic side population fraction. Stem Cells 26:1818–1830

252. Mathieu J, Zhou W, Xing Y, Sperber H, Ferreccio A, Agoston Z et al (2014) Hypoxia inducible factors have distinct and stage-specific roles during reprogramming of human cells to pluripotency. Cell Stem Cell 14:592–605
253. Prigione A, Rohwer N, Hoffmann S, Mlody B, Drews K, Bukowiecki R et al (2014) HIF1alpha modulates cell fate reprogramming through early glycolytic shift and upregulation of PDK1-3 and PKM2. Stem Cells 32:364–376
254. Luo W, Hu H, Chang R, Zhong J, Knabel M, O'Meally R et al (2011) Pyruvate kinase M2 is a PHD3-stimulated coactivator for hypoxia-inducible factor 1. Cell 145:732–744
255. Palsson-McDermott EM, Curtis AM, Goel G, Lauterbach MA, Sheedy FJ, Gleeson LE et al (2015) Pyruvate kinase M2 regulates Hif-1alpha activity and IL-1beta induction and is a critical determinant of the Warburg effect in LPS-activated macrophages. Cell Metab 21:65–80
256. Covello KL, Kehler J, Yu H, Gordan JD, Arsham AM, Hu CJ et al (2006) HIF-2alpha regulates Oct-4: effects of hypoxia on stem cell function, embryonic development, and tumor growth. Genes Dev 20:557–570
257. Johansson E, Grassi ES, Pantazopoulou V, Tong B, Lindgren D, Berg TJ et al (2017) CD44 interacts with HIF-2α to modulate the hypoxic phenotype of perinecrotic and perivascular glioma cells. Cell Rep 20:1641–1653
258. Qin J, Liu Y, Lu Y, Liu M, Li M, Li J et al (2017) Hypoxia-inducible factor 1 alpha promotes cancer stem cells-like properties in human ovarian cancer cells by upregulating SIRT1 expression. Sci Rep 7:10592
259. Zhang C, Zhi WI, Lu H, Samanta D, Chen I, Gabrielson E et al (2016) Hypoxia-inducible factors regulate pluripotency factor expression by ZNF217- and ALKBH5-mediated modulation of RNA methylation in breast cancer cells. Oncotarget 7:64527–64542
260. Shiraishi A, Tachi K, Essid N, Tsuboi I, Nagano M, Kato T et al (2017) Hypoxia promotes the phenotypic change of aldehyde dehydrogenase activity of breast cancer stem cells. Cancer Sci 108:362–372
261. Maeda K, Ding Q, Yoshimitsu M, Kuwahata T, Miyazaki Y, Tsukasa K et al (2016) CD133 modulate HIF-1α expression under hypoxia in EMT phenotype pancreatic cancer stem-like cells. Int J Mol Sci 17:1025
262. Lee G, Auffinger B, Guo D, Hasan T, Deheeger M, Tobias AL et al (2016) Dedifferentiation of glioma cells to glioma stem-like cells by therapeutic stress-induced HIF signaling in the recurrent GBM model. Mol Cancer Ther 15:3064–3076
263. Malanchi I, Santamaria-Martinez A, Susanto E, Peng H, Lehr HA, Delaloye JF et al (2011) Interactions between cancer stem cells and their niche govern metastatic colonization. Nature 481:85–89
264. Baccelli I, Schneeweiss A, Riethdorf S, Stenzinger A, Schillert A, Vogel V et al (2013) Identification of a population of blood circulating tumor cells from breast cancer patients that initiates metastasis in a xenograft assay. Nat Biotechnol 31:539–544
265. Dieter SM, Ball CR, Hoffmann CM, Nowrouzi A, Herbst F, Zavidij O et al (2011) Distinct types of tumor-initiating cells form human colon cancer tumors and metastases. Cell Stem Cell 9:357–365
266. Rumman M, Dhawan J, Kassem M (2015) Concise review: quiescence in adult stem cells: biological significance and relevance to tissue regeneration. Stem Cells 33:2903–2912
267. Cheung TH, Rando TA (2013) Molecular regulation of stem cell quiescence. Nat Rev Mol Cell Biol 14:329–340
268. Kusumbe AP, Bapat SA (2009) Cancer stem cells and aneuploid populations within developing tumors are the major determinants of tumor dormancy. Cancer Res 69:9245–9253
269. Chen J, Li Y, Yu T-S, McKay RM, Burns DK, Kernie SG et al (2012) A restricted cell population propagates glioblastoma growth after chemotherapy. Nature 488:522
270. Liau BB, Sievers C, Donohue LK, Gillespie SM, Flavahan WA, Miller TE et al (2017) Adaptive chromatin remodeling drives glioblastoma stem cell plasticity and drug tolerance. Cell Stem Cell 20:233–46.e7
271. Kreso A, O'Brien CA, van Galen P, Gan OI, Notta F, Brown AMK et al (2013) Variable clonal repopulation dynamics influence chemotherapy response in colorectal cancer. Science 339:543–548

272. Oshimori N, Oristian D, Fuchs E (2015) TGF-β promotes heterogeneity and drug resistance in squamous cell carcinoma. Cell 160:963–976
273. Kurtova AV, Xiao J, Mo Q, Pazhanisamy S, Krasnow R, Lerner SP et al (2014) Blocking PGE2-induced tumour repopulation abrogates bladder cancer chemoresistance. Nature 517:209
274. Buczacki SJA, Zecchini HI, Nicholson AM, Russell R, Vermeulen L, Kemp R et al (2013) Intestinal label-retaining cells are secretory precursors expressing Lgr5. Nature 495:65
275. van Es JH, Sato T, van de Wetering M, Lyubimova A, Yee Nee AN, Gregorieff A et al (2012) Dll1+ secretory progenitor cells revert to stem cells upon crypt damage. Nat Cell Biol 14:1099
276. Sato T, van Es JH, Snippert HJ, Stange DE, Vries RG, van den Born M et al (2010) Paneth cells constitute the niche for Lgr5 stem cells in intestinal crypts. Nature 469:415
277. Tetteh Paul W, Basak O, Farin Henner F, Wiebrands K, Kretzschmar K, Begthel H et al (2016) Replacement of lost Lgr5-positive stem cells through plasticity of their enterocyte-lineage daughters. Cell Stem Cell 18:203–213
278. Tammela T, Sanchez-Rivera FJ, Cetinbas NM, Wu K, Joshi NS, Helenius K et al (2017) A Wnt-producing niche drives proliferative potential and progression in lung adenocarcinoma. Nature 545:355–359
279. Takebe N, Miele L, Harris PJ, Jeong W, Bando H, Kahn M et al (2015) Targeting Notch, Hedgehog, and Wnt pathways in cancer stem cells: clinical update. Nat Rev Clin Oncol 12:445–464
280. Pattabiraman DR, Weinberg RA (2014) Tackling the cancer stem cells [mdash] what challenges do they pose? Nat Rev Drug Discov 13:497–512
281. Li Y, Atkinson K, Zhang T (2017) Combination of chemotherapy and cancer stem cell targeting agents: preclinical and clinical studies. Cancer Lett 396:103–109
282. Pan Q, Li Q, Liu S, Ning N, Zhang X, Xu Y et al (2015) Concise review: targeting cancer stem cells using immunologic approaches. Stem Cells 33:2085–2092
283. Putzer BM, Solanki M, Herchenroder O (2017) Advances in cancer stem cell targeting: how to strike the evil at its root. Adv Drug Deliv Rev
284. Colak S, Medema JP (2014) Cancer stem cells – important players in tumor therapy resistance. FEBS J 281:4779–4791
285. de Goeij BE, Lambert JM (2016) New developments for antibody-drug conjugate-based therapeutic approaches. Curr Opin Immunol 40:14–23
286. Maccalli C, De Maria R (2015) Cancer stem cells: perspectives for therapeutic targeting. Cancer Immunol Immunother 64:91–97
287. Gupta PB, Onder TT, Jiang G, Tao K, Kuperwasser C, Weinberg RA et al (2009) Identification of selective inhibitors of cancer stem cells by high-throughput screening. Cell 138:645–659
288. Lu D, Choi MY, Yu J, Castro JE, Kipps TJ, Carson DA (2011) Salinomycin inhibits Wnt signaling and selectively induces apoptosis in chronic lymphocytic leukemia cells. Proc Natl Acad Sci U S A 108:13253–13257
289. Friend C, Scher W, Holland JG, Sato T (1971) Hemoglobin synthesis in murine virus-induced leukemic cells in vitro: stimulation of erythroid differentiation by dimethyl sulfoxide. Proc Natl Acad Sci U S A 68:378–382
290. Breitman TR, Selonick SE, Collins SJ (1980) Induction of differentiation of the human promyelocytic leukemia cell line (HL-60) by retinoic acid. Proc Natl Acad Sci U S A 77:2936–2940
291. Fenaux P, Chastang C, Chevret S, Sanz M, Dombret H, Archimbaud E et al (1999) A randomized comparison of all transretinoic acid (ATRA) followed by chemotherapy and ATRA plus chemotherapy and the role of maintenance therapy in newly diagnosed acute promyelocytic leukemia. Blood 94:1192–1200
292. Abaza Y, Kantarjian HM, Garcia-Manero G, Estey E, Borthakur G, Jabbour E et al (2016) Long-term outcome of acute promyelocytic leukemia treated with all-trans-retinoic acid, arsenic trioxide, and gemtuzumab. Blood 129:1275–1283

293. Rustighi A, Zannini A, Tiberi L, Sommaggio R, Piazza S, Sorrentino G et al (2014) Prolyl-isomerase Pin1 controls normal and cancer stem cells of the breast. EMBO Mol Med 6:99–119
294. Luo M-L, Gong C, Chen C-H, Lee DY, Hu H, Huang P et al (2014) Prolyl isomerase Pin1 acts downstream of miR-200 to promote cancer stem-like cell traits in breast cancer. Cancer Res 74:3603–3616
295. Munster PN, Troso-Sandoval T, Rosen N, Rifkind R, Marks PA, Richon VM (2001) The histone deacetylase inhibitor suberoylanilide hydroxamic acid induces differentiation of human breast cancer cells. Cancer Res 61:8492–8497
296. Uchida H, Maruyama T, Nagashima T, Asada H, Yoshimura Y (2005) Histone deacetylase inhibitors induce differentiation of human endometrial adenocarcinoma cells through up-regulation of glycodelin. Endocrinology 146:5365–5373
297. Kim KH, Roberts CWM (2016) Targeting EZH2 in cancer. Nat Med 22:128
298. Mangraviti A, Raghavan T, Volpin F, Skuli N, Gullotti D, Zhou J et al (2017) HIF-1α- targeting acriflavine provides long term survival and radiological tumor response in brain cancer therapy. Sci Rep 7:14978
299. Cheloni G, Tanturli M, Tusa I, Ho DeSouza N, Shan Y, Gozzini A et al (2017) Targeting chronic myeloid leukemia stem cells with the hypoxia-inducible factor inhibitor acriflavine. Blood 130:655–665
300. Chen K, Huang YH, Chen JL (2013) Understanding and targeting cancer stem cells: therapeutic implications and challenges. Acta Pharmacol Sin 34:732–740
301. Ng SWK, Mitchell A, Kennedy JA, Chen WC, McLeod J, Ibrahimova N et al (2016) A 17-gene stemness score for rapid determination of risk in acute leukaemia. Nature 540:433
302. Li W, Ma H, Zhang J, Zhu L, Wang C, Yang Y (2017) Unraveling the roles of CD44/CD24 and ALDH1 as cancer stem cell markers in tumorigenesis and metastasis. Sci Rep 7:13856
303. He K, Xu T, Goldkorn A (2011) Cancer cells cyclically lose and regain drug-resistant highly tumorigenic features characteristic of a cancer stem-like phenotype. Mol Cancer Ther 10:938–948

Chapter 13
The Relevance of Induced Pluripotent Stem Cells for the Study of Physiological and Premature Aging

Dido Carrero and Carlos López-Otín

Abstract Aging is a process characterized by several alterations, such as accumulation of macromolecular damage, genomic instability, and loss of heterochromatin, which leads to a decline in stem cell function and reduced regenerative capacity. Recently, reprogramming of aged cells has been achieved, which illustrates the reversible potential of aging and evokes the idea of rejuvenation. In this chapter, we summarize the main advances in the field of reprogramming in aging, both in vitro and in vivo, which have contributed to further understand the molecular mechanisms underlying physiological and pathological aging. We also describe candidate therapeutic strategies that demolish age-associated reprogramming barriers and alleviate normal and premature aging.

Keywords Aging · Progeria · Stem cells · iPSCs

13.1 Introduction

Aging is a condition linked to the development of human pathologies such as cancer, diabetes, cardiovascular disorders, and neurodegenerative diseases [1]. This process is characterized by nine cellular and molecular characteristics, known as the hallmarks of aging, which include genomic instability, loss of proteostasis, epigenetic alterations, telomere attrition, mitochondrial dysfunction, deregulated nutrient sensing, stem cell exhaustion, cellular senescence, and altered intercellular communication [2].

D. Carrero
Departamento de Bioquímica y Biología Molecular, Facultad de Medicina, Instituto Universitario de Oncología (IUOPA), Universidad de Oviedo, Oviedo, Spain

C. López-Otín (✉)
Departamento de Bioquímica y Biología Molecular, Facultad de Medicina, Instituto Universitario de Oncología (IUOPA), Universidad de Oviedo, Oviedo, Spain

Centro de Investigación en Red de Cáncer, Spain
e-mail: clo@uniovi.es

R. Delgado-Morales (ed.), *Stem Cell Genetics for Biomedical Research*,
https://doi.org/10.1007/978-3-319-90695-9_13

Premature aging diseases, also known as progeroid syndromes, are a group of uncommon genetic disorders characterized by clinical features that resemble those observed in physiological aging, such as hair loss, short stature, cardiovascular affections, or osteoporosis. However, these features appear at an early age, triggered by mechanisms that are comparable to those described for normal aging, which suggests that research focused on pathological aging may facilitate the study of different aspects of physiological aging [3]. In the past years, several animal and cellular models of progeroid syndromes have proved useful not only to study aging but also to develop new therapeutic approaches for curing these dramatic diseases [4–6].

Research on biology and medicine has undergone a radical change with the introduction of a novel technique that allows the generation of induced pluripotent stem cells (iPSCs) from adult cells in vitro. This is achieved through retroviral gene transfer and overexpression of four transcriptional factors (*OCT4*, *SOX2*, *KLF4*, and *c-MYC*, OSKM), known as the Yamanaka factors [7]. Over recent years, this methodology has been widely used by the scientific community, since the pluripotency and self-renewal abilities of iPSCs constitute an endless source of patient-specific cell lines that can be used for regenerative medicine, disease modelling, and drug screening. However, the efficiency of this technique can decrease depending on the tissue source of the donor cells or the chronological age of cell donors [8].

In this chapter, we outline the main discoveries achieved through the generation of iPSCs from physiologically and prematurely aged cells and discuss their contribution to the development of rejuvenation strategies and therapeutic approaches that will help ameliorating normal and premature aging.

13.2 Premature Aging Syndromes

Progeroid syndromes show the main features of physiological aging but appear earlier and with a more aggressive course. Such disorders can be classified in two main groups according to the mechanisms that cause them, as they can be induced by defects in the nuclear envelope or in DNA repair mechanisms [3].

13.2.1 Nuclear Envelope Alterations

The nuclear lamina is a complex membrane barrier that isolates the nucleus from the cytoplasm in eukaryotic cells and contains many proteins involved in chromatin organization and gene regulation, such as lamins [9, 10]. There are two major types of lamin proteins, the A-type, encoded by the gene *LMNA*, which includes lamins A and C, and the B-type, encoded by the genes *LMNB1* and *LMNB2*, and includes lamins B1, B2, and B3 [11]. Mutations in genes that encode lamins, in proteins involved in their posttranslational processing, or in other components of the nuclear lamina can lead to severe accelerated aging pathologies known as progeroid laminopathies, such as Hutchinson-Gilford progeria syndrome (HGPS),

Néstor-Guillermo progeria syndrome (NGPS), atypical neonatal progeria syndromes (ANPS), restrictive dermopathy (RD), and mandibuloacral dysplasia (MAD) [12–19].

HGPS is the most prevalent and widely studied accelerated aging syndrome. Most cases of HGPS originate from a de novo heterozygous silent mutation in the *LMNA* gene (p.G608G). This mutation activates a cryptic splicing site and leads to the accumulation of a toxic protein called progerin, which disrupts the integrity of the nuclear envelope [14, 15]. Within the first year of life, patients suffering from HGPS display alopecia, atherosclerosis, lipodystrophy, heart infarction, osteoporosis, and death during puberty [4]. Cells obtained from these individuals present nuclear shape abnormalities, shortened telomeres, cell senescence, and genome instability [20]. Progerin also accumulates during physiological aging, reinforcing the similarities between normal and pathological aging [21].

Additionally, mutations in the gene *ZMPSTE24* - which encodes an enzyme involved in lamin A processing - are responsible for RD and MAD type B, both characterized by facial abnormalities and bone development problems [12, 16]. Alterations in other proteins of the nuclear envelope can also cause accelerated aging, such as the mutation p.A12T in the *BANF1* gene, that encodes the protein BAF, involved in chromatin organization and nuclear lamina assembly [18]. NGPS shares many clinical features with HGPS, except for the absence of cardiovascular alterations and a longer lifespan [22].

13.2.2 Defects in DNA Repair and Premature Aging

Aging is strictly related to the accumulation of DNA damage, since DNA repair mechanisms decline with age [23]. Defects in DNA repair machinery can also cause premature aging syndromes, such as Werner syndrome (WS), Bloom syndrome (BS), Rothmund-Thomson syndrome (RTS), Cockayne syndrome (CS), xeroderma pigmentosum (XP), thiotrichodystrophy (TTD), Fanconi anemia (FA), Seckel syndrome (SS), ataxia telangiectasia (AT), and dyskeratosis congenita (DC), among others.

In humans, there are five RecQ helicases involved in double-strand break repair during DNA replication and in the maintenance of genomic stability and telomere integrity [24]. Mutations in three of them (*WRN*, *BLM*, and *RECQ4*) are associated with WS, BS, and RTS progeroid syndromes, respectively. Patients with WS exhibit growth retardation, hair graying, cataracts, dermal atrophy, atherosclerosis, osteoporosis, and a high incidence of cancer related to chromosomal instability, telomere dysfunction, and the senescent phenotype observed in WS cells [25, 26]. BS and RTS patients are characterized by immunodeficiency and increased predisposition to cancer [27, 28].

Other DNA-repair proteins altered in progeroid syndromes include components of the nucleotide excision repair response, such as the excision repair cross-complementing (ERCC) family, the Fanconi anemia proteins, and the XP proteins. Mutations in genes encoding these and other proteins involved in DNA repair are responsible for different premature aging disorders, such as CS, TTD, and XP [29]. CS is a premature aging disorder characterized by neural abnormalities and microcephaly [30]. XP is caused by an impaired ability to repair damage caused by UV light, which leads to skin pigmentation abnormalities, neurological degeneration, and increased cancer susceptibility [31]. TTD patients show skin photosensitivity and growth retardation [32].

Another set of progeroid syndromes are mainly characterized by defects in DNA double-strand break repair and telomere dynamics, such as FA, SS, AT, and DC. FA is characterized by a higher cancer susceptibility, bone marrow failure, short stature, and skin abnormalities [33]. FA cells produce high levels of reactive oxygen species (ROS), which damage telomeric regions and induce accelerated telomere shortening [34, 35]. SS is caused by mutations in *ATR* and is characterized by intrauterine growth retardation and postnatal dwarfism [36]. AT is caused by mutations in *ATM*, and patients suffering from this syndrome show immunodeficiency, neurodegeneration, pigmentary abnormalities, hair graying, and increased cancer susceptibility [37]. DC is caused by mutations in several components of the telomerase complex. Patients with DC present bone marrow failure, pulmonary alterations, cancer predisposition, hair graying, and osteoporosis [38].

13.3 Adult Stem Cell Exhaustion in Normal and Premature Aging

During aging, tissue regeneration ability decreases due to an important decline in stem cell functions as a result of DNA damage, changes in tissue environment, and alterations in gene expression [3, 39]. Since viability of stem cells is a key factor in regeneration of adult tissues, changes in these cells presumably contribute to the development of age-related diseases and premature aging syndromes.

In normally aged mice, quiescent muscle stem cells show a pre-senescent state associated with the loss of epigenetic silencing of the INK4a locus. These cells lose their regenerative capacity and undergo full senescence upon regenerative pressure [40]. A similar process, also associated with increased levels of p16/INK4a, is observed in aged hematopoietic stem cells (HSCs) as a consequence of telomere shortening and accumulation of DNA damage [41]. It is also known that lamins play an important role in controlling adult stem cell function and that progerin accumulation reduces the proliferative and differentiation capacity of pluripotent and multipotent cells [42–46].

Changes in the epigenetic landscape are a common feature of aged stem cells [47, 48]. Accordingly, mesenchymal stem cells from aged individuals show a marked reduction in the heterochromatin-associated H3K9me3 mark, as well as a downregulation of proteins involved in heterochromatin maintenance [49]. This can also be observed in an embryonic stem cell model of Werner syndrome, supporting the idea that heterochromatin disorganization may cause cellular dysfunction with age [49].

HSC senescence has also been observed in many progeroid syndromes, especially in those related to defects in telomere dynamics. For example, a novel *TERT* mutation (T1129P) causing dyskeratosis congenita leads to cellular senescence and loss of CD34+ HSCs [50]. Furthermore, TNF-α induces premature senescence in bone marrow HSCs and progenitor cells in Fanconi anemia mouse models, which correlates with the accumulation of ROS and oxidative DNA damage [51]. Altogether, these studies demonstrate the close relation between organismal aging and stem cell exhaustion.

13.4 Cellular Models of Progeroid Syndromes

Cellular models are a basic experimental tool for understanding the mechanisms that cause diseases and for developing new drugs and therapies. However, obtaining tissue-specific cell lines from aged donors or from patients suffering from pathological aging is a challenging task due to their frailty. Since 2006, a new efficient approach for the generation of induced pluripotent stem cells was developed [7]. Hence, developing patient-derived iPSCs constitutes a valuable tool for the study of specific tissues affected by a disease and the development of tissue-specific drugs. Such approaches have also been explored for the study of different progeroid syndromes.

Through the generation of iPSCs from centenarian donors, it has been observed that aging constitutes a major barrier for somatic cell reprogramming, since this process is less effective than that of cells from young donors [8, 52] (Fig. 13.1). Despite this, reprogrammed aged cells show elongated telomeres, reduced oxidative stress, and pluripotency [53, 54]. Accordingly, differentiated cells derived from them do not show age-associated features such as telomere shortening, increased cellular senescence, and metabolic defects [2, 55]. However, that is not the case of differentiated iPSCs derived from progeroid syndromes, as they still possess the genetic alterations that cause these diseases [46, 56, 57]. Here, we discuss the most relevant results obtained from the generation of iPSCs from patients with premature aging syndromes and highlight their importance for a better understanding of the molecular basis that trigger aging (Table 13.1).

13.4.1 *iPSC Generation from Progeria Cells*

Reprogramming has allowed the generation of iPSCs from several progeroid syndromes and their differentiation along multiple lineages. Notably, all these progeroid iPSCs show lower reprogramming efficiency than control cells. Nonetheless, they are identical to normal iPSCs and lack disease-specific features that characterize progeroid cells, such as an aberrant nuclear envelope or

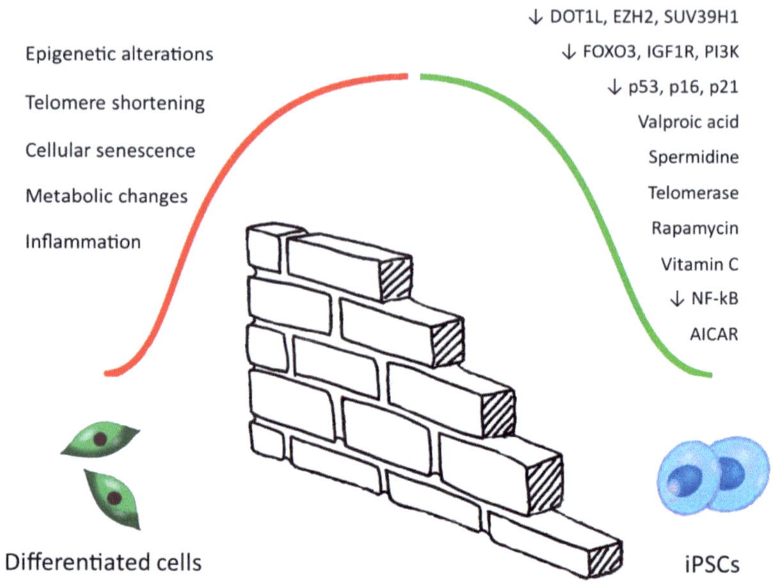

Fig. 13.1 Barriers and enhancers of somatic cell reprogramming during aging. Aging represents a crucial barrier that reduces somatic cell reprogramming efficiency due to the presence of characteristics such as cell senescence, inflammation, telomere shortening, metabolic changes, or epigenetic alterations. On the contrary, several molecules, such as vitamin C, rapamycin, spermidine, or inhibition of DOT1L, can defeat these barriers and increase aged cell reprogramming efficiency

Table 13.1 iPSCs derived from prematurely aged patients

Syndrome	Gene	Donor cells	Differentiated into	References
HGPS	*LMNA*	Fibroblasts	MSCs, dermal cells, and neural cells	[60]
HGPS	*LMNA*	Fibroblasts	Fibroblasts	[56]
HGPS	*LMNA*	Fibroblasts	VSMCs, fibroblasts	[64]
HGPS	*LMNA*	Fibroblasts	Neural cells, fibroblasts, endothelial cells, VSMCs, MSCs	[46]
HGPS	*LMNA*		Endothelial cells	[59]
HGPS	*LMNA*	Fibroblasts	Adipocytes	[158]
NGPS	*BANF1*	Fibroblasts	MSCs	[52]
aWS	*LMNA*	Fibroblasts	Fibroblasts	[56]
WS	*WRN*	Fibroblasts	Embryoid bodies	[61]
WS	*WRN*	Fibroblasts	MSCs and NSCs	[26]
CS	*ERCC6*	Fibroblasts	Embryoid bodies	[58]
CS	*ERCC6*	Fibroblasts	Neural lineage	[69]
XP	*XPA*, *XPB*, *XPC*, *XPG*, *XPV*	Fibroblasts	Neural lineage	[68, 67]
DKC	*TERT*, *DKC1*, *TCAB1*	Fibroblasts		[77]
FA	*FANCA*, *FANCC*	Fibroblasts	Hematopoietic cells	[70, 73, 71, 72]
AT	*ATM*	Fibroblasts	Neurons	[75, 76, 74]

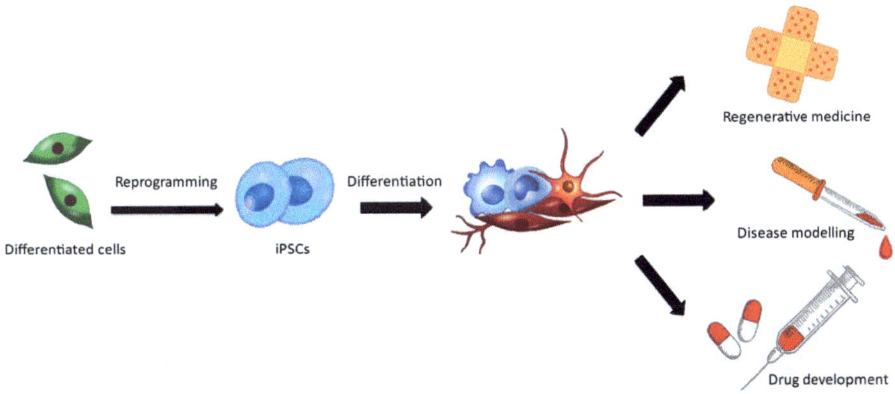

Fig. 13.2 Use of patient induced pluripotent stem cells (iPSCs) in research. The generation of patient-specific iPSCs constitutes a potentially unlimited source of cells that can be differentiated in multiple cell types for their use in regenerative medicine and cell transplantation. Moreover, iPSCs can also be used for disease modelling and developing new therapeutic approaches for genetic diseases

defects in DNA repair, which proves the absence of age-associated marks [26, 46, 52, 56–62]. However, when these cells are differentiated in several cell types, such as vascular smooth muscle cells, mesenchymal stem cells, or fibroblasts, they display again age-associated alterations like premature senescence or altered gene expression due to the activation of progerin expression. This demonstrates that iPSC models represent a very reliable source of specific cell lines affected by a disease, and therefore, they are useful tools for the study of the molecular mechanisms that cause these syndromes, for regenerative medicine, and for developing drug screenings that could be used in therapy [3, 54, 63] (Fig. 13.2).

As discussed above, HGPS-iPSCs are indistinguishable from control iPSCs due to the absence of *LMNA* expression. Moreover, when differentiated into neural progenitor cells (NPCs) and neurons, they do not show their progeroid phenotype, due to the high levels of microRNA miR-9 in neural cells, that specifically prevents lamin A expression [60]. However, HGPS-derived cells differentiated into smooth muscle cells (SMCs) or mesenchymal stem cells (MSCs) present several alterations as a result of progerin accumulation [46, 56, 64]. The vascular pathogenesis and lipodystrophy phenotype associated with this disease has also been studied through the generation of HGPS-iPSCs [62]. Moreover, expression of progerin in iPSC-derived fibroblasts and neurons resulted in several features related to aging, such as neuromelanin accumulation, which demonstrated the usefulness of progerin-induced aging for the study of late-onset age-related diseases [65]. Aged iPSCs have also been used as a model of cellular aging, since they show nuclear dysmorphism, imbalance in the components of the nuclear envelope, mitochondrial dysfunctions, *SIRT7* downregulation, and NF-κB p65 hyperactivation, among other aging-related features [66].

iPSCs from NGPS patients have also been generated, albeit at a low efficiency due to the hyperactivation of the NF-κB signalling pathway, which is mediated by the histone methyltransferase DOT1L [52]. Through in vitro differentiation experiments, it has also been found that NGPS MSCs have an impaired capacity to generate the bone, cartilage, and adipose tissue. Moreover, NGPS MSCs show profound alterations in nuclear envelope structure, as well as premature cell senescence. Together, these results demonstrate the utility of this cellular model to recapitulate age-associated alterations and NGPS pathogenesis [52].

Reprogramming of cells derived from WS patients has allowed the identification of the role of the WRN helicase in maintaining telomerase and DNA stability in the pluripotent state. These cells exhibited pluripotency abilities and genomic stability but recapitulated the cellular phenotype of WS cells during differentiation [61]. Differentiation into MSCs and NPCs restored their premature senescence phenotype and defective synthesis of the lagging strand of telomeres in MSCs, but not in NPCs. This may be regulated by telomerase, since expression of hTERT or p53 knockdown ameliorated the accelerated aging phenotype in MSCs, while inhibition of telomerase sensitized NPCs to DNA damage [26].

Xeroderma pigmentosum patient-derived iPSCs have been generated to study neurological disorders and to perform drug screening. Such iPSCs are able to differentiate into neural lineage cells, which suggest that the expression of wild-type *XPA* alleles is not necessary for iPSC generation and differentiation [67]. Differentiated neural stem cells (NSCs) or neurons showed severe DNA damage repair defects and an increased susceptibility to DNA damage-induced apoptosis [68]. The generation of iPSCs from CS patient fibroblasts and their differentiation into neural cells showed that one main component of the NER-based repair mechanisms, *ERCC6,* controls cell viability, ROS production, and normal neural function [58, 69].

Additionally, iPSCs from fibroblasts of patients with FA have been generated and show defective differentiation ability toward the hematopoietic and endothelial lineages in early hemoangiogenic progenitors, as well as decreased survival, due to defective genome stability maintenance [70–72]. Upon correction of the genetic defect, somatic cells from FA patients can be efficiently reprogrammed to pluripotency and differentiated into hematopoietic progenitors of the myeloid and erythroid lineages [73].

Reprogramming of fibroblasts from patients with AT has also been achieved, even though at a reduced efficiency. These iPSCs display hypersensitivity to ionizing radiation, defective radiation-induced signalling, alterations in DNA damage signalling pathways, changes in mitochondrial and pentose phosphate pathways, and cell cycle checkpoint defects [74]. Notably, AT iPSCs do not show any chromosomal instability in vitro for at least 80 passages [75]. AT iPSCs were also differentiated into functional neurons in order to study AT-associated neurodegeneration and drug screening [74]. This work has allowed the discovery of a small molecule that partly recovers ATM activity, a pharmacological approach that improved the DNA damage response of AT-iPSC-derived neurons [76]. Finally, generation of iPSCs from dyskeratosis congenita patients show a 50% reduction in telomerase levels and mislocalization of telomerase from Cajal bodies to nucleoli, which prevents natural telomere elongation that accompanies reprogramming [77]. This defect can be overcome by endogenous upregulation of *TERC* during the reprogramming process, which allows for telomere length restoration in DC cells [78].

Usually, the reprogramming efficiency of cells from progeroid syndromes is as low as that from aged donors. Considering that alterations in proteins involved in DNA damage and in the nuclear envelope have also been related to normal aging [2], the remarkable parallelism between these two processes supports the usefulness of these experimental models to study the mechanisms that cause physiological and premature aging. Remarkably, no iPSCs from other progeroid syndromes such as restrictive dermopathy, mandibuloacral dysplasia, Bloom syndrome, Rothmund-Thomson syndrome, trichothiodystrophy, or Seckel syndrome have been generated to date. This could be due to the low reprogramming efficiency of progeria cells, which in turn may be caused by a series of reprogramming barriers that are activated in prematurely aged cells.

13.5 Reprogramming In Vivo

The use of iPSCs as a therapeutic approach arises two major difficulties, which are the administration route and their stability upon tissue injection. Therefore, pluripotent cell dedifferentiation and reprogramming into iPSCs in vivo [79, 80] would avoid cell transplantation and supply specific iPSCs in situ that would contribute to tissue repair [81], while inducing rejuvenation of tissues and organs, ameliorating age-associated phenotypes, and extending organismal lifespan. However, this process appears to be more arbitrary than cell reprogramming in a dish, since iPSCs can produce teratomas if they are not properly differentiated [79], so the possibility of tumor generation from iPSC transplant in human patients should be evaluated before they are used in regenerative medicine. Several studies have also reported that incomplete in vitro reprogramming can result in a dedifferentiated state with advantageous properties [82–84], which could represent an appealing therapeutic and regenerative path that needs to be studied.

Reprogramming in vivo was first achieved in 2013, when Abad et al. reported that forced expression of Yamanaka factors resulted in teratoma formation in several organs, which implies that full reprogramming can occur in vivo [79]. Moreover, Nanog-positive cells were detected in multiple tissues, including stomach, intestine, pancreas, and kidney, while mouse bone marrow transplantation experiments indicate that the hematopoietic system is also reprogrammable in vivo [79]. Besides, in vivo-generated mouse iPSCs contribute to the trophectoderm lineage upon blastocyst injection, which suggests that iPSCs can drive cell differentiation and development and that they show totipotent features that are absent in standard iPS or ES cells [79]. Similarly, premature termination of reprogramming in vivo leads to cancer development through the accumulation of epigenetic alterations [80]. In this context, induction of the Yamanaka factors in mice for 4 days resulted in the proliferation of dysplastic cells, which then reverted to cells with normal morphologies. However, this was not the case for cells that overexpressed the Yamanaka factors for 7 days [80]. Reprogrammed cells showed epigenetic patterns that resembled those of ESCs. In the kidney, reprogramming in vivo led to a proliferation similar to Wilms tumor, a common pediatric kidney type of cancer [80].

Unfortunately, reprogramming in vivo is related to a high mortality in mice, partly as a result of teratoma development in different tissues, but in other cases, the cause of death is unknown [79, 80]. This fact suggests that loss of cellular identity upon dedifferentiation may impair organ or tissue functionality, leading to organismal death [85]. In this context, in vivo reprogramming in the absence of c-Myc delays the development of tumors [80], and mice expressing *Oct4* alone suffer from dysplastic growth that reverts back to normal when *Oct4* expression is concluded [86].

Recent studies have shown that partial reprogramming by short-term cyclic expression of the Yamanaka factors in vivo ameliorates cellular and physiological alterations related to aging, prolongs lifespan of progeroid mice, prevents tumor formation, and improves recovery from metabolic disease and muscle injury in aged wild-type mice [87]. Moreover, short-term induction of Yamanaka factors in progeroid cells reduced the expression of markers related to DNA damage - such as γH2AX, senescence, and ROS production - without causing the loss of cellular identity. Nevertheless, 8 days after OSKM induction, progeroid features slowly came back [87]. This was achieved as a result of restoration of the levels of two epigenetic marks of heterochromatin commonly associated with aging (H3K9me3 and H4K20me3), which reinforces the role of epigenetic alterations in aging [87].

It has also been described that overexpression of the Yamanaka factors in vivo exposes tissues to extensive damage, driving some other cells into cellular senescence [88]. Certain levels of tissue damage and senescence seem to be necessary for OSKM-driven reprogramming, as tissues lacking INK4a/ARF do not undergo senescence, and their ability to reprogram is extensively compromised, while tissues lacking p53 present higher levels of senescence, intensified cytokine production, and increased in vivo reprogramming [88]. This cross talk between senescence and reprogramming is mediated by the cytokine-rich microenvironment associated with senescent cells, since pharmacological inhibition of NF-κB or IL-6 reduces in vivo reprogramming [88]. Therefore, OSKM-driven reprogramming in both progeroid and physiologically aged mice seems to be more efficient than in young mice [88].

In addition, acute and chronic injury enables reprogramming in skeletal muscle, since it is associated with accumulation of senescent cells [89]. Similarly, advanced aging or local irradiation further enhanced in vivo reprogramming, while selective depletion of senescent cells reduced reprogramming efficiency. This effect appears to result from the release of IL-6, which suggests a potential link with the senescence-associated secretory phenotype (SASP) [89].

13.6 Barriers to Cell Reprogramming in Aging and Progeria

It is now widely acknowledged that both physiological and pathological aging share most of their underlying mechanisms [2, 3]. Recently there has been an increasing effort to improve the efficiency of the generation of iPSCs, which can be achieved through the identification and repression of reprogramming barriers and activation of enhancers that improve the global efficiency of the process [54].

The reprogramming process is indeed a slow and inefficient process, in part because forced expression of the Yamanaka factors is a stressful mechanism that activates apoptosis and cellular senescence by upregulating molecules such as p53, p16 (INK4a), and p21 (CIP1) [90, 91]. Accordingly, inhibition of these factors improves the efficiency of reprogramming in human and mouse somatic cells [92–94]. Moreover, several features displayed by aged cells, such as epigenetic alterations, telomere shortening, or cell senescence, represent strong barriers that greatly reduce the efficiency of cell reprogramming (Fig. 13.1). HGPS cells, as well as cells from *Zmpste24*-deficient mice, already show hyperactivation of p53 and p16 (INK4a), which makes somatic cell reprogramming even more difficult [95]. Moreover, MSCs differentiated from NGPS iPSCs show premature cell senescence characterized by senescence-associated-β-galactosidase (SA-β-Gal)-positive staining, senescence-associated heterochromatin foci, reduced bromodeoxyuridine (BrdU) incorporation, increased levels of p16, and significant loss of lamin B1 [52]. Furthermore, recent studies have shown that the E6 protein from the human papillomavirus 16 (HPV16) recovers FA iPSC colony formation through p53 inhibition. Consequently, it is demonstrated that the FA pathway is required for reprogramming through p53-dependent mechanisms [96].

Transcriptional repressor Patz1 also plays an inhibitory role in the reprogramming process since more iPSC colonies can be induced from *Patz1*$^{+/-}$ than from wild-type murine embryonic fibroblasts (MEFs). In this sense, *Patz1*$^{+/-}$ murine fibroblasts can overcome the senescence barrier of the INK4a/ARF locus, while *Patz1*$^{-/-}$ cells show the lowest reprogramming efficiency, which may be due to cell senescence caused by the upregulation of INK4a/ARF locus. Altogether, these results demonstrate that the expression of Patz1 can modulate reprogramming [97]. Telomere dynamics also play a major role in regulating cell senescence. Cell reprogramming requires the reactivation of telomerase activity in mammals, and while telomere shortening represents a major barrier to this process, ectopic expression of telomerase increases global reprogramming efficiency [94].

Regulation of cell cycle is crucial for the survival of the cell, including the repair of genetic damage and the prevention of uncontrolled cell division. Moreover, breakdown and reorganization of the nuclear envelope during cell division is a critical step in the cell cycle, in which lamins play a major role. This is the reason why prematurely aged cells present many cell cycle alterations, particularly those derived from patients that suffer from progeroid syndromes caused by defects in the nuclear envelope [3, 98].

Aging is also characterized by defects in main metabolic pathways [99], and the age-related decline in stem cell function is characterized by metabolic changes [48]; therefore it is likely that several metabolic regulators may play a relevant role in enhancing reprogramming efficiency. Accordingly, inhibitors of the insulin/IGF1 signalling pathway, such as the PI3K inhibitor LY294002 and the IGF1R inhibitor PQ401, are able to increase reprogramming efficiency in mice [100]. Moreover, mTOR inhibitors and autophagy activators such as PP242, rapamycin, or resveratrol notably improve the speed and efficiency of iPSC generation [100], decelerating the cellular senescence imposed by DNA damage-like responses and identifying mTOR as another important reprogramming repressor [101]. Activation of autophagy with spermidine also enhances the reprogramming efficiency of mice and human somatic cells [100] through the activation of SIRT1 and SIRT6 histone deacetylases [102]. Sirt6 is downregulated in iPSCs from old mice, in which nonhomologous end-joining (NHEJ) is less efficient than homologous recombination (HR). Introduction of a combination of Sirt6 and the Yamanaka factors during reprogramming promotes DSB repair in these iPSCs, which allows for the improvement of the quality of iPSCs derived from old donors by activating NHEJ and stabilizing the genome [103]. In this context, addition of *ZSCAN10* to the four Yamanaka factors during reprogramming of aged cells normalizes ROS-glutathione homeostasis and DNA response and corrects the genomic instability of IPSCs from aged donors [104, 105]. Moreover, the AMPK activator AICAR benefits mouse somatic cell reprogramming in an autophagy-independent Atg5-based way [106]. Lastly, recent studies demonstrate that FoxO3 plays a central role in neuronal reprogramming, since knockout or knockdown of this gene increased the reprogramming efficiency of aging but not of embryonic mouse fibroblasts [107].

During aging, mammals suffer from a chronic pro-inflammatory state known as "inflammaging" [108], when senescent cells secrete a growing combination of factors that alter their microenvironment and the neighboring cells in a paracrine way, reinforcing the senescence arrest, and activating immune surveillance. These factors conform the senescence-associated secretory phenotype (SASP), which is regulated by NF-κB [109, 110]. Inflammaging is also a main feature of accelerated aging [3]. Mouse models of progeroid laminopathies (*Zmpste24*-deficient and *Lmna*^{G609G} knock-in mice) show aberrant overactivation of NF-κB by ATM, as well as expression of several pro-inflammatory cytokines, such as IL-6, CXCL-1, and TNF-α, which contributes to the pathogenesis of accelerated aging syndromes [111]. Likewise, chronic inflammation induced by knockout of *nfkb1* in mice induces premature aging, while fibroblasts from these mice exhibit faster cell senescence. Moreover, accumulation of senescent cells in *nfkb1*$^{-/-}$ tissues can be blocked by the administration of anti-inflammatories to these mice, which also restores the regenerative potential of their tissues [112]. NF-κB signalling also plays an important role in lymphopoiesis, and accordingly, *Zmpste24*$^{-/-}$ mice show reduced numbers of early B-cell progenitors due to uncontrolled NF-κB activation [113]. An inflammatory phenotype associated with aging has also been observed in WS patients [114, 115]. Through derivation of iPSCs from individuals with premature or physiological aging, we recently found that age-associated NF-kB activation constitutes an important barrier for generation of iPSCs. Thus, pharmacological inhibition of NF-κB significantly increased the reprogramming efficiency of fibroblasts from NGPS and HGPS patients, as well as from normal aged donors [52].

Aged cells also show several epigenetic alterations, such as increased levels of active histone modifications and global loss of histones, those lead to chromatin relaxation and deregulation of gene expression [116]. Similarly, these epigenetic changes have also been found in cells from patients with progeroid syndromes [3]. Former studies have also pointed at chromatin-modifying proteins such as DOT1L and YYI as reprogramming repressors [52, 117]. NGPS fibroblasts also showed a noticeable overexpression of DOT1L, suggesting a direct regulation by NF-kB. Accordingly, DOT1L inhibitors strikingly increased the reprogramming efficiency of NGPS fibroblasts, and the treatment of progeroid mice with these inhibitors during lifetime resulted in a significant increase in longevity and an amelioration of their progeroid phenotype. These facts established a new relationship between rejuvenation and reprogramming improvement and suggest DOT1L as a novel target for rejuvenation-based approaches [52]. Other histone methyltransferases, such as Ezh2 or SUV39H1, have also been identified as reprogramming barriers [117, 118]. Furthermore, vitamin C treatment enhances mouse somatic cell reprogramming through a *Jhdm1b*-dependent mechanism, which represses the *Ink4/Arf* locus by removing H3K36me2/3 marks [119]. Moreover, administration of valproic acid, a histone deacetylase inhibitor, decreases cell senescence in reprogrammed human and mouse fibroblasts, enhancing the efficiency of iPSC generation [120].

13.7 Therapeutic and Rejuvenation Strategies

In the past few years, research on physiological and pathological aging has allowed to elucidate the basic mechanisms underlying these two processes, which undoubtedly contributed to the development of therapeutic strategies for these pathologies, especially for HGPS. Treatment of this progeroid syndrome has been explored from many different perspectives, being the first approach the one that involved the use of farnesyltransferase inhibitors (FTIs) such as lonafarnib, which proved efficient in cell culture and animal models. However, when tested in HGPS patients, lonafarnib administration only led to limited improvements in weight gain, vascular stiffness, and bone structure in some patients [121], likely due to the geranylgeranylation of progerin after FTI treatment, which produced another toxic form of prelamin A [6]. Consequently, inhibition of farnesylation and geranylgeranylation through combination of statins (pravastatin) and aminobisphosphonates (zoledronic acid) proved efficient in *Zmspte24*-deficient mice and is currently being tested in HGPS patients [4, 6].

Since these first strategies for HGPS treatment were described, new therapies have been tested in animal and cellular models with promising results for this and other progeroid syndromes, such as the administration of rapamycin [122–124]; morpholino antisense oligonucleotides [5, 125]; recombinant IGF-1 [126]; sodium salicylate, which inhibits the IKK complex [111]; pyrophosphate [127]; methylene blue [128]; Icmt inhibitors [129]; β-hydroxybutyrate; PARP inhibition; or NAD(+) supplementation [130]; serine protease inhibitors [131], danazol [132]; and p38 MAP-kinase inhibitors [133, 134].

Another strategy for treating HGPS patients is based on the administration of the SIRT1 activator resveratrol, which strengthens the binding between SIRT1 and lamin A, increasing its deacetylase activity. As a result, resveratrol treatment rescues adult stem cell decline, slows down body weight loss, improves trabecular bone structure, and significantly extends the lifespan in *Zmpste24*$^{-/-}$ mice [135]. Resveratrol also proved efficient in diminishing some of the features observed in *WRN*-deficient mice, such as hyperglycemia and insulin resistance, although it did not extend their lifespan [136]. Moreover, the proteasome inhibitor MG132 contributed to reducing progerin production through splicing regulation and promoted its degradation through macroautophagy, which improved the characteristic HGPS cellular phenotype, while injection in skeletal muscle of *Lmna*$^{G609G/G609G}$ mice locally reduced progerin levels Insertar cita [137]. Likewise, DOT1L inhibitors have been found to facilitate the generation of iPSCs, being also an effective rejuvenation strategy for both physiologically and prematurely aged human cells [52]. Additionally, treatment of *Zmpste24*-deficient mice during lifetime with DOT1L inhibitors resulted in a significant increase in their longevity and a remarkable amelioration of their progeroid features [52].

Loss of stem cell regenerative capacity and pluripotency is a major characteristic of organismal aging. Therefore, maintenance of these features can contribute to achieve rejuvenation of normally or prematurely aged individuals. In this context, several compounds have been observed to enhance the regenerative function of aged stem cells. For example, fibronectin is involved in the regenerative capacity of aged muscle stem cells (MuSCs). Deletion of *FN1* from young muscles replicates the aging phenotype and leads to a loss of MuSCs through deregulation of the p38 pathway. Moreover, reconstitution of FN levels in aged cells restores muscle regeneration ability [138]. Other compounds also alleviated aged satellite cell (SC) loss of self-renewal ability through inhibition of p38 MAPK activity [139, 140], which may contribute to enhance the skeletal muscle function and regenerative capacity, as well as to avoid loss of skeletal muscle mass and sarcopenia, both processes common during aging.

Similarly, β1-integrin also plays an important role in maintaining SC self-renewal and regeneration capacity in skeletal muscle, since it is deregulated in aged mouse SCs, and its hyperactivation improves regeneration after experimentally induced muscle injury, as well as in mouse models for Duchenne muscular dystrophy [141]. Inhibition of the JAK-STAT pathway also improves adult satellite stem cell function and muscle regeneration in aged mice [142]. Administration of the hormone oxytocin also enabled muscle regeneration by enhancing aged SC proliferation through activation of MAPK/ERK signalling pathway, while inhibition of oxytocin lead to premature sarcopenia [143]. Moreover, silencing of p16/INK4a expression restored quiescence and muscle SC function of aged mice [40]. Pharmacological inhibition of CDC42 activity functionally rejuvenates aged HSCs [144].

In the past years, senolytic drugs are becoming increasingly important as a successful strategy for the amelioration of aging. In this context, selective depletion of senescent muscle cells and aged HSCs was achieved through oral administration of a senolytic compound known as ABT263, which inhibited BCL-2/BCL-xL, contributing to the rejuvenation of the hematopoietic tissue [145]. Moreover, crossing the fast-aging mouse model *Bubr1*$^{H/H}$ with a systemic model for clearance of p16-positive cells (INK-ATTAC) resulted in the improvement of the disease phenotype shown by *Bubr1*$^{H/H}$ mice [146, 147]. In naturally aged mice, clearance of senescent cells successfully prolonged healthspan and lifespan [148].

Autophagy can maintain quiescence in SCs, leading to a muscle regenerative decline and stem cell loss of function. Accordingly, pharmacological and genetic reestablishment of autophagy restored homeostasis and regenerative functions in senescent satellite cells [149]. Moreover, NAD^+ repletion improved mitochondrial activity and muscle stem cell function and enhanced the lifespan of a mouse model of muscular dystrophy [150]. Furthermore, the inhibition of chromatin activation through deletion of Hoxa9 improved satellite cell function and muscle regeneration in old mice, which suggests that epigenetic stress response may be related to functionality of stem cells of aged mice [151].

These facts highlight the reversible potential of aging both in vitro and in vivo, which elicits the idea of amelioration of aging and organismal rejuvenation.

13.8 Concluding Remarks

Modelling aging in a dish can be a hard task due to the inability of reproducing an aged environment and to the reduced reprogramming efficiency of aged cells. Nevertheless, generation of iPSCs brings a new opportunity for improving regenerative medicine by avoiding the ethical problems related to the use of hESCs while preserving the benefits of autologous transplantation and self-renewal. According to this, promising results have been obtained using iPSCs derived from several age-related disorders and healthy aged donors. Moreover, iPSCs have also contributed to enlighten the mechanisms that cause aging [54] and to simplify drug screening for the identification of novel therapeutic targets that could be used as treatments for age-related diseases [152]. Somatic cell reprogramming is capable of reversing some cellular and molecular hallmarks of aging, which highlights the fact that rejuvenation is possible, at least at a cellular level [55, 153]. This is evidenced through the increase of telomere length [78, 94], the gain of mitochondrial function [154, 155], and the loss of senescence markers [156] observed in iPSCs derived from physiologically and pathologically aged patients. Furthermore, the development of gene editing tools such as the CRISPR/Cas9 technique [157] may contribute in the future to obtain corrected and safe iPSCs that could be used to treat age disorders with a genetic origin.

iPSCs represent a promising strategy for modelling age-related human diseases, since they could help to explore the molecular basis of such pathologies. Even though recent studies have obtained reprogrammed cells from patients suffering from premature aging syndromes and from centenarian donors, reprogramming cells from age-related diseases is still a challenging task. Herein, we have analyzed the barriers of somatic cell reprogramming in aging, in order to improve the reprogramming efficiency of this process. Overcoming age-related barriers of reprogramming is also a valid approach for developing rejuvenation strategies, which could inaugurate a new era in the study of aging.

Acknowledgments This work was supported by grants from European Union (DeAge, ERC Advanced Grant), Ministerio de Economía y Competitividad, Instituto de Salud Carlos III; CIBERONC, Plan Feder, and EDP Foundation. We also thank the generous support by J. I. Cabrera and Associazione Italiana Progeria Sammy Basso. The Instituto Universitario de Oncología is supported by Fundación Bancaria Caja de Ahorros de Asturias.

Bibliography

1. Khan SS, Singer BD, Vaughan DE (2017) Molecular and physiological manifestations and measurement of aging in humans. Aging Cell 16:624–633
2. Lopez-Otin C, Blasco MA, Partridge L, Serrano M, Kroemer G (2013) The hallmarks of aging. Cell 153:1194–1217
3. Carrero D, Soria-Valles C, Lopez-Otin C (2016) Hallmarks of progeroid syndromes: lessons from mice and reprogrammed cells. Dis Model Mech 9:719–735
4. Gordon LB, Rothman FG, Lopez-Otin C, Misteli T (2014) Progeria: a paradigm for translational medicine. Cell 156:400–407
5. Osorio FG, Navarro CL, Cadinanos J, Lopez-Mejia IC, Quiros PM, Bartoli C, Rivera J, Tazi J, Guzman G, Varela I et al (2011) Splicing-directed therapy in a new mouse model of human accelerated aging. Sci Transl Med 3:106ra107
6. Varela I, Pereira S, Ugalde AP, Navarro CL, Suarez MF, Cau P, Cadinanos J, Osorio FG, Foray N, Cobo J et al (2008) Combined treatment with statins and aminobisphosphonates extends longevity in a mouse model of human premature aging. Nat Med 14:767–772
7. Takahashi K, Yamanaka S (2006) Induction of pluripotent stem cells from mouse embryonic and adult fibroblast cultures by defined factors. Cell 126:663–676
8. Phanthong P, Raveh-Amit H, Li T, Kitiyanant Y, Dinnyes A (2013) Is aging a barrier to reprogramming? Lessons from induced pluripotent stem cells. Biogerontology 14:591–602
9. Turgay Y, Eibauer M, Goldman AE, Shimi T, Khayat M, Ben-Harush K, Dubrovsky-Gaupp A, Sapra KT, Goldman RD, Medalia O (2017) The molecular architecture of lamins in somatic cells. Nature 543:261–264
10. Ungricht R, Kutay U (2017) Mechanisms and functions of nuclear envelope remodelling. Nat Rev Mol Cell Biol 18:229–245
11. Cau P, Navarro C, Harhouri K, Roll P, Sigaudy S, Kaspi E, Perrin S, De Sandre-Giovannoli A, Levy N (2014) Nuclear matrix, nuclear envelope and premature aging syndromes in a translational research perspective. Semin Cell Dev Biol 29:125–147
12. Agarwal AK, Fryns JP, Auchus RJ, Garg A (2003) Zinc metalloproteinase, ZMPSTE24, is mutated in mandibuloacral dysplasia. Hum Mol Genet 12:1995–2001
13. Barthelemy F, Navarro C, Fayek R, Da Silva N, Roll P, Sigaudy S, Oshima J, Bonne G, Papadopoulou-Legbelou K, Evangeliou AE et al (2015) Truncated prelamin A expression in HGPS-like patients: a transcriptional study. Eur J Hum Genet 23:1051–1061

14. De Sandre-Giovannoli A, Bernard R, Cau P, Navarro C, Amiel J, Boccaccio I, Lyonnet S, Stewart CL, Munnich A, Le Merrer M et al (2003) Lamin a truncation in Hutchinson-Gilford progeria. Science 300:2055
15. Eriksson M, Brown WT, Gordon LB, Glynn MW, Singer J, Scott L, Erdos MR, Robbins CM, Moses TY, Berglund P et al (2003) Recurrent de novo point mutations in lamin A cause Hutchinson-Gilford progeria syndrome. Nature 423:293–298
16. Navarro CL, Cadinanos J, De Sandre-Giovannoli A, Bernard R, Courrier S, Boccaccio I, Boyer A, Kleijer WJ, Wagner A, Giuliano F et al (2005) Loss of ZMPSTE24 (FACE-1) causes autosomal recessive restrictive dermopathy and accumulation of lamin A precursors. Hum Mol Genet 14:1503–1513
17. Osorio FG, Ugalde AP, Marino G, Puente XS, Freije JM, Lopez-Otin C (2011) Cell autonomous and systemic factors in progeria development. Biochem Soc Trans 39:1710–1714
18. Puente XS, Quesada V, Osorio FG, Cabanillas R, Cadinanos J, Fraile JM, Ordonez GR, Puente DA, Gutierrez-Fernandez A, Fanjul-Fernandez M et al (2011) Exome sequencing and functional analysis identifies BANF1 mutation as the cause of a hereditary progeroid syndrome. Am J Hum Genet 88:650–656
19. Soria-Valles C, Carrero D, Gabau E, Velasco G, Quesada V, Barcena C, Moens M, Fieggen K, Mohrcken S, Owens M et al (2016) Novel LMNA mutations cause an aggressive atypical neonatal progeria without progerin accumulation. J Med Genet
20. Gonzalo S, Kreienkamp R (2015) DNA repair defects and genome instability in Hutchinson-Gilford Progeria Syndrome. Curr Opin Cell Biol 34:75–83
21. Scaffidi P, Misteli T (2006) Lamin A-dependent nuclear defects in human aging. Science 312:1059–1063
22. Cabanillas R, Cadinanos J, Villameytide JA, Perez M, Longo J, Richard JM, Alvarez R, Duran NS, Illan R, Gonzalez DJ et al (2011) Nestor-Guillermo progeria syndrome: a novel premature aging condition with early onset and chronic development caused by BANF1 mutations. Am J Med Genet A 155A:2617–2625
23. Vijg J, Suh Y (2013) Genome instability and aging. Annu Rev Physiol 75:645–668
24. Bernstein KA, Gangloff S, Rothstein R (2010) The RecQ DNA helicases in DNA repair. Annu Rev Genet 44:393–417
25. Crabbe L, Jauch A, Naeger CM, Holtgreve-Grez H, Karlseder J (2007) Telomere dysfunction as a cause of genomic instability in Werner syndrome. Proc Natl Acad Sci U S A 104:2205–2210
26. Cheung HH, Liu X, Canterel-Thouennon L, Li L, Edmonson C, Rennert OM (2014) Telomerase protects werner syndrome lineage-specific stem cells from premature aging. Stem Cell Reports 2:534–546
27. Ellis NA, Sander M, Harris CC, Bohr VA (2008) Bloom's syndrome workshop focuses on the functional specificities of RecQ helicases. Mech Ageing Dev 129:681–691
28. Puzianowska-Kuznicka M, Kuznicki J (2005) Genetic alterations in accelerated ageing syndromes. Do they play a role in natural ageing? Int J Biochem Cell Biol 37:947–960
29. Marteijn JA, Lans H, Vermeulen W, Hoeijmakers JH (2014) Understanding nucleotide excision repair and its roles in cancer and ageing. Nat Rev Mol Cell Biol 15:465–481
30. Tan WH, Baris H, Robson CD, Kimonis VE (2005) Cockayne syndrome: the developing phenotype. Am J Med Genet A 135:214–216
31. Cleaver JE (2005) Cancer in xeroderma pigmentosum and related disorders of DNA repair. Nat Rev Cancer 5:564–573
32. Itin PH, Sarasin A, Pittelkow MR (2001) Trichothiodystrophy: update on the sulfur-deficient brittle hair syndromes. J Am Acad Dermatol 44:891–920. quiz 921-894
33. Deakyne JS, Mazin AV (2011) Fanconi anemia: at the crossroads of DNA repair. Biochemistry (Mosc) 76:36–48
34. Bogliolo M, Surralles J (2015) Fanconi anemia: a model disease for studies on human genetics and advanced therapeutics. Curr Opin Genet Dev 33:32–40
35. Uziel O, Reshef H, Ravid A, Fabian I, Halperin D, Ram R, Bakhanashvili M, Nordenberg J, Lahav M (2008) Oxidative stress causes telomere damage in Fanconi anaemia cells - a possible predisposition for malignant transformation. Br J Haematol 142:82–93

36. Shanske A, Caride DG, Menasse-Palmer L, Bogdanow A, Marion RW (1997) Central nervous system anomalies in Seckel syndrome: report of a new family and review of the literature. Am J Med Genet 70:155–158
37. Sahin E, Depinho RA (2010) Linking functional decline of telomeres, mitochondria and stem cells during ageing. Nature 464:520–528
38. Stanley SE, Armanios M (2015) The short and long telomere syndromes: paired paradigms for molecular medicine. Curr Opin Genet Dev 33:1–9
39. Signer RA, Morrison SJ (2013) Mechanisms that regulate stem cell aging and life span. Cell Stem Cell 12:152–165
40. Sousa-Victor P, Gutarra S, Garcia-Prat L, Rodriguez-Ubreva J, Ortet L, Ruiz-Bonilla V, Jardi M, Ballestar E, Gonzalez S, Serrano AL et al (2014) Geriatric muscle stem cells switch reversible quiescence into senescence. Nature 506:316–321
41. Wang J, Lu X, Sakk V, Klein CA, Rudolph KL (2014) Senescence and apoptosis block hematopoietic activation of quiescent hematopoietic stem cells with short telomeres. Blood 124:3237–3240
42. Espada J, Varela I, Flores I, Ugalde AP, Cadinanos J, Pendas AM, Stewart CL, Tryggvason K, Blasco MA, Freije JM et al (2008) Nuclear envelope defects cause stem cell dysfunction in premature-aging mice. J Cell Biol 181:27–35
43. Pacheco LM, Gomez LA, Dias J, Ziebarth NM, Howard GA, Schiller PC (2014) Progerin expression disrupts critical adult stem cell functions involved in tissue repair. Aging (Albany NY) 6:1049–1063
44. Rosengardten Y, McKenna T, Grochova D, Eriksson M (2011) Stem cell depletion in Hutchinson-Gilford progeria syndrome. Aging Cell 10:1011–1020
45. Scaffidi P, Misteli T (2008) Lamin A-dependent misregulation of adult stem cells associated with accelerated ageing. Nat Cell Biol 10:452–459
46. Zhang J, Lian Q, Zhu G, Zhou F, Sui L, Tan C, Mutalif RA, Navasankari R, Zhang Y, Tse HF et al (2011) A human iPSC model of Hutchinson Gilford Progeria reveals vascular smooth muscle and mesenchymal stem cell defects. Cell Stem Cell 8:31–45
47. Neves J, Sousa-Victor P, Jasper H (2017) Rejuvenating strategies for stem cell-based therapies in aging. Cell Stem Cell 20:161–175
48. Ren R, Ocampo A, Liu GH, Izpisua Belmonte JC (2017) Regulation of stem cell aging by metabolism and epigenetics. Cell Metab 26:460–474
49. Zhang W, Li J, Suzuki K, Qu J, Wang P, Zhou J, Liu X, Ren R, Xu X, Ocampo A et al (2015) Aging stem cells. A Werner syndrome stem cell model unveils heterochromatin alterations as a driver of human aging. Science 348:1160–1163
50. Stockklausner C, Raffel S, Klermund J, Bandapalli OR, Beier F, Brummendorf TH, Burger F, Sauer SW, Hoffmann GF, Lorenz H et al (2015) A novel autosomal recessive TERT T1129P mutation in a dyskeratosis congenita family leads to cellular senescence and loss of CD34+ hematopoietic stem cells not reversible by mTOR-inhibition. Aging (Albany NY) 7:911–927
51. Zhang X, Sejas DP, Qiu Y, Williams DA, Pang Q (2007) Inflammatory ROS promote and cooperate with the Fanconi anemia mutation for hematopoietic senescence. J Cell Sci 120:1572–1583
52. Soria-Valles C, Osorio FG, Gutierrez-Fernandez A, De Los Angeles A, Bueno C, Menendez P, Martin-Subero JI, Daley GQ, Freije JM, Lopez-Otin C (2015) NF-kappaB activation impairs somatic cell reprogramming in ageing. Nat Cell Biol 17:1004–1013
53. Rohani L, Johnson AA, Arnold A, Stolzing A (2014) The aging signature: a hallmark of induced pluripotent stem cells? Aging Cell 13:2–7
54. Soria-Valles C, Lopez-Otin C (2016) iPSCs: on the road to reprogramming aging. Trends Mol Med 22:713–724
55. Freije JM, Lopez-Otin C (2012) Reprogramming aging and progeria. Curr Opin Cell Biol 24:757–764
56. Ho JC, Zhou T, Lai WH, Huang Y, Chan YC, Li X, Wong NL, Li Y, Au KW, Guo D et al (2011) Generation of induced pluripotent stem cell lines from 3 distinct laminopathies bearing heterogeneous mutations in lamin A/C. Aging (Albany NY) 3:380–390

57. Krishnan V, Chow MZ, Wang Z, Zhang L, Liu B, Liu X, Zhou Z (2011) Histone H4 lysine 16 hypoacetylation is associated with defective DNA repair and premature senescence in Zmpste24-deficient mice. Proc Natl Acad Sci U S A 108:12325–12330
58. Andrade LN, Nathanson JL, Yeo GW, Menck CF, Muotri AR (2012) Evidence for premature aging due to oxidative stress in iPSCs from Cockayne syndrome. Hum Mol Genet 21:3825–3834
59. Lo CY, Tjong YW, Ho JC, Siu CW, Cheung SY, Tang NL, Yu S, Tse HF, Yao X (2014) An upregulation in the expression of vanilloid transient potential channels 2 enhances hypotonicity-induced cytosolic Ca(2)(+) rise in human induced pluripotent stem cell model of Hutchinson-Gillford Progeria. PLoS One 9:e87273
60. Nissan X, Blondel S, Navarro C, Maury Y, Denis C, Girard M, Martinat C, De Sandre-Giovannoli A, Levy N, Peschanski M (2012) Unique preservation of neural cells in Hutchinson- Gilford progeria syndrome is due to the expression of the neural-specific miR-9 microRNA. Cell Rep 2:1–9
61. Shimamoto A, Kagawa H, Zensho K, Sera Y, Kazuki Y, Osaki M, Oshimura M, Ishigaki Y, Hamasaki K, Kodama Y et al (2014) Reprogramming suppresses premature senescence phenotypes of Werner syndrome cells and maintains chromosomal stability over long-term culture. PLoS One 9:e112900
62. Xiong Z, Lu Y, Xue J, Luo S, Xu X, Zhang L, Peng H, Li W, Chen D, Hu Z et al (2013) Hutchinson-Gilford progeria syndrome accompanied by severe skeletal abnormalities in two Chinese siblings: two case reports. J Med Case Rep 7:63
63. Brunauer R, Kennedy BK (2015) Medicine Progeria accelerates adult stem cell aging. Science 348:1093–1094
64. Liu GH, Barkho BZ, Ruiz S, Diep D, Qu J, Yang SL, Panopoulos AD, Suzuki K, Kurian L, Walsh C et al (2011) Recapitulation of premature ageing with iPSCs from Hutchinson-Gilford progeria syndrome. Nature 472:221–225
65. Miller JD, Ganat YM, Kishinevsky S, Bowman RL, Liu B, Tu EY, Mandal PK, Vera E, Shim JW, Kriks S et al (2013) Human iPSC-based modeling of late-onset disease via progerin-induced aging. Cell Stem Cell 13:691–705
66. Petrini S, Borghi R, D'Oria V, Restaldi F, Moreno S, Novelli A, Bertini E, Compagnucci C (2017) Aged induced pluripotent stem cell (iPSCs) as a new cellular model for studying premature aging. Aging (Albany NY) 9:1453–1469
67. Ohnishi H, Kawasaki T, Deguchi T, Yuba S (2015) Generation of xeroderma pigmentosum-A patient-derived induced pluripotent stem cell line for use as future disease model. Cell Reprogram 17:268–274
68. Fu L, Xu X, Ren R, Wu J, Zhang W, Yang J, Ren X, Wang S, Zhao Y, Sun L et al (2016) Modeling xeroderma pigmentosum associated neurological pathologies with patients-derived iPSCs. Protein Cell 7:210–221
69. Vessoni AT, Herai RH, Karpiak JV, Leal AM, Trujillo CA, Quinet A, Agnez Lima LF, Menck CF, Muotri AR (2016) Cockayne syndrome-derived neurons display reduced synapse density and altered neural network synchrony. Hum Mol Genet 25:1271–1280
70. Muller LU, Milsom MD, Harris CE, Vyas R, Brumme KM, Parmar K, Moreau LA, Schambach A, Park IH, London WB et al (2012) Overcoming reprogramming resistance of Fanconi anemia cells. Blood 119:5449–5457
71. Suzuki NM, Niwa A, Yabe M, Hira A, Okada C, Amano N, Watanabe A, Watanabe K, Heike T, Takata M et al (2015) Pluripotent cell models of fanconi anemia identify the early pathological defect in human hemoangiogenic progenitors. Stem Cells Transl Med 4:333–338
72. Yung SK, Tilgner K, Ledran MH, Habibollah S, Neganova I, Singhapol C, Saretzki G, Stojkovic M, Armstrong L, Przyborski S et al (2013) Brief report: human pluripotent stem cell models of fanconi anemia deficiency reveal an important role for fanconi anemia proteins in cellular reprogramming and survival of hematopoietic progenitors. Stem Cells 31:1022–1029
73. Raya A, Rodriguez-Piza I, Guenechea G, Vassena R, Navarro S, Barrero MJ, Consiglio A, Castella M, Rio P, Sleep E et al (2009) Disease-corrected haematopoietic progenitors from Fanconi anaemia induced pluripotent stem cells. Nature 460:53–59

74. Nayler S, Gatei M, Kozlov S, Gatti R, Mar JC, Wells CA, Lavin M, Wolvetang E (2012) Induced pluripotent stem cells from ataxia-telangiectasia recapitulate the cellular phenotype. Stem Cells Transl Med 1:523–535
75. Fukawatase Y, Toyoda M, Okamura K, Nakamura K, Nakabayashi K, Takada S, Yamazaki-Inoue M, Masuda A, Nasu M, Hata K et al (2014) Ataxia telangiectasia derived iPS cells show preserved x-ray sensitivity and decreased chromosomal instability. Sci Rep 4:5421
76. Lee P, Martin NT, Nakamura K, Azghadi S, Amiri M, Ben-David U, Perlman S, Gatti RA, Hu H, Lowry WE (2013) SMRT compounds abrogate cellular phenotypes of ataxia telangiectasia in neural derivatives of patient-specific hiPSCs. Nat Commun 4:1824
77. Batista LF, Pech MF, Zhong FL, Nguyen HN, Xie KT, Zaug AJ, Crary SM, Choi J, Sebastiano V, Cherry A et al (2011) Telomere shortening and loss of self-renewal in dyskeratosis congenita induced pluripotent stem cells. Nature 474:399–402
78. Agarwal S, Loh YH, McLoughlin EM, Huang J, Park IH, Miller JD, Huo H, Okuka M, Dos Reis RM, Loewer S et al (2010) Telomere elongation in induced pluripotent stem cells from dyskeratosis congenita patients. Nature 464:292–296
79. Abad M, Mosteiro L, Pantoja C, Canamero M, Rayon T, Ors I, Grana O, Megias D, Dominguez O, Martinez D et al (2013) Reprogramming in vivo produces teratomas and iPS cells with totipotency features. Nature 502:340–345
80. Ohnishi K, Semi K, Yamamoto T, Shimizu M, Tanaka A, Mitsunaga K, Okita K, Osafune K, Arioka Y, Maeda T et al (2014) Premature termination of reprogramming in vivo leads to cancer development through altered epigenetic regulation. Cell 156:663–677
81. Pareja-Galeano H, Sanchis-Gomar F, Perez LM, Emanuele E, Lucia A, Galvez BG, Gallardo ME (2016) iPSCs-based anti-aging therapies: recent discoveries and future challenges. Ageing Res Rev 27:37–41
82. Kurian L, Sancho-Martinez I, Nivet E, Aguirre A, Moon K, Pendaries C, Volle-Challier C, Bono F, Herbert JM, Pulecio J et al (2013) Conversion of human fibroblasts to angioblast-like progenitor cells. Nat Methods 10:77–83
83. Pulecio J, Nivet E, Sancho-Martinez I, Vitaloni M, Guenechea G, Xia Y, Kurian L, Dubova I, Bueren J, Laricchia-Robbio L et al (2014) Conversion of human fibroblasts into monocyte-like progenitor cells. Stem Cells 32:2923–2938
84. Thier M, Worsdorfer P, Lakes YB, Gorris R, Herms S, Opitz T, Seiferling D, Quandel T, Hoffmann P, Nothen MM et al (2012) Direct conversion of fibroblasts into stably expandable neural stem cells. Cell Stem Cell 10:473–479
85. Ocampo A, Reddy P, Izpisua Belmonte JC (2016) Anti-aging strategies based on cellular reprogramming. Trends Mol Med 22:725–738
86. Hochedlinger K, Yamada Y, Beard C, Jaenisch R (2005) Ectopic expression of Oct-4 blocks progenitor-cell differentiation and causes dysplasia in epithelial tissues. Cell 121:465–477
87. Ocampo A, Reddy P, Martinez-Redondo P, Platero-Luengo A, Hatanaka F, Hishida T, Li M, Lam D, Kurita M, Beyret E et al (2016) In vivo amelioration of age-associated hallmarks by partial reprogramming. Cell *167*:1719–1733.e12
88. Mosteiro L, Pantoja C, Alcazar N, Marion RM, Chondronasiou D, Rovira M, Fernandez-Marcos PJ, Munoz-Martin M, Blanco-Aparicio C, Pastor J et al (2016) Tissue damage and senescence provide critical signals for cellular reprogramming in vivo. Science 354
89. Chiche A, Le Roux I, von Joest M, Sakai H, Aguin SB, Cazin C, Salam R, Fiette L, Alegria O, Flamant P et al (2017) Injury-induced senescence enables in vivo reprogramming in skeletal muscle. Cell Stem Cell 20:407–414.e4
90. Banito A, Rashid ST, Acosta JC, Li S, Pereira CF, Geti I, Pinho S, Silva JC, Azuara V, Walsh M et al (2009) Senescence impairs successful reprogramming to pluripotent stem cells. Genes Dev 23:2134–2139
91. Hong H, Takahashi K, Ichisaka T, Aoi T, Kanagawa O, Nakagawa M, Okita K, Yamanaka S (2009) Suppression of induced pluripotent stem cell generation by the p53-p21 pathway. Nature 460:1132–1135
92. Kawamura T, Suzuki J, Wang YV, Menendez S, Morera LB, Raya A, Wahl GM, Izpisua Belmonte JC (2009) Linking the p53 tumour suppressor pathway to somatic cell reprogramming. Nature 460:1140–1144

93. Li H, Collado M, Villasante A, Strati K, Ortega S, Canamero M, Blasco MA, Serrano M (2009) The Ink4/Arf locus is a barrier for iPS cell reprogramming. Nature 460:1136–1139
94. Marion RM, Strati K, Li H, Tejera A, Schoeftner S, Ortega S, Serrano M, Blasco MA (2009) Telomeres acquire embryonic stem cell characteristics in induced pluripotent stem cells. Cell Stem Cell 4:141–154
95. Varela I, Cadinanos J, Pendas AM, Gutierrez-Fernandez A, Folgueras AR, Sanchez LM, Zhou Z, Rodriguez FJ, Stewart CL, Vega JA et al (2005) Accelerated ageing in mice deficient in Zmpste24 protease is linked to p53 signalling activation. Nature 437:564–568
96. Chlon TM, Hoskins EE, Mayhew CN, Wikenheiser-Brokamp KA, Davies SM, Mehta P, Myers KC, Wells JM, Wells SI (2014) High-risk human papillomavirus E6 protein promotes reprogramming of Fanconi anemia patient cells through repression of p53 but does not allow for sustained growth of induced pluripotent stem cells. J Virol 88:11315–11326
97. Ma H, Ow JR, Tan BC, Goh Z, Feng B, Loh YH, Fedele M, Li H, Wu Q (2014) The dosage of Patz1 modulates reprogramming process. Sci Rep 4:7519
98. Camozzi D, Capanni C, Cenni V, Mattioli E, Columbaro M, Squarzoni S, Lattanzi G (2014) Diverse lamin-dependent mechanisms interact to control chromatin dynamics. Focus on laminopathies. Nucleus 5:427–440
99. Lopez-Otin C, Galluzzi L, Freije JM, Madeo F, Kroemer G (2016) Metabolic control of longevity. Cell 166:802–821
100. Chen T, Shen L, Yu J, Wan H, Guo A, Chen J, Long Y, Zhao J, Pei G (2011) Rapamycin and other longevity-promoting compounds enhance the generation of mouse induced pluripotent stem cells. Aging Cell 10:908–911
101. Menendez JA, Vellon L, Oliveras-Ferraros C, Cufi S, Vazquez-Martin A (2011) mTOR-regulated senescence and autophagy during reprogramming of somatic cells to pluripotency: a roadmap from energy metabolism to stem cell renewal and aging. Cell Cycle 10:3658–3677
102. Sharma A, Diecke S, Zhang WY, Lan F, He C, Mordwinkin NM, Chua KF, Wu JC (2013) The role of SIRT6 protein in aging and reprogramming of human induced pluripotent stem cells. J Biol Chem 288:18439–18447
103. Chen W, Liu N, Zhang H, Qiao J, Jia W, Zhu S, Mao Z, Kang J (2017) Sirt6 promotes DNA end joining in iPSCs derived from old mice. Cell Rep 18:2880–2892
104. Bárcena C, López-Otín C (2017) A fruitful liaison of ZSCAN10 and ROS on the road to rejuvenation. Nat Cell Biol 19:1012–1013
105. Skamagki M, Correia C, Yeung P, Baslan T, Beck S, Zhang C, Ross CA, Dang L, Liu Z, Giunta S et al (2017) ZSCAN10 expression corrects the genomic instability of iPSCs from aged donors. Nat Cell Biol 19(9):1037–1048
106. Ma T, Li J, Xu Y, Yu C, Xu T, Wang H, Liu K, Cao N, Nie BM, Zhu SY et al (2015) Atg5-independent autophagy regulates mitochondrial clearance and is essential for iPSC reprogramming. Nat Cell Biol 17:1379–1387
107. Ahlenius H, Chanda S, Webb AE, Yousif I, Karmazin J, Prusiner SB, Brunet A, Sudhof TC, Wernig M (2016) FoxO3 regulates neuronal reprogramming of cells from postnatal and aging mice. Proc Natl Acad Sci U S A 113:8514–8519
108. Salminen A, Kaarniranta K, Kauppinen A (2012) Inflammaging: disturbed interplay between autophagy and inflammasomes. Aging (Albany NY) 4:166–175
109. Acosta JC, Banito A, Wuestefeld T, Georgilis A, Janich P, Morton JP, Athineos D, Kang TW, Lasitschka F, Andrulis M et al (2013) A complex secretory program orchestrated by the inflammasome controls paracrine senescence. Nat Cell Biol 15:978–990
110. Chien Y, Scuoppo C, Wang X, Fang X, Balgley B, Bolden JE, Premsrirut P, Luo W, Chicas A, Lee CS et al (2011) Control of the senescence-associated secretory phenotype by NF-kappaB promotes senescence and enhances chemosensitivity. Genes Dev 25:2125–2136
111. Osorio FG, Barcena C, Soria-Valles C, Ramsay AJ, de Carlos F, Cobo J, Fueyo A, Freije JM, Lopez-Otin C (2012) Nuclear lamina defects cause ATM-dependent NF-kappaB activation and link accelerated aging to a systemic inflammatory response. Genes Dev 26:2311–2324
112. Jurk D, Wilson C, Passos JF, Oakley F, Correia-Melo C, Greaves L, Saretzki G, Fox C, Lawless C, Anderson R et al (2014) Chronic inflammation induces telomere dysfunction and accelerates ageing in mice. Nat Commun 2:4172

113. Liu B, Zhou S, Liu X, Zhou K, Zhang F, Zhou Z (2013) Accumulation of prelamin A compromises NF-kappaB-regulated B-lymphopoiesis in a progeria mouse model. Longev Healthspan 2:1
114. Davis T, Kipling D (2006) Werner syndrome as an example of inflamm-aging: possible therapeutic opportunities for a progeroid syndrome? Rejuvenation Res 9:402–407
115. Goto M, Sugimoto K, Hayashi S, Ogino T, Sugimoto M, Furuichi Y, Matsuura M, Ishikawa Y, Iwaki-Egawa S, Watanabe Y (2012) Aging-associated inflammation in healthy Japanese individuals and patients with Werner syndrome. Exp Gerontol 47:936–939
116. Sen P, Shah PP, Nativio R, Berger SL (2016) Epigenetic mechanisms of longevity and aging. Cell 166:822–839
117. Onder TT, Kara N, Cherry A, Sinha AU, Zhu N, Bernt KM, Cahan P, Marcarci BO, Unternaehrer J, Gupta PB et al (2012) Chromatin-modifying enzymes as modulators of reprogramming. Nature 483:598–602
118. Pour M, Pilzer I, Rosner R, Smith ZD, Meissner A, Nachman I (2015) Epigenetic predisposition to reprogramming fates in somatic cells. EMBO Rep 16:370–378
119. Wang T, Chen K, Zeng X, Yang J, Wu Y, Shi X, Qin B, Zeng L, Esteban MA, Pan G et al (2011) The histone demethylases Jhdm1a/1b enhance somatic cell reprogramming in a vitamin-C-dependent manner. Cell Stem Cell 9:575–587
120. Chen X, Zhai Y, Yu D, Cui J, Hu JF, Li W (2016) Valproic acid enhances iPSC induction from human bone marrow-derived cells through the suppression of reprogramming-induced senescence. J Cell Physiol 231:1719–1727
121. Gordon LB, Kleinman ME, Miller DT, Neuberg DS, Giobbie-Hurder A, Gerhard-Herman M, Smoot LB, Gordon CM, Cleveland R, Snyder BD et al (2012) Clinical trial of a farnesyltransferase inhibitor in children with Hutchinson-Gilford progeria syndrome. Proc Natl Acad Sci U S A 109:16666–16671
122. Cao K, Graziotto JJ, Blair CD, Mazzulli JR, Erdos MR, Krainc D, Collins FS (2011) Rapamycin reverses cellular phenotypes and enhances mutant protein clearance in Hutchinson-Gilford progeria syndrome cells. Sci Transl Med 3:89ra58
123. Cenni V, Capanni C, Columbaro M, Ortolani M, D'Apice MR, Novelli G, Fini M, Marmiroli S, Scarano E, Maraldi NM et al (2011) Autophagic degradation of farnesylated prelamin A as a therapeutic approach to lamin-linked progeria. Eur J Histochem e36:55
124. Cenni V, Capanni C, Mattioli E, Columbaro M, Wehnert M, Ortolani M, Fini M, Novelli G, Bertacchini J, Maraldi NM et al (2014) Rapamycin treatment of Mandibuloacral dysplasia cells rescues localization of chromatin-associated proteins and cell cycle dynamics. Aging (Albany NY) 6:755–770
125. Scaffidi P, Misteli T (2005) Reversal of the cellular phenotype in the premature aging disease Hutchinson-Gilford progeria syndrome. Nat Med 11:440–445
126. Marino G, Ugalde AP, Fernandez AF, Osorio FG, Fueyo A, Freije JM, Lopez-Otin C (2010) Insulin-like growth factor 1 treatment extends longevity in a mouse model of human premature aging by restoring somatotroph axis function. Proc Natl Acad Sci U S A 107:16268–16273
127. Villa-Bellosta R, Rivera-Torres J, Osorio FG, Acin-Perez R, Enriquez JA, Lopez-Otin C, Andres V (2013) Defective extracellular pyrophosphate metabolism promotes vascular calcification in a mouse model of Hutchinson-Gilford progeria syndrome that is ameliorated on pyrophosphate treatment. Circulation 127:2442–2451
128. Xiong ZM, Choi JY, Wang K, Zhang H, Tariq Z, Wu D, Ko E, LaDana C, Sesaki H, Cao K (2016) Methylene blue alleviates nuclear and mitochondrial abnormalities in progeria. Aging Cell 15:279–290
129. Ibrahim MX, Sayin VI, Akula MK, Liu M, Fong LG, Young SG, Bergo MO (2013) Targeting isoprenylcysteine methylation ameliorates disease in a mouse model of progeria. Science 340:1330–1333
130. Scheibye-Knudsen M, Mitchell SJ, Fang EF, Iyama T, Ward T, Wang J, Dunn CA, Singh N, Veith S, Hasan-Olive MM et al (2014) A high-fat diet and NAD(+) activate Sirt1 to rescue premature aging in cockayne syndrome. Cell Metab 20:840–855

131. Chatre L, Biard DS, Sarasin A, Ricchetti M (2015) Reversal of mitochondrial defects with CSB-dependent serine protease inhibitors in patient cells of the progeroid Cockayne syndrome. Proc Natl Acad Sci U S A 112:E2910–E2919
132. Zlateska B, Ciccolini A, Dror Y (2015) Treatment of dyskeratosis congenita-associated pulmonary fibrosis with danazol. Pediatr Pulmonol 50:E48–E51
133. Saadatzadeh MR, Bijangi-Vishehsaraei K, Kapur R, Haneline LS (2009) Distinct roles of stress-activated protein kinases in Fanconi anemia-type C-deficient hematopoiesis. Blood 113:2655–2660
134. Tivey HS, Rokicki MJ, Barnacle JR, Rogers MJ, Bagley MC, Kipling D, Davis T (2013) Small molecule inhibition of p38 MAP kinase extends the replicative life span of human ATR-Seckel syndrome fibroblasts. J Gerontol A Biol Sci Med Sci 68:1001–1009
135. Liu B, Ghosh S, Yang X, Zheng H, Liu X, Wang Z, Jin G, Zheng B, Kennedy BK, Suh Y et al (2012) Resveratrol rescues SIRT1-dependent adult stem cell decline and alleviates progeroid features in laminopathy-based progeria. Cell Metab 16:738–750
136. Labbe A, Garand C, Cogger VC, Paquet ER, Desbiens M, Le Couteur DG, Lebel M (2011) Resveratrol improves insulin resistance hyperglycemia and hepatosteatosis but not hypertriglyceridemia, inflammation, and life span in a mouse model for Werner syndrome. J Gerontol A Biol Sci Med Sci 66:264–278
137. Harhouri K, Navarro C, Depetris D, Mattei MG, Nissan X, Cau P, De Sandre-Giovannoli A, Levy N (2017) MG132-induced progerin clearance is mediated by autophagy activation and splicing regulation. EMBO Mol Med 9:1294–1313
138. Lukjanenko L, Jung MJ, Hegde N, Perruisseau-Carrier C, Migliavacca E, Rozo M, Karaz S, Jacot G, Schmidt M, Li L et al (2016) Loss of fibronectin from the aged stem cell niche affects the regenerative capacity of skeletal muscle in mice. Nat Med 22:897–905
139. Bernet JD, Doles JD, Hall JK, Kelly Tanaka K, Carter TA, Olwin BB (2014) p38 MAPK signaling underlies a cell-autonomous loss of stem cell self-renewal in skeletal muscle of aged mice. Nat Med 20:265–271
140. Cosgrove BD, Gilbert PM, Porpiglia E, Mourkioti F, Lee SP, Corbel SY, Llewellyn ME, Delp SL, Blau HM (2014) Rejuvenation of the muscle stem cell population restores strength to injured aged muscles. Nat Med 20:255–264
141. Rozo M, Li L, Fan CM (2016) Targeting beta1-integrin signaling enhances regeneration in aged and dystrophic muscle in mice. Nat Med 22:889–896
142. Price FD, von Maltzahn J, Bentzinger CF, Dumont NA, Yin H, Chang NC, Wilson DH, Frenette J, Rudnicki MA (2014) Inhibition of JAK-STAT signaling stimulates adult satellite cell function. Nat Med 20:1174–1181
143. Elabd C, Cousin W, Upadhyayula P, Chen RY, Chooljian MS, Li J, Kung S, Jiang KP, Conboy IM (2014) Oxytocin is an age-specific circulating hormone that is necessary for muscle maintenance and regeneration. Nat Commun 5:4082
144. Florian MC, Dorr K, Niebel A, Daria D, Schrezenmeier H, Rojewski M, Filippi MD, Hasenberg A, Gunzer M, Scharffetter-Kochanek K et al (2012) Cdc42 activity regulates hematopoietic stem cell aging and rejuvenation. Cell Stem Cell 10:520–530
145. Chang J, Wang Y, Shao L, Laberge RM, Demaria M, Campisi J, Janakiraman K, Sharpless NE, Ding S, Feng W et al (2016) Clearance of senescent cells by ABT263 rejuvenates aged hematopoietic stem cells in mice. Nat Med 22:78–83
146. Baker DJ, Perez-Terzic C, Jin F, Pitel KS, Niederlander NJ, Jeganathan K, Yamada S, Reyes S, Rowe L, Hiddinga HJ et al (2008) Opposing roles for p16Ink4a and p19Arf in senescence and ageing caused by BubR1 insufficiency. Nat Cell Biol 10:825–836
147. Baker DJ, Wijshake T, Tchkonia T, LeBrasseur NK, Childs BG, van de Sluis B, Kirkland JL, van Deursen JM (2011) Clearance of p16Ink4a-positive senescent cells delays ageing-associated disorders. Nature 479:232–236
148. Baker DJ, Childs BG, Durik M, Wijers ME, Sieben CJ, Zhong J, Saltness RA, Jeganathan KB, Verzosa GC, Pezeshki A et al (2016) Naturally occurring p16(Ink4a)-positive cells shorten healthy lifespan. Nature 530:184–189

149. Garcia-Prat L, Martinez-Vicente M, Perdiguero E, Ortet L, Rodriguez-Ubreva J, Rebollo E, Ruiz-Bonilla V, Gutarra S, Ballestar E, Serrano AL et al (2016) Autophagy maintains stemness by preventing senescence. Nature 529:37–42
150. Zhang H, Ryu D, Wu Y, Gariani K, Wang X, Luan P, D'Amico D, Ropelle ER, Lutolf MP, Aebersold R et al (2016) NAD(+) repletion improves mitochondrial and stem cell function and enhances life span in mice. Science 352:1436–1443
151. Schworer S, Becker F, Feller C, Baig AH, Kober U, Henze H, Kraus JM, Xin B, Lechel A, Lipka DB et al (2016) Epigenetic stress responses induce muscle stem-cell ageing by Hoxa9 developmental signals. Nature 540:428–432
152. Lo Cicero A, Jaskowiak AL, Egesipe AL, Tournois J, Brinon B, Pitrez PR, Ferreira L, de Sandre-Giovannoli A, Levy N, Nissan X (2016) A high throughput phenotypic screening reveals compounds that counteract premature osteogenic differentiation of HGPS iPS-derived mesenchymal stem cells. Sci Rep 6:34798
153. Mahmoudi S, Brunet A (2012) Aging and reprogramming: a two-way street. Curr Opin Cell Biol 24:744–756
154. Prigione A, Hossini AM, Lichtner B, Serin A, Fauler B, Megges M, Lurz R, Lehrach H, Makrantonaki E, Zouboulis CC et al (2011) Mitochondrial-associated cell death mechanisms are reset to an embryonic-like state in aged donor-derived iPS cells harboring chromosomal aberrations. PLoS One 6:e27352
155. Suhr ST, Chang EA, Tjong J, Alcasid N, Perkins GA, Goissis MD, Ellisman MH, Perez GI, Cibelli JB (2010) Mitochondrial rejuvenation after induced pluripotency. PLoS One 5:e14095
156. Lapasset L, Milhavet O, Prieur A, Besnard E, Babled A, Ait-Hamou N, Leschik J, Pellestor F, Ramirez JM, De Vos J et al (2011) Rejuvenating senescent and centenarian human cells by reprogramming through the pluripotent state. Genes Dev 25:2248–2253
157. Cong L, Ran FA, Cox D, Lin S, Barretto R, Habib N, Hsu PD, Wu X, Jiang W, Marraffini LA et al (2013) Multiplex genome engineering using CRISPR/Cas systems. Science 339:819–823
158. Xiong ZM, LaDana C, Wu D, Cao K (2013) An inhibitory role of progerin in the gene induction network of adipocyte differentiation from iPS cells. Aging (Albany NY) 5:288–303

Part IV
Technical Challenges and Future

Chapter 14
Pluripotent Stem Cell Banks

Begoña Aran, Ángel Raya, and Anna Veiga

Abstract Embryonic stem cells (ESCs) are pluripotent stem cells that can be obtained from the inner cell mass (ICM) of blastocyst-stage pre-implantation embryos. The first lines of human ESC (hESC) were derived by Thomson et al. in 1998. These cells can be indefinitely cultured in vitro and differentiated into different cell types. These cell lines constitute an excellent source of cells for the study of human genetics and gene expression patterns and as a tool to understand the events that take place during human embryo development. These cell lines can also be useful for drug screening and for the disease modelling. In 2006, Takahashi and Yamanaka first introduced the technology of induced reprogramming to pluripotency in the mouse and later in human cells. These induced pluripotent stem cells (iPSCs) are generated through the reprogramming of somatic cells back to an embryonic-like state. The reprogram-

B. Aran
Center of Regenerative Medicine in Barcelona (CMRB), Barcelona, Spain

National Stem Cell Bank-Barcelona Node, Biomolecular and Bioinformatics Resources Platform PRB2, ISCIII, CMRB, Barcelona, Spain

Á. Raya
Institució Catalana de Recerca I Estudis Avançats (ICREA), Barcelona, Spain

Center of Regenerative Medicine in Barcelona (CMRB), Barcelona, Spain

Center for Networked Biomedical Research on Bioengineering, Biomaterials and Nanomedicine (CIBER-BBN), Madrid, Spain

A. Veiga (✉)
Center of Regenerative Medicine in Barcelona (CMRB), Barcelona, Spain

National Stem Cell Bank-Barcelona Node, Biomolecular and Bioinformatics Resources Platform PRB2, ISCIII, CMRB, Barcelona, Spain

Center for Networked Biomedical Research on Bioengineering, Biomaterials and Nanomedicine (CIBER-BBN), Madrid, Spain
e-mail: aveiga@cmrb.eu

R. Delgado-Morales (ed.), *Stem Cell Genetics for Biomedical Research*,
https://doi.org/10.1007/978-3-319-90695-9_14

ming process is induced by the addition of exogenous reprogramming factors. Human iPSC (hiPSC) technology avoids the destruction of human embryos and the ethical debate involved. Moreover, patient-specific iPSC can be obtained by this technology providing unprecedented opportunities, not only in regenerative medicine but also in basic research as disease models, drug discovery and toxicology. The emerging demands of stem cell research and applications require the establishment and cooperation of centralized banks at a translational and even global scale. This will ensure the availability of high-quality hPSC using standardized state-of-the-art methods and strategies to deal with a heterogeneous regulatory, ethical and legal landscape. Furthermore, stem cell banks are essential for the distribution of cell lines among research centres, promoting scientific collaboration and facilitating widespread use of the cells for research and clinical applications.

Keywords Embryonic stem cells · Induced pluripotent stem cells · Banking · Clinical trials · Disease models

Introduction

- Human embryonic stem cells (hESCs)
- Induced pluripotent stem cells (hiPSCs)

Banking Pluripotent Stem Cells

- Informed consent
- Provenance and selection of donor tissue
- Donor selection and risk factors
- Traceability and identifiability
- Characterization and quality control
- Preservation and storage
- Banks and registries
  - The International Stem Cell Banking Initiative (ISCBI)
  - EBiSC – European Bank for induced pluripotent Stem Cells
  - The human pluripotent stem cell registry (hPSCreg)
  - International Stem Cell Registry at the University of Massachusetts Medical School

hiPSC Applications

- Toxicology tests
- Disease models
- Cell therapy
  - Clinical-grade lines and GMP conditions
  - Banks of hiPSCs of frequents HLA
  - Clinical trials

14.1 Introduction

Embryonic stem cells (ESCs) are pluripotent stem cells that can be obtained from the inner cell mass (ICM) of blastocyst-stage pre-implantation embryos. Culture of cells from the inner cell mass of human blastocysts and subsequent growth of embryonic stem (ES)-like cells for two passages with partial characterization were first described by Bongso et al. [9]. The first lines of human ESC (hESC) were derived by Thomson et al. in 1998. These cells can be indefinitely cultured in vitro and differentiated into different cell types [78, 104]. From the first report of hESC derivation, numerous hESC lines have been generated by different laboratories [2, 10, 86]. These cell lines constitute an excellent source of cells for the study of human genetics and gene expression patterns and as a tool to understand the events that take place during human embryo development. These cell lines can also be useful for drug screening and for the disease modelling.

In 2006, Takahashi and Yamanaka first introduced the technology of induced reprogramming to pluripotency in the mouse and later in human cells [97, 113]. These induced pluripotent stem cells (iPSCs) are generated through the reprogramming of somatic cells back to an embryonic-like state. The reprogramming process is induced by the addition of exogenous reprogramming factors. Human iPSC (hiPSC) technology avoids the destruction of human embryos and the ethical debate involved. Moreover, patient-specific iPSC can be obtained by this technology providing unprecedented opportunities, not only in regenerative medicine but also in basic research as disease models, drug discovery and toxicology.

The research landscape for human pluripotent stem cells (hPSC) is changing rapidly. Thousands of both hESC and hiPSC lines have been generated since their establishment in 1998 and 2007, respectively. The expansion of national and international efforts in stem cell research has led to the establishment of stem cell banks and registries [37]. The emerging demands of stem cell research and applications require the establishment and cooperation of centralized banks at a translational and even global scale. This will ensure the availability of high-quality hPSC using standardized state-of-the-art methods and strategies to deal with a heterogeneous regulatory, ethical and legal landscape [17]. The principle of sharing quality-controlled research materials has been promoted in the hPSC field. Utilizing the knowledge of the existing well-established hPSC banks and their experience in standardization to promote consistent and robust quality control in hPSC is recommended [92]. Furthermore, stem cell banks are essential for the distribution of cell lines among research centres, promoting scientific collaboration and facilitating widespread use of the cells for research and clinical applications.

14.2 Human Embryonic Stem Cells

Most frequently, hESC lines are derived from surplus embryos from assisted reproduction cycles, usually frozen and specifically donated for human ESC research with informed patient consent. hESC are permanent cell lines derived from pre-implantation human embryos, commonly from the ICM of human blastocysts.

Today, the methodology of hESC derivation is widely established, and various protocols are used in the different steps of the process, including feeder cell preparation (if feeder cells are used), embryo culture, ICM isolation and initial steps of derivation. Originally, Thomson et al. [104] isolated the embryo ICM using antibodies (immunosurgery). Other authors, in order to avoid the use of components of animal origin that would limit the therapeutic use of these cells, proposed to isolate the ICM mechanically [94], by means of laser [105] or, even, seeding the entire blastocyst [33, 77]. However, there are other options for the derivation of hESC, such as the use of early embryos [114], in morula stage [93], as well as isolated blastomeres [41].

Usually, the ICM is grown on human or animal feeder cells, and the ensuing hESCs are further continuously propagated on feeder cells, in the presence of a special nourishing medium, until a cell line is established [57, 74, 80]. It is worth mentioning that an important goal in derivation attempts is to achieve derivation and culture of hESC under feeder-free conditions, using extracellular matrices [3, 32, 42, 56] and animal-free [25, 27, 52, 76] components, as well as under chemically defined culture conditions [52, 55, 57]. Moreover, performing the entire protocol following good manufacturing practice (GMP) conditions would be necessary for the safe clinical use of hESC in human therapy.

Some publications have revealed differences among hES cell lines related to the environment to which the cells have been exposed after embryo culture and derivation [2, 36]. The derivation efficiency varies between different authors according to the method used [5, 15], embryo quality [50, 114] or whether fresh or frozen embryos had been used [64, 88].

It is currently estimated that around 1000 hESC lines have been established [11, 54, 86], and more than 700 hESC lines have been registered in the hPSCreg (human pluripotent stem cell registry; www.hpscreg.eu).

14.3 Induced Pluripotent Stem Cells

Takahashi and Yamanaka [98] screened a set of 24 genes, including some differentially expressed in mouse ESC, by co-expression in foetal mouse fibroblasts. Subsequently, they narrowed down the really critical factors for induction of the pluripotency to four nuclear factors. The first reports of hiPSC generation used Oct4, Sox2, Klf4 and c-Myc (Yamanaka factors) to reprogram human dermal fibroblasts using retroviruses [97]. Nowadays, hiPSC technology has evolved enormously, and different reprogramming factors have been used with diverse reprogramming methods to reprogram a variety of somatic cells. Shortly after the first hiPSC paper was published, another group reported similar results using Oct4, Sox2, Nanog and Lin28 (Thompson factors) [113]. Other groups began experimenting with new strategies, and within a couple of years, non-integrative reprogramming methods were introduced, including Sendai virus [26], episomal plasmids [71, 112], recombinant proteins [38], PiggyBac transposons [109] and mRNA [107]. Nowadays, hiPSCs are readily generated from several types of starting somatic cells such as fibroblasts [97, 113], keratinocytes [1], umbilical cord blood cells [28], peripheral blood cells [16] or even urine sediment epithelial cells [115].

As new reprogramming methods were developed, reports begun to emerge showing that iPSC were not equivalent to ESC. Genetic and epigenetic analyses showed different DNA and histone modifications and gene expression patterns in iPSC compared with ESC. Differences in the somatic cell type, reprogramming method or even the extent of culture are thought to have an influence on these differences. However, improvements and modifications made to reprogramming methods over the past decade have helped improve the safety and quality of hiPSC [40].

14.4 Banking Pluripotent Stem Cellls

14.4.1 Informed Consent

Repositories and banks should guarantee the provenance of the embryos and human samples from which the hPSC were derived. Informed consent is mandatory in this regard. Consent requirements for embryo, cell or tissue donation are subject to national regulation.

Most of the hESC lines have been derived from frozen embryos arising from in vitro fertilization (IVF). Donating couples should be informed about stem cell derivation and about the specific consent. Possible future unspecified research use must be included and even potential clinical or commercial applications. The creation of human embryos for research and hESC derivation is allowed in very few countries such as United Kingdom and Sweden.

The donation of tissues or cells for the generation of iPSC must be in accordance with the requirements for tissue use for research (European Tissue and Cells Directive, EUTCD). Cell and tissue donors should also be informed about the iPSC generation process and about the specificities of the research project.

The informed consent must not replace a dialogue between research staff and donors of tissues and cells. Both embryo and tissue donors must receive oral information supported by relevant written material. Whenever possible, the person giving the information must not be involved in the research to avoid influencing donor decision. They have to provide the potential donor all ethic and scientific information that they need, in a transparent and understandable manner. The signature should be free and voluntary, and donor must be informed that the participation or rejection will not directly influence their personal treatment.

The informed consent should cover the following statements:

- That the donation is free and voluntary consented, after being informed that participation will not directly influence their personal treatment.

- That the research is not intended to provide direct medical benefit to the donor, except in the sense that it benefits the community.
- The ability of the donor to withdraw original specimens, derived cell lines or personal data to the participation in research.
- That derived cell lines might be deposited and stored in a repository for an unlimited time.
- The donor agreement could be used for unforeseen future research and/or even clinical applications. In this case, donor might be recontacted as consent form might be different for research or for clinical application.
- Disclosure of the possibility that resulting cell lines have commercial potential and might be used for the development of commercial products.
- That the donor and/or derived cell lines will be screened for infectious and possibly genetic diseases.
- That the resulting cell lines derived would carry the DNA of the donor and therefore be partially or completely genetically matched to the donor.
- That nucleic acid sequencing of the resulting cell lines might be performed and this data may be stored in databases available to the scientific community with the needed confidentiality, and this may compromise the capacity of donor identification. The implementation of mechanisms for the donor data protection and confidentiality with the limitations to assure traceability for safety reasons.
- If there is a need to request additional materials (such as blood), or if the information discovered during the research is relevant for his/her health, the donor must agree to be recontacted.

In the case that tissues and cells are procured from a child or an incapacitated adult, consent must be provided by a parent or person authorized. Researchers must justify why the research cannot be done using cells from adults with capacity.

Before accepting an hPSC line, a stem cell bank should determine its provenance and should guarantee that the original cells or tissue was collected consistently with international standards of ethics and with the regulation of the country of origin.

The human pluripotent stem cell registry (hPSC) proposes the questionnaire regarding informed consent described in Table 14.1.

14.5 Provenance and Selection of Donor Tissue.

14.5.1 Donor Selection and Risk Factors

Donor selection for generating hPSC is especially important in terms of clinical application. As a general rule, determining donor eligibility requires screening for risk factors associated with infection and communicable diseases [4]. This involves viral blood pathogens such as HIV, hepatitis B and C virus, malaria, Chagas disease and human T-cell lymphotropic viruses I and II.

For hESC, usually coming from assisted reproduction clinics, there are a number of considerations taking into account in the screening protocols of gamete and embryo donors. IVF couples that donate embryos for hESC are submitted to general

Table 14.1 Informed consent questionnaire proposed by the human pluripotent stem cell registry (hPSCreg)

Donor consent form and donor information sheet		Yes	No	Unk.
Has consent been obtained from the donor of the tissue from which iPSC have been derived?				
Was the consent voluntary given by the donor, custodian or parents?				
Was the donor been informed that participation will not directly influence their personal treatment?				
Confirm that consent was obtained by a qualified professional				
Has the donor agreed to be recontacted?				
Has the donor been informed about her/his data will be protected?				
Indicate whether the donated material has been	Pseudonymised			
	Anonymised			
Research				
Does consent expressly prevent derivation of iPSC?				
Does consent pertain to a specific research?				
Does consent permit unforeseen future research, without further consent?				
Does the consent permit uses of donated material or derived cell line intended for clinical treatment or human applications?				
Commercial exploitation				
Does consent expressly prevent development of commercial products?				
Does consent expressly prevent financial gain from any use of the donated material including any product made from it?				
Storage of and access to material				
Does consent expressly permit storage of donated material for an unlimited time?				
Does consent expressly permit storage of cells derived from the donated material for an unlimited time?				
Does consent prevent the donated biosample from being made available to researchers anywhere in the world?				
Does consent prevent cells derived from the donated biosample from being made available to researchers anywhere in the world?				
Does consent permit research by an academic institution?				
Does consent permit research by a public organization?				
Does consent permit research by a non-profit company?				
Does consent permit research by a for-profit corporation?				
Data and information				
Does consent expressly permit collection of genetic information?				
Does consent expressly permit storage of genetic information?				
Does consent expressly prevent dissemination of genetic information?				
Has the donor been informed that their donated biosample or derived cells may be tested for the presence of microbiological agents/pathogens?				
Has the donor consented to receive information discovered during use of donated material that has significant health implications for the donor?				
How may genetic information associated with the cell line accessed?	Open access			
	Controlled access			
	No information			

(continued)

Table 14.1 (continued)

Donor consent form and donor information sheet	Yes	No	Unk.
Benefits			
Will the donor expect to receive financial benefit, beyond reasonable expenses, in return for donating the biosample?			
Does the consent anticipate that the donor will be notified of results or outcomes of any research involving the donated samples or derived cells?			
Withdrawal			
Does the consent permit the donor, upon withdrawal of consent, to stop the use of the derived cell line(s) that have already been created from donated samples?			
Does the consent permit the donor, upon withdraw of consent, to stop delivery or use of information and data about the donor?			
Medical records			
Does consent permit access to medical records of the donor?			
Does consent permit access to any other source of information about the clinical treatment or health of the donor?			

screenings to avoid infections. It is recommended a medical history screening. Repositories and banks should have medical information of the donors and their families (anonymized or de-identified) that may be associated with risk of infection or transmission of disease. This will include cancer, drugs abuse, familial Creutzfeldt-Jakob disease and inherited diseases. Obviously, while hiPSC derived from patients have an invaluable utility as disease models, they are unlikely to be suitable for development of clinical therapies [4].

14.5.2 *Traceability and Identifiability*

Traceability has been defined as "tracking an individual through their medical history" (Medical Research Council, CoP for the use of human stem cell lines). In terms of hPSC generation and culture, traceability is the mechanism that allows a link to be made between donors and cell lines. Each cell line must be correctly identified at all times. It provides a system for tracking of handling and storage conditions to assure a safe and ethical provenance. It allows quality validation.

Traceability begins with the pseudoanonymization of the donated biosample and donor informed consent. It maintains donor anonymity but still allows for traceability from donor to recipient or vice versa in the event of a discovery which might significantly affect the health of the donor. Complete anonymization impedes the use of the samples because it is impossible to trace the sample back to the donor. If the procurement institution, the generation institution and/or the bank are distinct entities, enough data must be recorded to allow traceability, but donor details must be kept in the procurement institution [66].

Traceability is especially relevant in cases of hPSC generation for clinical application, as mandatory request in good manufacturing practice (GMP) production.

Careful evaluation associated with a stem cell line is necessary to determine its suitability for developing a clinical product. Stem cell banks should request history information from the depositor and continue to build a documented history relating to each cell line processed and banked. This compiled documentation is called a cell line "history file". This history file provides the necessary information to enable traceability of the cell line from derivation, culture, establishment, transport, testing, banking and storage. The history file should also include evidence that the cell banking was performed under principles of GMP.

Identifiability can be defined as information that may reasonably be expected to identify an individual alone or in combination with other available information (Interagency Secretariat on Research Ethics: Provisional final drafts of tri-council policy statement). In the context of hESC research, the possibility of donor identification based solely on the single nucleotide polimorphism (SNP) genotype of the cell line is extremely remote. The genotype of an hESC line does not correspond directly to the genotype of the individuals who donated the embryo. The publication of all genotypic information for these lines in banks and registries does not seem to pose a threat to the privacy and confidentiality of donors. Nevertheless, it may be possible to infer some genomic information about the gamete donors through hESC. Only if additional information were available regarding donation (e.g. date, place of embryo donation, etc.), it could be possible to identify a donor.

For hiPSC, the potential for donor identifiability seems easier. There is a concern that current technical and ethical safeguards to protect the identity of donors could be insufficient [44]. Novel methods and associated tools permitting individual identification in publicly accessible SNP databases have become an issue of debate. A study demonstrated that reidentification was possible even in the absence of a reference sample [30].

However, identifiability allows for the ongoing updating of clinical data, making the samples more interesting for research. Traceability and identifiability serve to consolidate the scientific validity and utility of research involving human tissues and can do the same for stem cell banking [43].

14.6 Characterization and Quality Control

The characterization of hPSC lines is mandatory in order to demonstrate self-renewal capacity and pluripotent differentiation ability. As recommended by the International Stem Cell Banking Initiative (ISCBI) [17], several tests must be performed before banking a new hESC or hiPSC line. The standard characterization before banking hPSC lines may include [36, 59]:

- Alkaline phosphatase (AP) activity: AP activity can be easily assayed because of the capacity of this enzyme to change the conformation of a colorimetric reagent from a soluble to a precipitated state. Commercial detection kits are available.
- Expression of hESC markers: Human pluripotency markers can be assessed by immunostaining or cytometry analysis. This panel of markers includes transcrip-

tion factors such as OCT4, NANOG and SOX2 and surface markers such as Tra-1-60, Tra-1-81, SSEA3 and SSEA4.

- Gene expression: RT-PCR is also commonly used to assay the expression of a series of genes and antigen expression patterns of undifferentiated cells.
- Differentiation tests:

Differentiation in vitro*:* hPSC colonies are first cultured in suspension forming large aggregates called embryoid bodies (EB), and later the EBs are cultured in specific conditions to promote differentiation to cell derivatives of the three main embryo germ layers. The differentiation is confirmed by immunofluorescence analysis of cell-type-specific markers.

Differentiation in vivo: Consists of injecting hPSC into immunocompromised mice, where they will proliferate and differentiate into teratomas, which contain tissues from the three primordial germinal layers. The differentiation is also confirmed by immunocytochemistry for differentiated markers. The requisite of performing the in vivo test is highly debated, and some groups argue that the in vitro test is enough.

- Karyotype: To confirm chromosome integrity. Checking karyotype stability every 20–25 passages and/or before banking is recommended as cells in culture can develop karyotype alterations.
- DNA fingerprinting: Cell line identification providing a genetic sample identifier from unique patterns on their DNA, such as short tandem repeat (STR) analysis.
- Microbiological tests: Mycoplasma testing should be routinely performed in the laboratory. Furthermore, a microbiological test is recommended to assure that the hPSC line is free of any possible contaminant gram-positive and gram-negative bacteria, fungi and yeast by standard methods.
- Reprogramming factors: In the case of hiPSC, tests are performed to ensure silencing of reprogramming vectors or elimination of non-integrative vectors. This may be assayed by RT-qPCR or antibody-based detection.

If the cell line is intended for clinical application, the characterization should be much more thorough for obvious safety reasons. In this case quality and safety controls must be included [4]:

- Sterility testing: It must be more exhaustive than routine microbiological tests performed for research. Standard methods for sterility testing for clinical applications are published by national authorities including US Pharmacopeia (USP) and the European Pharmacopoeia (EP).
- Virological testing: Detection of viral agents as HBsAg antigen, anti-HBc antibodies, anti-hepatitis C antibodies, anti-ViH antibodies, DNA VHB, RNA VIH-1, RNA VHC, TPHA (syphilis antibodies), Chagas and HTLV-I/II.
- DNA methylation analysis: In the case of hiPSC, DNA methylation patterns provide information as to whether cells have been completely reprogrammed from parental lines.
- Genetic screening: Whole-genome sequencing of cell lines intended for clinical use is generally agreed to be desirable to develop our scientific understanding of these cell types, and repositories should seek to contain such data.

14.7 Preservation and Storage

Satisfactory cryopreservation technologies are crucial in hPSC research and clinical applications, and it is important to establish effective freezing/thawing protocols to preserve early-passage cells. Moreover, efficient cryopreservation is essential for the establishment of hPSC banks and the transfer of cells among research centres.

The survival of hPSC after thawing ranges between 10% and 50%. hPSC cryopreservation has been extensively studied, and many reports comparing methods, containers and cryoprotectants have been published [34, 51, 110]. Vitrification has been suggested by some authors for hPSC cryopreservation [79] in order to avoid ice formation. This technique requires specialized skills and is not readily suitable for large amounts of hPSC. However, Li et al. [51] reported an ingenious method for vitrifying large amounts of hESC using a cell strainer. At any rate, slow freezing/rapid thawing protocols are the method of choice in most centres. Programmable freezers are also commonly used with controlled freezing rate around 0.2–0.5 °C/min. Several cryoprotectants (CPAs) have also been proposed [34, 53, 73]. CPAs such as the dimethyl sulfoxide (DMSO) and ethylene glycol (EG), the most commonly used, can be toxic to cells, and consideration must be given to the intrinsic toxicity of standard compounds that are time-, temperature- and concentration-dependent. Additives to the solution (e.g. serum) should be assessed for their ability to mitigate these effects. CPAs exert an osmotic effect during their addition to the cells. Such effects can be damaging and may affect cell survival. Osmotic damage can be reduced in stepwise addition protocols. Cells also can be damaged through inadequate thawing. Rapid warming is recommended for preventing cell damage from intracellular ice formation. This is especially important for vitrified cells. However, care must be taken to prevent temperature-dependent toxic effects of the CPAs, and their elution must be quick too [4].

Robust cryopreservation protocols are especially important when hPSC are intended for clinical application. hPSC must be banked under good manufacturing practices (GMP) before all the battery of tests that verify the integrity and sterility of the cell bank. The cells must "wait" frozen for the validation before their differentiation into a clinical relevant cell type [110]. Moreover current CPAs and additives must be replaced by xeno-free compounds manufactured under GMP conditions. Bovine serum can be replaced by human serum. Xeno-free commercial available freezing solution as FreSR™-S (Stemcell technologies) or GMP compliant as such CryoStor® (Stemcell technologies) can be used.

Regarding containers for cell store, the most generally used are vials, straws and bags. The use of open systems is not recommended due to the hazard to stored cells.

Scientific evidence suggests that storage cells at ultralow temperatures do not result in significant damage, as long as temperature remains uniform. Storage at −80 °C is acceptable for short periods of time if the sample has been preserved by freezing, but not for vitrified material. The most stable conditions for cell storage are below −160 °C, under liquid nitrogen (LN2). However, several reports indicate some contaminants as viruses can survive at LN2, and a fatal viral transmission has been reported [103]. Storage in the gas phase above LN2 has been recommended in order to reduce the risk of cross-contamination.

Cells cryopreserved by slow cooling may be transported in dry ice, but vitrified cells should not to avoid devitrification and cell damage. These cells must be transported in LN2, the most secure way of transport for any cryopreserved material.

Viability tests after thawing are recommended. This is mandatory when new freezing protocols are employed. Neutral red, propidium iodide or alamar blue staining may be used [4].

14.8 Banks and Registries

As previously discussed, the expansion of national and international efforts in stem cell research has led to the establishment of stem cell banks and registries. The emerging demands of stem cell research and therapeutics require the establishment of centralized banks at a translational and even global scale. International initiatives have been put in place. Below are the more prominent:

The International Stem Cell Banking Initiative (ISCBI)
http://www.stem-cell-forum.net/initiatives/international-stem-cell-banking-initiative

The ISCBI is funded by the International Stem Cell Forum (ISCF), and it is coordinated by the UK Stem Cell Bank. ISCBI comprises representatives from 17 countries, including national and institutional stem cell banks, registries collections, research centres and advisory bodies. ISCBI aims to create a global network of stem cell banks to facilitate best practice in stem cell research and clinical cell delivery, primary objectives of national and local governments worldwide. The ISCBI promotes, among other things, minimum standards for cell line quality control and comparability of data produced in different centres around the world.

The first ISCBI commitment was to produce a consensus document on principles of best practice for banking and distributing hESC and hiPSC for research purposes [35]. This document provided guidance on how a stem cell bank should operate from the point of obtaining deposits of new cell lines through to the process of shipping samples of cell lines to researchers. Crook et al. [17] present an overview of ISCBI, its activities, participants and goals. A second aim of ISCBI is the preparation of guidelines for banking clinical-grade hPSC lines. It will describe an agreed set of international minimum standard for clinically compliant hESC and hiPSC procurement, testing, maintenance and shipment. Kim et al. [39] report on the last two ISCBI meetings in 2016: California, 26th June and Korea 19–20 October. These meetings involved delegates from up to 24 countries) to reach consensus on standards for the field of stem cell research and clinical application.

ICSBI provides leadership for the unification of governance and scientific practices to support stem cell research and translational application. ISCBI aims to establishing an international banking network that encourages self-regulation to a global standard and ensures the legitimacy of the banks to the global stem cell effort. ISCBI members have also held regular workshops.

EBiSC, European Bank for Induced Pluripotent Stem Cells
http://www.ebisc.org

EBiSC is a project funded by the Innovative Medicines Initiative (IMI), resources of which are composed of financial contribution from the European Union's Seventh Framework Programme (FP7/2007–2013) and European Federation of Pharmaceutical Industries and Associations (EFPIA). EBiSC has been designed to address the increasing demand for quality-controlled, disease-relevant, research-grade iPSC lines, data and cell services. EBiSC is a centralized, not-for-profit iPSC bank providing researchers across academia and industry with access to scalable, cost-efficient and consistent, high-quality tools for new medicine development. EBiSC has established a robust, reliable supply chain from the generation of cell lines, over the specification, to internationally accepted quality criteria and their worldwide distribution to any qualified user. The main facility is located at the Babraham Research Campus (Cambridge, UK) and undertakes cell expansion, quality control and characterization. The European Cell Culture Collection (ECACC) of Public Health England (Department of Health, UK) coordinates cell line distribution, and the Fraunhofer IBMT (Saarbrücken, Germany) provides comprehensive operational backup.

EBiSC's capacity will be 10,000 cell lines; it will eventually process over 1000 lines per year. In a phased business strategy, EBiSC hot-started distribution of lines contributed by iPSC Centres in 2014 and lines collected based on specified user demand, reached full-scale operations in 2016, and with extended funding will become self-sustaining as a not-for-profit banking operation by 2019.

EBiSC aims at spearheading Europe in the international standardization of iPSC banking by forging collaborative links with similar endeavours in the USA and Asia. Through market research, stakeholder engagement, incentives and training, EBiSC promotes wider participation and use, maximize the value of the research resource, and secure its longevity.

The Human Pluripotent Stem Cell Registry (hPSCreg)
https://hpscreg.eu

hPSCreg was established in 2007 as the European human embryonic stem cell registry and is supported by the European Commission. It is a public registry and data portal for hPSC lines. It offers the research community, legislators, regulators and the general public at large an in-depth overview on the current status of hPSC research. The registry provides comprehensive and standardized biological and legal information as well as tools to search and compare information from multiple hPSC sources. The online registration tool provides more than 740 possible data fields for hPSC, which allow in-depth provision of detailed information on each registered cell line. This information includes data regarding provider information, donor information (including ethics), generation and culture cell line information, characterization, genotype and publications related. The registry allows searching for cell line and available cell line information. It allows to register new lines and to add new information of already registered lines. hPSCreg has put in place an automatic system for naming hPSC lines. The automatically generated and standardized name is limited to 15 characters. This restriction allows readability and manageability, not only for stem cell banks and literature search but also in daily lab use. All data required to generate the name are submitted to the database during cell line registration. The tool will assign these data into an exclusive and understandable cell line name, answering the call for standardization.

EU-funded research projects using hPSC are also collected in hPSCreg. Project information and hPSC lines associated to the project can be found in the registry. In 2017, a third edition of the project has been started, in which it is planned to gather information on the clinical-grade hPSC lines and a registry of clinical trials with hiPSC-derived cells. hPSCreg is led by the Berlin-Brandenburg Center for Regenerative Therapies (BCRT).

The Global Alliance for iPSC Therapies (GAiT)
http://www.globalallianceforipsctherapies.com

The Global Alliance for iPSC Therapies (GAiT) is a brand-new initiative to support the implementation and clinical application of therapies derived from pluripotent stem cells to the benefit of patients globally. GAiT is supported by an international consortium of organizations including the Cell and Gene Therapy Catapult (London, UK), the Centre for Commercialization of Regenerative Medicine (Toronto, Canada), the Korea HLA-Typed iPSC Banking Initiative (Seoul, Korea) and the New York Stem Cell Foundation (New York, USA). GAiT's mission is to serve as a central, international resource for organizations developing therapies from clinical-grade induced pluripotent stem cells and to support the expansion of this nascent field. With the support of its international partners, GAiT already has an early position on manufacturing, regulatory and quality standards, and GAiT's intention is to support the delivery of the next generation of cell therapies.

StemBANCC
http://stembancc.org/

StemBANCC is a large-scale, 5-year academic-industry partnership in the area of stem cell research. It brings together a consortium of 37 partners from pharmaceutical companies, research institutions and small and medium enterprises (SMEs) who collaborate and share their experience to exploit the rich expertise across sectors and enhance knowledge transfer between academia and industry for patient benefit. StemBANCC is coordinated by Hoffmann-La Roche Ltd.

StemBANCC aims to provide well-characterized, patient-derived induced pluripotent stem cell lines and associated biomaterials in an accessible and sustainable biobank. StemBANCC also aims to demonstrate proof of concept for the utility of induced pluripotent stem cells in drug discovery for hard-to-treat disorders. The main goal of the StemBANCC project is to generate and characterize high-quality hiPSC lines to study a range of chronic diseases and test for drug efficacy and safety. The five major disease groups of interest are peripheral neuropathies, neurodegenerative disorders, neurodysfunctional disorders, diabetes and adverse drug responders.

14.9 hPSC Applications

hPSC are powerful tools for research and drug discovery and may provide regenerative therapies for diseases that conventional medicine cannot currently cure. Thousands of hPSC lines have been established in laboratories around the world, and many research programs are focused on investigating their properties and their potential not only for therapy but for other applications as well, such as toxicology and drug discovery, developmental biology or disease modelling.

14.10 Toxicology Assays

The development of new drugs is enormously costly, particularly, because of failures and unanticipated side effects in clinical trials. There is considerable interest in developing more effective approaches that could predict the likelihood of candidate drugs to cause serious effects. The capability of hPSC lines to create tissue-like cultures in vitro could provide valuable information on the toxicity of drug compounds and hopefully avoid some of the serious chronic toxic effects of drugs that were not detected by standard assays. The use of hPSC in drug discovery and predictive toxicology allows limiting animal testing, as motivated by ethical and economic reasons, but also for relevance as physiological interspecies differences can be most confusing. Animal models have sometimes failed to detect critical toxic effects, and in some cases this was recognized only once the drug was on the market and had given rise to adverse effects in patients and customers [75]. hPSCs provide an opportunity to establish new in vitro human cell-based assays that could accelerate drug screening and improve safety assessment of pharmaceuticals, chemical products and agrochemicals.

The main challenge for the application of hPSCs in toxicology is to demonstrate that they can generate a significant range of toxicology relevant cell types with characteristics and responses typical of human cell types. A further goal would be to obtain data from stem cell-based systems that can indicate chronic toxicity effects (steatosis, cholestasis, cardiomyopathies, etc.) as well as acute responses (apoptosis, necrosis, etc.) [75].

hPSCs at undifferentiated state constitute a unique tool to perform developmental toxicology and genotoxicity studies. By differentiation, hPSC can give rise to all of the different cell types developing miniaturized models to explore many aspects of the biological activity of multiple compounds. Among the more specialized cell types, hepatocytes are of particular interest to the pharmaceutical industry because of their involvement in drug metabolism. Functional 3D liver organ buds have been developed which may result in better drug screening [87, 99]. Cardiomyocytes can be used in vitro to detect molecules that induce a delay in ventricular repolarization, in order to prevent the commercialization of drugs that could lead to ventricular arrhythmia in patients. hPSC-derived keratinocytes, through their capacity to form a

Table 14.2 Some examples of screening for toxicity with hPSC

hPSC	Cell type	Activity	Action
Undifferentiated state		Developmental toxicity	Proliferation
			Differentiation
		Genotoxicity	Chromosome stability
			Genetic stability
Differentiated state	Hepatocytes	Metabolism and hepatoxicity	Drugs interactions cytotoxicity
	Cardiomyocytes	Cardiotoxicity	Ventricular repolarization delay
	Neurons	Neurotoxicity	Gene expression Crosstalk between neurons
	Keratinocytes	Local tolerance	Pluriestratified epidermis

Based on [49]

pluristratified epidermis, could also provide models for local tolerance studies (absorption, corrosion, irritation, etc.) [49]. Regarding the nervous system, the analysis of gene expression alterations of cells such as neurons, mesenchymal stromal cells and vascular endothelial cells derived from hESC in a culture dish has been proposed [85]. Neuronal and glial cells from iPSC are a valuable tool to assess the neurotoxic effects of chemicals. Mixed cultures of neuronal cells and astrocytes may prove suitable for studying the crosstalk between neurons and glia in neurodegenerative diseases [75]. Table 14.2 shows some examples of screening for toxicity with hPSC.

hPSC may also enable the evaluation of identified candidates for different parameters in preclinical studies, in particular for their safety which represents a crucial issue [49]. It is recommended that researchers using hPSC for toxicology and drug screening should adopt the principles of good cell culture practice (GCCP). They should obtain cell with consistent quality control and characterization but also evidence of legal and ethical acceptability. It is important to take into account that the basic knowledge of cell culture is not enough to use this so special kind of cells. Some expertise is needed for the handling and culture of hPSC, as well as to carry out the differentiation protocols.

14.11 Disease Models

A comprehensive view of the biological processes underlying human pathologies is crucial to devising strategies for their prevention and treatment. As the molecular basis of many diseases can be narrowed down to defined genetic loci, modelling these diseases can be facilitated by studying specific genotypes in a suitable experimental context. The availability to model human diseases using hPSC allows the study of monogenic, complex and epigenetic disorders and even late-onset diseases.

Several strategies are used to generate such "diseases in a dish" using either hESC or patient-specific hiPSC.

hESC may be used as model human diseases in two different ways, by obtaining hESC lines derived from embryos carrying specific mutations or chromosomal disorders that have been identified by preimplantation genetic diagnosis (PGD) or preimplantation genetic screening (PGS), respectively. These embryos, which would otherwise be discarded, can be used as a source of hESC models for these disorders. Another option would be to induce a mutation in an established and healthy hESC line.

Currently, 79 hESCs coming from PGD or PGS discarded embryos are registered in the hPSCreg (human pluripotent stem cell registry) (www.hpscreg.eu).

However, only a small range of disorders can be traced by PGD and PGS, limiting the potential to generate disease models in hESC. This limitation has been overcome by the ability to reprogramme somatic patient-derived cells to pluripotency. So, iPSC generation is the most common way to achieve an hPSC for pathological model from patients whose disease is to be studied. Patient-specific lines are relevant models for any pathology including rare diseases that are difficult to analyse in vitro. This technology is especially useful for generating cell lines in particular when genetically inherited diseases affect tissues that cannot be easily accessed [8]. Six hundred and seven hiPSC lines coming from patients with more than 60 diseases are registered in the hPSCreg.

hPSC-based disease models are ideal when studying monogenic diseases, diseases showing high penetrance, have an early onset during development and are associated with a clear cellular phenotype. By contrast, diseases that originate from a set of genes (complex disorders) show low penetration, have a late onset or are associated with phenotypes involving aberrant pattern formation of target organs which are more challenging to model using hPSC. Monogenic, chromosomal and complex disorders have been successfully modelled using hiPSCs [6]. Complex diseases are still being successfully modelled using hiPSC. However, in cases where the same disorder can arise from mutations in a single gen or multiple gens, many investigators choose to model the rarer monogenic rather the more prevalent complex trait, for practical reasons [6]. Even so, ongoing studies also compare monogenic and complex cases of the same disease. Studies of the most common neurodegenerative disorder, Alzheimer disease, have identified mutations, as well as haplotypes that predispose to such disease [47, 111].

Late-onset disorders are more difficult to be modelled using hiPSC than early-onset ones. However, some strategies have been established to solve this. In some cases, early onset disease is available. For example, for the usually late-onset disease, Parkinson's, some patients have a variant that develop early aggressive symptoms. hiPSC from these patients allow researchers to identify disease-specific phenotypes in vitro following hiPSC differentiation towards dopaminergic neurons [20]. Another approach is to induce cellular ageing. Several strategies have been established to induce premature ageing [63, 87].

Many diseases have been modelled using hiPSC, but more frequent are cardiovascular and neurological diseases (long QT, Brugada and LEOPARD syndromes)

[65, 95]. hiPSC-derived neurons have been used to model Alzheimer disease [45, 90], Parkinson's disease [69, 82, 116] or autism [14]. Many authors have reported to model schizophrenia using patient hiPSC-derived neurons [13, 108] or Gaucher's disease [83]. For complex disorders of the nervous system, it is also of great importance to uncover whether the pathology is limited to neurons or whether other cell types are also affected. In these cases, coculture could be a useful platform to obtain these other cell types [8, 87]. For example, astrocyte-neuron coculture has been used to model amyotrophic lateral sclerosis [31, 67]. The coculture system allows the investigation of non-cell-autonomous aspects of the disease which would otherwise be impossible with single cell types as neurons.

The interactions between different cell types can be better modelling using organoids. Organoids have been generated for multiple organs, including the brain, kidney, retina, liver, lung and stomach using tissue stem cells and hiPSC [48]. hiPSC-derived organoids have been developed for several applications. They are particularly useful to study cell-cell interactions in a cellular context that mimics human physiology and development. Indeed, 3D organoids have been used to model human organ development and diseases to test therapeutic compounds in cell transplantation [21, 60, 96, 99]. Even though, 3D organoids for disease modelling has still limitations. Efficiency and reproducibility must be improved, and vascularization must be achieved. The transplantation approach may be applied when organoids with increase size and improved maturation will be solved.

14.12 Cell Therapy

A fundamental requirement for successful clinical translation is to create consistent high-quality hPSCs and their derivatives that assure safety and efficiency to the patient. As "living products", cells present a range of biomanufacturing challenges. The challenges for creating hiPSC-based drug therapeutics are particularly significant because there are multiple methodologies for reprogramming cells to create hiPSCs [88]. Differences in embryos or cell types of origin, reprogramming methodologies, reagents, culture conditions, differentiation protocols, etc., create significant variability in the generated products. A robust and standardized methodology would reduce the impact of technical variability. So, strict standards are needed to develop and validate manufacturing processes and the associated hPSC banks [106].

The translation of a discovery into a product for clinical evaluation involves preclinical development, which follows established processes that are guided by the appropriate regulatory body, such as Food and Drug Administration (FDA) in the USA or the European Medicines Agency (EMA) in Europe. In each clinical translation process, safety and efficacy are established by basic and preclinical research. Both FDA and EMA classify hPSC derivatives as advanced therapy medicinal products (ATMPs). This implies, among other requirements, hPSC and their derivatives must be produced under current good manufacturing practice (cGMP) conditions. The regulation for cell-based therapies varies across countries. As cell therapy products are being developed, manufacturers should market their products in different countries according the regulatory frameworks of each of them.

14.12.1 Clinical-Grade hPSC Lines and Good Manufacturing Practice (GMP)

Clinical-grade lines need to be established according to GMP conditions in facilities with a relevant product manufacturing licence under strict quality assurance. But these lines also must be generated with all ethical and legal requirements compliant. The hPSC for clinical applications should be developed under stringent ethical guidelines from anonymised traceable and tested donors who signed informed consent, as already explained.

Generation of hPSCs under GMP conditions includes tissue sourcing, hESC derivation or hiPSC generation, expansion, testing and storage. Although GMP-compliant hESCs have been generated [18, 19, 99] and also GMP-compliant hiPSC [7, 70], some authors adapt hPSC lines established in standard conditions to culture in GMP conditions [24]. In fact, most of the cells used in the current clinical trials were derived in no-GMP conditions and then qualified for GMP by conversion in GMP culture conditions and additional testing [58, 61, 84].

Modifications of specific components for the derivation and culture of hESC have been reported. Xenobiotics must be avoided. To avoid immunosurgery, strategies as mechanical or laser dissection has been described. Mouse fibroblast feeders have been replaced for human fibroblasts, and even derivation and culture in feeder-free conditions have been developed using recombinant or synthetic extracellular matrices [81]. Nowadays, commercially serum-free and produced under GMP condition cell growth media are available as SCGM (CellGenix™) or mTeSR™-1 (Stemcell technologies). For cell dissociation GMP collagenase as well as other commercially GMP enzymes as TrypLE™ Select (Gibco) are also available.

Regarding hiPSC, integration-free methods of reprogramming are mandatory for cell therapy. Although non-integrative methods have been developed with episomal vectors [22] or Sendai virus [62], it is difficult to exclude the possibility of integration of very small fragments of DNA, and sometimes extensive passaging is required to remove exogenous vectors or viral expression. It has been shown that modified mRNA can be used to efficiently derive footprint-free hiPSC [107].

GMP conditions include five fundamental pillars:

- Traceability: Set of measures, actions and procedures that allow identifying and registering each product from its origin to the final destination. The "history file" of each cell line includes information about depositor, storage (with a proper labeling system), shipping, provenance, culture, quality control, facility

and equipment details, environmental monitory records, deviations of standard operating procedures (SOPs) and records of staff training or illness among others [4]
- Documentation: All procedures should be documented as SOPs. Description of activities and processes. Registration and verification of activities. Compliance with standards. Reproducibility and standardization.
- Validation: Verify that each aspect of manufacturing works correctly. Security and guarantee that what is validated will produce adequate results. Equipment as freezers, LN2 tanks, refrigerators and incubators should be monitored with systems of recorded temperature and alarms. Validation is the documented act of ensuring that any procedure, equipment, material or activity gives the expected results with adequate reproducibility.
- Quality control: A quality management system (QMS) must be implemented that describes organization structure, policies, responsibilities, producers and resources required for quality assurance (QA). Systems for recording follow-up and corrective actions. System for reporting and investigating defects and accidents. Quality audit programme (independent audits and inspection from the legal authorities) must be performed.
- Training: Training and continuous qualification for all staff members in production processes and quality regulations.

It is recommended for a manufacturer producing and banking a hPSC line to create a first stock of vials called master cell bank (MCB) and focus the testing and characterization on vials of this MCB. One vial of this MCB will be used for producing future working cell banks (WCB). Additional testing of WCB should be considered based on risk assessment.

As a final point, it has to be taken into account that undifferentiated hPSC will not be used for cell therapy; all the security and quality considerations must be considered for the final products and differentiation protocols made through.

14.12.2 Banks of hiPSC of Frequent HLA Haplotypes

Although hPSC lines appear to demonstrate low expression of HLA antigens [23], differentiated cells generate an allogenic response as seen by T-cell proliferation and cellular apoptosis against IPSC-derived cells. The use of patient-specific cells for iPSC generation and future use in the replace of diseased and damaged tissues in patients assures immunological compatibility and avoids immunological rejection. However, the time and cost necessary for the production of clinical-grade iPSC, the production of differentiated cells in this conditions, quality controls and validation of the procedures could be unapproachable in a wide clinical practice. It seems unlikely that these will be used as a source for large numbers of patients.

An alternative to the use of patient-specific iPSC would be the creation of a bank, or a network of banks, of allogenic clinical GMP iPSC lines that could be expanded and differentiated to treat different patients. Such a bank should collect lines with diverse and sufficient HLA haplotypes to ensure the immune matching of most part of the population and minimize the risk of immunological rejection. This idea was already proposed by Bradley et al. [12] and Taylor et al. [101] using embryonic stem cells (hESCs). The iPSC generation methodology facilitates the procurement of PSC lines of specific HLA haplotypes, since somatic cell donors can be selected based on their HLA system [68]. The ability to preselect the donor genotype of iPSC lines provides important opportunities for immune matching in cell therapy. Selection of homozygous donors for HLA haplotypes for generation of iPSC would simplify compatibility with possible receptors [106].

Nakatsuji et al. [68] suggested that a cell bank size of only 30 iPSC lines will be able to find a three-locus match (HLA-A, HLA-B and HLA-DR) in 82.2% of Japanese population. If the bank size were increased to 50 lines, it would cover 90.7% of the population. However, the authors suggest that to find at least 1 homozygote for each of 50 different haplotypes, it would be necessary to study a database of 24,000 individuals. Okita et al. [72] estimated that 140 homozygous donors for HLA haplotypes would cover 90% of the Japanese population, requiring screening in 160,000 possible donors.

Gourraud et al. [29] developed a mathematical model whereby they calculated that a bank comprising iPSC lines representing the 20 most frequent haplotypes in each population would request quite different number of donors to screen between 26,000 for European Americans and 110,000 African-Americans. It would also match different fractions of the recipient population, more than 50% of the European Americans and 22% of African-Americans. Authors suggest that a bank comprising the 100 iPSC lines with the most frequent HLA in each population would leave out only 22% of the European Americans but 37% of the Asians, 48% of the Hispanic and 55% of the African-Americans. So, it seems that relatively few, but very well selected, donors would result in iPSC lines with very important clinical utility. HLA homozygote donors could be searched between the potential donors of bone marrow and the samples preserved in the umbilical cord banks, since both are already typed for the HLA system [68, 102].

Umbilical cord blood cells are an optimal cell type for generating iPSC in order to create a bank of homozygous iPSCs for HLA haplotypes for possible clinical translation. They are the cell type of choice for the following reasons:

- The method of obtaining during childbirth is not invasive for the mother nor for the foetus. Therefore the risk is null for both.
- There are already umbilical cord blood unit (UCBU) preserved in cord blood banks and with known HLA typing, which facilitates the screening of donors.
- It is a cellular type widely used to obtain iPSC and large experience has been achieved.

- Umbilical cord stem cells have shown some degree of non-haematopoietic potential and express transcription factors typically expressed in hESCs such as Oct4, Nanog, Sox2, Rex1 and SSEA-4 [46].

However, in order to carry out screening and to identify among the large number of potential donors, it would be necessary to collaborate with multiple centres worldwide [102]. Nevertheless, although matching of the three HLA loci would greatly reduce immunological rejection in future cell transplantation therapy, it is important to know that mismatching of other HLA loci would still likely necessitate the use of supplementary immunosuppression [68].

14.12.3 Clinical Trials

Cell therapies are emerging as the next major development in human medicine. The potential of hPSC is unique and extraordinary. Around 30 clinical trials are ongoing with hESC-derived cells, and first patients have been treated with hiPSC-derived cells.

The first clinical trial with cells from human pluripotent stem cells (hPSC) was launched in 2010. Geron Corp. initiated a Phase I study to evaluate the safety of the use of oligodendrocyte precursors derived from human embryonic stem cells (hESCs) (GRNOPC1), in patients with spinal cord injury. One year later, the company cancelled the trial for economic reasons, reporting that no significant side effects were observed in any of the five patients treated. In 2014, Asterias Biotherapeutics recovered the assay (by renaming the product as AST-OPC1), initiating a Phase I/II trial in which it is intended to treat 35 patients with different cell doses. Preliminary data presented by the company report reduced cavitation, improved myelin coating, neovascularization and the production of neuronal growth stimulating factors, in addition to the absence of relevant side effects.

In 2011, Advanced Cell Technologies initiated another Phase I trial, for the treatment of age-related macular degeneration (AMD) and Stargardt's disease by transplantation of retinal pigmented epithelium (RPE) derived from hESC (MA09-hRPE). ACT has published preliminary results [84] and medium term results [85] without side effects and improved vision in 17 of 18 treated patients. Currently, ACT (which was renamed Ocata Therapeutics and later Astellas Institute for Regenerative Medicine) has four more clinical trials underway for the treatment of macular degeneration (MD) in the USA and UK, registered at Clinical Trials.gov (https://clinicaltrials.gov). In addition, MA09-hRPE is also being used in two Phase I/II trials in South Korea with similar positive results [91]. Other clinical trials are registered at Clinical trials Trials.gov for the treatment of MD by RPE derived from hESC. They are carried out by Pfizer (UK), Regenerative Patch Technologies (USA), Cell Cure Neurosciences (Israel), Federal University of São Paulo (Brazil), Chinese Academy of Sciences (China) and Southwest Hospital, Chongqing (China) and two more are registered in the Chinese Clinical Trial Register (www.chictr.org) by the Chinese Academy of Sciences (China). The Eye Institute of Xiamen University is conducting a trial with corneal epithelium derived from hESC for severe ocular surface disease.

In 2013, the first Phase I clinical trial for the treatment of six patients with cardiac ischaemia with cardiac progenitors derived from hESC was started in the Assistance Publique Hopitaux in Paris. The preliminary results of the first treated patient were published by Menasche et al. [61] reporting improvement of cardiac function and absence of important side effects.

In 2014, Viacyte initiated a Phase I/II study for the treatment of 65 patients with type 1 diabetes. For that, they designed a device (VC-01) consisting in a semipermeable membrane that encapsulates the β-cell progenitor cells. This device, implanted subcutaneously, allows the entry of oxygen and nutrients, as well as the release of insulin and other hormones, while protecting the cells from the autoimmune reaction that causes type 1 diabetes. This year, the company has announced two more trials with a modified device (VC-02) that allows direct vascularization into itself.

In 2017, the Australian company Cyto Therapeutics Pty Limited has initiated a study in 12 patients with Parkinson's disease with neuronal progenitors from parthenogenetic hESC (ISC-hpNSC). After 6 months of treatment, a reduction of the off period was observed in the treated patients, as well as an improvement in the motor and cognitive abilities. Another clinical trial for Parkinson's disease has been announced by the Chinese Academy of Sciences by transplanting neuronal precursors derived from hESC (50 patients; Phase I).

In 2014, the RIKEN Institute in Japan treated a patient with cells from iPSC for the first time. These were RPEs derived from autologous iPSCs to avoid immune rejection. The patient did not present side effects. However, the trial was suspended when trying to treat a second patient, as some mutations were detected in the iPSC generated [58]. Due to these findings and the fact that the generation of patient-specific iPSC is long and expensive, Riken has opted for the use of tested and safe allogenic iPSC. They have started a new clinical trial, betting on the idea of creating an iPSC bank with HLA matching for the most frequent haplotypes. RIKEN has already announced the treatment of the first patient with RPE derived from allogenic iPSC.

Three other clinical trials have been announced with iPSC-derived cells. The Moorfields Eye Hospital NHS Foundation Trust (UK) will initiate a new trial for the treatment of AMD. Cynata Therapeutics (Australia) will use mesenchymal cells from allogeneic iPSC (CYP-001) for the treatment of graft versus host disease. And finally, the Center for iPS Cell Research (Japan) will conduct a trial for the treatment of Parkinson's with allogeneic iPSCs. Clinical trials registered in Clinical Trials.gov and other registries are shown in Table 14.3.

In summary, hPSC have become an essential tool for drug screening, developing disease models and regenerative medicine. The rapid advance in stem cells research have led to the creation of stem cell banks and registries in order to collect, organize and disseminate cell lines and cell lines specific information.

Table 14.3 Clinical trials with human pluripotent stem cells

n°	Disease	Cell origin	Device	Derived cells	Sponsor	Country	Phase	Start date	Final date	n° pac.	Status
NCT01217008	SCI	ESC	GRNOPC1	Oligodendrocytes	Asterias biotherapeutics	USA	I	Oct 10	Jul 13	5	Completed
NCT01344993	AMD	ESC	MA09-hRPE	Retinal pigmented epith.	Astellas Inst. Regen. Med.	USA	I/II	Apr 11	Aug 15	13	Completed
NCT01345006	SMD	ESC	MA09-hRPE	Retinal pigmented epith.	Astellas Inst. Regen. Med.	USA	I/II	Apr 11	Aug 15	13	Completed
NCT01469832	SMD	ESC	MA09-hRPE	Retinal pigmented epith.	Astellas Inst. Regen. Med.	UK	I/II	Nov 11	Sep 15	12	Completed
NCT01625559	SMD	ESC	MA09-hRPE	Retinal pigmented epith.	CHABiotech CO., Ltd	Korea	I	Sep 12	Jun 15	3	Unknown
NCT01674829	AMD	ESC	MA09-hRPE	Retinal pigmented epith.	CHABiotech CO., Ltd	Korea	I/IIa	Sep 12	Apr 16	12	Unknown
NCT01691261	AMD	ESC	PF-05206388	Retinal pigmented epith.	Pfizer	UK	I	Jun 15	Nov 16	10	Ongoing
NCT02057900	IHD	ESC		CD15+ Isl-1+ progen.	Assistance Pub. Hôp. Paris	France	I	Jun 13	Jun 18	6	Recruiting
NCT02239354	T1DM	ESC	VC-01™	β-cell progenitors	ViaCyte	USA	I/II	Sep 14	Jan 21	40	Ongoing
NCT02286089	AMD	ESC	OpRegen	Retinal pigmented epith.	Cell cure neurosciences	Israel	I/II	Mar 15	Sep 19	15	Recruiting
NCT02302157	SCI	ESC	AST-OPC1	Oligodendrocits progenit.	Asterias biotherapeutic.	USA	I/II	Mar 15	Dec 18	35	Recruiting
NCT02445612	SMD	ESC	MA09-hRPE	Retinal pigmented epith.	Astellas Inst. Regen. Med.	USA	I/II	Jul 12	Dec 29	13	Ongoing
NCT02452723	PD	pESC	ISC-hpNSC	Neural stem cells	Cyto therapeutics Pty Lim.	Australia	I	Mar 16	Mar 19	12	Recruiting
NCT02463344	AMD	ESC	MA09-hRPE	Retinal pigmented epith.	Astellas Inst. Regen. Med.	USA	I/II	Jul 12	Dec 29	11	Ongoing

NCT02464956	AMD	iPS autol		Retinal pigmented epith.	Moorfields eye Hosp. NHS	UK	I	Jul 15	Apr 16	10	Unknown
NCT02590692	AMD	ESC	CPCB-RPE1	Retinal pigmented epith.	Regenerative Patch Tech.	USA	I/II	Oct 15	Sep 22	20	Recruiting
NCT02749734	AMD; SMD	ESC	ESC-RPE	Retinal pigmented epith.	Southwest Hospital, China	China	I	May 15	Dec 17	15	Recruiting
NCT02755428	AMD	ESC	MA09-hRPE	Retinal pigmented epith.	Chinese Academy Sc.	China	I	Apr 17	Dec 20	10	Recruiting
NCT02903576	AMD; SMD	ESC		Retinal pigmented epith.	Federal Univ. São Paulo	Brazil	I/II	Aug 15	Jun 19	18	Recruiting
NCT02923375	GVHD	iPSC allog	CYP-001	MSC	Cynata therapeutics	Australia	I	Mar17	Sep 19	16	Recruiting
NCT02941991	SMD	ESC	MA09-hRPE	Retinal pigmented epith.	Astellas Inst. Regen. Med.	UK	I/II	Jun 13	Dec 19	11	Ongoing
NCT03046407	AMD	ESC	ESC-RPE	Retinal pigmented epith.	Chinese Academy Sc.	China	I	Mar 17	Dec 20	10	Recruiting
NCT03119636	PD	ESC		Neural precursors	Chinese Academy Sc.	China	I/II	May 17	Dec 20	50	Recruiting
NCT03162926	T1DM	ESC	VC-02	β-cell progenitors	ViaCyte	USA	I	Jul 17	Jun 18	15	Recruiting
NCT03163511	T1DM	ESC	VC-02	β-cell progenitors	ViaCyte	USA	I/II	Jul 17	Dec 20	55	Recruiting
ChiCTR-OCB-15005968	SOSD	ESC		Corneal-epithelium	Eye Institute Xiamen Univ.	China	I/II	Oct 15	Dec 18	20	Recruiting
ChiCTR-OCB-15007054	AMD	ESC	ESC-RPE	Retinal pigmented epith.	Chinese Academy Sc.	China	I	Jun 16	Jun 17	10	Recruiting
ChiCTR-OCB-15007055	RPD	ESC	ESC-RPE	Retinal pigmented epith.	Chinese Academy Sc.	China	I	Sep 15	Dec 17	10	Recruiting
	AMD	iPSC allog		Retinal pigmented epith.	RIKEN Center for Dev. Biol	Japan	I	Feb 17		5	Recruiting
UMIN000011929	AMD	iPS autol		Retinal pigmented epith.	RIKEN Center for Dev. Biol	Japan	I	Sep 14	Sep 15	1	Suspended
	PD	iPSC allog		Dopamine-secreting nerve	Center for iPS Cel Res.	Japan	I	2018			

SCI Spinal cord blood, *AMD* Age-related macular degeneration, *SMD* Stargardt's macular degeneration, *IHD* Ischemic heart disease, *T1DM* T1 diabetes mellitus, *PD* Parkinson's disease, *GVHD* Graft versus host disease, *SOSD* Severe ocular surface disease

References

1. Aasen T, Raya A, Barrero MJ, Garreta E, Consiglio A, Gonzalez F et al (2008) Efficient and rapid generation of induced pluripotent stem cells from human keratinocytes. Nat Biotechnol 26(11):1276–1284
2. Allegrucci C, Young LE (2007) Differences between human embryonic stem cell lines. Hum Reprod Update 13(2):103–120
3. Amit M, Shariki C, Margulets V, Itskovitz-Eldor J (2004) Feeder layer- and serum-free culture of human embryonic stem cells. Biol Reprod 70(3):837–845
4. Andrews PW, Baker D, Benvinisty N, Miranda B, Bruce K, Brustle O et al (2015) Points to consider in the development of seed stocks of pluripotent stem cells for clinical applications: International Stem Cell Banking Initiative (ISCBI). Regen Med 10(2 Suppl):1–44
5. Aran B, Rodriguez-Piza I, Raya A, Consiglio A, Munoz Y, Barri PN et al (2010) Derivation of human embryonic stem cells at the Center of Regenerative Medicine in Barcelona. In Vitro Cell Dev Biol Anim 46(3–4):356–366
6. Avior Y, Sagi I, Benvenisty N (2016) Pluripotent stem cells in disease modelling and drug discovery. Nat Rev Mol Cell Biol 17(3):170–182
7. Baghbaderani BA, Tian X, Neo BH, Burkall A, Dimezzo T, Sierra G et al (2015) cGMP-manufactured human induced pluripotent stem cells are available for pre-clinical and clinical applications. Stem Cell Rep 5(4):647–659
8. Bellin M, Marchetto MC, Gage FH, Mummery CL (2012) Induced pluripotent stem cells: the new patient? Nat Rev Mol Cell Biol 13(11):713–726
9. Bongso A, Fong CY, Ng SC, Ratnam S (1994) Isolation and culture of inner cell mass cells from human blastocysts. Hum Reprod 9(11):2110–2117
10. Borstlap J, Luong MX, Rooke HM, Aran B, Damaschun A, Elstner A et al (2010) International stem cell registries. In Vitro Cell Dev Biol Anim 46(3–4):242–246
11. Borstlap J, Stacey G, Kurtz A, Elstner A, Damaschun A, Aran B et al (2008) First evaluation of the European hESCreg. Nat Biotechnol 26(8):859–860
12. Bradley JA, Bolton EM, Pedersen RA (2002) Stem cell medicine encounters the immune system. Nat Rev Immunol 2(11):859–871
13. Brennand KJ, Simone A, Jou J, Gelboin-Burkhart C, Tran N, Sangar S et al (2011) Modelling schizophrenia using human induced pluripotent stem cells. Nature 473(7346):221–225
14. Brito A, Russo FB, Muotri AR, Beltrao-Braga PCB. Autism spectrum disorders and disease modeling using stem cells. Cell Tissue Res 2018;371(1):153–60
15. Chen AE, Egli D, Niakan K, Deng J, Akutsu H, Yamaki M et al (2009) Optimal timing of inner cell mass isolation increases the efficiency of human embryonic stem cell derivation and allows generation of sibling cell lines. Cell Stem Cell 4(2):103–106
16. Chou BK, Mali P, Huang X, Ye Z, Dowey SN, Resar LM et al (2011) Efficient human iPS cell derivation by a non-integrating plasmid from blood cells with unique epigenetic and gene expression signatures. Cell Res 21(3):518–529
17. Crook JM, Hei D, Stacey G (2010) The International Stem Cell Banking Initiative (ISCBI): raising standards to bank on. In Vitro Cell Dev Biol Anim 46(3–4):169–172
18. Crook JM, Peura TT, Kravets L, Bosman AG, Buzzard JJ, Horne R et al (2007) The generation of six clinical-grade human embryonic stem cell lines. Cell Stem Cell 1(5):490–494
19. De Sousa PA, Downie JM, Tye BJ, Bruce K, Dand P, Dhanjal S et al (2016) Development and production of good manufacturing practice grade human embryonic stem cell lines as source material for clinical application. Stem Cell Res 17(2):379–390
20. Devine MJ, Ryten M, Vodicka P, Thomson AJ, Burdon T, Houlden H et al (2011) Parkinson's disease induced pluripotent stem cells with triplication of the alpha-synuclein locus. Nat Commun 2:440
21. DiStefano T, Chen HY, Panebianco C, Kaya KD, Brooks MJ, Gieser L, et al. Accelerated and Improved differentiation of retinal organoids from pluripotent stem cells in Rotating-Wall vessel bioreactors. Stem cell reports. 2018;10(1):300–13

22. Dowey SN, Huang X, Chou BK, Ye Z, Cheng L (2012) Generation of integration-free human induced pluripotent stem cells from postnatal blood mononuclear cells by plasmid vector expression. Nat Protoc 7(11):2013–2021
23. Drukker M, Katchman H, Katz G, Even-Tov Friedman S, Shezen E, Hornstein E et al (2006) Human embryonic stem cells and their differentiated derivatives are less susceptible to immune rejection than adult cells. Stem Cells 24(2):221–229
24. Durruthy-Durruthy J, Briggs SF, Awe J, Ramathal CY, Karumbayaram S, Lee PC et al (2014) Rapid and efficient conversion of integration-free human induced pluripotent stem cells to GMP-grade culture conditions. PLoS One 9(4):e94231
25. Ellerstrom C, Strehl R, Moya K, Andersson K, Bergh C, Lundin K et al (2006) Derivation of a xeno-free human embryonic stem cell line. Stem Cells 24(10):2170–2176
26. Fusaki N, Ban H, Nishiyama A, Saeki K, Hasegawa M (2009) Efficient induction of transgene-free human pluripotent stem cells using a vector based on Sendai virus, an RNA virus that does not integrate into the host genome. Proc Jpn Acad Ser B Phys Biol Sci 85(8):348–362
27. Genbacev O, Krtolica A, Zdravkovic T, Brunette E, Powell S, Nath A et al (2005) Serum-free derivation of human embryonic stem cell lines on human placental fibroblast feeders. Fertil Steril 83(5):1517–1529
28. Giorgetti A, Montserrat N, Aasen T, Gonzalez F, Rodriguez-Piza I, Vassena R et al (2009) Generation of induced pluripotent stem cells from human cord blood using OCT4 and SOX2. Cell Stem Cell 5(4):353–357
29. Gourraud PA, Hollenbach JA, Barnetche T, Single RM, Mack SJ (2012) Standard methods for the management of immunogenetic data. Methods Mol Biol 882:197–213
30. Gymrek M, McGuire AL, Golan D, Halperin E, Erlich Y (2013) Identifying personal genomes by surname inference. Science 339(6117):321–324
31. Haidet-Phillips AM, Hester ME, Miranda CJ, Meyer K, Braun L, Frakes A et al (2011) Astrocytes from familial and sporadic ALS patients are toxic to motor neurons. Nat Biotechnol 29(9):824–828
32. Hasegawa K, Pomeroy JE, Pera MF (2010) Current technology for the derivation of pluripotent stem cell lines from human embryos. Cell Stem Cell 6(6):521–531
33. Heins N, Englund MC, Sjoblom C, Dahl U, Tonning A, Bergh C et al (2004) Derivation, characterization, and differentiation of human embryonic stem cells. Stem Cells 22(3):367–376
34. Imaizumi K, Nishishita N, Muramatsu M, Yamamoto T, Takenaka C, Kawamata S et al (2014) A simple and highly effective method for slow-freezing human pluripotent stem cells using dimethyl sulfoxide, hydroxyethyl starch and ethylene glycol. PLoS One 9(2):e88696
35. International Stem Cell Banking I (2009) Consensus guidance for banking and supply of human embryonic stem cell lines for research purposes. Stem Cell Rev 5(4):301–314
36. International Stem Cell I, Adewumi O, Aflatoonian B, Ahrlund-Richter L, Amit M, Andrews PW et al (2007) Characterization of human embryonic stem cell lines by the international stem cell initiative. Nat Biotechnol 25(7):803–816
37. Isasi RM, Knoppers BM (2009) Governing stem cell banks and registries: emerging issues. Stem Cell Res 3(2–3):96–105
38. Kim D, Kim CH, Moon JI, Chung YG, Chang MY, Han BS et al (2009) Generation of human induced pluripotent stem cells by direct delivery of reprogramming proteins. Cell Stem Cell 4(6):472–476
39. Kim JH, Kurtz A, Yuan BZ, Zeng F, Lomax G, Loring JF et al (2017) Report of the international stem cell banking initiative workshop activity: current hurdles and progress in seed-stock banking of human pluripotent stem cells. Stem Cells Transl Med 6(11):1956–1962
40. Kimbrel EA, Lanza R (2016) Pluripotent stem cells: the last 10 years. Regen Med 11(8):831–847
41. Klimanskaya I, Chung Y, Becker S, Lu SJ, Lanza R (2006) Human embryonic stem cell lines derived from single blastomeres. Nature 444(7118):481–485
42. Klimanskaya I, Chung Y, Meisner L, Johnson J, West MD, Lanza R (2005) Human embryonic stem cells derived without feeder cells. Lancet 365(9471):1636–1641

43. Knoppers BM, Isasi R (2010) Stem cell banking: between traceability and identifiability. Genome Med 2(10):73
44. Knoppers BM, Isasi R, Benvenisty N, Kim OJ, Lomax G, Morris C et al (2011) Publishing SNP genotypes of human embryonic stem cell lines: policy statement of the international stem cell forum ethics working party. Stem Cell Rev 7(3):482–484
45. Kondo T, Asai M, Tsukita K, Kutoku Y, Ohsawa Y, Sunada Y et al (2013) Modeling Alzheimer's disease with iPSCs reveals stress phenotypes associated with intracellular Abeta and differential drug responsiveness. Cell Stem Cell 12(4):487–496
46. Kucia M, Halasa M, Wysoczynski M, Baskiewicz-Masiuk M, Moldenhawer S, Zuba-Surma E et al (2007) Morphological and molecular characterization of novel population of CXCR4+ SSEA-4+ Oct-4+ very small embryonic-like cells purified from human cord blood: preliminary report. Leukemia 21(2):297–303
47. Lambert JC, Ibrahim-Verbaas CA, Harold D, Naj AC, Sims R, Bellenguez C et al (2013) Meta-analysis of 74,046 individuals identifies 11 new susceptibility loci for Alzheimer's disease. Nat Genet 45(12):1452–1458
48. Lancaster MA, Knoblich JA (2014) Generation of cerebral organoids from human pluripotent stem cells. Nat Protoc 9(10):2329–2340
49. Laustriat D, Gide J, Peschanski M (2010) Human pluripotent stem cells in drug discovery and predictive toxicology. Biochem Soc Trans 38(4):1051–1057
50. Lerou PH, Yabuuchi A, Huo H, Takeuchi A, Shea J, Cimini T et al (2008) Human embryonic stem cell derivation from poor-quality embryos. Nat Biotechnol 26(2):212–214
51. Li T, Zhou C, Liu C, Mai Q, Zhuang G (2008) Bulk vitrification of human embryonic stem cells. Hum Reprod 23(2):358–364
52. Li T, Zhou CQ, Mai QY, Zhuang GL (2005) Establishment of human embryonic stem cell line from gamete donors. Chin Med J 118(2):116–122
53. Li Y, Tan JC, Li LS (2010) Comparison of three methods for cryopreservation of human embryonic stem cells. Fertil Steril 93(3):999–1005
54. Loser P, Schirm J, Guhr A, Wobus AM, Kurtz A (2010) Human embryonic stem cell lines and their use in international research. Stem Cells 28(2):240–246
55. Lu J, Hou R, Booth CJ, Yang SH, Snyder M (2006) Defined culture conditions of human embryonic stem cells. Proc Natl Acad Sci U S A 103(15):5688–5693
56. Ludwig T, J AT. Defined, feeder-independent medium for human embryonic stem cell culture. Current protocols in stem cell biology. 2007;Chapter 1:Unit 1C 2
57. Ludwig TE, Levenstein ME, Jones JM, Berggren WT, Mitchen ER, Frane JL et al (2006) Derivation of human embryonic stem cells in defined conditions. Nat Biotechnol 24(2):185–187
58. Mandai M, Watanabe A, Kurimoto Y, Hirami Y, Morinaga C, Daimon T et al (2017) Autologous induced stem-cell-derived retinal cells for macular degeneration. N Engl J Med 376(11):1038–1046
59. Marti M, Mulero L, Pardo C, Morera C, Carrio M, Laricchia-Robbio L et al (2013) Characterization of pluripotent stem cells. Nat Protoc 8(2):223–253
60. McCracken KW, Cata EM, Crawford CM, Sinagoga KL, Schumacher M, Rockich BE et al (2014) Modelling human development and disease in pluripotent stem-cell-derived gastric organoids. Nature 516(7531):400–404
61. Menasche P, Vanneaux V, Hagege A, Bel A, Cholley B, Cacciapuoti I et al (2015) Human embryonic stem cell-derived cardiac progenitors for severe heart failure treatment: first clinical case report. Eur Heart J 36(30):2011–2017
62. Merling RK, Sweeney CL, Choi U, De Ravin SS, Myers TG, Otaizo-Carrasquero F, et al. Transgene-free iPSCs generated from small volume peripheral blood nonmobilized CD34+ cells. Blood 2013;121(14):e98–107
63. Miller JD, Ganat YM, Kishinevsky S, Bowman RL, Liu B, Tu EY et al (2013) Human iPSC-based modeling of late-onset disease via progerin-induced aging. Cell Stem Cell 13(6):691–705

64. Mitalipova M, Calhoun J, Shin S, Wininger D, Schulz T, Noggle S et al (2003) Human embryonic stem cell lines derived from discarded embryos. Stem Cells 21(5):521–526
65. Moretti A, Bellin M, Welling A, Jung CB, Lam JT, Bott-Flugel L et al (2010) Patient-specific induced pluripotent stem-cell models for long-QT syndrome. N Engl J Med 363(15):1397–1409
66. Murdoch A, Braude P, Courtney A, Brison D, Hunt C, Lawford-Davies J et al (2012) The procurement of cells for the derivation of human embryonic stem cell lines for therapeutic use: recommendations for good practice. Stem Cell Rev 8(1):91–99
67. Nagai M, Re DB, Nagata T, Chalazonitis A, Jessell TM, Wichterle H et al (2007) Astrocytes expressing ALS-linked mutated SOD1 release factors selectively toxic to motor neurons. Nat Neurosci 10(5):615–622
68. Nakatsuji N, Nakajima F, Tokunaga K (2008) HLA-haplotype banking and iPS cells. Nat Biotechnol 26(7):739–740
69. Nguyen HN, Byers B, Cord B, Shcheglovitov A, Byrne J, Gujar P et al (2011) LRRK2 mutant iPSC-derived DA neurons demonstrate increased susceptibility to oxidative stress. Cell Stem Cell 8(3):267–280
70. Ohmine S, Dietz AB, Deeds MC, Hartjes KA, Miller DR, Thatava T et al (2011) Induced pluripotent stem cells from GMP-grade hematopoietic progenitor cells and mononuclear myeloid cells. Stem Cell Res Ther 2(6):46
71. Okita K, Matsumura Y, Sato Y, Okada A, Morizane A, Okamoto S et al (2011a) A more efficient method to generate integration-free human iPS cells. Nat Methods 8(5):409–412
72. Okita K, Nagata N, Yamanaka S (2011b) Immunogenicity of induced pluripotent stem cells. Circ Res 109(7):720–721
73. Ota A, Matsumura K, Lee JJ, Sumi S, Hyon SH (2017) StemCell keep is effective for cryopreservation of human embryonic stem cells by vitrification. Cell Transplant 26(5):773–787
74. Pera MF, Trounson AO (2004) Human embryonic stem cells: prospects for development. Development 131(22):5515–5525
75. Pistollato F, Bremer-Hoffmann S, Healy L, Young L, Stacey G (2012) Standardization of pluripotent stem cell cultures for toxicity testing. Expert Opin Drug Metab Toxicol 8(2):239–257
76. Rajala K, Hakala H, Panula S, Aivio S, Pihlajamaki H, Suuronen R et al (2007) Testing of nine different xeno-free culture media for human embryonic stem cell cultures. Hum Reprod 22(5):1231–1238
77. Raya A, Rodriguez-Piza I, Aran B, Consiglio A, Barri PN, Veiga A et al (2008) Generation of cardiomyocytes from new human embryonic stem cell lines derived from poor-quality blastocysts. Cold Spring Harb Symp Quant Biol 73:127–135
78. Reubinoff BE, Pera MF, Fong CY, Trounson A, Bongso A (2000) Embryonic stem cell lines from human blastocysts: somatic differentiation in vitro. Nat Biotechnol 18(4):399–404
79. Reubinoff BE, Pera MF, Vajta G, Trounson AO (2001) Effective cryopreservation of human embryonic stem cells by the open pulled straw vitrification method. Hum Reprod 16(10):2187–2194
80. Richards M, Fong CY, Chan WK, Wong PC, Bongso A (2002) Human feeders support prolonged undifferentiated growth of human inner cell masses and embryonic stem cells. Nat Biotechnol 20(9):933–936
81. Rodin S, Domogatskaya A, Strom S, Hansson EM, Chien KR, Inzunza J et al (2010) Long-term self-renewal of human pluripotent stem cells on human recombinant laminin-511. Nat Biotechnol 28(6):611–615
82. Sanchez-Danes A, Richaud-Patin Y, Carballo-Carbajal I, Jimenez-Delgado S, Caig C, Mora S et al (2012) Disease-specific phenotypes in dopamine neurons from human iPS-based models of genetic and sporadic Parkinson's disease. EMBO Mol Med 4(5):380–395
83. Santos DM, Tiscornia G (2017) Induced pluripotent stem cell modeling of Gaucher's disease: what have we learned? Int J Mol Sci 18(4):888
84. Schwartz SD, Hubschman JP, Heilwell G, Franco-Cardenas V, Pan CK, Ostrick RM et al (2012) Embryonic stem cell trials for macular degeneration: a preliminary report. Lancet 379(9817):713–720

85. Schwartz SD, Regillo CD, Lam BL, Eliott D, Rosenfeld PJ, Gregori NZ et al (2015) Human embryonic stem cell-derived retinal pigment epithelium in patients with age-related macular degeneration and Stargardt's macular dystrophy: follow-up of two open-label phase 1/2 studies. Lancet 385(9967):509–516
86. Seltmann S, Lekschas F, Muller R, Stachelscheid H, Bittner MS, Zhang W et al (2016) hPSCreg–the human pluripotent stem cell registry. Nucleic Acids Res 44(D1):D757–D763
87. Shi Y, Inoue H, Wu JC, Yamanaka S (2017) Induced pluripotent stem cell technology: a decade of progress. Nat Rev Drug Discov 16(2):115–130
88. Silva M, Daheron L, Hurley H, Bure K, Barker R, Carr AJ et al (2015) Generating iPSCs: translating cell reprogramming science into scalable and robust biomanufacturing strategies. Cell Stem Cell 16(1):13–17
89. Sjogren A, Hardarson T, Andersson K, Caisander G, Lundquist M, Wikland M et al (2004) Human blastocysts for the development of embryonic stem cells. Reprod Biomed Online 9(3):326–329
90. Soldner F, Stelzer Y, Shivalila CS, Abraham BJ, Latourelle JC, Barrasa MI et al (2016) Parkinson-associated risk variant in distal enhancer of alpha-synuclein modulates target gene expression. Nature 533(7601):95–99
91. Song WK, Park KM, Kim HJ, Lee JH, Choi J, Chong SY et al (2015) Treatment of macular degeneration using embryonic stem cell-derived retinal pigment epithelium: preliminary results in Asian patients. Stem Cell Rep 4(5):860–872
92. Stacey GN, Crook JM, Hei D, Ludwig T (2013) Banking human induced pluripotent stem cells: lessons learned from embryonic stem cells? Cell Stem Cell 13(4):385–388
93. Strelchenko N, Verlinsky O, Kukharenko V, Verlinsky Y (2004) Morula-derived human embryonic stem cells. Reprod Biomed Online 9(6):623–629
94. Strom S, Inzunza J, Grinnemo KH, Holmberg K, Matilainen E, Stromberg AM et al (2007) Mechanical isolation of the inner cell mass is effective in derivation of new human embryonic stem cell lines. Hum Reprod 22(12):3051–3058
95. Sun N, Yazawa M, Liu J, Han L, Sanchez-Freire V, Abilez OJ et al (2012) Patient-specific induced pluripotent stem cells as a model for familial dilated cardiomyopathy. Sci Transl Med 4(130):130ra47
96. Takahashi Y, Sato S, Kurashima Y, Yamamoto T, Kurokawa S, Yuki Y, Takemura N, Uematsu S, Lai C-Y, Otsu M, Matsuno H, Osawa H, Mizushima T, Nishimura J, Hayashi M, Yamaguchi T, Kiyono H (2018) A refined culture system for human induced pluripotent stem cell-derived intestinal epithelial organoids. Stem Cell Reports 10(1):314–328
97. Takahashi K, Tanabe K, Ohnuki M, Narita M, Ichisaka T, Tomoda K et al (2007) Induction of pluripotent stem cells from adult human fibroblasts by defined factors. Cell 131(5):861–872
98. Takahashi K, Yamanaka S (2006) Induction of pluripotent stem cells from mouse embryonic and adult fibroblast cultures by defined factors. Cell 126(4):663–676
99. Takebe T, Sekine K, Enomura M, Koike H, Kimura M, Ogaeri T et al (2013) Vascularized and functional human liver from an iPSC-derived organ bud transplant. Nature 499(7459):481–484
100. Tannenbaum SE, Turetsky TT, Singer O, Aizenman E, Kirshberg S, Ilouz N et al (2012) Derivation of xeno-free and GMP-grade human embryonic stem cells–platforms for future clinical applications. PLoS One 7(6):e35325
101. Taylor CJ, Bolton EM, Pocock S, Sharples LD, Pedersen RA, Bradley JA (2005) Banking on human embryonic stem cells: estimating the number of donor cell lines needed for HLA matching. Lancet 366(9502):2019–2025
102. Taylor CJ, Peacock S, Chaudhry AN, Bradley JA, Bolton EM (2012) Generating an iPSC bank for HLA-matched tissue transplantation based on known donor and recipient HLA types. Cell Stem Cell 11(2):147–152
103. Tedder RS, Zuckerman MA, Goldstone AH, Hawkins AE, Fielding A, Briggs EM et al (1995) Hepatitis B transmission from contaminated cryopreservation tank. Lancet 346(8968):137–140
104. Thomson JA, Itskovitz-Eldor J, Shapiro SS, Waknitz MA, Swiergiel JJ, Marshall VS et al (1998) Embryonic stem cell lines derived from human blastocysts. Science 282(5391):1145–1147

105. Turetsky T, Aizenman E, Gil Y, Weinberg N, Shufaro Y, Revel A et al (2008) Laser-assisted derivation of human embryonic stem cell lines from IVF embryos after preimplantation genetic diagnosis. Hum Reprod 23(1):46–53
106. Turner M, Leslie S, Martin NG, Peschanski M, Rao M, Taylor CJ et al (2013) Toward the development of a global induced pluripotent stem cell library. Cell Stem Cell 13(4):382–384
107. Warren L, Manos PD, Ahfeldt T, Loh YH, Li H, Lau F et al (2010) Highly efficient reprogramming to pluripotency and directed differentiation of human cells with synthetic modified mRNA. Cell Stem Cell 7(5):618–630
108. Wen Z, Nguyen HN, Guo Z, Lalli MA, Wang X, Su Y et al (2014) Synaptic dysregulation in a human iPS cell model of mental disorders. Nature 515(7527):414–418
109. Woltjen K, Michael IP, Mohseni P, Desai R, Mileikovsky M, Hamalainen R et al (2009) piggyBac transposition reprograms fibroblasts to induced pluripotent stem cells. Nature 458(7239):766–770
110. Wong KG, Ryan SD, Ramnarine K, Rosen SA, Mann SE, Kulick A et al (2017) CryoPause: a new method to immediately initiate experiments after cryopreservation of pluripotent stem cells. Stem Cell Rep 9(1):355–365
111. Young JE, Boulanger-Weill J, Williams DA, Woodruff G, Buen F, Revilla AC et al (2015) Elucidating molecular phenotypes caused by the SORL1 Alzheimer's disease genetic risk factor using human induced pluripotent stem cells. Cell Stem Cell 16(4):373–385
112. Yu J, Hu K, Smuga-Otto K, Tian S, Stewart R, Slukvin II, Thomson JA (2009) Human induced pluripotent stem cells free of vector and transgene sequences. Science 324(5928):797–801
113. Yu J, Vodyanik MA, Smuga-Otto K, Antosiewicz-Bourget J, Frane JL, Tian S et al (2007) Induced pluripotent stem cell lines derived from human somatic cells. Science 318(5858):1917–1920
114. Zhang X, Stojkovic P, Przyborski S, Cooke M, Armstrong L, Lako M et al (2006) Derivation of human embryonic stem cells from developing and arrested embryos. Stem Cells 24(12):2669–2676
115. Zhou T, Benda C, Dunzinger S, Huang Y, Ho JC, Yang J et al (2012) Generation of human induced pluripotent stem cells from urine samples. Nat Protoc 7(12):2080–2089
116. Reinhardt P, Schmid B, Burbulla LF, Schöndorf DC, Wagner L, Glatza M, Höing S, Hargus G, Heck SA, Dhingra A, Wu G, Müller S, Brockmann K, Kluba T, Maisel M, Krüger R, Berg D, Tsytsyura Y, Thiel CS, Psathaki OE, Klingauf J, Kuhlmann T, Klewin M, Müller H, Gasser T, Schöler HR, Sterneckert J (2013) Genetic correction of a LRRK2 mutation in human iPSCs links parkinsonian neurodegeneration to ERK-dependent changes in gene expression. Cell Stem Cell 12(3):354–367. doi: 10.1016/j.stem.2013.01.008

Chapter 15
Engineering of Human-Induced Pluripotent Stem Cells for Precise Disease Modeling

P. Lisowski

Abstract Stem cell technologies and gene editing techniques are two of the most promising recent developments in biomedicine. The ability to reprogram common human cells into induced pluripotent stem cells (hiPSCs) and turn them into the cells of interest has already become a powerful research tool, thus providing a unique platform for disease studies. In combination with the use of designer nucleases approach to repair or to introduce disease-causing mutations, both are valuable in developing personalized disease models. This chapter provides an overview on designer nucleases-based gene editing in hiPSCs, describing the principles of CRISPR/Cas systems along with consecutive methodological steps such as nucleases selection, isolation, and genotyping of modified hiPSC clones with emphasis on the crucial role of isogenic cell lines in disease modeling. Moreover, the production of rare or complex genotypes in patient cell lines requires efficient and streamlined gene editing technologies. However, precise genome editing applications rely on infrequent homology-directed repair (HDR), with the abundant nonhomologous end joining (NHEJ) formed indels presenting a barrier to achieving high rates of precise sequence modification. The methods presented here are supported by theoretical framework to allow for the incorporation of inevitable improvements to achieve either higher rates of gene editing by promotion of HDR over NHEJ or application of different CRISPR/Cas platforms for robust and multiplex gene editing, toward decoding of neurodevelopmental as well as for modeling of late onset disorders by fast-forwarding the biological clock. Due to easy in theory but laborious and inefficient in practice, the precise

P. Lisowski (✉)
Department of Molecular Biology, Institute of Genetics and Animal Breeding, Polish Academy of Sciences, Jastrzebiec, Magdalenka, Poland

Department of Medical Genetics, Centre for Preclinical Research and Technology (CePT), Warsaw Medical University, Warsaw, Poland

Mitochondria and Cell Fate Reprogramming Group, Department of Proteomics and Molecular Mechanisms of Neurodegenerative Diseases, Max Delbrück Center for Molecular Medicine (MDC) in the Helmholtz Association, Berlin, Germany
e-mail: pawel.lisowski@mdc-berlin.de

R. Delgado-Morales (ed.), *Stem Cell Genetics for Biomedical Research*,
https://doi.org/10.1007/978-3-319-90695-9_15

and efficient genome editing in hiPSCs could be only achieved by the proper combination of the described methods in the process. This eventually would lead to generation of wide range of disease models for decoding of sporadic, polygenic, undiagnosed, and rare disorders using the adequate experimental design following appropriate gene editing toolbox selection.

Keywords Human-induced pluripotent stem cells (hiPSCs) · RNA-guided designer Cas9 nucleases · CRISPR/Cas · Gene editing · Multiplex genome editing · Genome engineering · Personalized disease models · hiPS cell-based disease modeling

15.1 Genome Editing Before CRISPR/Cas9

Experiments conducted in the early 1990s demonstrated that double-strand breaks (DSBs) generated by nucleases at a defined genomic site can be repaired either by the endogenous homology-mediated repair (HDR) using an exogenous provided repair template or by the error-prone nonhomologous end-joining (NHEJ) DNA repair pathways [1, 2]. Repair events frequently resulted in small deletions after rejoining of the two DNA ends, some occurred by simple blunt-ended ligation, whereas others through annealing of short regions of terminal homology. The crucial observation made during these experiments was that the DSBs were apparently recombinogenic, stimulating gene targeting of a homologous fragment by more than two orders of magnitude. Importantly, this principle of employing a DSB to facilitate DNA repair-mediated editing of genomes proved to be universal and applies to hPSCs as well as other systems such as *Caenorhabditis elegans* [3, 4] and *Drosophila melanogaster* [5–7].

In 2005 study that coined the term "genome editing," Urnov et al. demonstrated that engineered zinc finger nucleases (ZFNs) can serve as a designer nuclease to correct X-linked SCID disease-relevant mutations in patient-specific cells [8]. Ten years later the first clinical trials based on ZFN technological platform are underway to disrupt CCR5 in T cells to treat HIV patients [9]. Based on these experiments, designer nucleases such as ZFNs and transcription activator-like effector nucleases (TALENs) were applied to engineer hiPSCs [10–16]. These experiments provided proof of principle for designer nucleases-mediated gene knockouts, for the insertion of transgenes to generate cell-type lineage reporters, for the overexpression of transgenes from genetically defined loci, and for the introduction or repair of point mutations in hiPSCs. The initial ZFN and TALEN platforms for genome editing in stem cells were costly and labor-intensive, and their implementation as research tools therefore developed comparatively slowly. However, extensive work with ZFNs and TALENs has demonstrated the power of genome editing and highlighted the importance for development of more universal, cheaper, and simpler genome editing platforms.

15.2 Genome Editing with CRISPR/Cas9

Naturally occurring in some bacteria and archaea, an RNA-mediated adaptive defense mechanism, named the CRISPR (Clustered Regularly Interspaced Short Palindromic Repeat) system, exists to defeat invasions by foreign plasmid DNA or bacteriophages. The Type II CRISPR system is unique in the sense that the CRISPR-associated 9 (Cas9) protein alone, rather than as a complex of Cas proteins, can mediate double-strand breaks (DSBs) on the target DNA when associated with CRISPR RNA (crRNA) and trans-activating crRNA (tracrRNA). CRISPR/Cas9 system was discovered in 1980s; however its function was properly described in 2007 in *Streptococcus thermophiles* [17]. The system was applied to editing of mammalian genomes several years ago, and the recent applications of this system in genome engineering leads to a significant progress in stem cells field. It includes efficient generation of simple gene knockouts for basic molecular studies, correction or introduction of mutations for gene variants testing and disease modeling, knockin reporter cell lines for optimizing differentiation protocols, and cell labeling by reporter genes for tracking transplanted cells in animal models, drug testing, and even personalized gene therapies [18–21].

The nuclease Cas9 can generate DSBs on the target DNA sequences in a site-specific way directed by a single-guide RNA (sgRNA) upon the existence of a proto-spacer adjacent motif (PAM) sequence [19–21]. CRISPR/Cas9 system gives the ability to induce double-strand breaks (DSB) directly into the DNA. It relieves major technical problems of previous approaches such as zinc finger nucleases (ZFNs) and TALE nuclease (TALEN) systems. To adapt the prokaryotic system to work in eukaryotic cells, several groups have attached a nuclear localization signal to the Cas9 protein and have demonstrated the efficient generation of DSBs in mammalian cell lines, such as HEK293 cells [22], mouse neuro2A (N2A) cells [19], and human ES/iPS cells [17, 23]. The CRISPR/Cas9 system can be introduced into mouse or other species embryos to disrupt a target gene of interest. Moreover the system could be applied to large and farm animals opening new possibilities for examining of physiological functions that cannot be recapitulated in mouse or cellular models and for efficient production of multi-modified animals for xenotransplantation [24].

15.3 The Need for Isogenic Lines Due to Genetic and Epigenetic Background Variability of hiPSCs

The technical advances in genome editing are highly complemented with the development and translational applications of iPSC technology. Genome editing in hPSCs overcame the issue of enormous genetic and epigenetic background

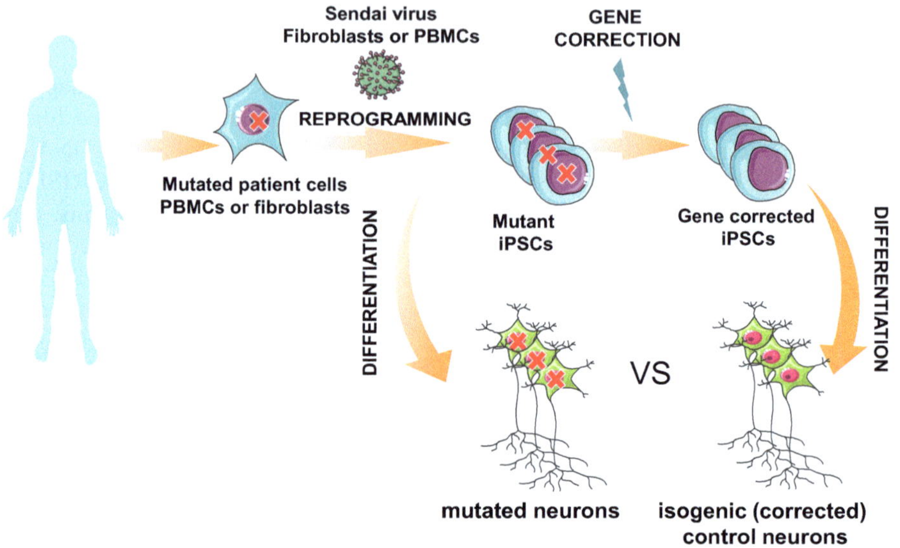

Fig. 15.1 The isogenic hiPSC line is engineered from a parental line through the correction of a targeted gene mutation. The parental cell line inherently becomes a control line to which the engineered line can be referenced. Isogenic hiPSC line pairs are crucial for studying the impact of genotype on cellular phenotype, as well as for parallel, high-throughput screening to enable the discovery of therapeutic compounds that exhibit genotype-specific toxicity

variability inherent in iPSC-based disease models [25, 26]. Aberrant epigenetic state after reprogramming leads to epigenetic defects in induced stem cells causing unpredictable long-term effects in differentiated cells [25]. Independent proof-of-concept studies demonstrated that designer nucleases can be used to repair or introduce disease mutations in hPSCs, and the resulting pairs of iPSC lines are isogenic, except for the disease-relevant mutation [15, 26, 27]. Parallel differentiation of such isogenic cells into disease-relevant cells and three-dimensional (3D) models such as brain organoids can be used to directly assess the contribution of a mutation to cellular pathology [27–30] under the experimental design as shown in the Fig. 15.1.

Thus, in fact genome engineering and human iPS cells are two powerful technologies, which can be combined to highlight phenotypic differences and identify pathological mechanisms of complex diseases only by providing the isogenic cellular material. Interestingly, even until now very few data are available regarding precise gene correction in human hiPS cells. This is due to technical limitations regarding precise HDR-based genome editing efficiency, clonal isolation of edited cells, assessment of editing events, and the effects of designer nucleases on engineered cells. All of these aspects are covered in below sections to highlight the currently available toolbox for genome editing in hiPSCs and to pinpoint the associated barriers which should be considered in each and every step in hiPS-cell based disease modeling.

15.4 Examples of CRISPR/Cas Genome Editing Applications: Reverse and Forward Genetics

15.4.1 Reverse Genetics

CRISPR/Cas system is an effective tool for both forward and reverse genetics [31]. As a reverse genetics tool, the system has facilitated germ line correction of a genetic mutation [32], somatic gene editing [33, 34], generation of inducible Cas9 knockin and knockout models [35], large genome modifications [36], and ex vivo genome editing in human cells [37, 38]. Since 2013, several groups have already demonstrated usefulness of the system in pluripotent stem cells genome editing. Genes can be inserted to generate reporter genes or to achieve ectopic expression. Genetic information can be deleted or inverted, and modifications as small as single base pair changes can be produced to introduce mutations or polymorphisms or to repair disease-relevant mutations. The resulting genetically engineered hPSCs differ from wild-type cells exclusively at the edited locus and are otherwise isogenic. Parallel differentiation of these isogenic cell lines into disease-relevant cell types can provide the basis for the phenotypic analysis of disease-specific cellular pathologies. Phenotypes found in these cells can be directly attributed to genetic manipulation. The system can simply induce disruption of both alleles, or using haplotype-specific target sites can induce changes in the allele of interest only and serve as a allele-specific editing strategy [39]. Additionally, as a combination of multiple sgRNAs, the system can cleave multiple genes simultaneously as multiplex genome editing strategy [21, 40]. Apart from genome editing, several groups have been investigating the modulation of gene expression or epigenetic status by taking advantage of the CRISPR system to bind to specific target sequences. Gene expression can be modulated (activated or repressed) using CRISPRi (CRISPR interference) and CRISPRa (CRISPR activation) by reversibly targeting their endogenous promoter. For instance, the Cas9 protein lacking nuclease activity (dCas9, with two point mutations, D10A and H840A) can be targeted to a desired transcriptional region to block transcriptional initiation or elongation [41]. To further modulate gene transcription, dCas9 can be also fused with a transcriptional repressor domain (KRAB) or activator domains (VP16 or VP64) to fine-tune gene transcription [42–46]. Such artificial transcription factors help in construction of synthetic gene circuits [47] or can be used to reprogram and/or to change the cell fate [48].

Without the need for ES cells, reverse genetics can now be applied to species where ES cells cannot be established and in mouse strains where germ line-competent ES cells are not available. Furthermore, the scalability of the CRISPR/Cas tool, the simplest among the programmable nuclease systems, can also be used as a forward genetics tool in constructing genome-wide mutation libraries to screen for new disease mutations. Unlike the previously used forward genetics approaches, such as mutagenic chemicals or radiation that induces random mutations and requires subsequent identification of these alterations, CRISPR/Cas represents the first-ever reported mutagenic tool that can be used for creating libraries of mutations at known sites in the genome [49].

15.4.2 Forward Genetics

In addition to editing of the hiPSCs, the advent of Cas9 is the development of a forward genetics methodology. This approach, to study hiPSCs, became available with the development of genome editing as a screening tool. The generation of bar-coded libraries of sgRNAs presently can cover almost each gene in the human genome and thus could be employed in gain and loss of function screens. Bulk transduction of hiPSCs with either Cas9 or nuclease-dead Cas9 (dCas9) in combination with genome-wide bar-coded sgRNA libraries (CRISPR cutting, CRISPRi, and CRISPRa) can be used to identify genes whose loss or gain of function changes the cellular representation within the specific cell pool. Enrichment or depletion of sgRNAs can be determined by screening of the sgRNAs, yielding candidate genes of interest [50–53].

15.5 Design of the Experiments

15.5.1 Selection and Design of sgRNA Using Guide Selection Tools

The guide sequence determines the efficiency of on-target cleavage. Recently high-throughput studies have found numerous off-targets not predicted by existing algorithms (up to 1-bp indels) [54, 55]. It is recommended to select guides to minimize potential off-target effects and test several to optimize on-target activity. The choice of the guide RNA sequence is facilitated by various tools and websites such as CHOPCHOP (http://chopchop.cbu.uib.no/), CRISP Design (http://crispr.mit.edu/), Genome Engineering 4.0 (http://www.genome-engineering.org/), or CRISPOR (http://crispor.tefor.net/). Among them the CRISPOR conducts independent evaluation of CRISP/Cas9 predictions [56].

15.5.2 Choosing the Right Cas9 (spCas9, St3Cas9, eCas9, HFCas9, Cpf1)

RNA-guided nucleases based on the Type II CRISPR-Cas9 system of *Streptococcus pyogenes* (Sp) have been most widely used for genome editing in experimental models. However the Cas9 nuclease component is available in multiple formats to allow for experimental flexibility and maximal success for particular cell types or specific applications. Since optimal activity of the RNA-guided Cas9 endonuclease is critical for successful gene editing, the considerations that can help in the type of Cas9 nuclease best suited for editing needs should be considered.

15.5.3 *Gene Editing Using "Standard" SpCas9 (Off-Target Activity)*

Editing of the cells is conducted by co-expressing of gRNA specific to the gene and/or sequence to be targeted and the endonuclease Cas9. The genomic target can be any of ~20 nucleotide DNA sequence, but the only requirement for the selection of Cas9 target sites should meet two conditions: 1) the sequence is unique compared to the rest of the genome; 2) the 20-bp target sequence is present directly upstream of a PAM sequence (the presence of a PAM sequence directly 3′ of the target sequence). Each Cas9 ortholog has a unique PAM sequence, while SpCas9 requires a 5′-NGG PAM sequence. SpCas9 PAM requirement does not severely limit the targeting range of SpCas9. Such target sites can be found on average every 8–12 bp in the human genome [19, 57]. In addition to the targeting range, another limitation of the SpCas9 is the potential for off-target mutagenesis. Cas9 can cleave off-target DNA targets in the genome at reduced frequencies. Off-target activity depends on a combination of factors including enzyme concentration and the abundance of similar sequences in the target genome. SpCas9 is a common choice when there is a suitable target site with a little concern for off-target effects and optimal gRNA design with minimal homologous sites throughout the genome. For routine application of Cas9, it is also important to consider ways to minimize the degree of off-target cleavage and to be able to detect the presence of off-target cleavage. In the sections below, we describe strategies on minimizing off-target modifications.

15.5.4 *Variants of the Cas9 with Lower Off-Target Activity (Cas9n, dCas9): Nickases*

The first reports of SpCas9 off-target activity appeared soon after the initial demonstrations of gene editing, particularly when Cas9 was expressed at high levels. Since then, the level of off-target activity in human cells hampers Cas9 applications in cell-based disease modeling as well as in clinical translation. Various strategies for mitigating off-target activity have been proposed to date [58]. Nucleases specificity depends on both the gRNA and the PAM recognized by the Cas9 protein. More stringent PAM requirements reduce the occurrence of off-target mutagenesis. The specificity of Cas9-mediated cleavage can be enhanced by properly designing of gRNA sequence. That is, choosing a target sequence that has minimal homology elsewhere in the genome. A variety of approaches have been used to further enhance

Cas9 specificity. For example, Cas9-nickase (Cas9n) takes advantage of the fact that Cas9 makes double-strand breaks (DSBs) through the combined activity of two nuclease domains, RuvC and HNH [59]. Converting one of the two critical enzymatic residues to an alanine (D10A or H840A) generates a "nicking" Cas9 that cannot generate a double-strand break. Thus, two properly targeted Cas9n molecules are required to efficiently create DSBs at the target locus, which greatly enhances specificity compared to wild-type SpCas9 [59]. As an example on nicking strategy, the pSpCas9n(BB), a D10A nickase mutant of Cas9 for HDR and for double-nicking applications, along with the 2AGFP and 2A-Puro fusion constructs (pSpCas9n(BB)-2A-GFP, pSpCas9n(BB)-2A-Puro) could be provided. It is worth to note that Cas9n is more specific than wild-type SpCas9; however DSBs could be still detectable at a target site when only one sgRNA is expressed within Cas9n. In other words Cas9n is able to bind and cause indels at one of the off-target sites for either of its sgRNAs.

This limitation could be overcome through the use of nuclease-dead Cas9 (dCas9) fused to the nonspecific endonuclease FokI which cleaves target DNA when dimerized. Therefore, dCas9-FokI essentially requires proper targeting of two dCas9-FokI molecules at the target site before any cleavage occurs; dCas9-FokI is much less likely to cut at an off target specified by a single sgRNA than Cas9n. A limitation of the Cas9n or dCas9-FokI approach is that they both necessitate two suitable target sequences in close proximity in order to efficiently generate a DSB.

15.5.5 High-Specificity Cas9 Variants with Minimal Off-Target Activity (eCas9, HFCas9)

Gene editing specificity could be improved by reducing the strength of the interactions of the Cas9-sgRNA complex with off-target sites to below the threshold necessary for binding while still retaining binding at the target site. Several labs, including Zhang's lab at the Broad Institute and Joung's group at Massachusetts General Hospital, have used structural biology to identify key residues that mediate Cas9's ability to cleave off-target sites. So-called enhanced Cas9 (eCas9, Zhang) and high-fidelity Cas9 (HFCas9, Joung) have comparable cleavage activity to wild-type SpCas9 at target loci but have greatly reduced off-target activity. These Cas9 variants enhance specificity without requiring two or more adjacent target sites within the target locus. The specific interactions occur between the Cas9 protein and a short DNA sequence known as the protospacer adjacent motif and through base pairing of the sgRNA and a DNA sequence of ~20 nucleotides. In addition, binding is stabilized by nonspecific interactions between the Cas9 protein and the DNA backbone [60]. Zhang [61] and Joung [62] showed that by decreasing the strength of these nonspecific interactions through amino acid substitutions, the resulting Cas9 variants are more dependent on correct sgRNA-DNA base pairing, leading to decreases in off-target activity with no significant decreases in on-target activity. Interestingly, the two groups approach this challenge by mutating different regions of *Streptococcus pyogenes* (Sp) Cas9. Zhang and colleagues focus on residues that interact with the nontarget DNA strand to generate eSpCas9, whereas Joung and colleagues alter amino acids that contact the DNA backbone on the target strand to produce SpCas9-HF1. Using unbiased, genome-wide assays, both groups showed a reduction of off-target effects to below detectable levels for several sgRNAs.

15.5.6 Non-S. Pyogenes Cas9s (SaCas9, St3Cas9)

Other Cas9 homologs have been isolated from a bacterial species. Many of them bind PAM sequences other than the common NGG PAM sequence, and it is important that non-SpCas9s are only compatible with the tracrRNA and crRNA (or synthetic gRNA) derived from the same species. Non-Sp Cas9s could be also more suitable for gene editing due to reasons other than the PAM sequence. For example, the coding sequence for Cas9 from *Staphylococcus aureus* (SaCas9) is about 1 kilobase smaller than SpCas9, which allows for packaging into adeno-associated virus (AAV), mostly used in current gene therapy strategies. It is worth to note that Müller and colleagues generated Cas9 based on two *Streptococcus thermophilus* CRISPR-3 Cas9 proteins (St3Cas9), which recognize longer PAMs, and performed a side-by-side comparison of the three nuclease systems targeted to matching sites in two endogenous human loci (PRKDC and CARD11) [63]. Results demonstrate that in samples with comparable on-target cleavage activities, significantly lower off-target mutagenesis was detected using St-based nucleases as compared to the standard Sp-nucleases. The St3Cas9 shares about 60% sequence identity with SpCas9. Although their crRNA repeats and tracrRNAs are interchangeable [64], SpCas9 and St3Cas9 require different PAM sequences (50-NGG-30 for SpCas9 and 50-NGGNG-30 for St3Cas9) for target DNA cleavage [64]. Moreover, similarly to SpCas9, the StCas9 proteins accepted truncated gRNAs, suggesting that the specificities of St-based nucleases can be further improved [63]. In conclusion, Cas9 proteins with longer or more restrictive PAM requirements could provide a more secure alternative to SpCas9-based nucleases and hence a potentially safe option for forthcoming usage.

15.5.7 Non-Cas9 CRISPR Endonuclease, Cpf1

A new CRISPR non-Cas9 endonuclease termed Cpf1 was recently identified by the Zhang [65]. A putative new class 2 nuclease, Cpf1 (CRISPR from *Prevotella* and *Francisella*), was classified as a novel, Type V CRISPR system. Like Cas9, Cpf1 contains a RuvC-like endonuclease domain, but it lacks Cas9's other HNH endonuclease domain, indicating that Cpf1 functions differently [65, 66]. Two enzymatic residues similar to Cas9 (D917 and E1006) residues are located in the RuvC domain, and mutation of either residue completely abolishes DNA cleavage. Cpf1 cleavage results in a 5 nucleotide 5′ overhang, 18 base pairs from the PAM sequence. This is different from Cas9 cutting, which results in blunt DNA ends three base pairs distal to the PAM sequence which can be converted into a nickase by mutating a single enzymatic residue. Cpf1 cleaves DNA in a staggered pattern and requires only one RNA rather than the two (tracrRNA and crRNA) needed by Cas9 for cleavage. For these reasons, Cpf1 may be even better suited for genome editing than Cas9, and the use of a different nuclease could potentially overcome some of Cas9's shortcomings, namely, its blunt double-stranded cleavage and G-rich PAM requirement. Cpf1's preferred PAM is 5'-TTN, differing from Cas9 PAM (3'-NGG), and mature crRNAs for Cpf1-mediated cleavage are 42–44 nucleotides in length, about the same size as Cas9's but with the direct repeat preceding the spacer rather than following it. The Cpf1 crRNA is also simpler in structure than Cas9's. Short stem-loop structure in the direct repeat region is necessary for cleavage of a target, and additional tracrRNA is not required in Cpf1. Moreover Cas9 generates blunt ends 3 nt upstream of the PAM site, but Cpf1 cleavage pattern creates 5 nucleotide 5′ overhang 18–23 bases away from the PAM. Among 16 diverse Cpf1 candidates, Zetsche et al. found two (AsCpf1 and LbCpf1; 1307 and 1228 amino acids long, respectively) that cleave in a staggered pattern similar to FnCpf1 [65].

15.5.8 Advantages of Cpf1 over Cas9

Introduction of Type V Cpf1-based systems has added another option to the CRISPR toolbox. Cleavage pattern of Cpf1 opens the possibility of directional gene transfer, analogous to traditional restriction enzyme cloning. Sticky-end-mediated gene transfer would be particularly helpful for targeting nondividing cells, which are difficult to modify through HDR. Cpf1 also expands the number of sites that can be targeted by CRISPR to AT-rich regions or AT-rich genomes that lack the 3'-NGG PAM sites favored by SpCas9 [65].

Since Cpf1 doesn't require a tracrRNA, guide RNAs are only ~42 nt long. Direct synthesis of these gRNAs should be significantly cheaper than that of the ~100 nt crRNA/tracrRNA hybrid guides needed for Cas9 function. Since both Cpf1 and its guide RNAs are smaller than their SpCas9 counterparts, they will also be easier to deliver in low-capacity vectors, such as adeno-associated viral (AAV) vectors [65, 66].

Cpf1 may also improve the frequency of HDR over NHEJ [65]. Cas9-mediated NHEJ usually destroys the PAM site due to its proximity to the cleavage site, preventing future edits. In contrast, since Cpf1 cleaves relatively far away from the PAM, NHEJ might retain the PAM site. Therefore, if HDR does not initially occur after Cpf1-mediated cleavage, the continued presence of the PAM may give Cpf1 the ability to cleave again and possibly mediate HDR. This double mechanism might improve the frequency of desired HDR edits, but the possibility has not yet been experimentally confirmed. To prevent new editing post-HDR, repair templates should remove the PAM sequence.

15.5.9 Synthetic Cas9s with Novel PAM Recognition (VQR, EQR, and VRER Cas9)

Mutants of *S. pyogenes* Cas9 (VQR, EQR, and VRER Cas9 variants) that recognize novel non-NGG PAM sequences were identified through a series of positive selection screens in bacteria [67, 68]. The VQR, EQR, and VRER Cas9 variants are capable of modifying genomic loci that cannot be modified using wild-type SpCas9, and their specificity for the PAM variants is similar to wild-type SpCas9 for several genomic targets in human cells. Including the VQR, EQR, and VRER SpCas9 variants effectively doubles the targeting range of CRISPR/Cas9 within the human genome.

15.6 Choosing the Right sgRNA/Cas9 Expression Plasmid

With the CRISPR/Cas system, the Cas9 and sgRNA must be expressed in the cell simultaneously. Plasmids containing both the gRNA and Cas9 act as an all-in-one vector, but their cutting functions are limited to the Cas9 present on the plasmid. sgRNA plasmids that do not express Cas9 require a separate Cas9 expression plasmid. Independent sgRNA plasmids can be paired with a wide variety of Cas9 plasmids and therefore are not limited to a single Cas9 function. Alternatively, independent gRNA plasmids can be used in cells in which Cas9 has been integrated into the genome of the panel of hPSCs which serve as genome-engineering platforms with doxycycline-inducible expression of Cas9 referred to iCas9 hPSCs [69].

15.6.1 sgRNA Delivery

Depending on the desired application, sgRNAs can be delivered as either PCR amplicons containing an expression cassette or sgRNA-expressing plasmids. PCR-based sgRNA delivery appends the custom sgRNA sequence onto the reverse PCR primer used to amplify a U6 promoter template. The resulting amplicon could be co-transfected with a Cas9 expression plasmid pSpCas9 [59]. This method is optimal for rapid screening of multiple candidate sgRNAs, as cell transfections for functional testing can be performed shortly after obtaining the sgRNA-encoding primers. Because this simple method counteracts the need for plasmid-based cloning and sequence verification, it is well suited for testing or co-transfecting a large number of sgRNAs for generating large knockout libraries. It is worth to note that the sgRNA-encoding primers are over 100 bp long, compared with the ~20-bp-long oligos required for plasmid-based sgRNA delivery [59].

Regarding the independent gRNA plasmids, many gRNA empty vectors have been developed to date. To choose the most proper gRNA vector for hiPSCs expression system, the application of Cas9 must be considered such as whether the

Cas9 will be used to cut, activate, interfere, or nick, then the selection such as puromycin or GFP/BFP/Venus, and finally the cloning enzyme preference used for insertion of gRNA sequence, such as BsbI or BsaI. Since the sgRNA scaffolds are specific for the type of Cas9Cpf, the vector to be used should be designed under consideration of *S. pyogenes*, *S. aureus*, *N. meningitidis*, *Prevotella/Francisella*, or more.

Construction of an expression plasmid for sgRNA is a straightforward step involving single cloning with a pair of partially complementary oligonucleotides. The oligo pairs encoding 20-nt guide sequences are annealed and ligated into a plasmid (pSpCas9(BB)) bearing both Cas9 and the residuum of the sgRNA as a scaffold directly following oligo cloning site [59].

The transfection plasmids can also be modified to enable virus production for in vivo delivery. The following plasmids could be used within a protocol: Cas9 alone (pSpCas9) or Cas9 with an sgRNA scaffold and cloning sites for inserting a guide sequence (pSpCas9(BB)). For the backbone cloning construct, 2A-GFP or 2A-Puro to Cas9 could be fused to allow screening or selection of transfected cells (pSpCas9(BB)-2AGFP or pSpCas9(BB)-2A-Puro, respectively) [59].

15.6.2 Cas9 Delivery

A wide variety of Cas9 containing plasmids are available that have been optimized for engineering of the human cells. Expression systems are useful in enrichment of Cas9-expressing cells or to establish a stable cell line. The selection of vector-based Cas9 nuclease comes to the choice between Cas9 expression plasmids with a multitude options for promoters and reporters and lentiviral Cas9 nuclease. The Cas9 nuclease expression plasmids are a good option for relatively difficult-to-transfect hiPS cell lines and to enrich the cells by antibiotic selection or fluorescent reporter cell sorting following a transient co-transfection of Cas9 plasmid with synthetic crRNA/tracrRNA.

15.6.3 Lentiviral-Based Cas9 Nuclease

Lentiviral-based Cas9 is an option for gene editing in difficult-to-transfect cells due to the wide dynamic range of lentiviral transduction. This option can also be used to establish a Cas9-expressing stable cell line for arrayed crRNA/tracrRNA or pooled sgRNA screening applications [69]. The activity of a particular promoter can result in variable Cas9 expression levels and potentially low editing efficiency. The lentiviral Cas9 nucleases are also available with an inducible promoter to support creation of stable cell lines with minimal background expression, or for temporal control over Cas9 expression for wide-ranging experimental applications.

15.6.4 DNA-Free Cas9: mRNA or Protein

DNA-free Cas9 is mainly used with synthetic crRNA/tracrRNA to avoid unwanted vector DNA integration into genomic DNA. Cas9 mRNA or protein is transiently available for editing events which means that these nuclease sources do not have the off-targeting potential of constitutive Cas9 expression. For that reason, CRISPR/Cas9 utilizing mRNA or protein is ideal for applications such as knockin of a fluorescent reporter using HDR or knockout cell line generation. Both are amenable to lipid-based transfection reagents or electroporation to support a wide variety of cell types. Finally, mRNA and protein are suited for experiments where available Cas9 plasmid or lentiviral particle promoters are not compatible in their cell line or model organism.

15.6.5 Cas9 mRNA

Purified, translation-ready Cas9 nuclease mRNA, in combination with crRNA/tracrRNA, is an economical option for applications involving DNA-free editing for generation of knockout or knockin clonal cell lines, or arrayed screening.

15.6.6 Cas9 Protein

Delivery of recombinant Cas9 protein and guide RNA (gRNA) as a preformed ribonucleoprotein (RNP) complex has recently emerged as a powerful and general approach to genome editing [70]. Once the Cas9 protein is available, the method involves no cloning and can be completed to the point of initial analysis in a matter of days. The main advantage of delivering Cas9 protein with crRNA/tracrRNA is that the translated protein is immediately available for complexing with the guide RNA and consequent editing in the cell. This is often a factor in applications where the cells lack robust translational machinery such as stem cells both induced and embryonic. Cas9 RNP delivery limits exposure to genome editing reagents, reduces off-target events, and drives high rates of homology-dependent repair. Delivery of RNP coupled with cell cycle arrest or modified donor DNA design can further be used to alter the balance of NHEJ and HDR products in human cells, leading to HDR frequencies greater than 30% [70–72]. RNP delivery thus minimizes some of the pitfalls of alternative editing modalities and is rapidly being adopted by the genome editing community.

15.6.6.1 Cas9 Protein Limitations

Cas9 RNPs are not without limitations. Cas9 RNPs are active immediately following transfection and are quickly degraded within the cell. These fast degradation kinetics enable Cas9 RNPs to modify target genes with reduced off-target effects.

A major drawback of using Cas9 RNPs is that expression is transient. Therefore, it may be best to use plasmid-based delivery or lentiviral-mediated delivery of CRISPR components in cases where stable or elevated expression of CRISPR components is necessary. Moreover, once cells are treated with Cas9 RNPs, the edits cannot be validated by fluorescence-activated cell sorting (FACS) duet, too, but rather should be validated either by isolating individual clones and screening target locus with Sanger sequencing or analyzing cleavage efficiency using a restriction digest-based assay (T7 endonuclease assay or Surveyor assay).

15.6.7 Cloning of Multiple sgRNA Expression Cassettes for Multiplex Genome Engineering

A unique capability of the CRISPR/Cas9 system is multiplex genome engineering by delivering a single Cas9 enzyme and two or more single-guide RNAs (sgRNAs) targeted to distinct genomic sites. Cas9 nucleases can be directed by short RNAs to induce precise cleavage at endogenous genomic loci of interest and can also be converted into a nicking enzyme to facilitate homology-directed repair with minimal mutagenic activity. Cas9 can be also used to simultaneously create multiple DNA breaks. In this approach multiple-guide sequences can be encoded into a single CRISPR array to enable simultaneous editing of several sites within the mammalian genome, demonstrating easy programmability and wide applicability of the RNA-guided nuclease technology [19]. To address the need for uniform and sustained delivery of multiplex CRISPR/Cas9-based genome engineering tools, a single lentiviral system to express a Cas9 variant, a reporter gene, and up to four sgRNAs from independent RNA polymerase III promoters that are incorporated into the vector by a convenient Golden Gate cloning method may be developed. Each sgRNA is efficiently expressed and can mediate multiplex gene editing in human and mouse cells. This delivery system will be significant to enabling the potential of CRISPR/Cas9-based multiplex genome engineering in diverse cell types.

15.6.8 Insertion or Regional Replacement by HDR

Generation of a knockin is either an insertion or a regional replacement achieved by HDR mechanism. HDR can occur either non-conservatively or conservatively. The non-conservative method is composed of the single-strand annealing (SSA) pathway. The conservative methods are characterized by the accurate repair of the DSB using homologous donor template such as plasmid but also with sister chromatid. The conservative HDRs are composed of the classical double-strand break repair (c-DSBR), synthesis-dependent strand annealing (SDSA), and break-induced repair (BIR). c-DSBR results in the formation of double Holliday junctions (dHJs) and crossover; SDSA results exclusively in noncrossover events when newly synthesized sequences are present on the same molecule. The BIR is not as well characterized, but one central feature of this pathway is the presence of only one invasive end at a DSB that can be used for repair (http://engelward-lab.mit.edu/animations/).

To generate targeted modification with HDR, the sgRNA target site should be located close (<100 bp) to the position of the intended mutation as the frequency of sequence conversion by HDR decreases with distance. Targeted mutations by HDR are guided by repair templates, being either single-stranded DNA oligonucleotides (ssODNs) or plasmid gene-targeting vectors [59, 73, 74].

15.6.8.1 Single-Stranded DNA Oligonucleotides (ssODNs)

ssODNs are convenient as they are synthetized for a reasonable price so that cloning work is not required. ssODNs used as HDR template contain a short sequence modification (deletion, insertion, or substitution) flanked by two homology sequences. Many manufacturers of custom ssODNs offer lengths up to 150 nt or more, enabling the insertion of sequences of up to 50 nt. When using ssODNs the selection of target sites that are less than 10 bp is distant to the planned mutation. In the standard design, as their length is limited, a desired mutation is located at the center of an ssODN, resulting in similarly sized homology regions (usually 40–60 nt each) that flank the desired mutation symmetrically on both sides, but these numbers may vary based on locus of interest and experimental design; thus 50–80 base homology arms are also commonly used. It is crucial to avoid the recognition and the re-cleaving of the recombined allele by Cas9 by the inclusion of one or more silent nucleotide replacements to lose the PAM recognition site. Silent mutations can be added to facilitate later screening for successful recombination, e.g., by the introduction of a recognition site for restriction enzymes.

15.6.8.2 Plasmid Gene-Targeting Vectors (dsDNA)

HDR efficiency with plasmid templates is generally low. However due to the difficulties associated with creating ssODNs longer than 200 bases, for larger and more sophisticated modifications such as the insertion of large DNA fragments and reporter and selection cassettes, targeting vectors with long homology arms are required. Such vectors include the insertion in between two regions homologous to sequences located up- and downstream of the DSB with no requirement for drug selection cassette, and the homology arms used are often shorter (500 bp) which facilitate cloning by PCR. In this “500 bp” approach, for vector construction, homology arms can be PCR-amplified from isogenic genomic DNA and ligated into a cloning plasmid backbone. In the simplest configuration, the selected Cas9 recognition sequence should cover the region of the intended mutation, bridging the homology arms of the targeting vector to prevent its cleaving by Cas9, being specific only to the wild-type genomic DNA.

The insert can be also generated by gene synthesis purchased from a commercial provider. These templates should have the homology arms of at least 800 bp. To increase the frequency of edits, researchers have designed self-cleaving plasmids that contain gRNA target sites flanking the template. When the Cas enzyme and the appropriate gRNAs are present, the template is liberated from the vector. To avoid plasmid cloning PCR-generated long dsDNA templates, but researchers from Integrated DNA Technologies, Inc. (IDT) suggest that these templates may be toxic to hiPSCs, thus lowering editing efficiency. Another disadvantage of dsDNA templates is their ability to bluntly integrate into the genome, duplicating the homology arm sequences.

15.7 Validation of Editing Events

Precise HDR-based genome editing in hiPSCs targets loci with low efficiency and in different way; thus the editing rate at a given locus needs to be evaluated. In principle, detailed determining of the sequences spectrum of all genomic targets in a population of hiPSCs will reveal the complete range of Cas9-mediated sequence modifications and their abundance. Thus, the analysis of mutants obtained using Cas9 requires specific methods for mutation detection. There are a variety of methodologies available to compare and validate Cas9 activity. Typically, the targeted exon is PCR amplified and cloned into a vector for bacterial single colony sequencing. Although PCR-based cloning is incredibly versatile and allows for integration of nearly any piece of DNA into a backbone vector of choice with minimal limitations, this method can be costly, time-consuming, and laborious, even for a limited number of clones [75]. The detailed editing rate could be evaluated by Sanger method or even by next-generation sequencing of cloned PCR products obtained from different hiPSC colonies. However, targeted, deep sequencing of required amplicons allows for precise evaluation of editing rate.

15.7.1 Decoding CRISPR/Cas-Mediated Editing Events by Sequencing

In order to identify the exact indel sequence of the resulting alleles in selected clones, most laboratories use Sanger sequencing. The PCR product can be sequenced directly by Sanger sequencing, but this results in a convoluted spectrum with overlapping peaks that is difficult to delineate with current methodologies. Indel distribution and even template-directed Cas9 editing events could be decoded by sequence trace decomposition analysis, but the precise modifications such as corrections and insertions and their accurate distribution have to be revealed by targeted deep sequencing.

15.7.1.1 Sequence Trace Decomposition Analysis of Non-template and Template Cas9 Editing

TIDE and TIDER

Since for many cell lines 5% of cutting efficiency in EMC methods is insufficient, an assay to accurately characterize and quantify quickly the induced mutations is needed. An alternative approach relies on algorithm developed to analyze Sanger sequencing trace and accurately determine the frequency of mutations at the projected editing site in a cell population down to 1–2%. The approach named TIDE and TIDER is available in a free-to-use Web portal (http://tide.nki.nl) as an interactive tool for automated decomposition of the sequence traces giving the quantitative spectrum of indels around the cut site (TIDE) and quantification of templated mutations plus the spectrum of non-templated indels (TIDER) (http://tide.nki.nl) [76]. The code was written in R, version 3.1.1, and requires as input a control sequence data file based on only a pair of PCR reactions (for non-templated Cas9 editing) and three sequence traces (for template-directed Cas9 editing) and standard capillary sequencing runs. The sequence traces are analyzed by a decomposition algorithm that identifies the major induced mutations in the editing site and accurately determines their frequency in a cell population. The sequencing data files (.abif or .scf format) are imported into TIDE using the R Bioconductor (www.bioconductor.org) package sangerseqR. The latter is a genotyping tool that can identify heterozygous short indels in sequence traces and templated mutations as well. TIDE first aligns the sgRNA sequence to the control sequence to determine the position of the expected Cas9 break site. Next, the control sequence region upstream of the break site is aligned to the experimental sample sequence in order to determine any offset between the two sequence reads. To model insertions, the TIDE software estimates the relative frequency with which each of the four nucleotides is introduced immediately after the break site. This is done by removal of the aggregate of estimated signals of mutants that have smaller number of insertions (including non-mutated and deletions). While this can be done for all insert sizes, TIDE currently only estimates the nucleotide composition of +1 insertions, which are the most frequently observed insertions [76]. TIDE method is cost-effective and quick, and it provides much more detailed information than enzyme-based assays. The method greatly facilitates the testing and rational design of genome editing strategies, but it cannot resolve sequences with complex indel mixtures [77]. TIDER as a modified version of TIDE estimates the frequency of targeted small nucleotide changes introduced by Cas9 in combination with homology-directed repair using a donor template. In addition, it determines the spectrum and frequency of non-templated indels. Compared to TIDE, the tool requires one additional sequencing trace (i.e., three instead of two). Preparation of this third "reference" DNA can be done with a simple two-step PCR protocol. TIDER reports the frequencies of the templated mutation and of all non-templated indels (http://tide.nki.nl).

CRISP-ID

CRISP-ID is a Web application which uses a unique algorithm for genotyping up to three alleles from a single Sanger sequencing trace, providing a robust and readily accessible platform to directly identify CRISPR-induced indels and significantly speed up the characterization of clones. PCR product can be sequenced directly by Sanger sequencing, but this results in a convoluted spectrum with overlapping peaks that is difficult to delineate with current methodologies. CRISP-ID directly reads sequencing trace files (.abi or .scf format) and is the first application with the ability to deconvolute the overlapping spectra from three different alleles, providing a robust and easy-to-use clone identification tool using direct standard Sanger sequencing of PCR products from cell line clones or patient material, without bacterial single-colony sequencing [78].

15.7.1.2 Next-Generation Sequencing of CRISPR/Cas-Mediated Editing Events

CRISPResso and BATCH-GE

Deep sequencing provides far more detail about editing outcomes than the other mentioned procedures due to missing of some products, such as large deletions and translocations with other methods. Initially purified PCR product (identical to that used in genotyping) should be then used to generate a smaller amplicon depending on Illumina system to be used (150–250 bp). Amplicons should be carefully prepared avoiding artifacts due to over-amplification. Importantly, if HDR template was used the forward primer should be designed in such way that cannot anneal to the template. The resulting products are prepared for sequencing using the Illumina TruSeq DNA library preparation kit according to manufacturer's protocols (https://www.illumina.com/). Sequencing could be performed on Illumina instruments such as MiSeq or MiniSeq (benchtop sequencer), with paired end reads length of 2 × 250 bp (MiSeq) or 2 × 150 (MiniSeq) following CRISPResso (http://crispresso.rocks/) [79] or BATCH-GE (https://github.com/WouterSteyaert/BATCH-GE.git) data analyses [80]. The advantage of benchtop MiniSeq instrument is that it could be used on demand even for low numbers of samples with no need to wait to batch samples for sequencing on a high-throughput instrument.

Such approach allows for precise characteristics of genome editing events and accurate calculation of the corresponding mutagenesis efficiency including detection of nucleotide substitutions ignored by other methods. Moreover CRISPResso allows for integration with other pipelines, tunable parameters of sequence quality and alignment fidelity, discrete measurement of insertions, deletions ignored by other methods, tunable windows around the cleavage site to minimize false-positive classification, and quantification of frameshift versus in-frame coding mutations [79].

These allow for distinction between NHEJ, HDR, and mixed events for a large number of samples in parallel with what is crucial during the gene editing and clonal isolation of hiPSCs procedures generating a large number of hiPSCs clones subjected for screening due to low efficiency of HDR in hiPSCs.

15.7.1.3 Fragment Analysis

When loading fluorescent-labeled PCR products of the edited sgRNA target site on a capillary sequencer, amplicons differing in size due to small insertions or deletions can be separated with accuracy in resolution, depending on the size of the PCR product. PCR products with a length of up to 650 bp can be separated with a resolution of a single base. For this approach, fluorescently labeled forward primers (HEX, TET, or 6-FAM) must be used to amplify the region of interest. Subsequently the PCR products are loaded on an agarose gel, purified using a PCR Clean-Up System and 60–150 ng of the purified samples are subjected to analysis using LIZ-500 size standard. It is worth to note that with this approach indel formations could be evaluated in single cell-derived clones as well as in cell populations [73].

15.7.1.4 Mismatch Cleavage Assays

Another strategy to identify unknown mutations relies on the identification of heteroduplex DNA formed after melting and hybridizing mutant and WT alleles. The identification of heteroduplex DNA can be done with chemicals, enzymes, or proteins that bind mismatches [81–87].

The enzyme mismatch cleavage (EMC) method is used commonly in genome editing for validation mutation rate obtained by engineered nucleases. This assay uses enzymes that cleave heteroduplex DNA at mismatches and extrahelical loops formed by single or multiple nucleotides. Bacteriophage resolvases and single-stranded nucleases are used commonly in these assays. Assays rely on cleavage of a mismatched PCR product as a proxy for the amount of disruption that has occurred in the genomic DNA of a pool of cells. The more cleavage means more disruption by Cas9 at the target site.

There are currently two commercially available and widely used assays T7E1 and Surveyor assays on deletions and point mutations obtained by nucleases in the hiPSC lines or lines for validation like HEK293T.

The EMC method takes advantages of enzymes able to cleave heteroduplex DNA at mismatches formed by single or multiple nucleotides. It is known that the cleavage activity of such enzymes is greater on mismatched than on Watson-Crick base pairs, but homoduplex DNA can be cleaved to a certain extent [84, 88].

T7 Endonuclease I

The structure, function, and substrate specificity of bacteriophage resolvase T7E1 have been well studied [89, 90]. The substrates of T7E1 are distorted dsDNA undergoing conformational changes [91]. Thus, T7E1 can recognize and cleave different dsDNA molecules if their structure is kinked and able to bend further [92, 93]. Typically, heteroduplex dsDNA containing bulges formed by extrahelical loops and even single-base mismatches can adopt such polymorphic structures [94]. Conversely, perfectly paired homoduplex dsDNA will not constitute good substrates for T7E1. The ability of T7E1 to discriminate between homoduplex and heteroduplex dsDNA has formed the basis of a mutation detection assay that worked with moderate success [92]. Yet the sequence and the number of mismatched nucleotides and the flanking sequence between the two DNA strands affect the heteroduplex structure and will therefore affect the cleavage efficiency by T7E1. This is why deletions are cleaved more efficiently than single-base mutations as noted by [92].

Routinely T7 endonuclease I digestion is performed at 37 for 30 min using 5 U of T7 endonuclease I (T7E1) on 20 mL of unpurified PCR products (~250 ng) in a reaction volume of 50 mL. The reaction is stopped by adding 4 mL of 0.5 methylenediaminetetraacetic acid.

Surveyor

Plant single-strand-specific endonucleases of the S1 nuclease family, such as CEL (CEL I and CEL II commercialized under the brand Surveyor), are single-stranded nucleases active on DNA or RNA. These nucleases do not form dimers, and they cleave DNA only one strand at a time on the 39 side of a mismatch [87, 95–98]. They have been described as useful for EMC assays, but they contain also 59 exonuclease activity [97, 99]. The specificity of CEL nucleases to identify single-base mismatches has been extensively and empirically proven by its use in the reverse genetics strategy of TILLING [99–106]. Yet there is limited evidence of their activity on extrahelical loops. The initial reports have shown an ability to detect extrahelical loops of up to 11 or 12 nucleotides [95, 97]. Despite the preference of CEL and ENDO in cleaving single-nucleotide mismatches, the Surveyor-based EMC assay is used commonly to scan mutations induced by engineered nucleases [5, 107–115].

To assess Cas9 cutting at a specific locus using a Surveyor assay, PCR primers are designed to produce a PCR amplicon of between 250 and 800 bp. The primers are designed such that the target cut region is located at approximately one third of the length of the amplicon (from either end). This will enable two distinct products to be resolved on a polyacrylamide or agarose gel and the efficiency of cutting to be assessed. In order to achieve consistent and sensitive Surveyor results, there are various steps in the experimental process that can be optimized.

Recently Vouillot L and colleagues [116] showed that T7E1 enabled the detection of 5% deletion mutant DNA. The exonuclease activity of Surveyor without Ampligase led to the appearance of many small DNA fragments that masked the specific signal coming from heteroduplex digestions. Authors showed that T7E1 outperforms the Surveyor nuclease in terms of sensitivity with deletion heteroduplex substrates and should be preferred here. In addition, they found that T7E1 outperforms the Surveyor nuclease in terms of sensitivity with deletion substrates, whereas Surveyor is better for detecting single-nucleotide changes. They conclude that T7E1 is the preferred enzyme to scan mutations triggered by engineered nucleases.

Routinely the Surveyor is performed using 20 mL of unpurified PCR products incubated with 1 mL of Surveyor nuclease (transgenomic Surveyor mutation detection kit for standard gel electrophoresis), 1 mL of Surveyor Enhancer, and 4 mL of 0.15 M MgCl2 in a 50-mL reaction. The mixture is then incubated for 1 h at 42°C, and the reaction is stopped by adding 4 mL of the stop solution provided in the kit.

15.7.1.5 Digital PCR

EMC assays are sensitive to ~5% cutting efficiency – which means that as long as 1 in 20 cells in the pool has undergone modification which should be detected. For many cell lines however this will be insufficient. Digital PCR offers not only a more sensitive test but a much more quantitative one. Digital PCR counts the number of non-modified versus modified amplicons. There are various ways to use digital PCR to quantify targeting efficiency, which vary in the design of the hydrolysis assay used. One example is a single PCR amplicon spanning the target site, one fluorophore-labeled hydrolysis probe that will bind every amplicon (thus quantifying total PCR product), and a differentially labeled probe that will bind the at the unmodified target site. In this manner, % modified amplicons can be extrapolated and efficiency quantified.

15.8 Modeling Sporadic Diseases: The Need for Isogenic Controls

hiPS cell-based modeling of sporadic and idiopathic diseases poses particular challenges due to expected subtle phenotypic changes. The genetics of sporadic diseases is thought to be a combination of multiple low-effect-size risk alleles, mostly in regulatory regions such as enhancers, which are identified by GWASs [117, 118]. Multiple risk variants with small effect size in combination with additional environmental factors are the drivers of sporadic diseases. Thus, a major challenge of using human-derived cells is that risk variants are not only present in patients but also in unaffected individuals, albeit with lower frequency. Thus, individual risk variants are not sufficient to cause disease-associated phenotypes in carrier individuals or in hiPSCs derived from carriers or patients [119].

The effect of a GWAS-identified risk regulatory allele on the target gene (or genes) would be predicted to be subtler than would be expected for monogenic diseases. Thus, it would be impossible to compare the disease-specific cells to a suitable control cell line because any control cells would have a different genetic background, which will affect the differentiation potential of the cells and thus would prevent a meaningful comparison. Thus, a major challenge for using hiPSCs for the study of sporadic diseases is how to generate pairs of isogenic cells that differ at one or multiple risk alleles.

Thus, the design of possible strategies to how the CRISPR/Cas9 gene editing approach could be used to generate isogenic cells that differ at multiple risk loci and thus would enable the mechanistic study of polygenic diseases. For example, once GWASs have identified genomic loci that may slightly increase the risk of developing a sporadic disease, then the key challenge of using patient-derived hiPSCs to get mechanistic insight into risk alleles is to create meaningful control cells by exchanging risk and protective alleles and generating appropriate control cells that differ exclusively at the risk loci under study [26]. Using this approach gene editing of noncoding distal enhancers resulted in deregulation of expression key genes implicated in the pathogenesis-studied disease [120]. In order to detect such subtle gene expression differences, an allele-specific assay should be developed that allowed the analysis of cis-acting effects of candidate variants on allele-specific gene expression as a consequence of deletion or exchange of disease-associated regulatory elements. Detailed analysis of isogenic cells with and without the risk allele could further demonstrate that a single base pair change causes loss of transcription factor binding sites for the transcription factors that otherwise function as a suppressor of associated risk variants transcription on a non-risk-associated allele. In addition, the cultures can be subjected to aging or environmental manipulation (e.g., oxidative stress, toxins). Figure 15.2 shows the general hiPSC technology platform and research workflow to study the interplay of genotypes and cellular environment experimentally.

15.9 Modeling of Undiagnosed and Rare Neurodevelopmental Diseases

Creating stem cell models for neurodevelopmental disorders (NDs) is time consuming and technically difficult, affecting the type of disorders that are studied. Specifically, common disorders are more likely to be studied because experimental outcomes will affect a greater number of patients. Completion of the Human Genome Sequencing Project [121] and the 1000 Genomes Project [122] has brought a great promise to revolutionize understanding of the genetic background of many human diseases, including NDs. Indeed, using whole-exome sequencing (WES), an almost exponential increase in knowledge on predisposition to Mendelian disorders, including NDs, has been observed. An even more powerful approach is represented by whole-genome sequencing (WGS) which covers all types of genetic

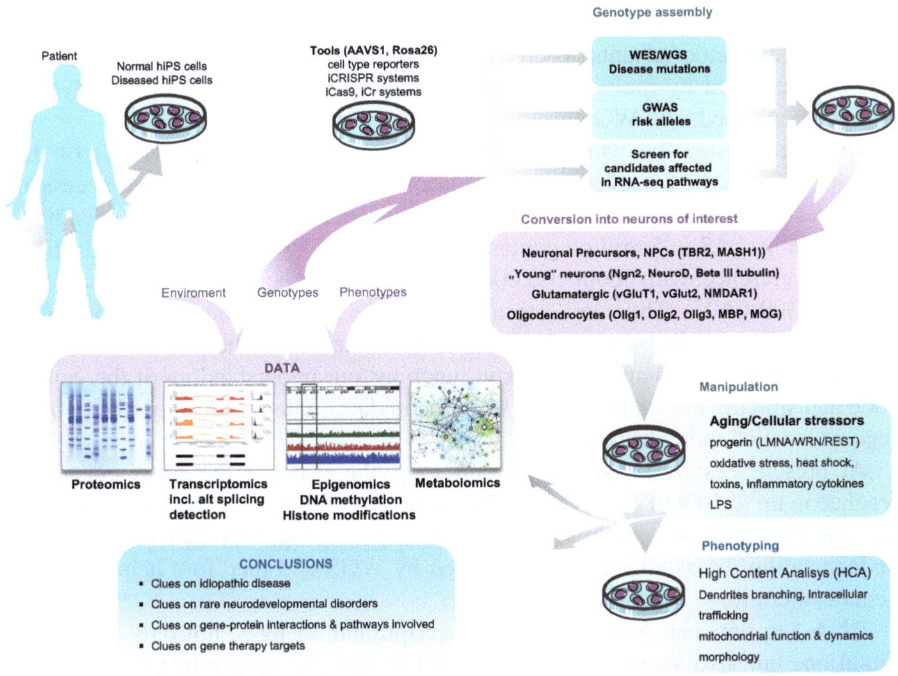

Fig. 15.2 hiPSC lines are derived by reprogramming of skin fibroblasts or peripheral blood mononuclear cells (PBMCs) from healthy or diseased individuals by the expression of the four transcription factors Oct4, Sox2, Klf2, and c-Myc (Yamanaka factors, OSKMs). Reprogrammed iPSCs represent self-renewing cell lines at early embryonic stage which can be propagated in vitro. Using designer nuclease technologies (Cas9), the candidate mutations could be introduced or corrected (isogenic cell lines) in one or more disease-related genes. These modified cells are then differentiated by expression of transcription factors into cells of interest and analyzed along the lineage for pathologic cellular phenotypes and genomic screening such as RNA-Seq and DNA methylation profiling. By the correlation of genotypes and phenotypes, this iPSC-based analysis platform delivers new insights into disease mechanisms. In addition, the cultures can be subjected to environmental manipulation or aged by fast-forwarding of the biological clock

variations, such as single-nucleotide variants (SNVs) and copy number variants (CNV) as well as structural variants (SVs) such as balanced chromosomal rearrangements. Moreover, WGS enables deciphering not only protein-coding sequence but also noncoding elements of human genome, including untranslated regions (UTRs), introns, regulatory elements, and diverse noncoding RNAs (ncRNAs) which may play an important role in genetic diseases. However, diagnosis based on WES did not result in detecting the causative variants, presumably due to the location of the causative variants. Similarly to sporadic disease, ND undiscovered variants possibly remain in the noncoding sequence, in regulatory motifs, and are reflected in transcriptome or epigenetic changes. Alterations in promoter, silencer/enhancer, and other regulatory sequences can cause reduced/increased amount of RNA, and these abnormalities can be revealed by using deep RNA-Seq. Changes in promoter sequences methylation can also be an indication and direction toward prioritizing possibly pathogenic variants. Therefore, causative variants in noncoding sequence can be uncovered by multi-omics study involving transcriptome and methylome analysis combined with WGS.

However, beside the undoubted benefits of WGS, several challenges of this approach have to be considered, such as prioritization and interpretation necessary for correct clinical variant annotation [123]. Indeed, during WGS "variant calling" provides a list of 3 to 5 million genomic positions (SNVs) where individuals differ from the reference, with about 100,000 of these variants being very rare or novel as well as ~300 CNVs. Thus, indication of causative gene(s)/variant(s) is not trivial, especially for rare diseases. It should be emphasized that whereas for variants in protein-coding regions and intron/exon junctions our understanding of the genetic code and splicing allows to identify likely candidates, interpreting variants in other regions is much more difficult.

At present despite progress in technology, the genetics diagnosis cannot be reached in up to 50% of cases of ND. The concept of combining the WGS data with the transcriptome and methylome study is a promising approach toward better understanding the extensive data generated by WGS and therefore help prioritize genetic alterations.

RNA-Seq data can reveal diverse transcriptional changes, not only splice site mutations but also silent mutations in coding regions, deep intronic mutations, and/or structural changes causing splicing aberrations, impossible to interpret using the WGS data only. Comparison and integrated analysis of the RNA-Seq results with the methylome analysis and whole-genome sequence (WGS) can provide new evidence that uncovered genomic mutations trigger those transcriptional changes and underlie ND. Therefore, by taking account of genomic mutations causing transcriptional aberrations, we can improve the sensitivity of deleterious mutation detection in the entire genome. Multi-omics approach has already been successfully introduced, predominantly focusing on particular gene panels which correspond with the particular disease [124–127]. However, due to the heterogeneous nature of the ND as well as the diverse phenotype it presents, integrated and comparative analyses of whole genomes and transcriptomes seem to be the most relevant to uncover novel molecular mechanisms contributing to ND. Identifying genomic variants based on WES or RNA-Seq only [128] limits the area of genome to the coding regions. On the other hand, WGS data allows for the screening of the noncoding regions however interpretation of such data is challenging. Combining the RNA-Seq results with the WGS is crucial for localization of the variants causing deregulations of the transcriptome.

The difficulties in obtaining of the human neural cells are the major challenge hiPSCs for the study of ND. Thus the experimental design should be expanded to isogenic control cells carrying the same genetic background as patient cells but differ at one or multiple risk alleles in the regions narrowed down by RNA-Seq of the transcriptome harvested from reprogrammed and differentiated hiPSCs into patient-relevant neurons of interest (Fig. 15.3).

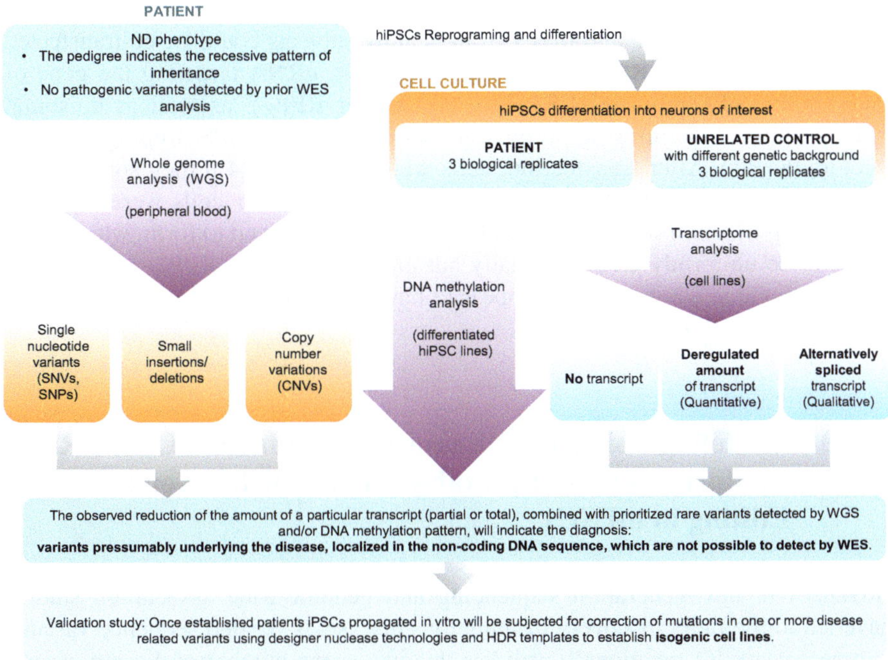

Fig. 15.3 Comparison and integrated analysis of the RNA-Seq results with the methylome analysis and whole-genome (exome) sequencing (WGS/WES) can provide new evidence that uncovered genomic mutations trigger those transcriptional changes and underlie ND. The observed reduction of the amount of a particular transcript (partial or total quantitative analyses, DEGs – Differentially Expressed Genes) or Alternative Isoform Regulation detection (AIR – qualitative analyses) combined with prioritized rare variants detected by WGS and/or DNA methylation pattern allows to narrow the WGS results and indicate the diagnosis, while variants underlying the disease are localized in the noncoding DNA sequence, which are not possible to detect by WGS/WES individually

15.9.1 Simultaneous Cas9 Genome Editing and iPSC Reprograming for ND Modeling

An interesting approach has been proposed recently by Bell et al. based on production of isogenic control lines generated by simultaneous CRIPSR/Cas9 genome editing and hiPSC reprograming [129]. This high-efficiency gene editing protocol can be performed in parallel to patient cell reprograming with reducing heterogeneity in final cell output. For ND research, this is an extremely important tool allowing for generation of isogenic controls directly from patient cells [129]. It is therefore desirable to make a CRIPSR/Cas9 gene edited cell line in which all cells are descended from a single gene edited cell omitting key concern when introducing CRIPSR/Cas9 gene editing results in heterogeneity repair due to double-strand breaks that introduce random mutations into the cut site, which can produce very different mutations in different cells, which consequently have different phenotypes. In this approach, fibroblasts from healthy subjects could be co-transfected with episomal CRIPSR/Cas9 construct containing gRNA targeting the gene of interest and RFP reporter (CRISPR/Cas9 Vector RFP+), as well as episomal iPSC vectors containing a puromycin resistance gene (iPS Reprogramming Vectors PurR+). After transfection, cells were plated for 24 h then sorted using FACS into RFP1+ and RFP2- single cell fractions. RFP1+ cells were replated, puromycin selection was applied, and colonies were allowed to form, which were found to be uniformly RFP1+ initially but gradually became RFP2- as induction proceeded. Gene editing did not appear to have any effect on transformation efficiency of cells [129].

15.10 Modeling Polygenic Diseases: The Inducible CRISPR Platforms for Multiplexable and Inducible Genome Editing in hiPSCs

Advances in next-generation sequencing and genome-wide association studies have led to the rapid discovery of numerous disease-associated sequence variants in recent years. To functionally validate this increasing number of disease-associated mutations, an ideal platform should not only closely recapitulate their human-specific context but also offer superior speed and capacity to meet the growing demand. The production of complex genotypes in multiple cell lines requires efficient and streamlined gene editing technologies. High rates of gene editing at up to three loci now could be achieved using modified inducible Cas9 systems (iCas9 and iCRISPR systems) [73, 130]. These systems include a doxycycline-inducible Cas9 and sgRNA/reporter plasmids for the enrichment of transfected cells by fluorescence-activated cell sorting (FACS) [73] or iCRISPR platform established through designer nuclease-mediated targeting of inducible Cas9 expression cassettes into the AAVS1 locus [130]. Depending on the experimental setup, both classical CRISPR/Cas9 and the iCRISPR platforms feature advantages and drawbacks. The former is applicable to wild-type cells if no further genomic modifications are desired. The latter was demonstrated to provide improved efficacy when applied to modify multiple loci, although prior cell line generation is required, and a modified AAVS1 locus will remain if no strategies for final Cas9 gene removal are applied [73]. These iCRISPR lines show a strong Cas9 expression when induced with doxycycline. These cells are transfected with a small vector for the expression, a fluorescent Venus reporter and of multiple sgRNAs, cloned into pU6-sgRNA-CAG-Venus plasmids. This allows the fraction of successfully transfected hiPSCs expressing sgRNAs and the reporter to be isolated by fluorescence-activated cell sorting (FACS). The sorted cells are then plated for the establishment of single cell-derived colonies [73]. To introduce specific nucleotide alterations, sgRNA with ssDNA as HDR template could be co-transfected with sgRNAs in close proximity to the genetic alteration locus [73]. ssDNA templates containing the genetic alternation are flanked by 40–80 nt of homology on each side of the space between the genetic alternation site and the Cas9 cutting site. Because of the close proximity of the sgRNAs designed, only one ssDNA template is necessary in most cases [73]. Moreover, introduction of silent mutations (e.g., G > A in the PAM) to eliminate the PAM sequence in the ssDNA template is highly recommended to avoid further editing of correctly modified alleles. If the genetic alternation introduces or disrupts a restriction site, HDR efficiency can be assessed through RFLP analysis (C > A introduces a BglII site).

15.10.1 Inducible Gene Knockouts Using iCas9 and iCr Systems

Gene-targeting strategy aimed at generating premature stop codons through frameshift mutations needs to be strategically chosen to maximize the possibility of disrupting the function of the corresponding protein. It is important to identify all possible splice variants of the gene of interest and design sgRNAs to target a functional region that is at least present in the isoform relevant for the study. Inducible gene knockout during differentiation of hiPSCs into cells of interest is of great importance for studying genes with pleiotropic effects. With the iCRISPR platform, inducible gene knockout could be achieved by inducible Cas9 expression and temporally regulated delivery of sgRNA due to the low toxicity of lipid-mediated sgRNA transfection [130]. Inducible gene knockout could also be achieved through generation of lines including a constitutive sgRNA expression module in addition to the doxycycline-inducible Cas9 expression cassette targeted into the AAVS1 locus [130]. In iCas9 systems the cells are first differentiated into cells or neurons of interest. Then, doxycycline induced expression of Cas9 together with sgRNAs transfection in differentiated cells result in inducible gene knockout. Whereas in iCr systems, a constitutive sgRNA expression module is inserted downstream into the Cas9 expression cassette under U6 promoter, and doxycycline treatment of differentiated iCr cells induces Cas9 expression, thus inducing gene knockout [130]. Generation of such lines allows inducible gene targeting in all the cells upon doxycycline treatment.

15.11 Modeling of Late-Onset Diseases: Aging of the Cells to Fast-Forward Disease Phenotypes by Modifications of DNA Repair Pathways

The aspect of aging as a necessary component to disease progression, i.e., the concern as to how well iPSC-derived cells can model late-onset diseases, where patients do not develop symptoms until later in life, will be covered by the modulation of genes associated with premature aging to induce multiple aging-related markers to study disease pathways of late-onset and aging-related disorders. hiPSC-derived neural cells also fail to recapitulate the major pathological hallmarks of the disease due to the rejuvenation of hiPSCs and production of neurons which represent years but not decades of neuronal age [131]. To date, one research group has artificially accelerated the aging of neurons through ectopic expression of progerin in hiPSCs achieving an exaggerated Parkinson's disease phenotype [132]. Cellular stressors have also been used to artificially induce a stressed phenotype in vitro [133–138]. However, there are serious drawbacks within both approaches. Prelamin A, mutated in progeria, is expressed at very low levels in the brain and may not actually have a large role in the brain [139, 140], while cellular stressors do not directly age the cells but elicit stress-response pathways. Aging is a gene-regulated process that is coupled in aged cells with tumor formation, cardiovascular disease, and neurodegenerative diseases. Major pathways associated with aging include the GH/IGF-1 axis, the TOR-S6K pathway, and sirtuin proteins [141]. However DNA metabolism is also linked to aging. One important gene that regulates DNA metabolism is the Werner syndrome, RecQ helicase-like gene (WRN). Loss of WRN function results in Werner's syndrome, which induces premature aging and early death. Another gene important in aging (particularly in the brain) is the neuron-restrictive silencer factor (REST). This gene has an important role in epigenetic modifications of histones and DNA [142] and has recently been identified to be depressed in the AD brain [143] compared to healthy age-matched brain tissue. By modulating gene expression of age-related proteins, it should therefore be possible to "age" a neuron and overcome the limitations of examining young neurons in a dish when studying diseases that affect "old" neurons. Independent gene editing approaches (LMNA/WRN/REST) could be applied to generate artificially aged neurons from patients hiPSCs carrying middle-age or late-onset phenotypes (middle such as > ~40 years old in Huntington's disease). This could be done by transfections of modified RNA inducing progerin expression generated by a silent point mutation (C1824T) in the lamin A gene, LMNA. Additionally, aging-related gene editing could be performed by the knockout (KO) of WRN (Werner protein repairing damaged DNA) and the knockdown (KD) of the REST using stable RNAi (neuron-restrictive silencer factor playing important role in epigenetic modifications of histones and DNA and recently identified as an epigenetic aging factor) [142, 143]. Constitutive and homozygous 70–80% reduction of REST knockdown could be introduced using shRNA vector designed with U6 RNA polymerase III promoter-driven shRNA to generate non-conditional hiPSC lines with an allover KD phenotype driving the expression of the sense, the loop, and the antisense sequence of the hairpin based on the insertion of shRNA vectors into the hiPSCs homolog of Rosa26 locus [144]. Using these alternative (LMNA/WRN/REST) aging strategies differentiating hiPSCs could be then subjected to assessment of aging rate by basic cell-type-specific parameters associated with neuronal aging in vivo such as degenerative changes in dendrite branching, dendrite degeneration, or enlarged mitochondria. Because aging of mitotic cells is associated with the shortening of telomeres to a critical point when the cell undergoes senescence, therefore, telomere lengths could be measured using quantitative fluorescence in situ hybridization. Since the ability to measure human aging from molecular profiles has practical implications in many fields, including disease prevention and treatment, such cell could be used to measure the impact of introduced mutations on the rate of methylome aging using measurements at genome-wide CpG markers. To build a model of aging based on the expression of genes that were associated with age, time course RNA-Seq could be also performed.

15.12 Future Perspectives: Increasing the Efficiency of Homology-Directed Repair (HDR) for Cas9-Induced Precise Gene Editing in hiPSCs

The precise genetic modifications by CRISPR/Cas9 is limited by the low efficiency of homology-directed repair (HDR) compared with the higher efficiency of the non-homologous end-joining (NHEJ) pathway in hiPSCs. Moreover, HDR normally acts exclusively during the S/G2 phases of the cell cycle, whereas the error-prone NHEJ pathway is dominant and active throughout the cell's different stages. The different activity pattern of the two pathways poses an efficiency problem for gene editing applications, for which precision when changing DNA sequences is crucial. Attempts to solve this problem involve mainly the shifting of the bias from the NHEJ pathway toward HDR-mediated mechanism, by timing control of CRISPR/Cas9 delivery, and optimizing the design of HDR donor templates with regard to the length and symmetry of the homology regions.

15.12.1 Enhancing the HDR Efficiency in hiPSCs by Consideration of the Repair Template Structure and Modifications

Recent studies suggest that choice of the target strand and an asymmetric design of ssODNs increase the knockin efficiency [72] and that phosphorothioate (PTO) modifications improve stability and HDR efficacy [145]. However, evaluation of ssODNs with symmetric or asymmetric structure with or without PTO modifications has shown no application of these rules to hiPSCs [73]. Interestingly, unmodified ssODNs without PTO modifications yield higher HDR rates compared to modified ones in hiPSCs [73]. Furthermore, in contrast to HEK293 cells as reported by Richardson et al. [72], a symmetric architecture of ssODNs resulted in higher HDR efficiencies than an asymmetrical in hiPSCs [73]. In particular they found that 36 bases distal to the PAM and 91 bases proximal to the PAM supported HDR efficiencies up to 60%. However, the efficacy of CRISPR/Cas9-mediated targeting of specific loci depends on the cell line subjected, the localization of the locus within the chromatin, and the sequence itself. Therefore, these observations cannot be generalized but may vary based on locus of interest architecture and experimental design.

15.12.2 Enhancing HDR Precision and Efficiency in hiPSCs by Controlled Timing of CRISPR/Cas9 Components Delivery: Cell Cycle Synchronization

NHEJ dominates DNA repair during G1, S, and G2 phases, whereas HDR is restricted to late S and G2 phases when DNA replication is completed and sister chromatids are available to serve as repair templates [146]. Impediments to HDR include competition with NHEJ in S and G2 phases and specific downregulation of HDR at M phase and early G1 to prevent telomere fusion [147]. Consequently, high cleavage activity of programmable nucleases does not correlate with efficient HDR [71]. Thus the robust approach that advances enhancement of HDR efficiency in human cells is highly recommended. The strategies are based on previous findings by Jinek et al. and combine well-established cell cycle synchronization techniques with direct nucleofection of preassembled Cas9 ribonucleoprotein (RNP) complexes to achieve controlled nuclease action at the phase of the cell cycle best for HDR (Fig. 15.4) [22]. In this approach Lin et al. applied chemical inhibitors to arrest cells at specific phases of cell cycle that included lovastatin (Lov), which blocks at early G1 and partially at G2/M phase; mimosine (Mim), aphidicolin (Aph), thymidine (Thy), and hydroxyurea (HU), which arrest cells at the G1-S border prior to onset of DNA replication; and nocodazole (Noc), which causes arrest at G2/M phase (shown in Fig. 15.4 cell cycle phases) [71]. These six reversible chemical inhibitors were used in parallel to synchronize cells at G1, S, and M phases of the cell cycle, followed by release from cell cycle synchronization prior to nucleofection with Cas9 RNP. As a result, human embryonic stem cells (hES) demonstrated robust HDR-mediated genome editing at levels up to 38% (~2% of HDR at 100 ρmol Cas9 RNP) with no detected off-target editing [71]. Cell cycle behavior was indistinguishable from unsynchronized control cultures, with no apparent changes in colony morphology, and hES colonies expressed high levels of alkaline phosphatase, a marker for pluripotency.

15.12.3 Enhancing HDR Precision and Efficiency in hiPSCs

Gene editing products generated by the native DSB repair do not align with experimental needs due to choice for the usage of alternative template-free annealing (a-NHEJ) and immediate protection of free ends through religation by classic

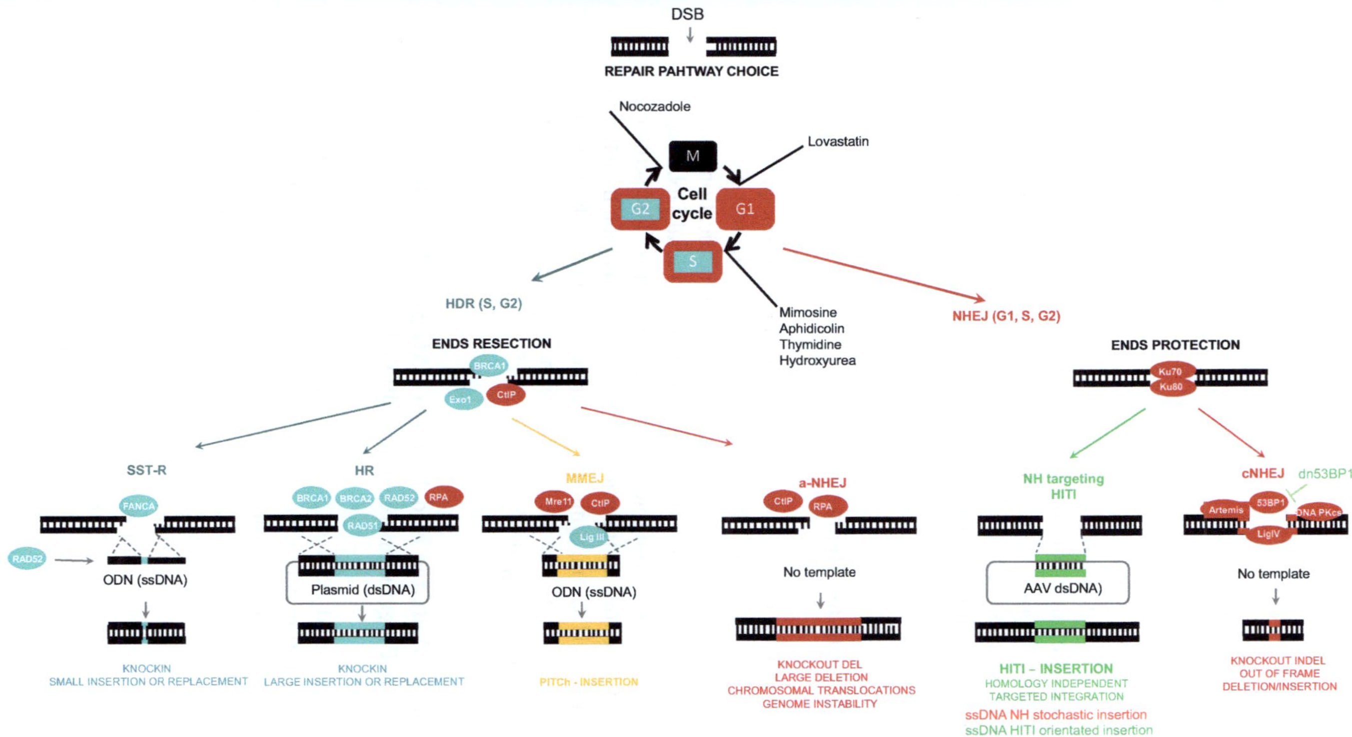

Fig. 15.4 The pathway used for double-strand break (DSB) repair depends on the cell-cycle phase and the presence of single-stranded DNA (ssDNA) oligodeoxynucleotides (ODNs) or double-stranded DNA (dsDNA) plasmids harboring homology sequences. 53BP1 initiates repair via the nonhomologous end-joining (NHEJ) pathway, which is active throughout the cell cycle and rejoins DSB ends without repair templates, frequently leading to the multiple nucleotide indels. HDR only occurs during the S/G2 phase of the cell cycle, whereas NHEJ is active throughout the cell cycle in a variety of adult cell types, making it accessible to nondividing cells. Homology-independent targeted integration (HITI) hijacks the NHEJ pathway and inserts ectopic DNA at target sites in both dividing and nondividing cells. The homologous recombination (HR) pathway, which depends on BRCA1, RAD51, and other proteins, repairs DSBs via recombination with dsDNA templates and enables the insertion or replacement of large sequence segments. SSTR is an alternative, RAD51-independent HDR mechanism mediated by core proteins of the Fanconi anemia repair pathway (FANCA) and accepts synthetic ODNs as templates, thus enabling the insertion or replacement of small-sequence segments flanked by 2 homology regions of 50–70 nucleotides. Precise gene editing via SSTR and ODNs is enhanced by ectopic expression of RAD52 and of a dominant-negative subfragment of 53BP1 (dn53BP1) counteracting the endogenous 53BP1. Alternatively, the PITCh approach utilizes MMEJ repair to achieve insertions with short homology sequences

nonhomologous end joining (c-NHEJ)-mediated DSB repair over precise homology-directed repair (HDR). c-NHEJ pathway operates throughout all phases of the cell cycle; HDR is restricted to the S and G2 phases [148]. The pathway choice is determined by the accumulation of 53BP1 as key molecule favoring c-NHEJ which blocks end resectioning by accessory protein CtIP, leading to a variety of edited alleles, with c-NHEJ as the dominant outcome. Ten percent of events of c-NHEJ result in a-NHEJ [149] leading to deletions of various sizes and extensive resection. These extensive resections, repaired by a-NHEJ, result in increased cancer-driving chromosomal translocations [150] and provide telomere dysfunction-induced genome instability [151]. Although HDR-based gene knockin is well established, it cannot necessarily be applied in every cell type duet to low HDR-mediated homologous recombination (HR) pathway frequencies especially in hiPSCs. Thus, current efforts in genome editing field aim in searching for further details of promotion of HDR over NHEJ or even using alternative repair pathways.

15.12.3.1 Single-Strand Template Repair (SST-R)

In proliferating cells, the native pathway uses the intact homologous sequences of sister chromatids as template for the repair of DSB sites and leads to the reconstitution of the wild-type allele [146, 152]. To achieve precise sequence modifications at targeted DSBs, the HR pathway can be co-opted by providing an artificial DNA repair template containing sequence regions homologous to the DSB ends. The sequence between the homologous ends, either an insertion or replacement, is then transferred into the targeted locus during HR, enabling the generation of precisely modified knockin alleles. Repair using ssODNs is mediated by a poorly defined mechanism designated as single-strand template repair (SST-R) (Fig. 15.4).

15.12.3.2 Microhomology-Mediated End Joining (MMEJ) and the Precise Integration into Target Chromosomes (PITCh)

Alternatively to HR, short resected DSB ends can be religated via stretches of a few complementary nucleotides through the mechanism of microhomology-mediated end joining (MMEJ) [153]. This repair pathway is more similar to HR, requiring Mre11 and CtIP for end resection and religating the ends (unlike c-NHEJ Ligase IV) using DNA ligase III (Fig. 15.4). MMEJ is harnessing independent machinery from HR and requiring an extremely short homologous sequence (5–25 bp) for DSB repair using the PITCh (Precise Integration into Target Chromosomes). Although PITCh is not yet widely used in hiPSCs, Sakuma et al. shows that PITCh results in precise gene knockin with easily constructed donor vector in human cell lines by transfection. The construction of the PITCh vectors requires only a few days, and the entire process takes ~1.5 months to establish knocked-in cells or ~1 week from injection to early genotyping [154].

15.12.3.3 Homology-Independent Targeted Integration (HITI)

NHEJ repair has been shown to frequently ligate exogenously provided DNA segments in between the ends of DSBs, in this approach, designated as nonhomology (NH) targeting (Fig. 15.4) [155]. The orientation of the inserted fragments is stochastic and cannot be controlled. This stochastic insertion has been recently overcome by a modification in the technique that allows a high degree of preference in the insertion orientation [156]. In homology-independent targeted integration (HITI) method, which uses NHEJ-mediated targeted integration if cleavage is repaired without gene integration or with integration in reverse direction, DNA undergoes additional cleavage until forward gene integration or gRNA can no longer bind to target cleavage sequence due to errors from NHEJ repair (Fig. 15.5). HITI donor vectors are constructed to ensure robust gene integration only when inserted in the forward direction, and if inserted in the reverse direction or is unintegrated, the DNA would undergo further cleavage by Cas9 until inserted correctly or gRNA is no longer able to bind to target sequences due to errors during NHEJ repair (Fig. 15.5). This (HITI) method was shown to be remarkably efficient in primary cells in vitro and in vivo and, more importantly, in vivo (in postnatal neurons), suggesting that c-NHEJ can be a method of choice for achieving targeted integrations in nondividing cells. As expected for c-NHEJ repair events, the ligation of the fragment into the genome frequently results in indels at one or both junctions [156, 157]. Effectiveness of HITI in nonproliferating cells was examined using mouse primary neurons, in which authors found gene insertion in 60% of transfected cells [156, 157]. Despite high efficiency in vitro, the in vivo HITI gene knockin efficiency is relatively low (1–5% of the cells) due to low DNA transduction efficiency [158]. Thus, to improve the HITI efficiency, more mechanistic

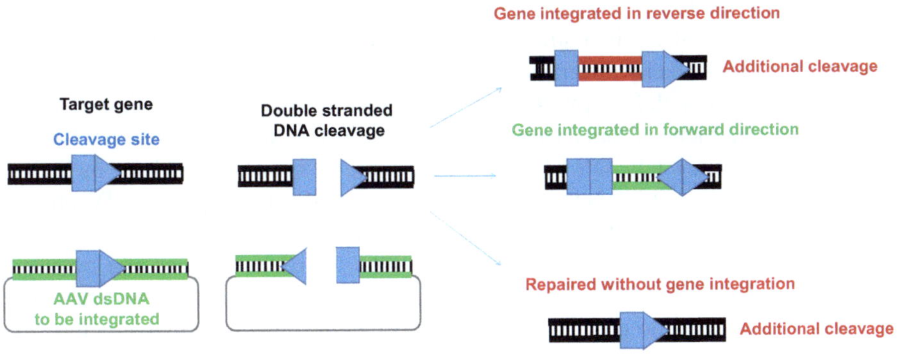

Fig. 15.5 Schematic of HITI using Cas9. The Cas9/sgRNA complex introduces DSBs resulting in two blunt ends. The exact same Cas9/sgRNA target sequence is loaded into the donor DNA in the reverse direction. Targeted genomic loci, as well as the donor DNA, are cleaved by Cas9/gRNA, and the linearized donor DNAs are integrated into target sites via the NHEJ DSB repair pathway. If donor DNA is integrated in the correct orientation, junction sequences are protected from further cleavage by Cas9/gRNA. If donor DNA integrates in the reverse orientation, Cas9 will excise the integrated donor DNA due to the presence of intact gRNA target sites

studies are important for identifying the major regulators of NHEJ as well as for development of the most suitable in vivo DNA delivery methods for specific cell types, tissues, and organs. Moreover, further unbiased deep sequencing analyses are needed to comprehensively evaluate HITI off-target effects. Finally, HITI can only insert ectopic DNA into the genome but cannot correct the mutation from the target locus. Thus, the range of genetic disorders that this technology is able to model or treat is limited. However, HITI has the potential to evolve into a comprehensive tool for precise disease modeling, in vivo genome corrections, and gene therapies.

15.12.4 Enhancing of Gene Editing in hiPSCs by Manipulation of DNA Repair Mechanisms

In cycling cells, sister chromatids are available as natural dsDNA templates for DSB repair. HDR can also occur with synthetic single-stranded oligodeoxynucleotides (ODNs) as templates, probably because they mimic natural recombination intermediates. This situation can be emulated by introducing plasmids with homology sequences into cells to serve as artificial repair templates together with further details of promotion of HDR over NHEJ.

15.12.4.1 Suppression of c-NHEJ Activity

Recently, the key molecules of the c-NHEJ pathway such as DNA Ligase-IV, Ku70/80, and DNA-dependent protein kinase catalytic subunits (DNA-PKcs) have been targeted to transiently suppress c-NHEJ activity (reviewed extensively by Danner et al. [157]). By transient suppression of c-NHEJ in cycling cells, DSB repair can be biased in favor of HDR. SCR7 is a small-molecule inhibitor that prevents the binding of Ligase IV to DNA and by this mean inhibits the ability of Ligase IV to join DSB ends [159]. Maruyama et al. [160] used SCR7 to inhibit c-NHEJ repair for Cas9-induced DSBs. The treatment of cells with SCR7 increased the rate of HDR events up to 19-fold. In a similar study, DNA Ligase IV was targeted in human cell lines by SCR7, shRNA-mediated gene silencing, or by adenoviral (Ad)-directed Ligase IV proteasomal degradation [161]. A combined knockdown of Ku70/Ligase IV, SCR7 treatment, or Ad protein expression was all effective for c-NHEJ suppression and increased HR events in HEK293 cells up to eightfold. However, SCR7 has shown unequal effectiveness when applied to other systems. It has been noted that its actual structure differs from the original description, and furthermore its selectivity toward other DNA ligases has been questioned [162]. The search for more potent and specific DNA Ligase IV inhibitors will certainly continue since Ligase IV is a promising target in cancer therapy. A step into this direction was reported by Menchon et al. who identified a lead compound disrupting the interaction of the DNA Ligase IV clamp domain to its binding partner XRCC4 [163].

Various small-molecule inhibitors against DNA-PK have been tested for suppression of c-NHEJ repair at Cas9-induced DSBs [157]. Robert et al. confirmed previous results [161] on the suppression of Ligase IV and Ku70/80 to increase HDR. They additionally identified the DNA-PK inhibitors NU7441 and KU-0060648 as effective for c-NHEJ suppression, resulting in increasing HDR twofold for both double- and single-stranded DNA repair templates [164]. Another approach antagonizes mammalian DNA-PK by depletion of its cofactor inositol hexakisphosphate (InsP6) using calmodulin inhibitors W7 and chlorpromazine [165]. InsP6 depletion by these compounds reduces the binding of the Ku complex to DSBs affecting the recruitment of other c-NHEJ factors. Although not yet tested for its effect on Cas9-induced DSBs in mammalian cells, both inhibitors phenocopy the positive effect (fivefold stimulation) on HDR seen in Ku deletion strains of *Cryptococcus* fungi [166]. 53BP1 is a key determinant of DSB repair pathway choice, promoting c-NHEJ by blocking DSB end resection. 53BP1 is therefore a prime target for interventions aiming to suppress c-NHEJ. Earlier studies of the role of 53BP1 in DSB repair used siRNA-mediated knockdown [167], the ectopic expression of dominant-negative (DN)53BP1 subdomains including the tandem Tudor domain [168, 169], or 53BP1-inactivated cell lines [170].

15.12.4.2 Manipulation of Single-Stranded Template Repair (SSTR) and Fanconi's Anemia (FA) Repair Pathway Components

Although chemical or genetic interruption of the NHEJ pathway can favor HDR [171], such manipulations can be difficult to employ, harmful to cells, or both [71]. The HDR substrates are recombined with DSBs through the classical homologous recombination (HR) pathway, which depends on RAD51 and on numerous other proteins. HDR can also occur with ODNs as templates, probably because they mimic natural recombination intermediates [172]. In these cases, ODN sequences are read into the genome in a process called single-stranded template repair (SSTR), which differs from HR and does not depend on RAD51 (Fig. 15.4). The extent of the practical uses of ODNs in precision gene editing, however, remains unclear, and the enzymatic machinery involved remains largely unexplored [172]. Recently Paulsen et al. showed that ectopic expression of RAD52 and the suppression of 53BP1 activity (by expressing a truncated fragment that occupies the signal sites on H2A and acts as a dominant-negative (dn) 53BP1 mutant) can be an effective means of enhancing the recombination of ODNs with Cas9-induced DSBs and these interventions lead to a two- to threefold increase in ODN recombination with various target genes [173]. The expression of dn53BP1 alone has a beneficial effect on ODN recombination, but its highest benefit is obtained when combined with the expression of RAD52 (but not of RAD51), EXO1, or BLM. Moreover, Richardson et al. found that SSTR requires multiple components of the Fanconi's anemia (FA) repair pathway but does not require RAD51-mediated homologous recombination, distinguishing SSTR from repair using double-stranded DNA donors [174]. These findings suggest a focus for future work aimed to decoding what pushes the FA pathway to repair Cas9-induced DSBs and how do RAD52 and related proteins act as recombination enhancers.

Acknowledgments This work was supported by the National Science Centre, Poland Grant No. 2016/22/M/NZ2/00548. I would like to thank Dr. Ralf Kühn (Max-Delbrück-Centrum für Molekulare Medizin) for invaluable support and guidance in development and application of hiPSC gene editing technologies described in this work. I would like to thank Ms. Aleksandra Golonko (Bialystok University of Technology) for excellent support in manuscript preparation.

References

1. Rouet P, Smih F, Jasin M (1994) Expression of a site-specific endonuclease stimulates homologous recombination in mammalian cells. Proc Natl Acad Sci U S A 91:6064–6068
2. Rouet P, Smih F, Jasin M (1994) Introduction of double-strand breaks into the genome of mouse cells by expression of a rare-cutting endonuclease. Mol Cell Biol 14:8096–8106
3. Morton J, Davis MW, Jorgensen EM, Carroll D (2006) Induction and repair of zinc-finger nuclease-targeted double-strand breaks in Caenorhabditis elegans somatic cells. Proc Natl Acad Sci 103:16370–16375
4. Wood AJ, Lo TW, Zeitler B, et al (2011) Targeted genome editing across species using ZFNs and TALENs. Science (80-) 333:307
5. Beumer KJ, Trautman JK, Bozas A, Liu J-L, Rutter J, Gall JG, Carroll D (2008) Efficient gene targeting in Drosophila by direct embryo injection with zinc-finger nucleases. Proc Natl Acad Sci 105:19821–19826
6. Bibikova M, Golic M, Golic KG, Carroll D (2002) Targeted chromosomal cleavage and mutagenesis in Drosophila using zinc-finger nucleases. Genetics 161:1169–1175
7. Bibikova M, Beumer K, Trautman JK, Carroll D (2003) Enhancing gene targeting with designed zinc finger nucleases. Science (80-) 300:764
8. Urnov FD, Miller JC, Lee YL, Beausejour CM, Rock JM, Augustus S, Jamieson AC, Porteus MH, Gregory PD, Holmes MC (2005) Highly efficient endogenous human gene correction using designed zinc-finger nucleases. Nature 435:646–651
9. Tebas P, Stein D, Tang WW et al (2014) Gene editing of *CCR5* in autologous CD4 T cells of persons infected with HIV. N Engl J Med 370:901–910
10. DeKelver RC, Choi VM, Moehle EA et al (2010) Functional genomics, proteomics, and regulatory DNA analysis in isogenic settings using zinc finger nuclease-driven transgenesis into a safe harbor locus in the human genome. Genome Res 20:1133–1142
11. Hockemeyer D, Jaenisch R (2010) Gene targeting in human pluripotent cells. Cold Spring Harb Symp Quant Biol 75:201–209
12. Hockemeyer D, Soldner F, Beard C et al (2009) Efficient targeting of expressed and silent genes in human ESCs and iPSCs using zinc-finger nucleases. Nat Biotechnol 27:851–857
13. Lombardo A, Genovese P, Beausejour CM et al (2007) Gene editing in human stem cells using zinc finger nucleases and integrase-defective lentiviral vector delivery. Nat Biotechnol 25:1298–1306
14. Sexton AN, Regalado SG, Lai CS, Cost GJ, O'Neil CM, Urnov FD, Gregory PD, Jaenisch R, Collins K, Hockemeyer D (2014) Genetic and molecular identification of three human TPP1 functions in telomerase action: recruitment, activation, and homeostasis set point regulation. Genes Dev 28:1885–1899
15. Soldner F, Laganière J, Cheng AW et al (2011) Generation of isogenic pluripotent stem cells differing exclusively at two early onset parkinson point mutations. Cell 146:318–331
16. Zou J, Maeder ML, Mali P et al (2009) Gene targeting of a disease-related gene in human induced pluripotent stem and embryonic stem cells. Cell Stem Cell 5:97–110
17. Mali P, Yang L, Esvelt KM, Aach J, Guell M, DiCarlo JE, Norville JE, Church GM (2013) RNA-guided human genome engineering via Cas9. Science (80-) 339:823–826
18. Xue X, Papusha A, Choi K et al (2016) Differential regulation of the anti-crossover and replication fork regression activities of mph1 by mte1. Genes Dev 30:687–699

19. Cong L, Ran FA, Cox D, et al (2013) Multiplex genome engineering using CRISPR/Cas systems. Science (80-) 339:819–823
20. Mali P, Aach J, Stranges PB, Esvelt KM, Moosburner M, Kosuri S, Yang L, Church GM (2013) CAS9 transcriptional activators for target specificity screening and paired nickases for cooperative genome engineering. Nat Biotechnol 31:833–838
21. Wang H, Yang H, Shivalila CS, Dawlaty MM, Cheng AW, Zhang F, Jaenisch R (2013) One-step generation of mice carrying mutations in multiple genes by CRISPR/cas-mediated genome engineering. Cell 153:910–918
22. Jinek M, East A, Cheng A, Lin S, Ma E, Doudna J (2013) RNA-programmed genome editing in human cells. elife 2013:e00471
23. Ding Q, Regan SN, Xia Y, Oostrom LA, Cowan CA, Musunuru K (2013) Enhanced efficiency of human pluripotent stem cell genome editing through replacing TALENs with CRISPRs. Cell Stem Cell 12:393–394
24. Fischer K, Kraner-Scheiber S, Petersen B et al (2016) Efficient production of multi-modified pigs for xenotransplantation by “combineering”, gene stacking and gene editing. Sci Rep 6:29081
25. Grzybek M, Golonko A, Walczak M, Lisowski P (2017) Epigenetics of cell fate reprogramming and its implications for neurological disorders modelling. Neurobiol Dis 99:84–120
26. Hockemeyer D, Jaenisch R (2016) Induced pluripotent stem cells meet genome editing. Cell Stem Cell 18:573–586
27. Yusa K, Rashid ST, Strick-Marchand H et al (2011) Targeted gene correction of α1-antitrypsin deficiency in induced pluripotent stem cells. Nature 478:391–394
28. Chung K, Wallace J, Kim SY et al (2013) Structural and molecular interrogation of intact biological systems. Nature 497:332–337
29. Ambasudhan R, Ryan SD, Dolatabadi N et al (2013) XIsogenic hman iPSC prkinson’s mdel shows ntrosative stress-induced dsfunction in MEF2-PGC1α tanscription. Cell 155:1351–1364
30. Wang Y, Li Y, Toth JI, Petroski MD, Zhang Z, Zhao JC (2014) N6-methyladenosine modification destabilizes developmental regulators in embryonic stem cells. Nat Cell Biol 16:191–198
31. Gurumurthy CB, Grati M, Ohtsuka M, Schilit SLP, Quadros RM, Liu XZ (2016) CRISPR: a versatile tool for both forward and reverse genetics research. Hum Genet 135:971–976
32. Mianné J, Chessum L, Kumar S et al (2016) Correction of the auditory phenotype in C57BL/6N mice via CRISPR/Cas9-mediated homology directed repair. Genome Med 8:16
33. Nelson CE, Hakim CH, Ousterout DG, et al (2016) In vivo genome editing improves muscle function in a mouse model of Duchenne muscular dystrophy. Science (80-) 351:403–407
34. Yang Y, Wang L, Bell P et al (2016) A dual AAV system enables the Cas9-mediated correction of a metabolic liver disease in newborn mice. Nat Biotechnol 34:334–338
35. Platt RJ, Chen S, Zhou Y et al (2014) CRISPR-Cas9 knockin mice for genome editing and cancer modeling. Cell 159:440–455
36. Yoshimi K, Kunihiro Y, Kaneko T, Nagahora H, Voigt B, Mashimo T (2016) SsODN-mediated knock-in with CRISPR-Cas for large genomic regions in zygotes. Nat Commun 7:10431
37. Claussnitzer M, Dankel SN, Kim K-H et al (2015) *FTO* obesity variant circuitry and adipocyte Browning in humans. N Engl J Med 373:895–907
38. Orack JC, Deleidi M, Pitt D, Mahajan K, Nicholas JA, Boster AL, Racke MK, Comabella M, Watanabe F, Imitola J (2015) Concise review: modeling multiple sclerosis with stem cell biological platforms: toward functional validation of cellular and molecular phenotypes in inflammation-induced neurodegeneration. Stem Cells Transl Med 4:252–260
39. Shin JW, Kim K-H, Chao MJ, Atwal RS, Gillis T, MacDonald ME, Gusella JF, Lee J-M (2016) Permanent inactivation of Huntington’s disease mutation by personalized allele-specific CRISPR/Cas9. Hum Mol Genet 25(20):4566–4576. ddw286
40. Yang H, Wang H, Shivalila CS, Cheng AW, Shi L, Jaenisch R (2013) XOne-step generation of mice carrying reporter and conditional alleles by CRISPR/cas-mediated genome engineering. Cell 154:1370–1379

41. Qi LS, Larson MH, Gilbert LA, Doudna JA, Weissman JS, Arkin AP, Lim WA (2013) Repurposing CRISPR as an RNA-γuided platform for sequence-specific control of gene expression. Cell 152:1173–1183
42. Cheng AW, Wang H, Yang H, Shi L, Katz Y, Theunissen TW, Rangarajan S, Shivalila CS, Dadon DB, Jaenisch R (2013) Multiplexed activation of endogenous genes by CRISPR-on, an RNA-guided transcriptional activator system. Cell Res 23:1163–1171
43. Gilbert LA, Larson MH, Morsut L et al (2013) XCRISPR-mediated modular RNA-guided regulation of transcription in eukaryotes. Cell. https://doi.org/10.1016/j.cell.2013.06.044
44. Maeder ML, Linder SJ, Cascio VM, Fu Y, Ho QH, Joung JK (2013) CRISPR RNA-guided activation of endogenous human genes. Nat Methods 10:977–979
45. Perez-Pinera P, Kocak DD, Vockley CM et al (2013) RNA-guided gene activation by CRISPR-Cas9-based transcription factors. Nat Methods 10:973–976
46. Tanenbaum ME, Gilbert LA, Qi LS, Weissman JS, Vale RD (2014) A protein-tagging system for signal amplification in gene expression and fluorescence imaging. Cell 159:635–646
47. Zalatan JG, Lee ME, Almeida R et al (2015) Engineering complex synthetic transcriptional programs with CRISPR RNA scaffolds. Cell 160:339–350
48. Chakraborty S, Ji H, Kabadi AM, Gersbach CA, Christoforou N, Leong KW (2014) A CRISPR/Cas9-based system for reprogramming cell lineage specification. Stem Cell Rep 3:940–947
49. Chen S, Sanjana NE, Zheng K et al (2015) Genome-wide CRISPR screen in a mouse model of tumor growth and metastasis. Cell 160:1246–1260
50. Peng J, Zhou Y, Zhu S, Wei W (2015) High-throughput screens in mammalian cells using the CRISPR-Cas9 system. FEBS J 282:2089–2096
51. Shen B, Zhang J, Wu H, Wang J, Ma K, Li Z, Zhang X, Zhang P, Huang X (2013) Generation of gene-modified mice via Cas9/RNA-mediated gene targeting. Cell Res 23:720–723
52. Wang T, Wei JJ, Sabatini DM, Lander ES (2014) Genetic screens in human cells using the CRISPR-Cas9 system. Science (80-) 343:80–84
53. Zhou Y, Zhu S, Cai C, Yuan P, Li C, Huang Y, Wei W (2014) High-throughput screening of a CRISPR/Cas9 library for functional genomics in human cells. Nature 509:487–491
54. Wang X, Wang Y, Wu X, Wang J, Wang Y, Qiu Z, Chang T, Huang H, Lin RJ, Yee JK (2015) Unbiased detection of off-target cleavage by CRISPR-Cas9 and TALENs using integrase-defective lentiviral vectors. Nat Biotechnol 33:175–179
55. Tsai SQ, Zheng Z, Nguyen NT et al (2015) GUIDE-seq enables genome-wide profiling of off-target cleavage by CRISPR-Cas nucleases. Nat Biotechnol 33:187–198
56. Haeussler M, Schönig K, Eckert H et al (2016) Evaluation of off-target and on-target scoring algorithms and integration into the guide RNA selection tool CRISPOR. Genome Biol 17:148
57. Hsu PD, Scott DA, Weinstein JA et al (2013) DNA targeting specificity of RNA-guided Cas9 nucleases. Nat Biotechnol 31:827–832
58. Bolukbasi MF, Gupta A, Wolfe SA (2015) Creating and evaluating accurate CRISPR-Cas9 scalpels for genomic surgery. Nat Methods 13:41–50
59. Ran FA, Hsu PD, Wright J, Agarwala V, Scott DA, Zhang F (2013) Genome engineering using the CRISPR-Cas9 system. Nat Protoc 8:2281–2308
60. Anders C, Niewoehner O, Duerst A, Jinek M (2014) Structural basis of PAM-dependent target DNA recognition by the Cas9 endonuclease. Nature 513:569–573
61. Slaymaker IM, Gao L, Zetsche B, Scott DA, Yan WX, Zhang F (2016) Rationally engineered Cas9 nucleases with improved specificity. Science (80-) 351:84–88
62. Kleinstiver BP, Pattanayak V, Prew MS, Tsai SQ, Nguyen NT, Zheng Z, Joung JK (2016) High-fidelity CRISPR-Cas9 nucleases with no detectable genome-wide off-target effects. Nature 529:490–495
63. Müller M, Lee CM, Gasiunas G, Davis TH, Cradick TJ, Siksnys V, Bao G, Cathomen T, Mussolino C (2016) Streptococcus thermophilus CRISPR-Cas9 systems enable specific editing of the human genome. Mol Ther 24:636–644

64. Fonfara I, Le Rhun A, Chylinski K, Makarova KS, Lécrivain AL, Bzdrenga J, Koonin EV, Charpentier E (2014) Phylogeny of Cas9 determines functional exchangeability of dual-RNA and Cas9 among orthologous type II CRISPR-Cas systems. Nucleic Acids Res 42:2577–2590
65. Zetsche B, Gootenberg JS, Abudayyeh OO et al (2015) Cpf1 is a single RNA-guided endonuclease of a class 2 CRISPR-Cas system. Cell 163:759–771
66. Fonfara I, Richter H, BratoviÄ M, Le Rhun A, Charpentier E (2016) The CRISPR-associated DNA-cleaving enzyme Cpf1 also processes precursor CRISPR RNA. Nature 532:517–521
67. Kleinstiver BP, Prew MS, Tsai SQ et al (2015) Engineered CRISPR-Cas9 nucleases with altered PAM specificities. Nature 523:481–485
68. Hirano S, Nishimasu H, Ishitani R, Nureki O (2016) Structural basis for the altered PAM specificities of engineered CRISPR-Cas9. Mol Cell 61:886–894
69. González F, Zhu Z, Shi ZD, Lelli K, Verma N, Li QV, Huangfu D (2014) An iCRISPR platform for rapid, multiplexable, and inducible genome editing in human pluripotent stem cells. Cell Stem Cell 15:215–226
70. DeWitt MA, Corn JE, Carroll D (2017) Genome editing via delivery of Cas9 ribonucleoprotein. Methods 121–122:9–15
71. Lin S, Staahl BT, Alla RK, Doudna JA (2014) Enhanced homology-directed human genome engineering by controlled timing of CRISPR/Cas9 delivery. elife 3:e04766
72. Richardson CD, Ray GJ, DeWitt MA, Curie GL, Corn JE (2016) Enhancing homology-directed genome editing by catalytically active and inactive CRISPR-Cas9 using asymmetric donor DNA. Nat Biotechnol 34:339–344
73. Yumlu S, Stumm J, Bashir S, Dreyer AK, Lisowski P, Danner E, Kühn R (2017) Gene editing and clonal isolation of human induced pluripotent stem cells using CRISPR/Cas9. Methods 121–122:29–44
74. Yang L, Guell M, Byrne S et al (2013) Optimization of scarless human stem cell genome editing. Nucleic Acids Res 41:9049–9061
75. Dahlem TJ, Hoshijima K, Jurynec MJ, Gunther D, Starker CG, Locke AS, Weis AM, Voytas DF, Grunwald DJ (2012) Simple methods for generating and detecting locus-specific mutations induced with TALENs in the zebrafish genome. PLoS Genet 8:e1002861
76. Brinkman EK, Chen T, Amendola M, Van Steensel B (2014) Easy quantitative assessment of genome editing by sequence trace decomposition. Nucleic Acids Res https://doi.org/10.1093/nar/gku936
77. Hill JT, Demarest BL, Bisgrove BW, Su YC, Smith M, Yost HJ (2014) Poly peak parser: method and software for identification of unknown indels using sanger sequencing of polymerase chain reaction products. Dev Dyn 243:1632–1636
78. Dehairs J, Talebi A, Cherifi Y, Swinnen JV (2016) CRISP-ID: decoding CRISPR mediated indels by Sanger sequencing. Sci Rep 6:28973
79. Pinello L, Canver MC, Hoban MD, Orkin SH, Kohn DB, Bauer DE, Yuan G-C (2016) Analyzing CRISPR genome-editing experiments with CRISPResso. Nat Biotechnol 34:695–697
80. Boel A, Steyaert W, De Rocker N, Menten B, Callewaert B, De Paepe A, Coucke P, Willaert A (2016) BATCH-GE: Batch analysis of next-generation sequencing data for genome editing assessment. Sci Rep 6:30330
81. Dodgson JB, Wells RD (1977) Action of single-strand specific nucleases on model DNA Heteroduplexes of defined size and sequence. Biochemistry 16:2374–2379
82. Bhattacharyya A, Lilley DMJ (1989) The contrasting structures of mismatched DNA sequences containing looped-out bases (bulges) and multiple mismatches (bubbles). Nucleic Acids Res 17:6821–6840
83. Wagner R, Debbie P, Radman M (1995) Mutation detection using immobilized mismatch binding protein (MutS). Nucleic Acids Res 23:3944–3948
84. Youil R, Kemper BW, Cotton RG (1995) Screening for mutations by enzyme mismatch cleavage with T4 endonuclease VII. Proc Natl Acad Sci U S A 92:87–91
85. Howard JT, Ward J, Watson JN, Roux KH (1999) Heteroduplex cleavage analysis using S1 nuclease. BioTechniques 27:18–19

86. Taylor GR, Deeble J (1999) Enzymatic methods for mutation scanning. Genet Anal Biomol Eng 14:181–186
87. Yeung AT, Hattangadi D, Blakesley L, Nicolas E (2005) Enzymatic mutation detection technologies. BioTechniques 38:749–758
88. Babon JJ, Youil R, Cotton RGH (1995) Improved strategy for mutation detection – a modification to the enzyme mismatch cleavage method. Nucleic Acids Res 23:5082–5084
89. Hadden JM, Déclais AC, Carr SB, Lilley DMJ, Phillips SEV (2007) The structural basis of Holliday junction resolution by T7 endonuclease I. Nature 449:621–624
90. Freeman ADJ, Déclais AC, Lilley DMJ (2013) The importance of the N-terminus of T7 endonuclease i in the interaction with DNA junctions. J Mol Biol 425:395–410
91. Déclais AC, Lilley DM (2008) New insight into the recognition of branched DNA structure by junction-resolving enzymes. Curr Opin Struct Biol 18:86–95
92. Mashal RD, Koontz J, Sklar J (1995) Detection of mutations by cleavage of DNA heteroduplexes with bacteriophage resolvases. Nat Genet 9:177–183
93. Déclais AC, Liu J, Freeman ADJ, Lilley DMJ (2006) Structural recognition between a four-way DNA junction and a resolving enzyme. J Mol Biol 359:1261–1276
94. Gohlke C, Murchie AI, Lilley DM, Clegg RM (1994) Kinking of DNA and RNA helices by bulged nucleotides observed by fluorescence resonance energy transfer. Proc Natl Acad Sci U S A 91:11660–11664
95. Oleykowski CA, Bronson Mullins CR, Godwin AK, Yeung AT (1998) Mutation detection using a novel plant endonuclease. Nucleic Acids Res 26:4597–4602
96. Yang B, Wen X, Kodali NS, Oleykowski CA, Miller CG, Kulinski J, Besack D, Yeung JA, Kowalski D, Yeung AT (2000) Purification, cloning, and characterization of the CEL I nuclease. Biochemistry 39:3533–3541
97. Qiu P, Shandilya H, D'Alessio JM, O'Connor K, Durocher J, Gerard GF (2004) Mutation detection using Surveyor™ nuclease. BioTechniques 36:702–707
98. Voskarides K, Deltas C (2009) Screening for mutations in kidney-related genes using SURVEYOR nuclease for cleavage at heteroduplex mismatches. J Mol Diagn 11:311–318
99. Till BJ, Burtner C, Comai L, Henikoff S (2004) Mismatch cleavage by single-strand specific nucleases. Nucleic Acids Res 32:2632–2641
100. Bentley A, Maclennan B, Calvo J, Dearolf CR (2000) Targeted recovery of mutations in drosophila. Genetics 156:1169–1173
101. Colbert T (2001) High-throughput screening for induced point mutations. Plant Physiol 126:480–484
102. Coghill EL, Hugill A, Parkinson N, Davison C, Glenister P, Clements S, Hunter J, Cox RD, Brown SDM (2002) A gene-driven approach to the identification of ENU mutants in the mouse. Nat Genet 30:255–256
103. Perry JA (2003) A TILLING reverse genetics tool and a web-accessible collection of mutants of the legume Lotus japonicus. Plant Physiol 131:866–871
104. Wienholds E, van Eeden F, Kosters M, Mudde J, Plasterk RHA, Cuppen E (2003) Efficient target-selected mutagenesis in zebrafish. Genome Res 13:2700–2707
105. Comai L, Young K, Till BJ et al (2004) Efficient discovery of DNA polymorphisms in natural populations by Ecotilling. Plant J 37:778–786
106. Slade AJ, Fuerstenberg SI, Loeffler D, Steine MN, Facciotti D (2005) A reverse genetic, nontransgenic approach to wheat crop improvement by TILLING. Nat Biotechnol 23:75–81
107. Geurts AM, Cost GJ, Freyvert Y, et al (2009) Knockout rats via embryo microinjection of zinc-finger nucleases. Science (80-) 325:433
108. Guschin DY, Waite AJ, Katibah GE, Miller JC, Holmes MC, Rebar EJ (2010) A rapid and general assay for monitoring endogenous gene modification. Methods Mol Biol 649:247–256
109. Miller JC, Tan S, Qiao G et al (2011) A TALE nuclease architecture for efficient genome editing. Nat Biotechnol 29:143–150
110. Tesson L, Usal C, Meqnoret S et al (2011) Knockout rats generated by embryo microinjection of TALENs. Nat Biotechnol 29:695–696

111. Isalan M (2012) Zinc-finger nucleases: how to play two good hands. Nat Methods 9:32–34
112. Sanjana NE, Cong L, Zhou Y, Cunniff MM, Feng G, Zhang F (2012) A transcription activator-like effector toolbox for genome engineering. Nat Protoc 7:171–192
113. Holkers M, Maggio I, Liu J, Janssen JM, Miselli F, Mussolino C, Recchia A, Cathomen T, Gonçalves MAFV (2013) Differential integrity of TALE nuclease genes following adenoviral and lentiviral vector gene transfer into human cells. Nucleic Acids Res 41:e63
114. Maier DA, Brennan AL, Jiang S et al (2013) Efficient clinical scale gene modification via zinc finger nuclease–targeted disruption of the HIV co-receptor CCR5. Hum Gene Ther 24:245–258
115. Van Rensburg R, Beyer I, Yao XY et al (2013) Chromatin structure of two genomic sites for targeted transgene integration in induced pluripotent stem cells and hematopoietic stem cells. Gene Ther 20:201–214
116. Vouillot L, Thélie A, Pollet N (2015) Comparison of T7E1 and Surveyor mismatch cleavage assays to detect mutations triggered by engineered nucleases. G3: Genes|Genomes|Genetics 5:407–415
117. Gibson DG (2011) Enzymatic assembly of overlapping DNA fragments. Methods Enzymol 498:349–361
118. Merkle FT, Eggan K (2013) Modeling human disease with pluripotent stem cells: from genome association to function. Cell Stem Cell 12:656–668
119. Soldner F, Jaenisch R (2012) iPSC disease modeling. Science (80-) 338:1155–1156
120. Soldner F, Stelzer Y, Shivalila CS, Abraham BJ, Latourelle JC, Barrasa MI, Goldmann J, Myers RH, Young RA, Jaenisch R (2016) Parkinson-associated risk variant in distal enhancer of α-synuclein modulates target gene expression. Nature 533:95–99
121. Collins FS, Lander ES, Rogers J, Waterson RH (2004) Finishing the euchromatic sequence of the human genome. Nature 431:931–945
122. Auton A, Abecasis GR, Altshuler DM et al (2015) A global reference for human genetic variation. Nature 526:68–74
123. Dewey FE, Grove ME, Pan C et al (2014) Clinical interpretation and implications of whole-genome sequencing. JAMA J Am Med Assoc 311:1035–1044
124. Mangino M, Cecelja M, Menni C, Tsai PC, Yuan W, Small K, Bell J, Mitchell GF, Chowienczyk P, Spector TD (2016) Integrated multiomics approach identifies calcium and integrin-binding protein-2 as a novel gene for pulse wave velocity. J Hypertens 34:79–87
125. Gerstung M, Pellagatti A, Malcovati L et al (2015) Combining gene mutation with gene expression data improves outcome prediction in myelodysplastic syndromes. Nat Commun 6:5901
126. Saykin AJ, Shen L, Yao X et al (2015) Genetic studies of quantitative MCI and AD phenotypes in ADNI: progress, opportunities, and plans. Alzheimers Dement 11:792–814
127. Shiraishi Y, Fujimoto A, Furuta M et al (2014) Integrated analysis of whole genome and transcriptome sequencing reveals diverse transcriptomic aberrations driven by somatic genomic changes in liver cancers. PLoS One 9:e114263
128. Piskol R, Ramaswami G, Li JB (2013) Reliable identification of genomic variants from RNA-seq data. Am J Hum Genet 93:641–651
129. Bell S, Peng H, Crapper L, Kolobova I, Maussion G, Vasuta C, Yerko V, Pan Wong T, Ernst C (2017) A rapid pipeline to model rare neurodevelopmental disorders with simultaneous CRISPR/Cas9 gene editing. Stem Cells Transl Med 6:886–896
130. Zhu Z, González F, Huangfu D (2014) The iCRISPR platform for rapid genome editing in human pluripotent stem cells. Methods Enzymol 546:215–250
131. Studer L, Vera E, Cornacchia D (2015) Programming and reprogramming cellular age in the era of induced pluripotency. Cell Stem Cell 16:591–600
132. Miller JD, Ganat YM, Kishinevsky S et al (2013) Human iPSC-based modeling of late-onset disease via progerin-induced aging. Cell Stem Cell 13:691–705
133. Byers B, Cord B, Nguyen HN, Schüle B, Fenno L, Lee PC, Deisseroth K, Langston JW, Pera RR, Palmer TD (2011) SNCA triplication parkinson's patient's iPSC-derived DA neurons accumulate α-Synuclein and are susceptible to oxidative stress. PLoS One 6:e26159

134. Cooper O, Seo H, Andrabi S et al (2012) Pharmacological rescue of mitochondrial deficits in iPSC-derived neural cells from patients with familial Parkinson's disease. Sci Transl Med 4:141ra90
135. Liu GH, Ding Z, Izpisua Belmonte JC (2012) IPSC technology to study human aging and aging-related disorders. Curr Opin Cell Biol 24:765–774
136. Nguyen HN, Byers B, Cord B et al (2011) LRRK2 mutant iPSC-derived da neurons demonstrate increased susceptibility to oxidative stress. Cell Stem Cell 8:267–280
137. Reinhardt P, Schmid B, Burbulla LF et al (2013) Genetic correction of a lrrk2 mutation in human iPSCs links parkinsonian neurodegeneration to ERK-dependent changes in gene expression. Cell Stem Cell 12:354–367
138. Seibler P, Graziotto J, Jeong H, Simunovic F, Klein C, Krainc D (2011) Mitochondrial Parkin recruitment is impaired in neurons derived from mutant PINK1 induced pluripotent stem cells. J Neurosci 31:5970–5976
139. Young SG, Jung H-J, Lee JM, Fong LG (2014) Nuclear Lamins and neurobiology. Mol Cell Biol 34:2776–2785
140. Baek JH, Schmidt E, Viceconte N et al (2015) Expression of progerin in aging mouse brains reveals structural nuclear abnormalities without detectible significant alterations in gene expression, hippocampal stem cells or behavior. Hum Mol Genet 24:1305–1321
141. Longo VD, Antebi A, Bartke A et al (2015) Interventions to slow aging in humans: are we ready? Aging Cell 14:497–510
142. Huang Y, Myers SJ, Dingledine R (1999) Transcriptional repression by REST: recruitment of Sin3A and histone deacetylase to neuronal genes. Nat Neurosci 2:867–872
143. Lu T, Aron L, Zullo J et al (2014) REST and stress resistance in ageing and Alzheimer's disease. Nature 507:448–454
144. Hitz C, Steuber-Buchberger P, Delic S, Wurst W, Kühn R (2009) Generation of shrna transgenic mice. Methods Mol Biol 530:101–129
145. Renaud JB, Boix C, Charpentier M et al (2016) Improved genome editing efficiency and flexibility using modified oligonucleotides with TALEN and CRISPR-Cas9 nucleases. Cell Rep 14:2263–2272
146. Heyer W-D, Ehmsen KT, Liu J (2010) Regulation of homologous recombination in eukaryotes. Annu Rev Genet 44:113–139
147. Sfeir A, De Lange T (2012) Removal of shelterin reveals the telomere end-protection problem. Science (80-) 336:593–597
148. Hustedt N, Durocher D (2017) The control of DNA repair by the cell cycle. Nat Cell Biol 19:1–9
149. Corneo B, Wendland RL, Deriano L et al (2007) Rag mutations reveal robust alternative end joining. Nature 449:483–486
150. Deriano L, Roth DB (2013) Modernizing the nonhomologous end-joining repertoire: alternative and classical NHEJ share the stage. Annu Rev Genet 47:433–455
151. Badie S, Carlos AR, Folio C, Okamoto K, Bouwman P, Jonkers J, Tarsounas M (2015) BRCA1 and CtIP promote alternative non-homologous end-joining at uncapped telomeres. EMBO J 34:828–828
152. San Filippo J, Sung P, Klein H (2008) Mechanism of eukaryotic homologous recombination. Annu Rev Biochem 77:229–257
153. Sfeir A, Symington LS (2015) Microhomology-mediated end joining: a back-up survival mechanism or dedicated pathway? Trends Biochem Sci 40:701–714
154. Sakuma T, Nakade S, Sakane Y, Suzuki KIT, Yamamoto T (2016) MMEJ-assisted gene knock-in using TALENs and CRISPR-Cas9 with the PITCh systems. Nat Protoc 11:118–133
155. He X, Tan C, Wang F, Wang Y, Zhou R, Cui D, You W, Zhao H, Ren J, Feng B (2016) Knock-in of large reporter genes in human cells via CRISPR/Cas9-induced homology-dependent and independent DNA repair. Nucleic Acids Res 44:e85–e85
156. Suzuki K, Tsunekawa Y, Hernandez-Benitez R et al (2016) In vivo genome editing via CRISPR/Cas9 mediated homology-independent targeted integration. Nature 540:144–149

157. Danner E, Bashir S, Yumlu S, Wurst W, Wefers B, Kühn R (2017) Control of gene editing by manipulation of DNA repair mechanisms. Mamm Genome 28:262–274
158. Suzuki K, Izpisua Belmonte JC (2018) In vivo genome editing via the HITI method as a tool for gene therapy. J Hum Genet 63:157–164
159. Srivastava M, Nambiar M, Sharma S et al (2012) An inhibitor of nonhomologous end-joining abrogates double-strand break repair and impedes cancer progression. Cell 151:1474–1487
160. Maruyama T, Dougan SK, Truttmann MC, Bilate AM, Ingram JR, Ploegh HL (2015) Increasing the efficiency of precise genome editing with CRISPR-Cas9 by inhibition of nonhomologous end joining. Nat Biotechnol 33:538–542
161. Chu VT, Weber T, Wefers B, Wurst W, Sander S, Rajewsky K, Kühn R (2015) Increasing the efficiency of homology-directed repair for CRISPR-Cas9-induced precise gene editing in mammalian cells. Nat Biotechnol 33:543–548
162. Greco GE, Matsumoto Y, Brooks RC, Lu Z, Lieber MR, Tomkinson AE (2016) SCR7 is neither a selective nor a potent inhibitor of human DNA ligase IV. DNA Repair (Amst) 43:18–23
163. Menchon G, Bombarde O, Trivedi M et al (2016) Structure-based virtual ligand screening on the XRCC4/DNA ligase IV Interface. Sci Rep 6:22878
164. Robert F, Barbeau M, Éthier S, Dostie J, Pelletier J (2015) Pharmacological inhibition of DNA-PK stimulates Cas9-mediated genome editing. Genome Med 7:93
165. Byrum J, Jordan S, Safrany ST, Rodgers W (2004) Visualization of inositol phosphate-dependent mobility of Ku: depletion of the DNA-PK cofactor InsP6 inhibits Ku mobility. Nucleic Acids Res 32:2776–2784
166. Arras SDM, Fraser JA (2016) Chemical inhibitors of non-homologous end joining increase targeted construct integration in cryptococcus neoformans. PLoS One 11:e0163049
167. Tang J, Cho NW, Cui G, Manion EM, Shanbhag NM, Botuyan MV, Mer G, Greenberg RA (2013) Acetylation limits 53BP1 association with damaged chromatin to promote homologous recombination. Nat Struct Mol Biol 20:317–325
168. Xie A, Hartlerode A, Stucki M et al (2007) Distinct roles of chromatin-associated proteins MDC1 and 53BP1 in mammalian double-strand break repair. Mol Cell 28:1045–1057
169. Yoo E, Kim BU, Seung YL, Chae HC, Chung JH, Lee CH (2005) 53BP1 is associated with replication protein a and is required for RPA2 hyperphosphorylation following DNA damage. Oncogene 24:5423–5430
170. Orthwein A, Noordermeer SM, Wilson MD et al (2015) A mechanism for the suppression of homologous recombination in G1 cells. Nature 528:422–426
171. Shrivastav M, De Haro LP, Nickoloff JA (2008) Regulation of DNA double-strand break repair pathway choice. Cell Res 18:134–147
172. Bashir S, Kühn R (2017) Enhanced precision and efficiency. Nat Biomed Eng 1:856–857
173. Paulsen BS, Mandal PK, Frock RL et al (2017) Ectopic expression of RAD52 and dn53BP1 improves homology-directed repair during CRISPR-Cas9 genome editing. Nat Biomed Eng 1:878–888
174. Richardson CD, Kazane KR, Feng SJ, Bray NL, Schaefer AJ, Floor S, Corn J (2017) CRISPR-Cas9 genome editing in human cells works via The Fanconi Anemia Pathway. DoiOrg 136028

Index

A
Acetylcholinesterase inhibitors, 4
Acidic pH, 290–291
Acinar cells, 203–204
Acriflavine, 294
Action potential (AP)
  CMs, 175
  drug effects and exposure, 180
  hERG inhibition, 175
  hiPSC embryoid bodies, 179
  patch clamp, 182
  phases, 181
  types, 179
  VSDs, 182
Action potential duration (APD), 179, 180, 182, 188
Acute myeloid leukaemia (AML), 272
Acute promyelocytic leukaemia (APL), 294
Adaptive immunity
  anti-IL-12p40 antibody, 248
  B-cell depletion therapy, 246
  CD8+ T cells, 247
  chemokine CCL2, 247
  culture medium, 247
  cytolytic granules, 247
  healthy donors and SLE, 246
  HLA class I polymorphisms, 247
  IL-12p35 subunit, 248
  immunosuppressive, 249
  in vitro differentiation, 249
  multi-organ autoimmune disease, 249
  murine models, 249
  OPG, 249
  osteoclasts, 250
  paracrine mechanisms, 249
  PD-1 molecular pathway, 249
  PGIA, 246
  psoriasis, 250
  soluble factors, 247
  T cells, 247
  Th cells, 248
  transcription factor Foxp3, 248–249
  Treg cells, 246
Adeno-associated virus (AAV), 377, 378
Adipocytes, 46
Adipose stem cells, 56
Adult progenitor cells
  acinar cells, 203–204
  ductal/centroacinar cells, 206–208
  endocrine and exocrine pancreatic compartments, 203
  endocrine cell, 204–206
  intestine and skin, 203
  in vivo regeneration, 208–209
  non-homeostatic conditions, 202
  proliferative tissues, 203
Adult stem cells
  adult tissues/organs, 226
  BM, 226
  MSC, 226
Advanced cell technologies (ACT), 358
Age-related macular degeneration (AMD), 358
Aging
  cell donors, 312
  human pathology, 311 (*see also* Induced pluripotent stem cells (iPSCs))
  transcriptional factors, 312
Aldehyde dehydrogenase (ALDH), 145
Aldehyde dehydrogenase 1 (ALDH1), 274

R. Delgado-Morales (ed.), *Stem Cell Genetics for Biomedical Research*,
https://doi.org/10.1007/978-3-319-90695-9

Alkaline phosphatase (AP) activity, 345
ALLogeneic Heart STem Cells to Achieve Myocardial Regeneration (ALLSTAR), 155
All-*trans* retinoic acid (ATRA), 294
α cells, 198, 201, 204–206, 208, 209
α-smooth muscle actin (α-SMA), 289
Alzheimer's disease (AD), 354
  APP gene, 4
  3D, 8
  dementia, 4
  gene editing, 6, 7
  hiPSCs, 5
  insights, 10
  iPSCs, 5
  neurodegeneration, 5
  PSEN1/PSEN2 gene, 4
  SAD, 8, 9
  γ-secretase-modulating drugs, 9
American Spinal Injury Association (ASIA), 61
Amyloid-beta (Aβ)
  human neurons, 9
  neuropathological hallmarks, 4, 10
  oligomeric, 6
  presenilin 2, 5
Aneuploidy, 55
Angiogenesis, 231
Antibody-drug conjugates (ADCs), 293
Anti-citrullinated protein antibodies (ACPA), 246
Apolipoprotein E4 allele (ApoE4), 8
APOLLON, 154
Arrhythmia, 177, 179, 181, 183, 185, 186
Arteries (arterioles), 229
Asterias OPC1 (AST-OPC1), 64
Astrocyte differentiation, 33–35
Astrocytes, 19, 58
Ataxia telangiectasia (AT), 313, 314
Atherosclerosis, 234
ATP-binding cassette (ABC), 275
Atypical neonatal progeria syndromes (ANPS), 313
Australian company Cyto Therapeutics Pty Limited, 359
Autism spectrum disorders (ASD), 14
AutoLogous Human CArdiac-Derived stem cell to treat Ischemic cArdiomyopathy (ALCADIA), 149, 150
Automatic label-free beating rate imaging, 182, 183
Avermectins, 9

B
BAM factors, 59
BAMN factors, 59
Basal forebrain cholinergic neurons (BFCNs), 7, 8
Basic fibroblast growth factor (BFGF), 47, 110, 150
B-cell depletion therapy, 246
B-cell lymphoma-2 (BCL-2), 287
β-cells, 202, 209–212
  acinar secretions, 198
  biomedical research, 214
  cell replacement therapy, diabetes, 202
  cell transplantation, 212
  diabetes mellitus, 198
  dysfunctional, 213
  embryogenesis, 213
  embryological origin, 213
  exocrine pancreas, 197
  fibroblasts, 213
  hormone-producing cells, 198
  human embryonic, 198
  iPSC-reprogramming factors, 213
  islet transplantation, 198
  pancreas development, 199, 200
  plastic pancreas, 201
  pluripotency, 213
  progenitor cells (*see* Adult progenitor cells)
  stem cells (*see* Stem cells)
  type 1 and 2 diabetes patients, 198
β-sarcoglycan-null mouse model, 235
Blastocyst, 225
Blood-brain barrier (BBB), 97
Blood-spinal cord barrier (BSCB), 44
Bloom syndrome (BS), 313, 319
B lymphocyte, 247
B lymphocyte-induced maturation protein-1 (Blimp-1), 247
Bone marrow (BM), 139, 226
Bone marrow-derived cardiac side population (BM-cSP), 146, 147
Bone marrow-derived cardiac stem cells (BMdCSC), 146
Bone marrow-derived stromal cells (BMSCs), 98
Bone marrow MSCs (BM-MSCs), 289
Bone morphogenetic protein 2 (BMP-2), 159
Bone morphogenetic protein 4 (BMP-4), 91
Bone morphogenetic proteins (BMPs), 33, 289
Brain-derived neurotrophic factor (BDNF), 48
Bruch's membrane, 76, 78

C

Calcium-sensitive detectors (CSDs), 181, 182
Cancer-associated fibroblasts (CAFs), 289
Cancer risk, 80
Cancer stem cells (CSCs), 288
  AML cells, 272
  epigenetic therapy (differentiation therapy), 293, 294
  heterogeneity
    classical theory (hierarchical model), 276
    classification, 276
    clonal evolution model, 276
    genetic and epigenetic abnormalities, 276
    plasticity theory, 278
  immunogenicity (*see* Immunogenicity)
  immunotherapy and ADCs, 293
  inhibition, regulatory pathways, 293
  isolation
    side population, 275
    sphere-forming assays, 275
    stem cell markers, 273, 275
    validation, 275
  markers, 274
  metabolic reprogramming, 286
  phenotypes, 272
  plasticity (*see* CSC plasticity)
  quiescence, 291
  regenerative power, 272
  therapeutic targeting, 292
  TME, 294 (*see also* Tissue microenvironment (TME))
  tumorigenesis and plasticity, 277
  tumour heterogeneity, 295
Capillaries, 229
Cardiac atrial appendage stem cells (CASCs), 145, 146
Cardiac colony-forming unit fibroblast (cCFU-F), 145
Cardiac excitation-contraction coupling, 185
Cardiac progenitor cells (CPCs), 140
Cardiac side population cells (cSP), 143, 144
Cardiac stem cells (CSC)
  ALCADIA trial, 149, 150
  ALLSTAR trial, 155
  APOLLON trial, 154
  CADUCEUS trial, 153
  cell handling and preparation, 160
  clinical trials, 151–152, 156–157
  CONCERT-HF trial, 154, 155
  DYNAMIC trial, 155
  endogenous
    CASCs, 145, 146
    cCFU-F, 145
    CDCs, 144, 145
    clonogenicity, self-renewal and multipotency, 140
    CPCs, 140–142
    cSP, 143, 144
    EPDC, 142, 143
    flow cytometry immunophenotype, 141
    Islet-1, 144
    $Sca1^+$, 143
  ESC and iPSC, 147–148
  HOPE-Duchenne trial, 156
  meta-analysis, 157–158
  non-endogenous
    BM-cSP, 146, 147
    BMdCSC, 146
  paracrine factors, 148–149
  PERSEUS trial, 154
  preclinical studies, 143, 145, 148, 160
  regenerative medicine, 158–160
  SCIPIO trial, 150, 153
  TICAP trial, 153, 154
CardioECR instrument, 181
Cardiomyocytes, 139, 140
Cardiomyogenesis, 139
CArdiosphere-Derived aUtologous stem CElls to reverse ventricUlar dySfunction (CADUCEUS), 153
Cardiosphere-derived cells (CDCs), 144, 145
Cardiovascular diseases (CVD), 136
Cas9 delivery, 380
Cas9 genome editing, 393–394
Cas9 mRNA, 381
Cas9 protein, 381–382
Catumaxomab, 293
Cell-based remyelinating therapy, 91–92
  brain biopsies, 91
  consequences, 90
  demyelinating/dysmyelinating disorders, 90
  EAE, 91
  endogenous process, 90
  ES cells, 91
  experimental models, 90
  in human, 90
  myelin-making oligodendrocytes, 90
  oligodendrocytic fate, 91
  OPCs (*see* Oligodendrocyte progenitor cells (OPCs))
  pathogenic mechanisms, 89
  postmitotic oligodendrocytes, 91
  in post-mortem, 89

Cell-based therapy
  cancer risk, 80
  delivery, 77
  ERG, 76
  eye, 75
  human embryonic stem, 91
  immune response, 79
  induction of tolerance, 79
  MS, 88, 89, 104, 113
  in myelin disorders, 96
  replacement and rescue, 74
  research, 28
  RP and GA, 75
  RPE cells, 75, 78
  scaffold-based approaches, 78
  sources, 76
  uncontrolled use of, 114
Cell cycle behavior, 398
Cell cycle synchronization, 398
Cell index monitoring, 183
Cell therapy, 354–359
Central memory cells, 250
Central nervous system (CNS), 44
Centroacinar cells (CACs), 206, 208
Chinese Academy of Sciences, 359
Chitosan, 50
Chondrocytes, 46
Chondroitin sulfate proteoglycans (CSPGs), 44
Ciliary neurotrophic factor (CNTF), 47, 103, 110
Citrullinated cyclic peptides (CCP), 252
C-kit⁺ cardiac progenitor cells (CPCs), 140–142
Classical theory (hierarchical model), 276
Clonal evolution model, 276
Clustered Regularly Interspaced Short Palindromic Repeats (CRISPR), 20
c-NHEJ pathway, 402
Cockayne syndrome (CS), 313, 314
Collagen-induced arthritis (CIA), 243
Combination of Mesenchymal and c-kit⁺ Cardiac Stem Cells as Regenerative Therapy for Heart Failure (CONCERT-HF), 154, 155
Copy number variations (CNVs), 14, 55
Coronary angioplasty, 137
Coronary artery bypass graft surgery (CABG), 137
Corrected field potential duration (cFPD), 181
Creutzfeldt-Jakob disease, 344
CRISPR/Cas genome editing applications
  forward and reverse genetics, 373
  forward genetics methodology, 374
CRISPR/Cas-mediated editing
  Cas9 cutting, 388
  CRISPResso and BATCH-GE, 386–387
  digital PCR, 389
  EMC method, 387
  fragment analysis, 387
  TIDE and TIDER, 385
CRISPR/Cas system, 373, 379
CRISPR/Cas9-based genome engineering tools, 382
CRIPSR/Cas9 gene editing, 390, 393
CRISPR/Cas9 system, 21, 36, 325, 382
CryoStor®, 347
CSC plasticity
  cellular identity, 278
  dedifferentiation, 278
  EMT, 280–282
  transcriptional and epigenetic regulation, 278
  transdifferentiation, 279
C-terminal domain, 31, 33
CXC-chemokine ligand 12 (CXCL12), 289
CXC chemokine receptor 4 (CXCR4), 48

D
Darwinian evolution, 276
Dedifferentiation, 278
Delayed after depolarizations (DADs), 179–180
δ cells, 198, 201, 204–206, 208
Dementia, 4
Demyelination, 49
Dendritic cells (DCs), 245, 287
Dendritic complexity, 29, 30
Dental pulp pluripotent-like stem cells (DPPSC)
  advantage, 227
  CD73, 227
  characteristics, 227
  embryonic layers, 227
  populations, 227
Dental pulp stem cells (DPSC)
  embryonic layers, 226
  fibroblasts, 227
  multipotent stem/progenitor cells, 226
  osteoblasts and chondroblasts, 226
Differentiation therapy, 293, 294
Dilated cardiomYopathy iNtervention with Allogeneic MyocardIally-regenerative Cells (DYNAMIC), 155
Direct reprogramming, 59
Discoidin domain receptor 2 (DDR2), 153
Disease-modifying treatments (DMTs), 88, 89, 104

Disrupted in schizophrenia 1 (DISC1), 17
Diuretics, 137
DNA-free Cas9, 381
Docosahexaenoic acid (DHA), 6
Donor tissue
  GMP, 345
  hESC, 342
  identifiability, 345
  pseudoanonymization, 344
  selection, 342
  SNP databases, 345
  traceability, 344
Doublecortin-like kinase-1 (Dclk1), 203
Double Holliday junctions (dHJs), 382
Double-strand breaks (DSBs), 370, 376, 399
Doxorubicin, 183
Draining lymph nodes (dLNs), 249
Drugs
  acetylcholinesterase inhibitors, 4
  AD treatment, 4
  NMDA, 4
  screening assays, 185
Duchenne muscular dystrophy (DMD), 232, 324
Dynamin-related protein 1 (DRP1), 285
Dyskeratosis congenita (DC), 313, 314
Dystrophin (DYS), 232
Dystrophin glycoprotein complex (DGC), 232

E
Early afterdepolarization (EAD), 179–181, 185
E-cadherin, 280
Effector memory cells, 250
Ejection fraction (EF), 143
Electrocardiogram (ECG), 175
Electrophysiological methods, 182–183
Electrophysiology studies
  fluorescent imaging, 181–182
  MEA, 181
  multielectrode arrays, 181
  patch clamp, 179–180
Electroretinogram (ERG), 76
Elegant animal models, 10
Embryoid bodies (EB), 346
Embryonic stem cells (ESCs), 28, 33, 91, 339
  blastocyst, 225
  hESC, 225
  inner cell mass, 225
  iPSC, 147–148
  morula, 225
  trophoblast, 225
  zygote, 225
Endogenous Sca1^{+} CPC, 143
Endomysium, 231
Endonuclease I, 388
Endothelial cells (EC), 229, 289
  autologous transplantation therapy, 233
  cell lineages, 233
  DPPSC, 234
  human pluripotent stem cells, 233
  tissue-engineered vessels, 233
  vascular grafts, 233
Engineered heart tissues (EHTs), 188
Enzyme mismatch cleavage (EMC) method, 387
Ependymal glia, 53
Epicardium-derived cells (EPDC), 142, 143
Epidermal growth factor (EGF), 47, 227
Epidermal growth factor receptor (EGFR), 51
Epigallocatechin gallate (EGCG), 293
Epigenetic memory, 55
Epigenetic therapy, 293, 294
Epigenome, 55
Epithelial-mesenchymal transition (EMT), 142
  embryonic morphogenesis, 280
  homeobox factor PRRX1, 281
  lineage tracing analysis, 282
  mesenchymal phenotype, 280
  MET, 280, 281
  metastatic tumours, 280
  microRNAs, 281
  non-coding RNAs, 280
  protein families, 280
  stem-like cancer cells, 281
  stemness, 282
  TFs, 281
  tumour growth and dissemination, 281
Ethidium bromide, 94
European bank for induced pluripotent stem cells (EBiSC), 349
The European Cell Culture Collection (ECACC), 349
European Federation of Pharmaceutical Industries and Associations (EFPIA), 349
Excision repair cross-complementing (ERCC), 313
Extracellular acidification rate (ECAR), 184
Extracellular flux (XF), 184
Extracellular matrix (ECM), 290
Ex vivo experiments, 177

F
Fanconi anemia (FA), 313, 314
FA repair pathway components, 403
Farnesyltransferase inhibitors (FTIs), 323

Fatty acid oxidation (FAO), 285
Fibroblast-associated protein (FAP), 289
Fibroblast growth factor receptor (FGFR), 159
Fibroblast-NPC reprogramming method, 60
Fibronectin, 324
Field potential duration (FPD), 181, 189
Field potential (FP), 181, 185
Fluo-4, 182
Fluorescence-activated cell sorting (FACS), 92, 394
Forward genetics methodology, 374

G
Gamma-aminobutyric acid (GABA) receptor, 205, 208
γ-secretase inhibitor (GSI), 256
γ-secretase-modulating drugs, 9
Gaucher's disease, 354
Gene and pharmacological therapy, 75
Genetic editing, 6, 20–21, 376
Genetic screening, 346
Genome editing, 370, 371
Genome-wide association studies (GWAS), 8
Geographic atrophy (GA), 74
Glial-derived neurotrophic factor (GDNF), 110
Glial fibrillary acidic protein (GFAP), 33, 108, 110
Glial scar, 44, 47
Glioblastoma multiforme (GBM), 60
Glioblastoma stem cells (GSCs), 279
Global alliance for iPSC therapies (GAiT), 350
Glucose-6-phopshate dehydrogenase, 284
Glycolysis, 282–284
Good manufacturing practice (GMP), 344, 347, 355–356
Green fluorescent protein (GFP), 144

H
Haematopoietic stem cells (HSCs), 88, 89, 98, 272, 288
Hallmarks of aging, 311
Halt cardiomyOPathy progrEssion (HOPE) in Duchenne trial, 156
HDR-based genome editing, 384
Heart attack, 137
Heart failure (HF), 137
Hematopoietic stem cells (HSCs), 47, 314
Hepatocyte growth factor (HGF), 48, 101, 255
High-throughput screening (HTS), 9, 293
Histone deacetylases (HDACs), 31, 294
HLA homozygote donors, 357
Homologous recombination (HR), 322
Homology-independent targeted integration (HITI), 401–402
Homology-mediated repair (HDR)
- CRISPR/Cas9, 397, 398
- dsDNA, 384
- hiPSCs, 397–402
- HITI, 401
- mechanism, 382
- MMEJ, 400
- PITCh, 400
- repair template structure and modifications, 397–398
- ssODNs, 383, 397
- SST-R, 400
- targeted mutations, 383

Hot-spot mutations, 33
hPSC-based disease models, 353
Human cardiac-derived stem cells (hCSC), 149
Human central nervous system stem cells (HuCNS-SCs), 62
Human embryonic stem cells (hESC), 96, 175, 209–211, 339–340
- characterization, 345
- human therapy, 340
- karyotype, 346
- methodology, 340
- pluripotent and undifferentiated cells, 225
- nourishing medium, 340
- RT-PCR, 346

Human *Ether-à-go-go-Related* Gene (hERG), 175–177, 180, 181, 185, 186
Human Genome Sequencing Project, 390
Human growth factor (HGF), 289
Human-induced pluripotent stem cell-derived cardiomyocytes (hiPSC-CMs)
- cardiotoxicity screening, 176–177
- comparison of phenotypical parameters, 187
- electrophysiological parameters, 180
- future perspectives, 186
- pros and cons, 176
- safety pharmacology assays, 179
- safety pharmacology screenings
  - bioenergetics and cell viability, 184–185
  - contraction and beating behavior, 182–183
  - electrophysiology studies, 179–182

Human-induced pluripotent stem cells (hiPSCs), 209, 211, 213
- AD cellular phenotypes, 9
- Cas9 homologs, 377
- Cpf1-based systems, 378
- CRISPR/Cas9 system, 371

epigenetic state, 372
γ-secretase-modulating drugs, 9
genome editing, 370
genome engineering, 372
HTS, 9
non-Cas9 CRISPR endonuclease, 377–378
sgRNA, 374
sgRNA-DNA base pairing, 376
SpCas9, 375
S. pyogenes Cas9, 379
therapeutic screening, 9, 10
three-dimensions, 10
type II CRISPR-Cas9 system, 374
ZFN and TALEN platforms, 370
Human leukocyte antigen (HLA), 356
Human papillomavirus 16 (HPV16), 321
Human pluripotent stem cell registry (hPSCreg), 342, 350
Human pluripotent stem cells (hPSC), 339, 351, 360–361
Huntington's disease, 396
Hutchinson-Gilford progeria syndrome (HGPS), 312, 313
Hyperphosphorylated tau protein, 4, 8
Hypertension, 234
Hypoplastic left heart syndrome, 153
Hypoxia-inducible factors (HIF), 290

I

IL-1 receptor antagonist (IL1RA), 244, 247
Immunoediting, 287
Immunogenicity, 53, 54
immune cell regulation, 287, 288
immunoediting, 287
immunological molecules, 287
Immunotherapy, 293
Indoleamine 2,3-dioxygenase (IDO), 101, 102, 106, 245
Induced neural stem/precursor cells (iNSPCs), 113
Induced oligodendrocyte precursor cells (iOPCs), 97
Induced pluripotent stem cells (iPSCs), 5, 14, 175, 339
advantages, 28
cardiac repair, 147–148
cell-based drug, 28
cellular models
cell types, 317
centenarian donors, 315
DNA damage, 319
dyskeratosis congenita patients, 319
fibroblasts, 318
hematopoietic and endothelial lineages, 318
myeloid and erythroid lineages, 318
neurological disorders, 318
neuromelanin accumulation, 317
NGPS, 318
nuclear dysmorphism, 317
telomerase and DNA stability, 318
vascular pathogenesis and lipodystrophy phenotype, 317
(*see also* Progeroid syndromes)
3D organoids, 36
gene editing, 36
genes, 225, 226
nervous system cells, 28
reprogramming in vivo, 319–321
reprogramming methods, 341
RTT, 33, 34, 36
technology, 340
uses, 36
Inducible nitric oxide synthase (iNOS), 256
Inflammaging, 322
Informed consent questionnaire, 343–344
Inner cell mass (ICM), 147, 225
Innovative medicines initiative (IMI), 349
Insulin growth factor 1 (IGF-1), 148
Integrated DNA Technologies, Inc. (IDT), 384
Integration-free methods, 355
Intercellular adhesion molecule 1 (ICAM-1), 100, 112
Interferon gamma (IFN-γ), 101, 287
International Council on Harmonization of Technical Requirements for Registration of Pharmaceuticals for Human Use (ICH), 175
International Society for Cellular Therapy (ISCT), 99, 106, 242
International Standards for Neurological Classification of Spinal Cord Injury (ISNCSCI), 62
The International stem cell banking initiative (ISCBI), 348
International Stem Cell Forum (ISCF), 348
Inter-tumour heterogeneity, 276
Intracerebroventricular (i.c.v.), 103
Intra-tumour heterogeneity, 275
In vitro fertilization (IVF), 341
In vivo QT assays, 177
Ischemic heart diseases (IHD), 136, 137, 143
Islet-1 CPC, 144
Isogenic hiPSC line, 372

K
Keratinocytes, 56
Krebs cycle, 284
Kruppel-like factor-4 (KLF4), 147

L
Lactate dehydrogenase (LDH), 284
Lamins, 312–314
Late-onset Alzheimer's disease (LOAD), 4
Late-onset and aging-related disorders, 396
Late-onset diseases, 353
hiPSC-derived neural cells, 396
KD phenotype, 396
Lentiviral-based Cas9, 380
Leukemia inhibitory factor (LIF), 227
Leukocyte function antigen 1 (LFA-1), 102
Limitations in applicability of hiPSC-CMs
phenotypic immaturity, 187–188
phenotypic variability, 188–189
Lipopolysaccharide (LPS), 101
Lonafarnib, 323
Lumbar punctures, 64
Lymphocyte function-associated antigen 1 (LFA-1), 112
Lymphocytes, 288

M
Macroautophagy, 324
Macrophage colony-stimulating factor 1 (M-CSF), 244
Macrophages, 243
Major depressive disorder (MDD), 14
Major histocompatibility complex (MHC), 79, 287
Mammalian heart, 138–139
Mammalian Müller glial cells, 77
Mandibuloacral dysplasia (MAD), 313, 319
Marrow-isolated adult multilineage inducible (MIAMI), 227
Master cell bank (MCB), 356
Mechanistic target of rapamycin (mTOR), 34
MeCP2-knockout (KO), 30
Mendelian disorders, 390
Mesenchymal-epithelial transition (MET), 280
MEsenchymal StEm Cells for Multiple Sclerosis (MESEMS), 105
Mesenchymal stem/stromal cells (MSCs), 44, 46, 58, 226, 289, 317
adaptive and innate immunity, 100
adherent fibroblast-like cells, 98
BMSCs, 98
CFU-Fs, 98
characterization, 99
clinical translation, 104–106
exogenous myoblasts, 99
intrinsic, 100
ISCT, 99
local delivery, 103–104
myogenic and skeletogenic CD146+ cells, 99 (*see also* Rheumatoid arthritis (RA))
skeletal tissues, 98
systemic delivery, 100–103
Mesodermal progenitor cells (MPC), 227
Mesoderm-derived cell types
angiogenesis, 229, 231
blood vessels, 229, 230
DGC, 232
DPPSC, 229
MD, 232
myofiber nuclei (myonuclei), 232
skeletal muscle, 231, 232
VSMC, 230
Metabolic reprogramming
glycolysis, 282, 284
metabolic flexibility, 285, 286
mitochondrial assets and behaviours, 284, 285
Methyl-CpG-binding domain (MBD), 31, 32
Methyl-CpG-binding protein 2 (MECP2), 31, 32
gene, 30, 31
glia, 30
5-hydroxymethylcytosine-containing DNA, 32
mutations, 32, 33
neuronal maturation, 33
NSCs, 33
postmitotic neurons, 30, 31
protein functions, 31, 32
C-terminal domain, 31
MBB, 31, 32
TRD, 31
symptoms, 30
Microhomology-mediated end joining (MMEJ), 400
MicroRNA processing genes, 18
Milk fat globule-EGF-8 (MFG-E8), 289
miRNA processing/mRNA splicing, 32
Mitochondrial DNA (mtDNA), 285
Mitochondrial functions, 184
Mitogen-activated protein kinase (MAPK), 245
Morula, 225
Mouse embryonic fibroblast (MEF), 60
Multidrug resistance (MDR), 275

Multielectrode arrays (MEA), 181, 183, 189
Multiplex genome editing, 373
Multipotent adult progenitor cells (MAPC), 227
Multipotent mesenchymal stromal cells, 99
Multipotent retinal stem cells, 77
Murine embryonic fibroblasts (MEFs), 321
Muscle fascicle/fiber bundle, 231
Muscle stem cells (MuSCs), 324
Muscular dystrophies (MD), 232, 235, 325
Mycoplasma testing, 346
Myelin basic protein (MBP), 92, 100
Myelin oligodendrocyte glycoprotein (MOG), 92, 100, 102
Myelogenous cells, 288
Myoblast Autologous Grafting in Ischemic Cardiomyopathy (MAGIC), 156
Myoblast determination (MYOD), 279
Myocardial infarction (MI), 137
Myocardial injury, 140
Myocytes, 46
Myopic macular degeneration, 75

N

Nanovesicles, 18
Natural killer (NK) cells, 244, 287
N-cadherin, 280
NCoR/SMRT interaction domain (NID), 31, 32
Néstor-Guillermo progeria syndrome (NGPS), 313
Neural cell, 14, 15, 17, 21
Neural cell adhesion molecule (NCAM), 59
Neural diversity, 18–20
Neural progenitor cells (NPCs), 44, 46, 58, 317
Neural stem cells (NSCs), 33–35, 46
  mammal, 108
  mesenchymal, embryonic and foetal, 114
  proliferation and differentiation, 103
  SVZ, 92
Neural stem/precursor cells (NSPCs)
  acute EAE rats, 108
  adult CNS, 106, 108
  adult mammalian brain, 108
  categorization, 106
  clinical translation, 112–113
  graft-to-host interactions, 109
  identification, 108
  intrinsic ability, 108
  local delivery, 111–112
  mammal neural stem cell, 108
  systemic delivery, 109–111
Neurodegeneration, 30
Neurodevelopment, 14, 20
Neurodevelopmental disorders (NDs), 390
  CRIPSR/Cas9 gene editing, 393
  hiPSCs, 392
  iPSC reprograming, 393–394
  molecular mechanisms, 392
  RFP1+ and RFP2- single cell fractions, 394
  RNA-Seq data, 392
  WGS, 392
Neuronal differentiation, 33, 34
Neuronal phenotypes, 34, 35
Neuroprotection, 47–49
Neuroregeneration
  agarose hydrogel, 49, 50
  bifunctionalized hydrogels, 50
  biomaterials, 49, 51
  C3 transferase gene, 50
  chitosan, 50
  EGFR, 51
  microenvironment, 51
  microinjected transcription factors and hormones, 50
  neurogenesis and axonal regeneration, 51
  oligodendro gliogenesis and functional remyelination, 50
  PEG, 50
  PgP, 50
  RhoA, 50
New chemical entities (NCE)
  cardiac toxicity, 177
  components, 175
  preclinical safety screening, 177
  TdP risk, 177
N-methyl-D-aspartate (NMDA), 4
NOGA angiogenesis revascularization therapy, 148
Non-haematopoietic stem cell (nHSC) transplantation, 98–100, 106–109
  CNS, 97, 98
  innate and adaptive immunity, 97
  microenvironment, 97
  migration and maturation, OPCs, 97
  MSCs (*see* Mesenchymal stem cells (MSCs))
  noncanonical properties, 98
  NSPCs (*see* Neural stem/precursor cells (NSPCs))
  pro-inflammatory cytokines, 97
Nonhomologous end-joining (NHEJ), 322

Non-stem autologous cell transplantation
clinical trials, 62–64
OECs, 61
Schwann cells, 61
stem cell fate, 61–62
Novel gene expression, 6
Nuclear factor-κB (NF-κB), 244

O

Olfactory ecto-MSCs (OE-MSCs), 256
Olfactory ensheathing cells (OECs), 61
Oligodendrocyte progenitor cells (OPCs), 47, 58
characteristics, 92
clinical translation, 95–97
disease/pathway models, 92–95
galactocerebroside, 92
immature oligodendrocyte precursors, 92
in mammals, 92
physiology, 91
Oligodendrocyte progenitors (OPCs), 19
Oligodendrocytes, 20
Organ regeneration, 228
Organoid, 36
Osteoblasts, 46
Osteoprotegerin (OPG), 249
Oxidative phosphorylation (OXPHOS), 282–284
Oxygen consumption rate (OCR), 184

P

Pancreatic duct ligation (PDL), 203, 207
Parkinson's disease, 354, 396
Pathogenesis-studied disease, 390
Patient cell lines
Aβ-specific antibodies, 6
hiPSCs, 5, 6
Percutaneous coronary intervention, 137
Peripheral blood mononuclear cells (PBMC), 251
Peroxisome proliferator-activated receptor-gamma (PPAR-γ), 245
Peroxisome proliferator-activated receptor (PPAR) β/δ, 257
Phosphorothioate (PTO) modifications, 397
Phosphorylated tau protein, 5
Phytohemagglutinin (PHA), 251
Placental MSCs (PL-MSCs), 104
Plasticity theory, 278
Platelet-derived growth factor (PDGF), 227
Platelet-derived growth factor-α (PDGFα), 289
Platelet-derived growth factor subunit A (PDGF-A), 47
Pleiotrophin (PTN), 289
Pluripotent stem cell banks
informed consent, 341–342
iPSC, 341
Polyethylene glycol (PEG), 50
Polygenic diseases
CRISPR/Cas9 and iCRISPR platforms, 394
iCas9 and iCr systems, 395
next-generation sequencing and genome-wide association studies, 394
sgRNA with ssDNA, 394
Polymerase chain reaction (PCR), 80
Polysialated form of neural cell adhesion molecule (PSA-NCAM), 108
Postmitotic neurons, 30
PP cells, 198, 201
Precise integration into target chromosomes (PITCh), 400
Preimplantation genetic diagnosis (PGD), 353
Premature aging diseases, *see* Progeroid syndromes
ProgEnitoR cell infuSion to treat hEart Univentricular heart diseaseS (PERSEUS), 154
Progerin, 313
Progeroid laminopathies, 312
Progeroid syndromes
ABT263, 325
adult stem cell exhaustion, 314–315
autophagy, 325
β1-integrin, 324
cell reprogramming
aged cells, 321, 323
cell cycle, 321
DOT1L inhibitors, 323
Ezh2/SUV39H1, 323
FoxO3, 322
inflammaging, 322
insulin/IGF1 signalling pathway, 322
metabolic pathways, 322
MSCs, 321
NF-κB signalling, 322
Sirt6, 322
telomere dynamics, 321
transcriptional repressor Patz1, 321
Yamanaka factors, 321, 322
DNA repair, 313–314
FTI treatment, 323
genetic disorders, 312
HGPS, 324 (*see also* Induced pluripotent stem cells (iPSCs))
JAK-STAT pathway, 324

MuSCs, 324
nuclear lamina, 312, 313
physiological aging, 312
p38 MAPK activity, 324
senolytic drugs, 325
stem cell regenerative capacity and pluripotency, 324
WRN-deficient mice, 324
Programmed cell death 1 (PD-1), 247
Progressive MS, 89–113
allogeneic stem cells, 114
cell-based remyelinating therapy (*see* Cell-based remyelinating therapy)
cell-based therapy, 88, 107
clinical cell therapy, 114
CNS, 88
core pathological features, 90
DMTs, 88
glioproliferative lesion, 114
in histopathology, 88
immuno-ablation, 89
MRI, 88
myelin-forming cells, 89
neural therapy space, 114
nHSCs, 89
non-haematopoietic stem cell (nHSC) (*see* Non-haematopoietic stem cell (nHSC) transplantation)
phase of, 88
preclinical research, 89
scientific progress, 113
therapeutic approach, 89
Proinflammatory cytokines, 243
Prostaglandin E2 (PGE2), 101, 102, 244, 289
Proteoglycan-induced arthritis (PGIA), 246
Proteolipid protein (PLP), 92, 100, 101
Psoriatic arthritis (PsA), 250
Psychiatric disorders, 14, 18, 21
Psychiatric Genomics Consortium (PGC), 14
Pyruvate dehydrogenase kinase-1 (PDK1), 284
Pyruvate kinase muscle (PKM), 284

Q

QT prolongation
cardiac events, 175
hERG inhibition, 175, 177
ICH, 175
MEA, 189
Quality management system (QMS), 356

R

Radial-glial cells, 52
Radical oxygen species (ROS), 137
Ratiometric and non-metric, 182
Reactive oxygen species (ROS), 284, 314
Recombination signal-binding protein for immunoglobulin kappa J (RBP-J), 257
Regenerative medicine, 96
dental pulp tissue, 229
donor-recipient immunocompatibility, 229
local anesthesia, 229
organ repair, 228
paracrine effect, 228
postnatal function, 228
stem-cell therapy, 228
transplantation and replacement therapies, 228
Reinfusion of Enriched Progenitor cells And Infarct Remodeling in Acute Myocardial Infarction (REPAIR-AMI), 156
Relapsing-remitting multiple sclerosis (RRMS), 88, 89
Remyelination, 49
Renin-angiotensin-aldosterone, 137
Reperfusion injury, 137
Repolarization reserve, 177
Repressor element 1 silencing transcription factor (REST), 52
Reprogramming in vivo, 319–321
Restenosis, 234
Restrictive dermopathy (RD), 313, 319
Resveratrol, 324
Retinal trauma, 80
Rett syndrome-induced pluripotent stem cells (RTT-iPSC), 34, 35
Rett syndrome (RTT)
dendritic complexity, 29, 30
de novo mutations, 28
iPSC technology, 33, 34, 36
MECP2, 29–32, 34, 36
mouse model, 29, 30, 36
neurological symptoms, 30
progressive neurological dysfunction, 29
neurodevelopmental disorder, 28, 33
neuronal phenotypes, 35
NSCs, 33
stem cell, 33
synaptic density, 29
TALEN gene editing technology, 36
Reverse genetics tool, 373

Rheumatoid arthritis (RA), 246–250
adaptive (*see* Adaptive immunity)
adaptive memory
acute kidney inflammation, 252
blood and nonlymphoid tissues, 250
CCP, 252
central and effector memory cells, 250
chemokine receptors, 251
ectopic lymphoid follicles, 253
γδ T cells, 252
genome-wide association studies, 253
human amniotic membrane, 251
immunological memory, 250
LT-α expression, 251
lymph nodes, 252
pathogenesis, 251
stem cell memory, 250
B lymphocytes, 255
cardiovascular and pulmonary systems, 242
chemokines, 255
CIA model, 256, 257
clinical trials, 253–254
hepatic ischemia/reperfusion injury model, 257
hypoxia, 255
immune cells, 242
immunosuppressive agents, 242–243
immunosuppressive phenotype, 254
innate immunity, 243–246
iNOS, 256
lymphoid organs, 256
notch signaling, 256
OE-MSCs, 256
PPARβ/δ, 257
proinflammatory cytokines, 256
RBP-J, 257
tissue degeneration, 255
RhoA pathway, 80
Ribonucleoprotein (RNP), 381
RNA-Seq data, 392
Robust cryopreservation protocols, 347
Rostral neurons, 6
Rothmund-Thomson syndrome (RTS), 313, 319

S

S100β, 33
Sanger sequencing, 385, 386
Satisfactory cryopreservation technologies, 347
Scaffold-based approaches, 78
Schizophrenia (SCZ)
allelic effects, 14
astrocytes, 19
complexity, 20
CRISPR-Cas9 system, 21
DISC1 and Ndel1/Nde1, 17
gene and protein, 14
genotype and phenotype, 15
glia cells, 15
iPSCs, 14–17, 20
microRNA processing genes, 18
MPB and MOG, 18
neurodevelopment, 20
neuropsychiatric models, 14
next-generation sequencing, 14
oligodendrocytes, 19
pathophysiology, 15
pluripotent stem cells, 20–21
postmortem brain samples, 17
TALENs, 21
Schwann cells, 61
Seckel syndrome (SS), 313, 314, 319
Senescence-associated secretory phenotype (SASP), 321, 322
sgRNA delivery, 379–380
Shared epitope (SE), 253
Signal regulatory protein-α (SIRPα), 153
Signal transducer and activator of transcription 3 (STAT3), 111
Single nucleotide polymorphism (SNP), 55, 345
Single-strand annealing (SSA) pathway, 382
Single-stranded DNA oligonucleotides (ssODNs), 383
Single-strand template repair (SST-R), 400, 403
Skin fibroblasts, 55–56
Smooth muscle cells (SMCs), 229, 317
cardiovascular diseases, 234
dystrophy, 235
embryonic development, 234
human embryonic stem cell, 234
in vitro model systems, 234
vascular constructs/bladder reconstruction substrates, 234
Somatic cell reprogramming, 325, 341
Somatic cells, 28, 34
Sox9^{CreER} transgenic mice, 208
Spinal cord injury (SCI), 61
adipose stem cells, 56
annual incidence, 44
astrocytes, 58
cell types, 57
differentiated cells, 57–58
direct reprogramming/transdifferentiation, 59–60
glial scar, 44, 47
human ESC genomes, 52

immune rejection, 53–54
induced pluripotency, 46
inducing pluripotency, 56–57
iPSC-derived cells, 54
keratinocytes, 56
long/short interspersed nuclear elements, 52
MSCs, 58
neocortex, 51
neuroinflammation, 52
neuroprotection, 47–49
neuroregeneration, 47, 49–51
neurorepair, 47
non-stem autologous (*see* Non-stem autologous cell transplantation)
NPCs, 52, 58
OPCs, 58
outer subventricular zone, 51
pathophysiology, 45
phases, 43
primary and secondary injury, 43
principals, 47
remyelination, 49
skin fibroblasts, 55–56
source cells, 54–56
stem cells, 44–47
translational neural repair, 53
umbilical cord blood cells, 56
visual cortex, 51
zebra fish, 52, 53
Sporadic Alzheimer's disease (SAD), 4, 5, 8, 9
AD-iPSC-derived neurons, 9
oligomeric Aβ, 6, 8
Sporadic diseases
hiPS cell-based modeling, 389
hiPSC technology platform, 390
isogenic cells, 390
risk regulatory allele, 390
Stargardt macular dystrophy (SMD), 75
Stem cell antigen-1 (Sca-1), 143
Stem Cell Infusion in Patients with Ischemic cardiOmyopathy (SCIPIO), 150, 153
Stem cells
challenges, 212
embryonic, 209–211
functions, 224
human pluripotent, 209
induced pluripotent, 211
types, 224
StemBANCC, 350
Sterility testing, 346
Stochastic model, 276
Stromal cell-derived factor-1 (SDF-1), 48
Stromal cell-derived factor-1α (SDF-1α), 247, 257
Structural and functional endpoints, 183
Suberoylanilide hydroxamic acid (SAHA), 294
Subretinal cell transplantation, 76
Subventricular zone (SVZ), 52, 92, 103, 108
Synaptic density, 29, 34
Synaptosome-associated protein 91 (SNAP91), 21
Systemic lupus erythematosus (SLE), 246

T

Tartrate-resistant acid phosphatase (TRAP), 250
T-cell lymphocytes, 287
T-helper (Th) cells, 248
Thiotrichodystrophy (TTD), 313, 314
3'-untranslated region (3'UTR), 30
Tight seal (giga seal), 180
Tissue microenvironment (TME), 288–290
components
ECM, 290
haematopoietic cells, 288
MSCs, 289, 290
hypoxia and acidic pH, 290–291
metastasis, 291
Tissue-resident memory (TRM), 251
TNF-α-stimulated gene 6 protein (TSG-6), 244
Tolerogenic DCs (tDCs), 245
Toll-like receptor 2 (TLR2), 244
Torsade des pointes (TdP), 175, 177, 181
Toxicology assays, 351–352
Transcoronary Infusion of Cardiac Progenitor (TICAP), 153, 154
Transcription activator-like effector nucleases (TALENs), 21, 36, 370
Transcription factors (TFs), 175, 278
Transcriptional repression domain (TRD), 31
Transdifferentiation, 279
Transforming growth factor beta (TGF-β), 48, 149, 234
Transnational AllianCe for regenerative Therapies In Cardiovascular Syndromes (TACTICS), 158
Transplantation of Progenitor Cells and Regeneration Enhancement in Acute Myocardial Infarction (TOPCARE-AMI), 157
Trichothiodystrophy, 319
Trophoblast, 225
T-tubules, 188
Tumorigenicity, 64
Tumour-associated MSCs (TA-MSCs), 289
Tumour spheres, 275
Type II CRISPR system, 371

U
Umbilical cord blood cells, 56, 357
Umbilical cord blood unit (UCBU), 357
Umbilical cord-derived MSCs (UC-MSCs), 249
Umbilical cord perivascular cells (UCPVCs), 100

V
Valproic acid, 323
Vascular aneurysm, 234
Vascular endothelial growth factor (VEGF), 109, 295
Vascular smooth muscle cells (VSMC), 229
Veins (venules), 229
Very small embryonic-like (VSEL) cells, 227
Viability tests, 348
Vimentin, 289
Virological testing, 346
Volga German mutation, 5
Voltage-sensitive dyes (VSDs), 181, 182

W
Warburg effect, 282
Werner syndrome (WS), 313, 314, 396
Whole-exome sequencing (WES), 390
Whole-genome sequencing (WGS), 390
Wnt/β-catenin pathway, 77
Wnt ligand, 77
Working cell banks (WCB), 356
World Health Organization (WHO), 136

X
Xenotransplantation, 371
Xeroderma pigmentosum (XP), 313, 318
X-linked gene, 28

Y
Yamanaka factors, 312, 319
Yes-associated protein (YAP), 279

Z
Zinc finger nucleases (ZFNs), 21, 36, 370, 371
ZMPSTE24 gene, 313
Zygote, 225

MIX
Papier aus verantwortungsvollen Quellen
Paper from responsible sources
FSC® C105338

If you have any concerns about our products,
you can contact us on
ProductSafety@springernature.com

In case Publisher is established outside the EU,
the EU authorized representative is:
Springer Nature Customer Service Center GmbH
Europaplatz 3, 69115 Heidelberg, Germany

Printed by Libri Plureos GmbH
in Hamburg, Germany